Conducting Polymer-Based Energy Storage Materials

Conducting Polymer-Based Energy Storage Materials

Edited by

Inamuddin
Rajender Boddula
Mohammad Faraz Ahmer
Abdullah M. Asiri

CRC Press
Taylor & Francis Group
Boca Raton London New York

CRC Press is an imprint of the
Taylor & Francis Group, an **informa** business

CRC Press
Taylor & Francis Group
6000 Broken Sound Parkway NW, Suite 300
Boca Raton, FL 33487-2742

First issued in paperback 2021

© 2019 by Taylor & Francis Group, LLC
CRC Press is an imprint of Taylor & Francis Group, an Informa business

No claim to original U.S. Government works

ISBN 13: 978-0-367-19394-2 (hbk)
ISBN 13: 978-1-03-223817-3 (pbk)

Visit the Taylor & Francis Web site at
http://www.taylorandfrancis.com

and the CRC Press Web site at
http://www.crcpress.com

Publisher's Note
The publisher has gone to great lengths to ensure the quality of this reprint but points out that some imperfections in the original copies may be apparent.

Contents

Preface

Electrochemical storage devices, for example, batteries and supercapacitors, have demonstrated extraordinary assurance in empowering the use of energy produced from wind and solar energy sources. It is necessary to think about the development of independent electrochemical energy storage systems. The most important aspect of an energy storage system is the designing of active electrode materials for excellent electrochemical performance. Among the various materials used in energy storage systems, conducting polymers have gained significant attention by researchers across the world due to their low cost, simple tunability of arrangements, rich redox chemistry, flexibility, structure, and morphology. Thus, conducting polymers seem to be the alternative for next-generation electrochemical devices. However, cycle stability is a bottle-neck limitation for practical applications of energy storage devices based on conducting polymers. Therefore, the research for the development of next-generation conducting polymers-driven energy storage has been in progress to get practically viable technologies. The conducting polymer-based energy storage devices seem to be the incredible future, but still more research and development studies are needed to commercialize them at a large scale.

This book explores the new aspects of conducting polymer processing, usable properties, nanostructures, and engineering strategies applicable to energy storage applications. It briefly surveys recent advances of conducting polymers and their composites, synthetic approaches, and applications in all-types electrochemical energy storage devices. It deliberates cutting-edge knowledge of energy storage technology based on synthetic metals and presents up-to-date literature coverage on a large, rapidly growing, and complex conducting polymer-based electrochemical energy storage system. This book is planned to provide the readers a clear thought regarding the execution of conducting polymeric materials as electrodes in electrochemical power sources. This book is a one-stop reference guide that overviews up-to-date literature in the field of conducting polymer-based energy storage systems. This book is an invaluable guide to students, professors, scientists, and R&D industrial specialists working in the field of advanced science, nanodevices, flexible electronics, and energy science.

Inamuddin, Rajender Boddula, Mohammad Faraz Ahmer, and Abdullah M. Asiri

Contributors

Mutayyab Afreen
Department of Physics
University of Agriculture
Faisalabad, Pakistan

Jazib Ali
Department of Physics and Astronomy
Shanghai Jiao Tong University
Shanghai, China

Faiza Altaf
Department of Environmental Sciences (Chemistry)
Fatima Jinnah women university
Rawalpindi, Pakistan

Muhammad Zeeshan Ashfaq
Department of Physics
University of Agriculture
Faisalabad, Pakistan

Sadia Zafar Bajwa
Nanobiotechnology Group National Institute for
    Biotechnology and Genetic Engineering (NIBGE)
Faisalabad, Pakistan

Fawzi Banat
Department of Chemical Engineering
Khalifa University
Abu Dhabi, United Arab Emirates

Prasun Banerjee
Department of Physics
GITAM Deemed to be University
Bangalore, India

D. Baba Basha
Department of Physics
College of Computer and Information Sciences,
    Majmaah University
Al'Majmaah, KSA

Zahid Bashir
Department of Polymer Engineering and Technology
University of the Punjab
Lahore, Pakistan

Younus Raza Beg
Department of Chemistry
Govt. Digvijay PG Autonomous College
Rajnandgaon, Chhattisgarh, India

G. Bharath
Department of Chemical Engineering
Khalifa University
Abu Dhabi, United Arab Emirates

Muhammad Bilal
School of Life Science and Food Engineering
Huaiyin Institute of Technology
Huaian, China

Swati Chaudhary
Amity Institute of Nanotechnology
Amity University
Noida, India

U. S. Chavan
Department of Mechanical Engineering
Sinhgad Institute of Technology Lonavala
Lonavala, India

Anirban Dandapat
Department of Chemistry
DSB Campus, Kumaun University
Nainital, India

Chittaranjan Das
Institute of Applied Materials-Energy Storage System
Karlsruhe Institute of Technology
Karlsruhe, Germany

Navaneethan Duraisamy
Department of Chemistry
Periyar University
Salem, India

Kalyan Gayen
Department of Chemical Engineering
NIT Agartala
West Tripura, India

Karl I Jacob
School of Materials Science and Engineering
Georgia Institute of Technology
Atlanta, GA, USA

C. Jahnke
Department of Materials and Metallurgical
    Engineering
South Dakota School of Mines and Technology
Rapid City, South Dakota

Nuzhat Jamila
Nanobiotechnology Group National Institute for
    Biotechnology and GeneticEngineering (NIBGE)
Faisalabad, Pakistan

Joon Ching Juan
Nanotechnology & Catalysis Research Centre
    (NANOCAT), Institute for Advanced Studies
    University of Malaya (UM)
Kuala Lumpur, Malaysia

Veerendra Kumar A. Kalalbandi
Department of Chemistry, Jain College of Engineering
Belagavi, Karnataka, India

Shankara S. Kalanur
Department of Materials Science and Engineering
Ajou University
Suwon, Republic of Korea

Manoj Karakoti
Prof. R. S. Nanoscience and Nanotechnology Centre
Department of Chemistry DSB Campus, Kumaun
    University
Nainital, India

Muhammad Khalil
Department of Polymer Engineering and Technology
University of the Punjab
Lahore, Pakistan

Waheed S. Khan
Nanobiotechnology Group National Institute for
    Biotechnology and Genetic Engineering (NIBGE)
Faisalabad, Pakistan

Rafi Ullah Khan
Department of Polymer Engineering and Technology
University of the Punjab
Lahore, Pakistan

Institute of Chemical Engineering and Technology
    (ICET)
University of the Punjab
Lahore, Pakistan

Asimananda Khandual
Department of Textile Engineering
College of Engineering & Technology (BPUT)
Bhubaneswar, India

Ashok Kumar
Department of Physics
Napaam, Tezpur
Assam, India

R. Jeevan Kumar
Department of Physics
Sri Krishnadevaraya University
Anantapuramu, India

A.B.V. Kiran Kumar
Nano world India, Buchireddy Plaem
Nellore, India

Amity Institute of Nanotechnology
Amity University
Noida, India

Chin Wei Lai
Nanotechnology & Catalysis Research Centre
    (NANOCAT), Institute for Advanced Studies
University of Malaya (UM)
Kuala Lumpur, Malaysia

P. E. Lokhande
Department of Mechanical Engineering
Sinhgad Institute of Technology Lonavala
Lonavala, India

Suman Mahendia
Department of Physics
Kurukshetra University
Kurukshetra, India

R.V. Mangalaraja
Advanced Ceramics and Nanotechnology Laboratory,
    Department of Materials Engineering, Faculty of
    Engineering
University of Concepcion
Concepcion, Chile

Technological Development Unit (UDT)
University of Concepcion
Coronel, Chile

Mriganka Sekhar Manna
Department of Chemical Engineering
NIT Agartala
West Tripura, India

Satyabadi Martha
Government (SSD) +2 Science College (HSS),
    Badabharandi
Umerkote, India

Priyadharshini Matheswaran
Smart Materials Interface Laboratory, Department of
    Physics
Periyar University
Salem, India

Dilip Kumar Mishra
Department of Physics, Faculty of Engineering and
    Technology (ITER)
Siksha 'O' Anusandhan Deemed to be University
Bhubaneswar, India

Nadhratun Naiim Mobarak
Centre for Advanced Materials and Renewable
    Resources (CAMARR) Faculty of Science and
    Technology
Universiti Kebangsaan Malaysia
Selangor, Malaysia

Mohamad Azuwa Mohamed
Centre for Advanced Materials and Renewable
    Resources (CAMARR) Faculty of Science and
    Technology
Universiti Kebangsaan Malaysia
Selangor, Malaysia

Dania Munir
Department of Polymer Engineering and Technology
University of the Punjab
Lahore, Pakistan

K. Chandra Babu Naidu
Department of Physics
GITAM Deemed to be University
Bangalore, India

U. Naresh
Department of Physics
Sri Krishnadevaraya University
Anantapuramu, India

G. Nasymov
Department of Materials and Metallurgical Engineering
South Dakota School of Mines and Technology
Rapid City, South Dakota

Gokul Ram Nishad
Department of Chemistry
Govt. Digvijay PG Autonomous College
Rajnandgaon, Chhattisgarh, India

A. M. Numan-Al-Mobin
Department of Materials and Metallurgical Engineering
South Dakota School of Mines and Technology
Rapid City, South Dakota

Sandeep Pandey
Prof. R. S. Nanoscience and Nanotechnology Centre
Department of Chemistry DSB Campus, Kumaun
    University
Nainital, India

Roshan Paul
Department of Textile Science and Technology
University of Beira Interior
Covilha, Portugal

N. Pugazhenthiran
Laboratorio de Tecnologías Limpias, Faculty of
    Engineering
University of Concepcion
Concepcion, Chile

Tahir Rasheed
Department of Chemistry and Chemical Engineering
Shanghai Jiao Tong University
Shanghai, China

C.H.V.V. Ramana
Department of Electrical and Electronics Engineering
    Science
University of Johannesburg, Auckland Park Campus
Johannesburg, South Africa

S. Ramesh
Department of Physics
GITAM Deemed to be University
Bangalore, India

Narendra Reddy
Center for Incubation, Innovation, Research and
    Consultancy
Jyothy Institute of Technology
Bengaluru, India

Asma Rehman
Nanobiotechnology Group National Institute for
    Biotechnology and Genetic Engineering
    (NIBGE)
Faisalabad, Pakistan

Aneela Sabir
Department of Polymer Engineering and Technology
University of the Punjab
Lahore, Pakistan

Nanda Gopal Sahoo
Prof. R. S. Nanoscience and Nanotechnology Centre
Department of Chemistry DSB Campus, Kumaun
    University
Nainital, India

Samina Saleem
Department of Polymer Engineering and
    Technology
University of the Punjab
Lahore, Pakistan

Devalina Sarmah
Department of Physics
Napaam, Tezpur
Assam, India

A. Dennyson Savariraj
Advanced Ceramics and Nanotechnology Laboratory,
    Department of Materials Engineering, Faculty of
    Engineering
University of Concepcion
Concepcion, Chile

Hyungtak Seo
Department of Materials Science and Engineering
Ajou University
Suwon, Republic of Korea

Department of Energy Systems Research
Ajou University
Suwon, Republic of Korea

Muhammad Shafiq
Department of Polymer Engineering and Technology
University of the Punjab
Lahore, Pakistan

Faiqa Shakeel
Department of Polymer Engineering and Technology
University of the Punjab
Lahore, Pakistan

Priyanka Singh
Department of Chemistry
Govt. Digvijay PG Autonomous College
Rajnandgaon, Chhattisgarh, India

A. Smirnova
Department of Materials and Metallurgical Engineering
South Dakota School of Mines and Technology
Rapid City, South Dakota

K. Srinivas
Department of Physics
GITAM Deemed to be University
Bangalore, India

J. Swanson
Department of Materials and Metallurgical Engineering
South Dakota School of Mines and Technology
Rapid City, South Dakota

Muhammad Hassan Tariq
Department of Polymer Engineering and
    Technology
University of the Punjab
Lahore, Pakistan

Pazhanivel Thangavelu
Smart Materials Interface Laboratory, Department
    of Physics
Periyar University
Salem, India

Christelle Pau Ping Wong
Nanotechnology & Catalysis Research Centre
    (NANOCAT), Institute for Advanced Studies
University of Malaya (UM)
Kuala Lumpur, Malaysia

Muhammad Yasir
Department of Chemistry
University of Lahore
Lahore, Pakistan

Muhammad Hamad Zeeshan
Institute of Chemical Engineering and Technology
    (ICET)
University of the Punjab
Lahore, Pakistan

Editors

Dr. Inamuddin is currently working as Assistant Professor in the Chemistry Department, Faculty of Science, King Abdulaziz University, Jeddah, Saudi Arabia. He is a permanent faculty member (Assistant Professor) at the Department of Applied Chemistry, Aligarh Muslim University, Aligarh, India. He obtained master of science degree in organic chemistry from Chaudhary Charan Singh (CCS) University, Meerut, India, in 2002. He received his master of philosophy and Ph.D. degrees in Applied Chemistry from Aligarh Muslim University (AMU), India, in 2004 and 2007, respectively. He has extensive research experience in multidisciplinary fields of analytical chemistry, materials chemistry, and electrochemistry and, more specifically, renewable energy and environment. He has worked on different research projects as project fellow and senior research fellow funded by University Grants Commission (UGC), Government of India, and Council of Scientific and Industrial Research (CSIR), Government of India. He has received Fast Track Young Scientist Award from the Department of Science and Technology, India, to work in the area of bending actuators and artificial muscles. He has completed four major research projects sanctioned by University Grant Commission, Department of Science and Technology, Council of Scientific and Industrial Research, and Council of Science and Technology, India. He has published 147 research articles in international journals of repute and eighteen book chapters in knowledge-based book editions published by renowned international publishers. He has published 60 edited books with Springer (U.K.), Elsevier, Nova Science Publishers, Inc. (U.S.A.), CRC Press Taylor & Francis Asia Pacific, Trans Tech Publications Ltd. (Switzerland), IntechOpen Limited (U.K.), and Materials Research Forum LLC (U.S.A). He is a member of various journals' editorial boards. He is also serving as Associate Editor for journals (*Environmental Chemistry Letter, Applied Water Science and Euro-Mediterranean Journal for Environmental Integration, Springer-Nature*), Frontiers Section Editor (*Current Analytical Chemistry*, Bentham Science Publishers), Editorial Board Member (*Scientific Reports-Nature*), Editor (*Eurasian Journal of Analytical Chemistry*), and Review Editor (*Frontiers in Chemistry*, Frontiers, U.K.) He also serves as guest editor for various special thematic special issues to the journals of Elsevier, Bentham Science Publishers, and John Wiley & Sons, Inc. He has attended as well as chaired sessions in various international and national conferences. He has worked as a Postdoctoral Fellow, leading a research team at the Creative Research Initiative Center for Bio-Artificial Muscle, Hanyang University, South Korea, in the field of renewable energy, especially biofuel cells. He has also worked as a Postdoctoral Fellow at the Center of Research Excellence in Renewable Energy, King Fahd University of Petroleum and Minerals, Saudi Arabia, in the field of polymer electrolyte membrane fuel cells and computational fluid dynamics of polymer electrolyte membrane fuel cells. He is a life member of the *Journal of the Indian Chemical Society*. His research interest includes ion exchange materials, a sensor for heavy metal ions, biofuel cells, supercapacitors, and bending actuators.

Dr. Mohammad Faraz Ahmer is presently working as Assistant Professor in the Department of Electrical Engineering, Mewat Engineering College, Nuh, Haryana, India, since 2012 after working as Guest Faculty in University Polytechnic, Aligarh Muslim University Aligarh, India, during 2009–2011. He completed M.Tech. (2009) and bachelor of engineering (2007) degrees in electrical engineering from Aligarh Muslim University, Aligarh, in the first division. He obtained a Ph.D. degree in 2016 on his thesis entitled "Studies on Electrochemical Capacitor Electrodes." He has published six research papers in reputed scientific journals. He has edited two books with Materials Science Forum, LLC (USA). His scientific interests include electrospun nanocomposites and supercapacitors. He has presented his work at several conferences. He is actively engaged in searching of new methodologies involving the development of organic composite materials for energy storage systems.

Prof. Abdullah M. Asiri is the Head of the Chemistry Department at King Abdulaziz University since October 2009, and he is the Founder and the Director of the Center of Excellence for Advanced Materials Research (CEAMR) since 2010 till date. He is a Professor of organic photochemistry. He graduated from King Abdulaziz University (KAU) with a B.Sc. in chemistry in 1990 and received a Ph.D. degree from University of Wales, College of Cardiff, Wales, in 1995. His research interest covers color chemistry, synthesis of novel photochromic and thermochromic systems, synthesis of novel coloring matters and dyeing of textiles, materials chemistry, nanochemistry and nanotechnology, and polymers and plastics. Prof. Asiri is the principal supervisor of more than 20 M.Sc. and six Ph.D. theses. He is the main author of ten books of different chemistry disciplines. Prof. Asiri is the Editor-in-Chief of *King Abdulaziz University Journal of Science*. His major achievement is the research of tribochromic compounds, a new class of compounds that change from slightly colored or colorless to deep colored when subjected to small pressure or when ground. This discovery was introduced to the scientific community as a new terminology published by International Union of Pure and Applied Chemistry (IUPAC) in 2000. This discovery was awarded a patent from European Patent office and from UK patent. Prof. Asiri has involved in many committees at both KAU level and national level. He played a major role in identifying the national plan for science and technology when he was part of the Advanced Materials Committee working for King Abdulaziz City for Science and Technology (KACST) in 2007 and in advancing the chemistry education and research in KAU. He has been awarded the best researchers from KAU for the past five years. He also received the Young Scientist Award from the Saudi Chemical Society in 2009 and the first prize for the distinction in science from the Saudi Chemical Society in 2012. He also received a recognition certificate from the American Chemical Society (Gulf Region Chapter) for the advancement of chemical science in the Kingdome. He received a Scopus certificate for the most publishing scientist in Saudi Arabia in chemistry in 2008. He is also a member of the editorial board of various journals of international repute. He is the Vice-President of Saudi Chemical Society (Western Province Branch). He holds four US patents, more than one thousand publications in international journals, several book chapters, and edited books.

Dr. Rajender Boddula is currently working for Chinese Academy of Sciences President's International Fellowship Initiative (CAS-PIFI) at National Center for Nanoscience and Technology (NCNST, Beijing). His academic honors includes University Grants Commission National Fellowship and many merit scholarships, and CAS-PIFI. He has published many scientific articles in international peer-reviewed journals and has authored twenty book chapters. He is also serving as an editorial board member and a referee for reputed international peer-reviewed journals. He has published edited books with Springer (UK), Elsevier, Materials Science Forum LLC (USA) and CRC Press Taylor & Francis Asia Pacific, Trans Tech Publications Ltd. (Switzerland). His specialized areas of research are energy conversion and storage, which include sustainable nanomaterials, graphene, polymer composites, heterogeneous catalysis for organic transformations, environmental remediation technologies, photoelectrochemical water-splitting devices, biofuel cells, batteries, and supercapacitors.

1 Polythiophene-Based Battery Applications

Younus Raza Beg and Gokul Ram Nishad
Department of Chemistry
Govt. Digvijay PG Autonomous College
Rajnandgaon, Chhattisgarh, India

Chittaranjan Das
Institute of Applied Materials-Energy Storage System
Karlsruhe Institute of Technology
Karlsruhe, Germany

Priyanka Singh
Department of Chemistry
Govt. Digvijay PG Autonomous College
Rajnandgaon, Chhattisgarh, India

CONTENTS

1.1 INTRODUCTION

An electrochemical cell that converts stored chemical energy into electrical energy is known as a battery. Conventionally, it consists of a cathode as the positive electrode and an anode as the negative electrode, a separator and an electrolyte which undergoes electrochemical reaction and allows the flow of ions, thus conducting current. We will discuss the application of polythiophenes (PTs) and their derivatives in the battery technology as electrode, binder, conduction-promoting agents, separator, electrolyte, coin-cell cases, and Li–O$_2$ catalyst. Until now, a lot would have been discussed about battery along with composition and functions of conductive polymers. Therefore, we would directly deal with the prime focus of this chapter, i.e., PTs. PTs are the polymers of a sulfur heterocycle, thiophene, that becomes conductive upon oxidation. Alan J. Heeger, Alan MacDiarmid, and Hideki Shirakawa received the 2000 Nobel Prize in Chemistry for the discovery and development of conductive polymers. PTs are also conductive polymers and their electrical conductivity is due to the delocalization of electrons along the polymer backbone. Therefore, the conducting polymers are also known as "synthetic metals."

Molecular structure is a major physical property that determines factor for conducting polymers. Synthesis of these materials helps in determining the magnitude of p overlap along the backbone and thus eliminating the structural defects. Interchain overlap and dimensionality depend on the material assembly and/or processing. Performance can be enhanced by planarization of the backbone and assembly of the backbone in the form of p stacks through improvement of the electronic and photonic properties of the resultant materials. A lot of work has been done on the selective engineering of the properties of PTs (Figure 1.1) through synthesis and assembly. Although processability is difficult, the probable high-temperature stability [1] and very high electrical conductivity of PT films make it a highly desirable material. Their optical properties depend on environmental stimuli. Remarkable color shifts are observed in response to changes in solvent, temperature, potential applied, and binding to other molecules.

Among different PT derivatives, the most widely used one is poly(3,4-ethylene dioxythiophene) (PEDOT) (Figure 1.2). It was first prepared by Bayer while attempting to fabricate an easily oxidized and stable conducting polymer without undesirable α–β′ couplings, often found in PT. But PEDOT films did possess high conductivity (3×10^4 S m^{-1}) and were found to be cathodically coloring and almost transparent in the oxidized state [2]. A water-soluble polyelectrolyte [poly(styrene sulfonate), PSS] was incorporated as the counter-ion in the doped state to yield the product poly(3,4-ethylene dioxythiophene):poly(styrenesulfonate) (PEDOT:PSS) (Figure 1.2) [3] in order to tackle its insolubility issues. PEDOT:PSS consists of a component, PSS, with deprotonated sulfonyl groups carrying negative charge and PEDOT carrying positive charges. PEDOT and PEDOT:PSS play key role as organic semiconductors in electronics because they are p- and n-dopable as they can transport either holes or electrons [4]. Several derivatives of PEDOT have been synthesized to produce high-contrast, fast-switching polymer films. It can withstand 200°C temperature without being degraded. Hence, it has superior thermal properties in ambient air as well as good electrochemical stability, charge capacity, and ionic conductivity [5,6]. It offers optical transparency in thin and oxidized films with very high stability, reasonable bandgap, and low redox potential [7]. PT is used in nonlinear optics, electrical supercapacitor, polymeric light-emitting diode (PLED)s, eletrochromics, antistatic coatings, sensors, batteries, solar cells, memory devices, transistors, and a lot more [8].

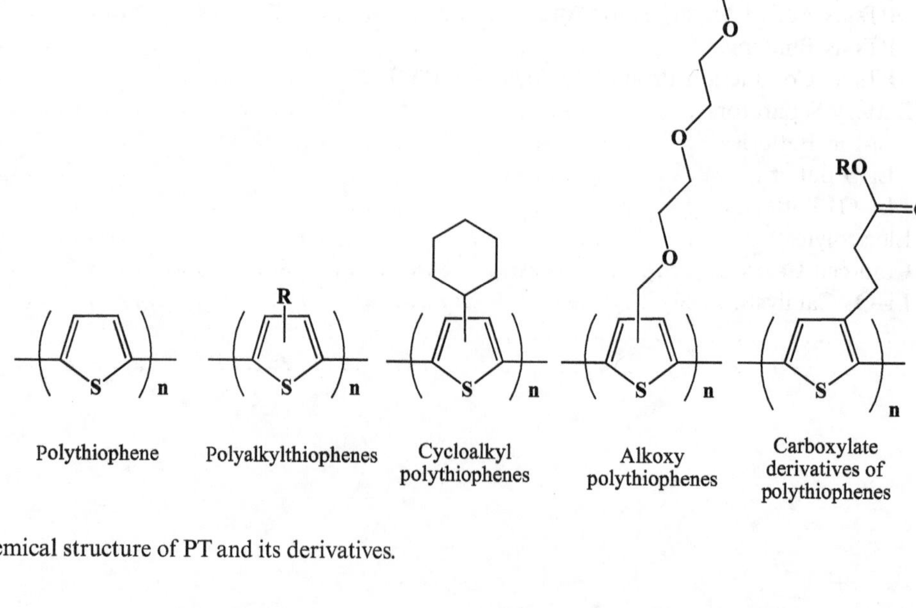

FIGURE 1.1   Chemical structure of PT and its derivatives.

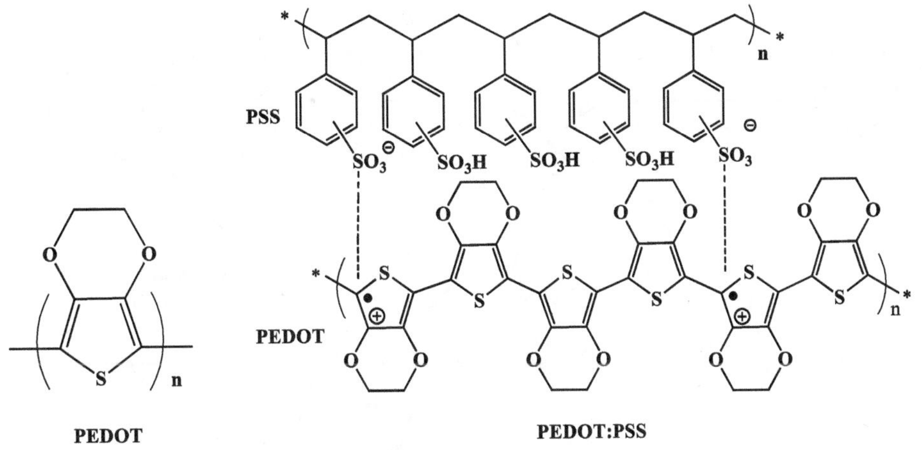

FIGURE 1.2 Chemical structure of PEDOT and PEDOT:PSS.

1.2 SYNTHESIS

PTs can be synthesized by either of the following methods:

1) *Electrochemical polymerization*: it involves the application of potential across a solution of the monomer.
2) *Chemical polymerization*: it involves the use of oxidants or cross-coupling catalysts.

Both the methods will be discussed below along with their advantages and disadvantages.

1.2.1 ELECTROCHEMICAL POLYMERIZATION

Thiophene and an electrolyte solution produce a conductive PT film on the anode in electrochemical polymerization. One of the advantages of electrochemical polymerization is its convenient nature because the polymer does not need to be isolated and purified. Its limitation is the formation of polymers with undesirable alpha–beta linkages and varying degrees of regioregularity. Oxidation of a monomer (educt) by removal of one electron produces a radical cation (polaron). It couples with another radical cation to form a dication dimer (Figure 1.3). The radical cation can also couple with another monomer to produce a radical cation dimer followed by removal of a proton. This process is repeated to give the polymer. The polymer chains are also oxidized that obtain electrical neutrality due to the presence of counter-ion from the electrolyte. The resulting polymer is a good conductor of electricity and acts as a new electrode. Polymerization conditions are responsible for the growth of either long, flexible chains or shorter and more crosslinked chains through deposition of long and well-ordered chains onto the electrode surface.

The quality of electrochemically prepared PT film depends on electrode material, current density, temperature, solvent, electrolyte, presence of water, monomer concentration, monomer structure, and applied potential [8]. The potential required for the oxidation of monomer depends on the electron density of thiophene ring π-system. The required oxidation potential is lowered due to the presence of electron-donating groups, but it gets increased in case of electron-withdrawing groups. Table 1.1 deals with the reaction conditions for electrochemical synthesis of PTs from monomers and their conductances.

1.2.2 CHEMICAL SYNTHESIS

Linear alkyl-substituted polymers have been studied extensively due to their ease of synthesis. Unsubstituted PTs were chemically prepared for the first time in 1980 by Yamamoto et al. (1980) and Lin and Dudek (1980) [17,18]. Both of them synthesized PT through metal-catalyzed polycondensation polymerization of 2,5-dibromothiophene. Since PT is insoluble in tetrahydrofuran (THF) even at low molecular weights, its precipitation under the proposed reaction conditions [17,18] limits the formation of higher-molecular-weight PTs.

Sugimoto et al. (1986) reported the FeCl$_3$ method for polymerization of polyalkylthiophenes (PATs) [19]. But this method gives variable results. Pomrantz et al. (1991) have studied the reproducibility of this reaction [20]. Still, it is the most widely used method for the preparation of PT and its derivatives, in spite of its limitations. Pomerantz et al. (1995) reported the Ullmann coupling for the synthesis of carboxylate derivatives of PTs [21]. Table 1.2 deals with the reaction conditions for the synthesis of some of the PT derivatives that are being widely used in battery technology and various other fields. PTs are used in different fields after mixing with different organic or inorganic additives to give nanoparticles [24], nanocomposites [25–27], nanowires [28,29], nanotubes [30], nanofibers [31,32], nanosheets [33,34], hybrid materials [35], etc. These different forms of PTs have been compared in Table 1.3.

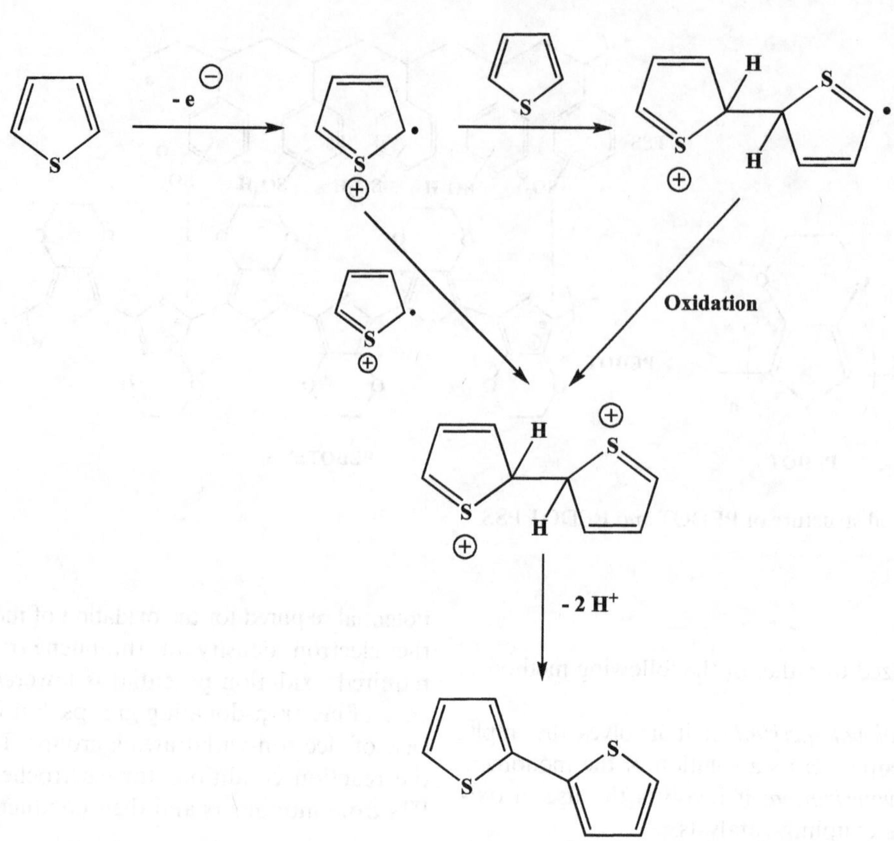

FIGURE 1.3 Initial steps involved in the electrochemical polymerization of thiophenes [8].

TABLE 1.1

Reaction conditions for electrochemical synthesis of PTs from monomers and their conductances

[Monomer] (M)	Solvent	Electrolyte	Electrical condition	σ (Sm^{-1})	Reference
Thiophene					
0.01	CH_3CN	Bu_4NClO_4	1.6 V/SCE	1×10^3–1×10^4	Tourillon and Gamier (1982) [9]
0.2	Propylene carbonate	Et_4NPF_6	1×10^4 mA m^{-2}	2.7×10^4	Sato et al. (1985) [10]
0.1	$PhNO_2$	Bu_4NPF_6	2.0×10^4 mA m^{-2}	3.7×10^4	Roncali et al. (1989) [11]
Methyl thiophene					
0.01	CH_3CN	Et_4NPF_6	1.7 V/SCE	1.0×10^2	Waltman et al. (1983) [12]
-	CH_3CN	$Bu_4NSO_3CF_3$	1.5 V/SCE	3.0×10^3–1.0×10^5	Tourillon and Gamier (1983) [13]
0.05	$PhNO_2$	Bu_4NPF_6	1.5×10^4 mA cm^{-2}	1.975×10^5	Roncali et al. (1988) [14] Yassar et al. (1989) [15]
Bithiophene					
0.1	CH_3CN	HSO_4^-	1.17 V/SCE	1.0×10^{-1}	Waltman et al. (1983) [12]
Trithiophene					
0.002	CH_3CN	Et_4NPF_6	1.7 V/SCE	1.0×10^1–1.0	Inganas et al. (1985) [16]

SCE: Standard calomel electrode.

TABLE 1.2

Various routes for the chemical synthesis of PTs along with the reaction conditions

S. no.	Route	Reactant	Reagent and reaction condition	Product	Reference
1	Yamamoto route	Br—thiophene—Br	Mg/THF/Ni(biyp)Cl$_2$	polythiophene	Yamamoto et al. (1980) [17]
2	Lin and Dudek route	Br—thiophene—Br	Mg/THF/M(acac)$_n$ where M = Pd, Ni, Co or Fe	polythiophene	Lin and Dudek (1980) [18]
3	Sugimoto route	thiophene	FeCl$_3$/CHCl$_3$	polythiophene	Sugimoto et al. (1986) [19]
4	Demercuration polymerization	ClHg—thiophene(R)—HgCl, R = alkyl or esters	Cu/PdCl$_2$/pyridine/Δ	poly(R-thiophene)	Yamamoto et al. (1992) [22]
5	Souto Maior route for synthesis of HH–TT PATs	bithiophene with R groups	FeCl$_3$ or electrochemical method	polymer	R. M. Souto Maior et al. (1990) [23]
6	Souto Maior route for synthesis of TT–HH PATs	bithiophene with R groups	FeCl$_3$ or electrochemical method	polymer	R. M. Souto Maior et al. (1990) [23]
7	Ullmann coupling route	thiophene-3-COOH	1. Br$_2$/AcOH 2. SOCl$_2$ 3. ROH, pyridine 4. Cu, DMF/150°C, 7 days	poly(COOR-thiophene)	Pomerantz et al. (1995) [21]

HH–TT PATs: head–head tail–tail polyalkylthiophenes.
TT–HH PATs: tail–tail head–head polyalkylthiophenes.
THF: tetrahydrofuran.

TABLE 1.3

Method of preparation of various PT-based polymeric materials along with their important properties

S. no.	Polymeric materials	Monomer	Method of preparation/oxidant	Properties	Reference
Nanoparticles					
1	Spherical PT nanoparticles	Thiophene	COP in the presence of SDS and H_2O_2/cupric nitrate, cupric sulfate and cupric chloride	• Regular spherical morphology • Relatively low surface resistivities and • High thermal stability	Wang et al. (2010a) [24]
Nanocomposites					
2	Nanophotoadduct of pentaammine chlorocobalt(III) chloride with hexamine and PT	Thiophene	COP/$FeCl_3$	• Acts as a Schottky diode material • Increases the conductance and thermal stability of PT • Operates at relatively high temperatures in electrical appliances	Najar and Majid (2013) [25]
3	PT/TiO_2 nanocomposites	Thiophene	COP/$FeCl_3$	• Amorphous structures • Semiconductor property was observed for PT/TiO_2-anionic system	Uygun et al. (2009) [26]
4	Graphene/Fe_3O_4@PT	Thiophene	Oxidative chemical polymerization/$FeCl_3$	• Magnetic nanocomposite • An adsorbent with the excellent extraction properties • Simple, fast, and efficient extraction technique for trace analysis of Polycyclic aromatic hydrocarbons (PAHs) in the seawater samples	Mehdinia et al. (2014) [254]
Nanowires					
5	Poly(3-hexylthiophene)-block-poly(3-(3-thioacetylpropyl)oxymethylthiophene) (P3HT)-b-(P3TT) diblock copolymers	2,5-Dibromo-3-hexylthiophene and 2,5-dibromo-3-(3-bromopropyl) oxylmethylthiophene	solvent-induced crystallization/$FeCl_3$	• Capable of maintaining their structural integrity in solvents that normally dissolve the polymers • Could be reduced to the fully solvated polymer	Hammer et al. (2014) [28]
6	Azide-functionalized poly(3-hexylthiophene) (P3HTazide) NWs	3-(Azidohexyl)thiophene and 3-hexylthiophene	Photo-crosslinking and click-functionalization/–	• Water-processable PT nanowires	Kim et al. (2015) [29]
Nanotubes					
7	Nanotube composites consisting of metal nanoparticles and PT	Terthiophene-functionalized metal (Au, Pd) nanoparticles	EOP/–	• Recyclable highly effective catalysts for the carbon–carbon coupling reaction • Hybrid nanomaterials with controllable surface chemistry	Umeda et al. (2008) [30]

Nanofibers

#	Material	Monomer/polymer	Method/oxidant	Features	Reference
8	Poly(3-alkylthiophene) (P3AT) nanofibers	Regioregular poly(3-butylthiophene-2,5-diyl; P3BT), poly(3-hexylthiophene-2,5-diyl; P3HT), poly(3-octylthiophene-2,5-diyl; P3OT), poly(3-decylthiophene-2,5-diyl; P3DT), Regiorandom poly(3-hexylthiophene-2,5-diyl; RRa-P3HT)	Whisker method using anisole solvent/–	• Anisotropic cross-section • Low-cost and mass production of nanofibers	Samitsu et al. (2008) [31]
9	PT nanofibers and carbonaceous PT nanofibers	Thiophene	Surfactant-assisted dilute polymerization method/$FeCl_3$	• Used as electrode materials for asymmetric supercapacitor applications	Balakrishnan et al. (2014) [32]

Nanosheets

#	Material	Monomer/polymer	Method/oxidant	Features	Reference
10	Graphene oxide–PT derivative hybrid nanosheets	Thiophene and benzaldehyde	In situ polymerization method/–	• Enhances the performance of supercapacitor • Long-term stability • Larger energy density, up to 1.48×10^2 Wh kg^{-1} at a power density of 4.16×10^1 W kg^{-1}	Alabadi et al. (2016) [33]
11	PT-coated ultrathin MnO_2 nanosheets	Thiophene	Nanopainting with a thin layer of conducting polymer/–	• Exhibits enhanced capacity retention upon high-rate charge/discharge cycling	Xiao et al. (2010) [34]

Hybrid materials

#	Material	Monomer/polymer	Method/oxidant	Features	Reference
12	Regioregular PT/gold nanoparticle hybrid materials	Poly(3-hexylthiophene)	One-pot synthesis/–	• Narrow size distribution • May find applications in single-electron tunneling studies as well as field effect transistors	Zhai et al. (2004) [35]

PT: polythiophene.
COP: chemical oxidative polymerization.
EOP: electrochemical oxidative polymerization.

1.3 BATTERY APPLICATIONS OF PTs

1.3.1 PTs AS CATHODIC MATERIALS

PTs have been used as the cathodic materials for batteries in various forms that will be discussed below. Electrochemical performance of the batteries is characterized by the charge/discharge test, cyclic voltammetry (CV), electrochemical impedance spectroscopic (EIS) studies, and galvanostatic intermittent titration technique (GITT). Use of PT enhances the electrical conductivity, cycling performance, and Li diffusion coefficient (in Li-ion batteries) along with decrease in the charge transfer resistance. PTs have generally been prepared by facile in-situ chemical oxidation polymerization (COP) method. Typically, a cathode is made up of

- active materials,
- conductive additives to ensure electrical conductivity, and
- binder supporting matrix.

1.3.1.1 PTs as Active Materials

1.3.1.1.1 Li-Ion Batteries (LIBs)

Lithium-ion batteries (LIBs) are the most promising and advanced secondary rechargeable devices due to high energy density (2.50×10^2 Wh kg^{-1}) [36]. Li ions move from the negative to the positive electrode during discharge and back while charging. LIBs use intercalated Li compound as one electrode material. LIBs find versatile applications from small portable electronic devices to electric vehicles or heavy electric vehicles. Handheld electronics use LIBs based on lithium cobalt oxide ($LiCoO_2$) with high energy density and safety risks upon damage [37]. Electric tools, medical equipment, etc., use lithium iron phosphate ($LiFePO_4$), lithium-ion manganese oxide battery ($LiMn_2O_4$, Li_2MnO_3), and lithium nickel manganese cobalt oxide ($LiNiMnCoO_2$) which possess comparatively low energy density but offer longer lives and are safer. Life extension, energy density, charge/discharge capacity, safety, cost reduction, and charging speed improvement are the major research areas for LIBs. Therefore, a lot of work is being done in order to improve the electrochemical performance of LIBs. Renewable organic electrodes based on organosulfur compounds, organic radical polymers, organic carbonyl compounds, and conducting polymers have been worked upon [38].

Olivine–lithium iron phosphates are inexpensive, ecofriendly, stable, and naturally abundant electrode materials and offer relatively large theoretical capacity (1.70×10^5 mAh kg^{-1}) [39], but suffer from poor electrical conductivity. Bai et al. (2010) synthesized a series of $LiFePO_4$/PT composites through in-situ polymerization of thiophene monomers on the surface of $LiFePO_4$ particles [39]. Electrochemical impedance spectra (EIS)

results prove that the PT coating decreases the charge transfer resistance of $LiFePO_4$ electrodes significantly. Charge/discharge testing was done to test the electrochemical performance and study Li insertion and extraction. Enhanced reversible capacity and better cycling ability were observed in comparison to the bare $LiFePO_4$.

Trinh et al. (2013) successfully prepared free-standing PEDOT–$LiFePO_4$ composite films by dynamic three-phase interline electropolymerization (D3PIE) and used as the positive electrode in standard LIBs without further modifications [40]. All the electrochemically inactive materials—such as carbon, polymer binder, and current collector—used in conventional composite cathodes were eliminated. A discharge capacity of 7.5×10^4 mAh kg^{-1} and high capacity retention were obtained. Xing et al. (2013) prepared a $LiFePO_4$/C composite having three-dimensional (3D) carbon network [41]. Cetyltrimethyl ammonium bromide (CTAB) and starch were used as carbon sources. It consisted of a full and uniform carbon coating having a continuous carbon film framework that acted as a good conductive path, thus improving the electron transfer efficiency. The porous structure facilitates diffusion of Li^+ ions that ultimately lead to high ionic and electrical conductivities. An excellent electrochemical performance with high-rate cyclic performance was obtained. A capacity of 9.5×10^4 mAh kg^{-1} at a current density of 3.4×10^3 A kg^{-1} (20 C) and a discharge capacity of 7.4×10^4 mAh kg^{-1} was obtained even after 1200 cycles. Capacity retention was observed to be 80%. This electrode system shows better cyclic performance than reported in previous works.

Lithium orthosilicate, Li_2MSiO_4 (where M = Fe, Mn and Co), based cathodic materials offer high theoretical capacity (4.330×10^6 mAh kg^{-1}) along with high thermal stability due to the presence of strong Si–O bond [42]. They are safe, cost effective, ecofriendly, and easy to synthesize but have poor conductivity that hampers its application in practical cells. While comparing with $LiFePO_4$, orthosilicates have about three to five times lower electronic conductivities. Kempaiah et al. (2012) synthesized monodispersed Li_2MnSiO_4 nanoparticles through supercritical solvothermal method [42]. Upon coating it with PEDOT, discharge capacity of 3.13×10^5 mAh kg^{-1} was achieved. Figure 1.4 depicts the powder X-ray diffraction pattern and Figure 1.5 shows the transmission electron microscopy (TEM) and high-resolution transmission electron microscopy (HRTEM) images as well as particle size distribution of Li_2MnSiO_4 and PEDOT/Li_2MnSiO_4 nanoparticles which reveal that as-synthesized and PEDOT-coated particles exhibit non-aggregated spherical morphology of diameter 5–20 nm (Figure 1.5a and b), with majority of particles around 10 nm (Figure 1.5c).

The lithium vanadium oxide (LiV_3O_8) is safe, cost effective, and offers high-specific capacity and structural stability [43,44]. Its nanomaterials have been tested extensively as it provides short diffusion pathways for

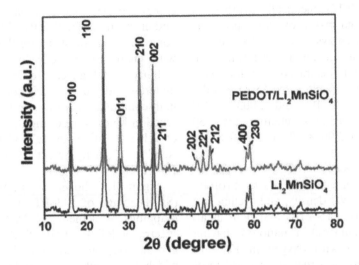

FIGURE 1.4 Powder X-ray diffraction pattern of as-synthesized Li_2MnSiO_4 and $PEDOT/Li_2MnSiO_4$ nanoparticles by supercritical method (Reprinted with permission from Royal Society of Chemistry, London, Great Britain) [42].

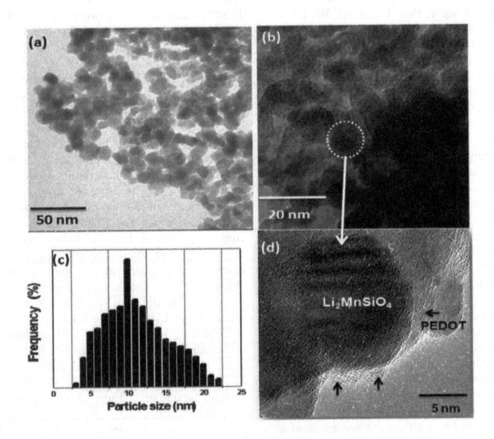

FIGURE 1.5 (a, b) TEM and HRTEM images of as-synthesized Li_2MnSiO_4 and $PEDOT/Li_2MnSiO_4$ nanoparticles, (c) particle size distribution, and (d) HRTEM image showing PEDOT-coated Li_2MnSiO_4 nanoparticles (Reprinted with permission from Royal Society of Chemistry, London, Great Britain) [42].

Li-ion insertion and extraction reactions. It also has high specific surface area that increases contact between the active material and electrolyte [45]. Poor rate capability and capacity that fade during cycling have been observed for bare LiV_3O_8 due to phase transformation during cycling and dissolution of small quantity of LiV_3O_8 in the electrolyte [46]. Hence, structural modification through coating has been tried to enhance its electrochemical performance. Guo et al. (2014) chemically synthesized LiV_3O_8/PT composite through in-situ oxidative polymerization [47]. The composite shows a single phase in the X-ray diffraction (XRD) pattern, but Fourier transform infrared spectroscopy (FTIR) spectra confirm the existence of PT. Initial discharge capacities for 15 wt.% LiV_3O_8/PT composite are 2.133×10^5 and 2.003×10^5 mAh kg^{-1}, with nearly no capacity retention after 50 cycles at current densities of 3.00×10^5 and 9.00×10^5 mA kg^{-1}, respectively.

Conversion-reaction cathodes have the potential to increase the energy density of current LIBs by twofold. But, low energy efficiency, power density, and cycling stability hamper their application to LIBs [48]. Fan et al. (2015) used core-shell FeOF@PEDOT nanorods having fluorine partially substituted with oxygen in FeF_3 in a conversion-reaction cathode in order to increase reaction kinetics and reduction of potential hysteresis [48]. Nanolayer PEDOT coating facilitates fast electronic connection and prevents side reactions. A capacity of 5.60×10^5 mAh kg^{-1} was delivered at 1.0×10^4 mA kg^{-1} with an energy density of higher than 1.10×10^3 Wh kg^{-1}.

Inorganic oxides, sulfides, and graphite allow intercalation of guests into this host using suitable procedures [49,50]. Different two-dimensional (2D) lamellar nanocomposites have also been chemically synthesized through delamination of a layered host into molecular single layers [51,52]. Transition metal dichalcogenides having lamellar structures are amenable to intercalation [53]. For example, MoS_2 has been employed as an inorganic host for polymers to be used as battery cathodes [54] and encapsulating support for magnetic materials [55]. Insertion of conducting polymers into layered host materials has been adopted for preparing hybrid materials with enhanced properties. Murugan et al. (2006) used exfoliation-induced PEDOT nanoribbon formed between the layers of MoS_2 nanosheets as cathode material for LIBs [56]. Significant improvement in the discharge capacity (1×10^5 mAh kg^{-1}) was observed as compared to MoS_2 (4×10^4 mAh kg^{-1}).

Organic electrodes such as aromatic polyimides (PIs) with theoretical capacity of about 400 mAh g^{-1} and working voltage of 2.5 V vs. Li/Li$^+$ have also been used [38]. Discharging/lithium intake by aromatic polyimides involves acceptance of two electrons, thus forming a delocalized radical anion and dianion [57]. However, it has low redox potential and poor structural stability, restricting its use [58,59]. Lithium storage mechanism is a simple redox reaction in aromatic polyimide [60]. Additives such as graphene or carbon nanotubes have been used to overcome this intrinsic electrical insulation of aromatic polyimides [61,62]. A polyimide derivative obtained from condensation polymerization of pyromellitic dianhydride and 2,6-diaminoanthraquinone (PMAQ) and single-walled carbon nanotube network (SWNT) delivered capacity of 1.90×10^5 mAh kg^{-1} at a current rate of 0.1 C [63]. Lyu et al. (2017) prepared an organic

composite electrode material based on an aromatic PI and electron conductive PT [38]. Both PI and PT possess common aromatic structure that allows intimate contacts with highly reversible redox reactions, good structural stability, and high electronic conductivity. PI composite material with 30 wt.% PT coating (PI30PT) shows optimal combination with good conductivity and fast Li reaction kinetics, resulting into a high reversible capacity of 2.168×10^5 mAh kg^{-1} along with high-rate cycling stability. Capacity retention of 94% was obtained after 1000 cycles.

Figure 1.6 shows the scanning electron microscopy (SEM) image of PI30PT, its total and individual EDS mappings of elements C, N, O, and S along with EDS spectra of PT, PI, PI10PT, PI30PT, and PI50PT. Nitrogen and oxygen peaks are clearly seen in the PI sample which are hardly observed in other composite samples. However, sulfur peak in the PI@PT composites increases with an increase in the content of PT. Figure 1.7 represents the cyclic voltammograms (CVs) and charge–discharge profiles of PI, PI10PT, PI30PT, and PI50PT. The PI30PT electrode shows two pairs of well-resolved redox peaks with two reduction peaks at 2.39 and 2.58 V and corresponding oxidation peaks at 2.52 and 2.77 V, respectively (Figure 1.7a). The appearance of doublets confirms a reversible two-electron redox reaction during both lithiation and de-lithiation processes, related to stepwise formation of radical anion and dianion. However, corresponding doublets for both PI and PI10PT are not well defined, owing to their low electronic conductivity caused by zero or lower concentration of highly conductive PT coating which results in slow charge transfer kinetics between the radical anion and the dianion. Higher potential during discharge and lower potential during charge for the PI30PT electrode were observed suggesting that it has the lowest polarization among these polymeric electrodes, resulting in high utilization efficiency of polymer cathode and high specific capacity (Figure 1.7b). The composition and electrochemical properties of different PT-based cathodes have been tabulated for LIBs in Table 1.4.

Organosulfur compounds with multiple thiol groups or disulfide moieties also possess high theoretical capacity, chemical tunability, and redox system to capture Li ions during discharge [64,65]. Despite of low cost and globally abundant nature of sulfur, these compounds have low electronic conductivity and sluggish kinetics of redox reactions at room temperature [66,67]. 2,5-Dimercapto1,3,4-thiadiazole (DMcT) is a promising organosulfur compound as cathode and Kiya et al. (2006) used DMcT–PEDOT composite cathodes for LIBs [68].

1.3.1.1.2 Li–S Batteries

Low energy density of conventional LIBs cannot fulfill the high energy requirement. Therefore, LSBs possessing high theoretical capacity (1.675×10^6 mAh kg^{-1}) and

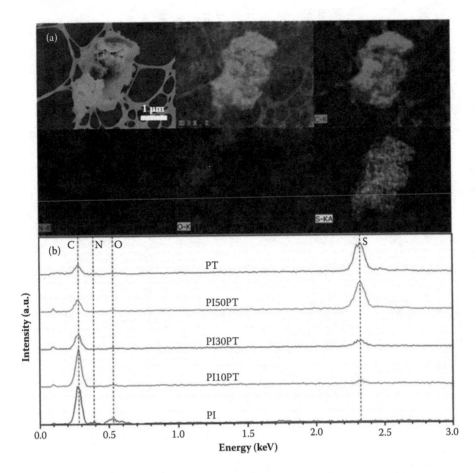

FIGURE 1.6   (a) SEM image of PI30PT, its total and individual EDS mappings of elements C, N, O, and S; (b) EDS spectra of PT, PI, PI10PT, PI30PT, and PI50PT (Reprinted with permission from Royal Society of Chemistry, London, Great Britain) [38].

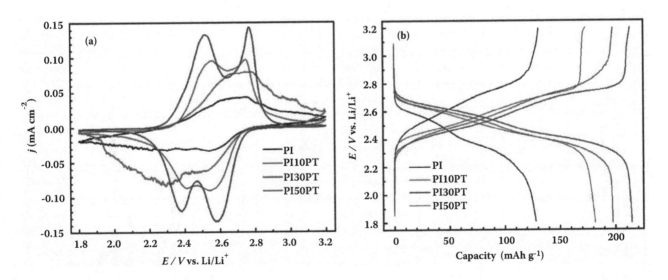

FIGURE 1.7   Cyclic voltammograms (CVs) of (a) PI, PI10PT, PI30PT, and PI50PT at a scan rate of 0.05 mV s^{-1}; charge–discharge profiles of (b) PI, PI10PT, PI30PT, and PI50PT at a current rate of C/10 (Reprinted with permission from Royal Society of Chemistry, London, Great Britain) [38].

TABLE 1.4

Composition and electrochemical properties of PT-based cathodes for Li-ion batteries (LIBs)

S. no.	Working electrode (cathodic material)			Counter electrode/Separator	Electrolyte/Solvent	Method of preparation/Oxidant	Working voltage (V)	Specific capacity (mAhkg⁻¹)/current density (mA kg⁻¹)	Charge capacity/discharge capacity (mAh kg⁻¹)	Reversible capacity/irreversible capacity (mAh kg⁻¹)	Coulombic efficiency (%)/capacity retention (%)/stable cycle life	Reference
	Active material	Conductive additive	Binder									
1	60 wt.% Polyimide@PT composites	30 wt.% C45	10 wt.% PVDF	Li foil/Celgard 2320	$LiPF_6$/EC: DEC: DMC (1:1:1 v/v/v)	In-situ chemical oxidation polymerization/–	1.8–3.2	–/8.0 × 10⁶ ᵃ	–/–	2.168 × 10⁵/–	100/94/1000	Lyu et al. (2017) [38]
2	75 wt.% $LiFePO_4$/PT composites	20 wt.% Acetylene black	5 wt.% PTFE	Li foil/Celgard 2400 microporous membrane	$LiClO_4$/EC: DMC (1:1 v/v)	In-situ polymerization/$FeCl_3$	2.5–4.1	1.569 × 10⁵/3.4 × 10⁴ ᵇ	–/4.998 × 10⁴	–/–	–/–	Bai (2010) [39]
3	PEDOT–$LiFePO_4$ films	–	–	RVC foam/Celgard 2500	$LiPF_6$/EC: DMC (1:1 v/v)	D3PIE/–	2.2–4.0	–/–	–/1.6 × 10⁵	–/–	–/69/>75	Trinh et al. (2013) [40]
4	80 wt.% $LiFePO_4$/C composite powders	10 wt.% acetylene black	10 wt.% PVDF	Li foil/polypropylene (Celgard 2500)	$LiPF_6$/EC: EMC:DEC (1:1:1 v/v/v)	Surfactant–mediated electropolymerization/–	2.3–4.5	9.5 × 10⁴/1.7 × 10⁴	–/1.66 × 10⁵	–/–	100/86/1200	Xing et al. (2013) [41]
5	PEDOT/Li_2MnSiO_4 nanoparticles	–	–	Li/–	$LiClO_4$/EC and DMC	Supercritical solvothermal method/–	1.5–4.5	–/2.165 × 10⁵ ᶜ	–/3.13 × 10⁵	–/–	100/42/20	Kempaiah et al. (2012) [42]
6	85 wt.% LiV_3O_8/PT composite	10 wt.% Carbon black	5 wt.% PVDF	Li/Celgard 2300 film	$LiPF_6$/EC: DMC (1:1 v/v)	In situ oxidative polymerization/$FeCl_3$	1.8–4.0	–/3.0 × 10⁵	–/2.133 × 10⁵	–/–	–/50	Guo et al. (2014) [47]
7	FeOF@PEDOT nanorods	–	–	Li/–	–	In-situ polymerization/–	1.2–4.0	8.85 x 10⁵/1.0 × 10⁴	–/5.50 × 10⁵	–/–	100/94/>150	Fan et al. (2015) [48]
8	Nanoribbon of PEDOT/MoS_2 nanocomposite	Carbon black	PTFE	Li foil/–	$LiClO_4$/E: DMC (1:1 v/v)	Refluxing followed by acidification/$FeCl_3$	2.0–4.4	–/1.5 × 10⁴	–/1.0 × 10⁵	–/–	–/–	Murugan et al. (2006) [56]

PT: polythiophene.
PEDOT: poly(3,4-ethylene dioxythiophene).
DMcT: 2,5-dimercapto-1,3,4-thiadiazole.
PTFE: polytetrafluoroethylene.
PVDF: poly(vinylidene)fluoride.
D3PIE: dynamic three-phase interline electropolymerization.
EC: ethylene carbonate.
DEC: diethyl carbonate.
EMC: ethylmethyl carbonate.
DMC: dimethyl carbonate.

[a]. Unit conversion has been done by following the relation, 1 C = 4.00 × 10⁵ mA kg⁻¹ [38].
[b]. Unit conversion has been done by following the relation, 1 C = 1.70 × 10⁵ mA kg⁻¹ [39].
[c]. Unit conversion has been done by following the relation, 1 C = 4.33 × 10⁵ mA kg⁻¹ [42].

energy density (2.567×10^3 Wh kg^{-1}) have gained attention of the scientific community as new-generation energy-storage devices [69,70]. However, LSBs face several issues such as

- low sulfur utilization,
- poor long-term cycling stability due to the low conductivity of sulfur,
- lithium polysulfide shuttle effect,
- polysulfide corrosion,
- large volume change during discharge–charge processes,
- Li dendrite growth, and
- less reactivity with liquid electrolytes than Li metal.

Conductive polymers act as capable matrices for confinement of lithium polysulfides. The role of different conductive polymers on electrochemical performances of S electrode is still poorly understood due to its vastly different structural configurations. The conductive polymer layer facilitates charge transport and prevents the dissolution of polysulfides. PEDOT composites achieve excellent reversibility, good stability, and fast kinetics. Li et al. (2013) systematically investigated the influence of three of the most well-known conductive polymers—polyaniline, polypyrrole, and PEDOT—on sulfur cathode based on polymer-coated hollow sulfur nanospheres with high uniformity [71]. Experimental observations and theoretical simulations revealed that the capability of these three polymers in improving long-term cycling stability and high-rate performance of sulfur cathode decreased in the order PEDOT > polypyrrole > polyaniline. Table 1.5 presents a comparison of the composition and electrochemical properties of different PT-based cathodes for LSBs.

Porous carbon support is very important in carbon–sulfur composite cathode materials for LSBs, but it is electrochemically inactive, thus lowering the specific capacity and overall energy density. Sulfur nanoparticles (SNPs) have been used to overcome the electrical insulating issue without any other supporting materials because SNPs offer high specific surface area, abundant electron transport active sites and require short electron transfer distance. Chen et al. (2013) reported the synthesis of ultrafine SNPs having diameter 10–20 nm through membrane-assisted precipitation technique followed by PEDOT coating to form S/PEDOT core/shell nanoparticles [72]. Ultrasmall size of SNPs assists in electrical conduction and increases sulfur utilization. Conducting PEDOT shell encapsulation restricts polysulfide diffusion and lessens self-discharging as well as shuttle effect, thus enhancing the cycling stability. Initial discharge capacity of 1.117×10^6 mAh kg^{-1} and stable capacity of 9.30×10^5 mAh kg^{-1} after 50 cycles were obtained. Li et al. (2014) prepared PEDOT–PSS-coated sulfur@activated porous graphene composite (PEDOT/S@aPG) by impregnation of S with aPG followed by encapsulation with PEDOT–PSS and used it as cathode for LSBs [73].

Abundant nanopores and large surface area of aPG provide intimate contact and strong interaction with S species. High specific discharge capacity (1.198×10^6 mAh kg^{-1} at 0.1 C) in the first cycle, excellent rate capability, and good cycling stability were obtained with reversible capacity of 8.45×10^5 mAh kg^{-1} after 200 cycles.

Raman spectra and thermogravimetric curves of sulfur, S@aPG, PEDOT/S@aPG, aPG, and PEDOT–PSS have been presented in Figures 1.8 and 1.9, respectively. Elemental sulfur shows three sharp peaks (155, 220, and 476 cm^{-1}) and the S@aPG composite shows the characteristic peaks of elemental sulfur and the aPG (G band at 1590 cm^{-1} and D band at 1353 cm^{-1}) and additional three new peaks at 318, 374, and 394 cm^{-1} attributable to C–S in plane bending, C–S deformation, and S–S stretching vibrations, respectively (Figure 1.8b). However, the S@aPG composite only shows characteristic peaks of the aPG (1590 and 1353 cm^{-1}) only and the peaks in the range of 150–500 cm^{-1} are not obvious due to the coating of PEDOT–PSS thin layer on the surface of S@aPG. Figure 1.9 indicates the existence of close interactions between sulfur and aPG. PEDOT/S@aPG composite shows the weight loss peak of sulfur that shifts to higher temperature range of 150–320°C with a much lower evaporation rate indicating strong confinement effect due to incorporation of PEDOT–PSS for improving the stability of sulfur, and the sulfur loading of PEDOT/S@aPG is 60.1%.

Oschmann et al. (2015) reported the copolymerization of sulfur- and allyl-terminated poly(3-hexylthiophene-2,5-diyl) (P3HT) for the first time through Grignard metathesis polymerization [74]. Homogeneous composite of S-P3HT forms a framework of interconnected charge transfer channels, and its assembly into electrodes increases the performance of LSBs. Nuclear magnetic resonance (NMR) spectroscopy, size exclusion chromatography, and near-edge X-ray absorption fine spectroscopy confirm the formation of a C–S bond in the copolymer. Even though P3HT is incompatible with elementary sulfur, S-P3HT copolymer can be well dispersed in sulfur on the sub-micrometeric level. Enhanced cycling performance was obtained for S-P3HT copolymer (7.99×10^5 mAh kg^{-1}) as compared to simple mixture lacking covalent link between sulfur and P3HT (5.44×10^5 mAh kg^{-1}) after 100 cycles at 0.5 C.

Sulfur is stored in the large interstitial sites of Prussian blue analogs (PBAs), and efficient reversible insertion/extraction of both Li$^+$ and electrons takes place due to the presence of well-trapped mobile dielectron redox centers in LSBs. $Na_2Fe[Fe(CN)_6]$ possesses a large open framework. As a cathode, it can store sulfur and act as a polysulfide diffusion inhibitor depending on the Lewis acid–base bonding effect. $S@Na_2Fe[Fe(CN)_6]@PEDOT$ composite prepared by Su et al. (2017) has excellent electrochemical properties due to the internal transport of Li$^+$/e$^-$, thus maximizing the use of sulfur [75]. The open metal centers serve as Lewis acid sites with high

TABLE 1.5

Composition and electrochemical properties of PT-based cathodes for Li–S batteries (LSBs) and LPB

S. no.	Working electrode (cathodic material)			Counter-electrode/Separator	Electrolyte/solvent	Method of preparation/oxidant	Working voltage (V)	Specific capacity (mAh kg⁻¹)/current density (mA kg⁻¹)	Charge capacity/discharge capacity (mAh kg⁻¹)	Reversible capacity/irreversible capacity (mAh kg⁻¹)	Coulombic efficiency (%)/capacity retention (%)/stable cycle life	Reference
	Active material	Conductive additive	Binder									
Li–S batteries												
1	Hollow sulfur nanospheres@PEDOT	–	–	Li/–	LiTSFI/DME:DOL (1:1 v/v) LiNO$_3$ (1 wt.%) as an additive	COP/(NH$_4$)$_2$S$_2$O$_8$	2.5–4	–/–	–/1.285×10^6	1.071×10^6 (after 100 cycles)/–	–/83 (after 100 cycles)/–	Li et al. (2013) [71]
2	70 wt.% S/PEDOT core/shell nanoparticles	20 wt.% carbon black	10 wt.% LA-132 latex	Li foil/porous polypropylene	LiTFSI/DOL:DME (1:1 v/v)	Membrane-assisted precipitation and COP/FeCl$_3$	1.0–2.8	9.3×10^5/4.0×10^5	–/1.117×10^6	9.3×10^5 (after 50 cycles)/–	90/–/50	Chen et al. (2013) [72]
3	80 wt.% PEDOT–PSS-coated sulfur@activated porous graphene composite	10 wt.% acetylene black	10 wt.% LA132 in deionized water–ethanol solution	Li foil/porous polypropylene membrane (Celgard 2400)	LiTFSI/DME:DOL (1:1 v/v) with 1% LiNO$_3$ as an additive	Ultrasonic stirring/–	1.5–3.0	1.198×10^6/1.675×10^{5} ᵃ	–/1.180×10^6	8.45×10^5 (after 200 cycles)/–	99/43.6/200	Li et al. (2014) [73]
4	70 wt.% S-P3HT copolymer	25 wt.% carbon (Super P)	5 wt.% polyethylene	Li foil/polyethylene	LiTFSI and LiNO$_3$/DOL:DME (1:1 v/v)	Grignard metathesis polymerization/–	1.7–2.8	1.154×10^6/8.375×10^{5} ᵃ	–/–	–/–	–/–	Oschmann et al. (2015) [74]
5	S@Na$_2$Fe[Fe(CN)$_6$]@PEDOT composite	–	–	–/–	–	Solution method and PEDOT layer coating/–	1.7–2.7	1.291×10^6/1.675×10^{6} ᵃ	–/1.020×10^6	7.7×10^5 (after 100 cycles)/–	–/–	Su et al. (2017) [75]
6	PEDOT-coated diamond-shaped sulfur/SWCNT composite	–	–	Li foil/porous polypropylene membrane (Celgard 2400)	LiTFSI/DOL:DME (1:1 v/v) with 0.1 M LiNO$_3$ as an additive	PEDOT coating through centrifugation/–	1.5–3.0	1.675×10^{5} ᵃ	–/1.52×10^6	–/–	–/76 (after 250 cycles)/–	Zhang et al. (2017a) [76]

Li primary battery

| 7 | 80 wt.% Polythiophene/graphite fluoride composites | 10 wt.% Super P carbon | 10 wt.% PVDF | Li disk/microporous polypropylene/polyethylene/polypropylene film | LiPF$_6$/EC:DMC (1:1 v/v) | COP/FeCl$_3$ | 1.5–3.0 | 7.15×10^5/$4.35 \times 10^{4\,b}$ | –/ | –/ | –/– | Yin et al. (2016) [77] |

PEDOT: poly(3,4-ethylene dioxythiophene).
PEDOT-PSS: poly(3,4-ethylene dioxythiophene): poly(styrenesulfonate).
P3HT: poly(3-hexylthiophene-2,5-diyl).
SWCNT: single-walled carbon nanotube.
PVDF: poly(vinylidene fluoride).
LiTFSI: lithium bis(trifluorosulfonyl imide).
COP: chemical oxidative polymerization.
EC: ethylene carbonate.
DMC: dimethyl carbonate.
DME: dimethyl ether.
DOL: 1,3-dioxolane.

n^a: Unit conversion has been done by following the relation, 1 C = 1.675×10^6 mA kg^{-1} [70].
n^b: Unit conversion has been done by following the relation, 1 C = 8.70×10^5 mA kg^{-1} [78].

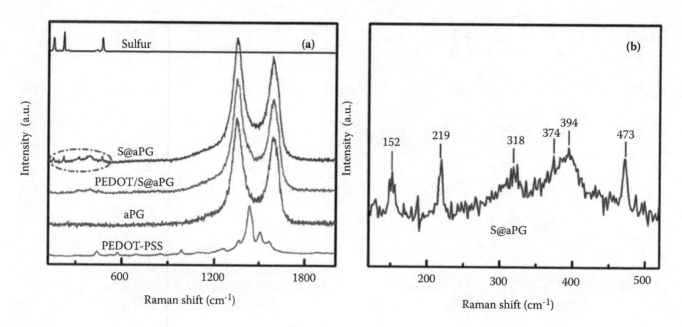

FIGURE 1.8 (a) Raman spectra of sulfur, S@aPG, PEDOT/S@aPG, aPG, and PEDOT–PSS. (b) The magnification of the area marked by the red frame (Reprinted with permission from Royal Society of Chemistry, London, Great Britain) [73].

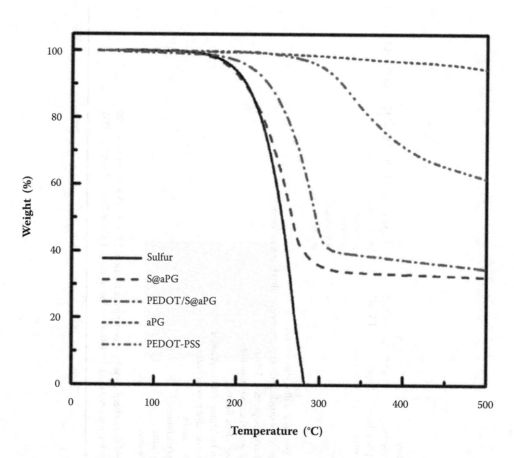

FIGURE 1.9 Thermogravimetric curves for sulfur, S@aPG, PEDOT/S@aPG, aPG, and PEDOT–PSS (Reprinted with permission from Royal Society of Chemistry, Great Britain) [73].

affinity to the negatively charged polysulfide anions, thus decreasing the diffusion of polysulfides from the cathode and minimizing the shuttling effect. Zhang et al. (2017a) designed a free-standing PEDOT-coated diamond-shaped sulfur (D-sulfur)/single-walled carbon nanotube (SWCNT) composite for flexible and binder-free cathode of LSBs [76]. A flexible network was fabricated and dissolution of lithium polysulfides was confined successfully. Initial discharge capacity reached 1.520×10^6 mAh kg^{-1} at 0.1 C, and capacity retention was observed to be 76% after 250 cycles at 0.5 C.

1.3.1.1.3 Li Primary Batteries (LPBs)

Fluorinated carbon (CFx) offers several advantages on being used as cathode materials in LPBs, such as high energy density (up to 1.5×10^3 Wh kg^{-1}), high and flat discharge potential (about 2.2–2.4 V) with broad working temperature range (from 40°C to 170°C) [79]. It has a theoretical capacity of 8.70×10^5 mA kg^{-1} in case of an LPB [78]. But still it lacks good electrical conductivity, leading to low rate capability and an initial potential delay, thus hampering its application in high-power devices. Rate capability increases upon decreasing the fluorine content [80]. Nanostructured conductive additives such as carbon nanotubes [81] or graphene [82] increase the rate capability of the CFx cathode due to the formation of conductive paths for charge transport. Surface coating is also an effective approach for improving the conductivity of CFx cathodes [83,84]. Yin et al. (2016) prepared a series of PT/graphite fluoride (PT/CFx) composites through in-situ polymerization of thiophene monomers on the surface of CFx [77]. PT/CFx of 22.94 wt.% is the most suitable ratio for composite preparation as revealed from TEM due to the formation of uniform and complete PT coating on the surface of CFx particles. It can be discharged at a high rate up to 4 C offering a maximum power density of 4.997×10^3 W kg^{-1} along with high energy density of 1.707×10^3 Wh kg^{-1} (Table 1.5). EIS measurements prove decrease in the charge transfer resistance of CFx cathode due to PT coating. Conductive PT serves the role of both conducting additive and porous adsorbing agent. The rate capability of PT/CFx composites increases remarkably as compared to pure CFx cathode.

1.3.1.1.4 Na-Ion Batteries

Cost effectiveness and natural abundance of sodium have led to increased attention toward rechargeable Na batteries. Different types of electrodes have been worked upon for increasing Na battery performances. Sodium and lithium, both being alkali-metals, have parallel chemical properties but differ in physical properties. Computational studies show that the Ni-ion battery (NIB) materials have working voltages lower than LIBs by 0.18–0.57 V [85]. Therefore, NIBs have lower energy density as compared to LIBs. Ni_3S_2 has low cost, natural abundance, and high theoretical capacity, rendering it demandable for sodium batteries, but its poor conductivity along with large volume expansion hampers its prospective applications. Shang et al. (2015) constructed monolithic Ni_3S_2–PEDOT electrodes with stable electrochemical performance for sodium batteries [86]. Ni_3S_2 was grown directly on Ni foam substrate having high electron transport efficiency. PEDOT layer was capable of protecting the Ni_3S_2 from wrecking due to severe volume expansion during charge–discharge. Ni_3S_2 breaks into Ni and gives Na_2S upon reaction with Na. This results in large volume expansion and structure destruction, causing severe capacity loss upon cycling. Hence, the interface between Ni_3S_2 and electrolyte needs to be stabilized for improving the cycling performance. Stable performance was observed for the monolithic Ni_3S_2–PEDOT electrode during cycling without any overcharging behavior, thus signifying the protective function of PEDOT layer. Stable cycling performance with a capacity of 2.8×10^5 mAh kg^{-1} after 30 cycles between 0.5 and 2 V was obtained due to the insertion/deinsertion of Na^+ ion during cycling. High reversible specific capacity of 4.0×10^5 mAh kg^{-1} even at 6.0×10^5 mA kg^{-1} with high initial coulombic efficiency has also been observed.

Na-incorporated olivine-type cathodes are difficult to prepare because the phase of $NaFePO_4$ grown from conventional chemical synthesis through precursors is an electrochemically inactive maricite phase [87]. It also suffers from low electronic and ionic conductivities, thus limiting the alkali ion insertion/extraction [88]. Ali et al. (2016) reported a discharge capacity of 1.42×10^5 mAh kg^{-1} and stable cycle life over 100 cycles with a capacity retention of 94% for PT-wrapped olivine $NaFePO_4$ [89]. The $NaFePO_4$/PT electrode exhibited appreciable performance at high current densities, and reversible capacities of 7.0×10^4 mAh kg^{-1} and 4.2×10^4 mAh kg^{-1} were observed at 1.50×10^5 mA kg^{-1} and 3.0×10^5 mA kg^{-1}, respectively. In-situ X-ray absorption spectroscopy (XAS) was used to study the related electrochemical reaction mechanism. A systematic change of Fe valence and reversible contraction/expansion of Fe−O octahedra takes place upon desodiation/sodiation. Ex-situ XRD studies show that the deintercalation in $NaFePO_4$/PT electrodes takes place through stable intermediate phase and the lattice parameters show reversible contraction/expansion of unit cell during cycling. Table 1.6 shows the details regarding the composition and electrochemical properties of different PT-based cathodes for NIBs.

Biopolymers like lignin contain quinone/hydroquinone redox systems for charge storage. Lignins are polyphenolic compounds without any regular structure having chemical and physical properties depending on the biological source. It requires conductive additive to tackle its non-conductive nature and allow electron transfer. Biopolymer and conjugated polymer composites have appreciable charge-storage properties but their working has been studied in acidic aqueous media, thus limiting its applications, primarily to

TABLE 1.6

Composition and electrochemical properties of PT-based cathodes for Na-ion batteries (NIBs) and alkaline super-iron battery

S. no.	Working electrode (cathodic material)			Counter electrode	Electrolyte/ solvent	Method of preparation/oxidant	Separator	Working voltage (V)	Specific capacity (mAh kg^{-1})/ current density (mA kg^{-1})	Specific capacity (mAh kg^{-1})/ current density (mA kg^{-1})	Specific capacity (mAh kg^{-1})/ current density (mA kg^{-1})	Coulombic efficiency (%)/capacity retention (%)/stable cycle life	Reference
	Active material	Conductive additive	Binder										
Na-ion batteries													
1	Ni$_3$S$_2$–PEDOT	–	–	Na pieces	NaClO$_4$/EC: DMC (1:1 v/v) with 10% FEC	Hydrothermal synthesis of Ni$_3$S$_2$ followed by surfactant-mediated electrodeposition of PEDOT/–	Glass-fiber and a polypropylene (Celgard 2400)	0.5–2.8	4.0 × 10^5/ 6.0 × 10^5	3.183 × 10^5/–	4.08 × 10^5/–	83.6/–/–	Shang et al. (2015) [86]
2	80 wt.% PT-coated NaFePO$_4$ powder	10 wt.% carbon black	10 wt.% PVDF	Na foil	NaClO$_4$/DEC: PC:EC (1:1:1 v/v/v)	In-situ polymerization followed by sodiation with NaI/FePO$_4$ ratio of 1.1:1/NO$_2$BF$_4$	Glass-fiber membrane	2.2–4.0	~1.0 × 10^4	~1.42 × 10^5	7.0 × 10^4/–	99/94/100	Ali et al. (2016) [89]
3	PEDOT/ Lignin (80/20 or 60/40)	–	–	Na	BMPyrTFSI or BMPyrFSI or EMImTFSI or EMImFSI ionic liquids	Oxidative polymerization/–	Solupor separator	−1.5 to 1.5	–/–	7.5 × 10^4/ 4.6 × 10^4 after 100 cycles	–	–/–	Casado et al. (2017) [90]
Alkaline super-iron battery													
4	K$_2$FeO$_4$ @P3HT	Acetylene black	–	Zinc foil	KOH	In-situ polymerization/ K$_2$FeO$_4$	Polyethylene	0.8–2.0	–/–	~3.51 × 10^5	–	–/–	Wang et al. (2016) [91]

PT: polythiophene.
PEDOT: Poly(3,4-ethylene dioxythiophene).
P3HT: Poly(3-hexylthiophene-2,5-diyl).
PVDF: poly(vinylidene)fluoride.
EC: ethylene carbonate.
PC: propylene carbonate.
DEC: diethyl carbonate.
DMC: dimethyl carbonate.
BMPyrTFSI: 1-butyl-1-methylpyrrolidinium bis(trifluoromethylsulfonyl)imide.
BMPyrFSI: 1-butyl-1-methylpyrrolidinium bis(fluorosulfonyl)imide.
EMImTFSI: 1-ethyl-3-methylimidazolium bis(trifluoromethylsulfonyl)imide.
EMImFSI: 1-ethyl-3-methylimidazolium bis(fluorosulfonyl)imide.

supercapacitors. PEDOT/lignin biopolymers were used as low-cost and sustainable electroactive materials in aprotic ionic liquid electrolytes—1-butyl-1-methylpyrrolidinium bis(trifluoromethylsulfonyl)imide (BMPyrTFSI), 1-butyl-1-methylpyrrolidinium bis(fluorosulfonyl)imide (BMPyrFSI), 1-ethyl-3-methylimidazolium bis(trifluoromethylsulfonyl)imide (EMImTFSI) and 1-ethyl-3-methylimidazolium bis(fluorosulfonyl)imide (EMImFSI)—in sodium full cell batteries by Casado et al. (2017) [90]. Effects of water and sodium salt were also investigated. Sodium batteries having PEDOT/lignin cathode with imidazolium-based ionic liquid electrolyte showed higher capacity values as compared to pyrrolidinium-based ionic liquid electrolyte (7.0×10^4 mAh kg^{-1}).

1.3.1.1.5 Alkaline Super-Iron Battery

K_2FeO_4 batteries possess high reduction potential, high specific capacity (4.06×10^5 mAh kg^{-1}), and high solid-state stability (<0.1% decomposition percent year), and the starting material is abundantly available [91]. Super-iron batteries are chemically instable in alkaline battery system. Therefore, Wang et al. (2016a) used poly(3-hexylthiophene)-coated K_2FeO_4 (K_2FeO_4@P3HT) in order to achieve enhanced capacity and stability [91]. High discharge capacity of 3.51×10^5 mAh kg^{-1} was obtained for K_2FeO_4@P3HT-1% electrode which is about 13% higher than the K_2FeO_4 electrode (Table 1.6). In-situ formation of a two-layer film on the surface of K_2FeO_4 crystal protects direct contact between the electrolyte and K_2FeO_4 and hence reduces the resistance of charge transfer.

1.3.1.1.6 Zn/PEDOT Batteries

Talking about the use of conducting polymers as cathode materials for rechargeable Zn batteries, polyaniline (PANi) has been investigated in aqueous and non-aqueous media, owing to its good redox reversibility and stability in contact with air and aqueous solutions. However, its performance was not satisfactory due to degradation of PANi during cycling [92–94]. Simons et al. (2015) used Zn metal as anode, PEDOT as cathode, and 1-ethyl-3-methylimidazolium dicyanamide ionic liquid as electrolyte for non-aqueous secondary battery in order to avoid dendritic growth, upon charge/discharge cycling, on the Zn surface (Table 1.7) [95]. Use of ionic liquid electrolytes is also beneficial in overcoming inherent flammability and volatility problems of rechargeable batteries. Figure 1.10 represents the schematic diagram of the coin-cell assembly used in this study. High efficiency, cycling ability, and performance over 320 cycles without any short circuit were obtained. Both Zn and PEDOT electrode surfaces had minimal degradation, thus portentous of Zn/PEDOT electrochemical device having extended cycle life under numerous charge/discharge cycles.

1.3.1.1.7 Zn–Mn₂ Batteries

The fiber-shaped Zn–Mn₂ battery delivered a discharge capacity of 1.58×10^5 mAh kg^{-1} at a current density of 7.0×10^1 A kg^{-1} [99]. However, both conventional aqueous and flexible Zn–Mn₂ batteries are non-rechargeable suffering from sharp capacity attenuation [100–102]. Use of bulk Zn foil/paste and binders causes resource wastage, high resistance, low capacity, and rate capability [99,103]. Zeng et al. (2017) prepared a high-performance and stable flexible rechargeable quasi-solid-state Zn–Mn₂ battery by engineering MnO_2 electrodes and gel electrolyte. They used a PEDOT buffer layer and an Mn^{2+}-based neutral electrolyte [96]. The Zn–Mn₂@PEDOT battery has a capacity of 3.666×10^5 mAh kg^{-1} and cycling performance of 83.7% after 300 cycles in aqueous electrolyte (Table 1.7). While using $PVA/ZnCl_2/MnSO_4$ gel as electrolyte, it remains highly rechargeable with more than 77.7% of its initial capacity and 100% coulombic efficiency after 300 cycles. An admirable energy density of 504.9 Wh kg^{-1} and a power density of 8.6 kW kg^{-1} were obtained.

1.3.1.1.8 Zinc–Carbon Batteries

Carbon Nanotubes (CNTs) act as efficient conductive additives for composite electrodes due to high electrical conductivity, strength, flexibility, and large surface area. Carbon nanotube films act as lightweight bendable mechanical support/current collector for a composite electrode [104]. SWCNTs are too expensive as conductive additives and/or materials for current collectors in commercial batteries. Wang et al. (2013) demonstrated the implementation of relatively inexpensive multiwalled carbon nanotubes (MWCNTs) and PEDOT:PSS as conductive composite electrode for flexible Zn/MnO_2 batteries [97] (Table 1.7). The MWCNTs were more efficient than graphite. Carboxylated MWCNTs increase the resistance of the electrode and decrease the electrochemical performance.

1.3.1.1.9 Al Batteries

Volumetric capacity of Al (8.0×10^6 Ah m^{-3}) is four times higher than that of Li. Its gravimetric capacity is 3.0×10^3 Ah kg^{-1} as compared to 3.9×10^3 Ah kg^{-1} for Li and 1.2×10^3 Ah kg^{-1} for Na. Also, Al is the most abundant metal in the earth's crust and has significantly low cost as compared to most of the other metals available for electrochemical energy storage [105]. Hudak (2013) demonstrated the use of conducting polymers as positive electrode-active materials in rechargeable aluminum-based batteries operating at room temperature [98]. Chloroaluminate ionic liquid electrolytes were used that allowed reversible stripping and plating of Al metal at the negative electrode. Stable galvanostatic cycling of polypyrrole and PT cells with capacities at near-theoretical levels (3.0×10^4–1.00×10^5 mAh kg^{-1}) and coulombic efficiencies of about 100% were observed. Energy density of

TABLE 1.7

Composition and electrochemical properties of PT-based cathodes for Zn/PEDOT batteries, zinc–carbon batteries, Zn–Mn₂ batteries, and Al batteries

S. no.	Working electrode (cathodic material)			Counter electrode	Electrolyte/solvent	Method of preparation	Separator	Working voltage (V)	Specific capacity (mAh kg⁻¹)/current density (mA kg⁻¹)	Charge capacity/discharge capacity (mAh kg⁻¹)	Reversible capacity/irreversible capacity (mAh kg⁻¹)	Coulombic efficiency (%)/capacity retention (%)/stable cycle life	Energy density (Wh kg⁻¹)/power density (mW cm⁻²)	Reference
	Active material	Conductive additive	Binder											
Zn/PEDOT battery														
1	85% PEDOT/PDDA	10% carbon black	5% PVDF	Zn	1-Ethyl-3-methylimidazolium dicyanamide ionic liquid	In-situ polymerization	Glass fiber	-0.8–0.7	$-/6.5 \times 10^5$	$-/2.85 \times 10^4$	$-/-$	$-/320$	$-/-$	Simons et al. (2015) [95]
Zn–Mn₂ battery														
2	MnO₂@PEDOT	–	–	Zn nanosheet	PVA/LiCl–ZnCl₂–MnSO₄ gel	Electrodeposition method	NKK separator	1.8	$-/4.0 \times 10^{4\,a}$	$-/3.666 - 10^5$	$2.194 - 10^5/-$	100 (after 300 cycles)/83.7 (after 300 cycles)/–	$5.049 - 10^2/8.6^{\,b}$	Zeng et al. (2017) [96]
Zinc–carbon battery														
3	MnO₂ powder, 1.3% wt. aqueous dispersion of PEDOT:PSS and MWCNTs /PEDOT:PSS/MnO₂/CNTs (12%, raw)	8.6% w/w Graphite	PVP	83.5% w/w Zn dust, 2% w/w PVP, 14.5% w/w PEDOT:PSS + Zn acetate	Zn(CH₃COO)₂ (+ PbCl₂ + HDTAB as inhibitors)	Ultrasonic homogenization	Craft paper separator (MUNKSJÖ paper)	0.5–1.5	$1.2 \times 10^5/-$	$-/-$	$-/-$	$-/-$	$-/-$	Wang et al. (2013) [97]

Al battery

| 4 | Polythiophene film on glassy carbon | Conductive carbon | PTFE | Al | Chloroaluminate ionic liquid AlCl$_3$: EMIC (1.5:1) | Oxidative electropolymerization | Whatman® GF/D glass microfiber mats | 0.6 – 1.6 1.2 (OCV) | –/1.0 × 10^1 [a] | –/5.66 × 10^4 [a] (10th cycle) 5.64 × 10^3 [a] (400th cycle) | –/– | 100.4 (10th cycle) 99.6 (400th cycle)/ – | 4.4 × 10^1/– | Hudak (2013) [98] |

PEDOT: poly(3,4-ethylene dioxythiophene).

PEDOT-PSS: poly(3,4-ethylene dioxythiophene): poly(styrenesulfonate).

PDDA: poly(diallyldimethyl) ammonium.

MWCNTs: multiwalled carbon nanotubes.

CNTs: carbon nanotubes.

PVDF: poly(vinylidene) fluoride.

PTFE: polytetrafluoroethylene.

PVP: polyvinylpyrrolidone.

PVA: poly(vinyl alcohol).

HDTAB: hexadecyltrimethylammonium bromide.

EMIC: 1-ethyl-3-methylimidazolium chloride.

OCV: open circuit voltage;

n[a]. The unit has been taken as mA m^{-2}

n[b]. The unit has been taken as kW kg^{-1}

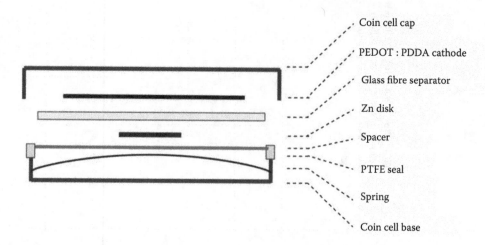

Coin cell cap

PEDOT : PDDA cathode

Glass fibre separator

Zn disk

Spacer

PTFE seal

Spring

Coin cell base

FIGURE 1.10 Schematic of the coin-cell assembly used in this study (Reprinted with permission from John Wiley and Sons Inc., New York, United States) [95].

4.4×10^1 Wh kg^{-1} was obtained for a sealed sandwich-type cell with PT as the positive electrode (Table 1.7).

1.3.1.1.10 Textile Batteries

Several textile-based devices such as transistors, light emitting diodes, biosensors, super capacitors, and energy-storage devices made from conductive polymers and yarns have been reported. These are also known as smart textiles or electronic textiles which are able to respond to environmental stimuli. Smart textile system consists of a sensor, actuator, data-processing and energy-storage devices and links connecting them, hence are bulky, rigid, and heavy leading to reduced comfort of clothing and increased weight. Therefore, thin, flexible batteries compatible with other textile aspects need to be developed.

Odhiambo et al. (2014) investigated textile batteries consisting of PEDOT:PSS as electro-active polymer. Conductive yarns sewn into a textile substrate and then coated with PEDOT:PSS were used as the electrodes (Figure 1.11) [106]. Cotton/polyester fabric was used because of its structure and appreciable wettability of the cotton counterpart. A thermoplastic polyurethane (TPU) layer was used to make the upper surface of the fabric hydrophobic except a central region. PEDOT:PSS was coated in seven layers. A comparison was done between the batteries made with silver-coated polybenzoxazol filament yarns and pure stainless steel filament yarns. The devices were charged and voltage-decay study was done to measure self-discharge. Effect of charging time and various load resistors on the voltage decay was also observed. Devices with pure stainless steel filaments yarn electrode had performed better.

1.3.1.2 PTs as Binder

Polyvinylidene fluoride (PVDF) is a commercially used binder for LIBs owing to its outstanding properties like high electrochemical window at room temperature and tolerance to 10% volume change of graphite upon Li intercalation. But, it requires *N*-methyl-2-pyrrolidone (NMP) as solvent, which is non-ecofriendly and is unable to accommodate large volume changes related to Si- or even Sn-based anodes. The content of conductive carbon additives and inactive polymer binders reaches up to 15–20 wt.% of electrode material. Therefore, electrode capacity can be increased by reducing the amount of inactive components. This can be achieved through their replacement with conducting polymers.

1.3.1.2.1 Li-Ion Batteries

Das et al. (2015) reported the behavior and properties of carbon black-free LiFePO$_4$ composite electrode, with PEDOT:PSS serving as binder and conducting additive [107]. Scanning electron microscopy (SEM), mercury porosimetry, and high-resolution X-ray computed tomography were used to study the effect of polymer amount on morphometric properties of the electrodes. They observed overvoltage decline along with rate capability improvement with increasing PEDOT:PSS content. Eliseeva et al. (2015) used PEDOT:PSS/carboxymethylcellulose (CMC) binder for the improvement of capacity of LiFePO$_4$-based cathode materials [108]. The carbon black-free active material with 96 wt.% C–LiFePO$_4$ and 4 wt.% binder displayed a discharge capacity of 1.48×10^5 mAh kg^{-1} at 0.2 C. Excellent rate capability with the discharge capacity of 1.26×10^5 mAh kg^{-1} at 5 C and good cycling stability at 1 C with <1% decay after 100 cycles were observed.

Large amounts of conductive carbon additives must be added to olivine-type LiFePO$_4$ cathodes for improving their electrical conductivity. Levin et al. (2015) used a small amount (0.5 wt.%) of PEDOT:PSS as a binder and conducting additive [109]. They reported that the gravimetric energy density of these composite electrodes is 15% higher than that of the conventional LiFePO$_4$-based

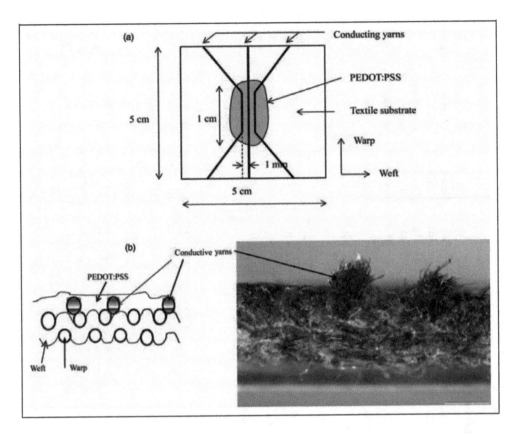

FIGURE 1.11 Schematic: (a) top view of the PEDOT:PSS device and (b) cross-sectional view of the PEDOT:PSS device (Permission granted under the STM Guidelines) [106].

electrodes, but cyclability and C-rate capability remained at the same level. The electrochemical properties of different PT binder-based LIBs and LSBs have been compared in Table 1.8.

Ma et al. (2016) used a sulfur heterocyclic quinine, dibenzo[b,i]thianthrene-5,7,12,14-tetraone (DTT) cathode, and PEDOT:PSS as multifunctional binder for a high-performance rechargeable LIB [110]. DTT has an initial Li-ion intercalation potential of 2.89 V that is 0.3 V higher than its carbon analog. This is caused by the reduction of lowest unoccupied molecular orbital (LUMO) energy level due to S atom introduction. Noncovalent interaction between DTT and PEDOT:PSS due to the electron transfer of extended p-conjugated configuration of DTT (an electron donor) and cationic PEDOT (an electron acceptor) also suppresses the dissolution of DTT; hence, its conductivity is enhanced, thus improving the electrochemical performance. Long-term cycling stability of 2.92×10^5 mAh kg^{-1} for the first cycle, 2.66×10^5 mAh kg^{-1} after 200 cycles at 0.1 C, and a high-rate capability of 2.20×10^5 mAh kg^{-1} at 1 C were obtained.

1.3.1.2.2 Li–S Batteries

High electrical conductivity and enhanced affinity of PEDOT for both sulfur and polysulfides have been used for improving the performance of LSBs. Wang et al.

(2014) used it as binder for sulfur electrode [111]. They used commercial micrometric sulfur particles and synthesized colloidal nanometric sulfur powders of different sizes as active materials in order to study the impact of particle size on cell performance. This electrode system with polyethylene glycol dimethyl ether (PEGDME) electrolyte showed the best cycle performance, with a capacity retention of 68% and specific capacity of 5.78×10^5 mAh kg^{-1} (after 100 cycles) attributable to the high conductivity of PEDOT and viscosity of PEGDME.

Pan et al. (2016b) used a mixture of polyacrylic acid (PAA) and PEDOT:PSS as binder for improving the specific capacity and cycling stability of LSBs [112]. PAA improves the solvent system of sulfur cathodes and promotes the transfer of Li ion, whereas PEDOT:PSS aids the electron transfer process and prevents the dissolution of polysulfides. Initial specific capacities of 1.121×10^6 mAh kg^{-1} and 8.30×10^5 mAh kg^{-1} (after 80 cycles) at 0.5 C are observed and the electrochemical performance was found better than either PAA or PEDOT:PSS single-component binders. Figure 1.12 represents the surface morphologies of sulfur cathodes with PAA, PAA/PEDOT:PSS (2:3), and PEDOT:PSS before cycling and after 80 cycles. Severe solid precipitation of Li$_2$S and Li$_2$S$_2$, formed from the deep discharge of sulfur and strong reduction of soluble polysulfides in the electrolyte, had

TABLE 1.8

Composition and electrochemical properties of PT binder-based Li-ion batteries (LIBs) and Li–S batteries (LSBs)

S. no.	Working electrode (cathodic material)			Counter electrode	Electrolyte	Separator	Method of preparation	Working voltage (V)	Specific capacity (mAh kg⁻¹)/current density (mA kg⁻¹)	Charge capacity/discharge capacity (mAh kg⁻¹)	Columbic efficiency (%)/energy density (mAh kg⁻¹)	Reference
	Active material	Conductive additive	Binder									
Li-ion batteries												
1	94/92/84 wt.% LiFePO$_4$	–	6/8/16 wt.% PEDOT:PSS	Li	LiPF$_6$/EC:EMC (1:1 v/v)	Glass microfiber separator	Coat preparation	2.8–4.0	1.7×10^5/ 3.4×10^4 [a]	–/–	–/–	Das et al. (2015) [107]
2	96 wt.% of C–LiFePO$_4$	–	4 wt.% PEDOT:PSS/CMC	Li	LiPF$_6$ (EC:DMC = 1:1 v/v)	Celgard 2325 membrane	Mechanical mixing	2.0–4.0	1.48×10^5/ 3.4×10^4 [a]	–/1.48×10^5	–/–	Eliseeva et al. (2015) [108]
3	(94/92/84)% LiFePO$_4$		(6/8/16)% PEDOT:PSS	Li	LiPF$_6$/ EC: DMC (1:1 v/v)	Celgard 2325 membrane	Electropolymerization	2.0–4.0	1.42×10^5/ 1.7×10^5 [a]	–/1.47×10^5	98/1.47×10^5	Levin et al. (2015) [109]
4	Sulfur heterocyclic quinone DTT		PEDOT:PSS	Li	LiTFSI/DOL: DME with 1 wt.% LiNO$_3$ additive	–	–	1.6–3.6	2.85×10^5/ 2.85×10^4	–/2.92×10^5 (1st cycle) 2.66×10^5 (after 200 cycles)	99/–	Ma et al. (2016) [110]
Li-S batteries												
5	50% S	40% AB	10% PVDF/ PEDOT	Li disk	LiTFSI/DOL/DME or PEGDME containing LiNO$_3$ (1% wt.)	Polypropylene film (Celgard 2400)	Motorized film application	1.5–2.6	5.78×10^5 (after 100 cycles)/–	–/5.20×10^5 (DOL/DME) 290 (PEGDME)	–/–	Wang et al. (2014) [111]

6 | 70 wt.% ketjen-black-sulfur (KJC/S) composite | 20 wt.% AB | 10 wt.% PAA/ PEDOT: PSS (2:3) | Li foil | LiTFSI/DME:DOL (1:1 v/v) with LiNO$_3$ | Porous polypropylene membrane (Celgard 2400) | Magnetic stirring with DMSO solvent | 1.7–3.0 | 1.121×10^6 6.1×10^5 (after 100 cycles)/ 8.375×10^5 [b] | $-/9.0 \times 10^5$ | $-/-$ | Pan et al. (2016b) [112]

PEDOT: poly(3,4-ethylene dioxythiophene).
PEDOT-PSS: poly(3,4-ethylene dioxythiophene): poly(styrenesulfonate).
PAA: polyacrylic acid.
PVDF: poly(vinylidene)fluoride.
DTT: dibenzo[b,j]thianthrene-5,7,12,14-tetraone.
EMC: ethylmethyl carbonate.
EC: ethylene carbonate.
DME: dimethyl ether.
DOL: 1,3-dioxolane.
CMC: carboxymethyl cellulose.
LiTFSI: lithium bis(trifluorosulfonyl imide).
PEGDME: polyethylene glycol dimethyl ether.

n^a: Unit conversion has been done by following the relation, 1 C = 1.70×10^5 mA kg^{-1} [39].
n^b: Unit conversion has been done by following the relation, 1 C = 1.675×10^6 mA kg^{-1} [70].

FIGURE 1.12 Surface morphologies of sulfur cathodes with (a) PAA, (b) PAA/PEDOT:PSS (2:3), and (c) PEDOT:PSS before cycling and surface morphologies of sulfur cathodes with (d) PAA, (e) PAA/PEDOT:PSS (2:3), and (f) PEDOT:PSS after 80 cycles (Reprinted with permission from Royal Society of Chemistry, London, Great Britain) [112].

been observed in the PAA cathode that led to denser surface, active materials loss, electronic and ionic transport disturbances, and capacity fading. However, in case of the cathode containing PEDOT:PSS, only massive particles appeared on the surface.

1.3.1.3 PTs as Conduction-Promoting Agents

The conducting agents used with water-soluble binders for cathode and anode materials include conventional carbonaceous materials such as acetylene black (AB) [27], carbon black [113], graphite powder [114], grapheme [115], carbon nanotubes [116], and nanofibers [117]. These conducting materials form conducting bridges for connecting the active material particles. But, they have a tendency to aggregate, leading to discontinuous conducting bridges and increased electrode resistance, thus decreasing the electrochemical performance of the cells [118]. PTs are conducting polymers and improve the conductivity of PT-based electrodes. But, in this section we are particularly dealing with electrodes, using only PT as conduction-promoting agents (CPAs) in place of acetylene black, super carbon P, etc. The combination of conductive PEDOT:PSS and a water-soluble binder results into:

- the formation of homogeneous and continuous conducting bridges throughout the electrode and
- enhancement of the compaction density of electrode by reduction of conventional conducting agent content.

Das et al. (2015) have used PEDOT:PSS as a binder for carbon black-free LiFePO$_4$ (LFP) cathode [107].

PEDOT:PSS served twin role of binder and conductive agent. Long-term cycling performances of the electrode are affected due to its weak adhesion ability. Zhong et al. (2016) also used the conducting polymer (PEDOT:PSS) as a conduction-promoting agent and carboxymethyl chitosan (CCTS) as a binder for LiFePO$_4$ cathode in LIBs [119]. They investigated the replacement ratios for conventional acetylene black with PEDOT:PSS through specific resistance and compaction density testing of LFP cathodes in LIBs. The electrical conductivity, peel strength, and compaction density of the electrode sheets were measured. Better cycling and rate performances were also observed. The capacity of 89.7% is retained at 1 C/2 C (charge/discharge) rate, which is comparable with 90% capacity retention for commercial PVDF−LFP over 1000 cycles. Yan et al. (2018) used PEDOT:PSS Mg^{2+} as a conduction-promoting agent to improve the conductivity of the cathode composite [120] and obtained high initial specific capacity up to 1.097 × 10^6 mAh kg^{-1} and high capacity retention of up to 74% after 250 cycles at 0.5 C with a sulfur content of 70 wt.% in the cathode (Table 1.9).

1.3.2 PTs as Air Cathode

1.3.2.1 Li–Air Batteries

Li–air batteries are considered to be the next-generation battery systems due to energy-storage capabilities far exceeding the LIBs but is hampered due to significant overpotential, low-rate capability and limited cyclic performance [121,122]. In aprotic Li–air batteries, air electrodes are charged by reversible formation and dissociation of Li$_2$O$_2$. Side reactions, such as electrolyte decomposition and Li$_2$CO$_3$ formation, also occur that

TABLE 1.9

Composition and electrochemical properties of PT conduction-promoting agent-based Li-ion battery (LIB)

S. no.	Cathode			Counter electrode	Electrolyte	Separator	Working voltage (V)	Current density (mA kg⁻¹)	Specific capacity (mAh kg⁻¹)	Reference
	Active material	Conductive additive	Binder							
1	$LiFePO_4$	PEDOT: PSS	CCTS	Li	$LiPF_6$/EC: DEC: DMC (1:1:1 v/v/v)	Celgard 2400	2.5–4.0	$3.40 \times 10^{4\,a}$	1.55×10^5	Zhong et al. (2016) [119]

PEDOT-PSS: poly(3,4-ethylene dioxythiophene):poly(styrenesulfonate).

CCTS: carboxymethyl chitosan.

EC: ethylene carbonate.

DEC: diethyl carbonate.

DMC: dimethyl carbonate.

n^a: Unit conversion has been done by following the relation, $1\ C = 1.70 \times 10^5\ mA\ kg^{-1}$ [39].

get accumulated on the air-electrode surface leading to increased overpotential and limited Li–air cell-cycle life [123,124]. These side reactions are promoted by carbon, a base material used due to its high conductivity, low weight, and wide surface area [125]. Use of carbon-free electrodes [126,127] or carbon surface coated with conducting polymers has been opted as a remedy [128]. Yoon et al. (2016) used PEDOT:PSS, a multifunctional composite material coated on the surface of carbon (graphene) as an air electrode for enhanced Li–air battery (Figure 1.13) [128]. The PEDOT:PSS layer was deposited using a simple method. The air electrode was prepared without binder like PVDF due to high adhesion of PEDOT:PSS. Considerable discharge and charge capacity

were obtained at all current densities. PEDOT:PSS also served the role of a redox reaction matrix and conducting binder. The accumulation of reaction products formed after cycling due to side reaction in the electrode was effectively reduced by using PEDOT:PSS coating layer as it is capable of suppressing unwanted side reactions between carbon and the electrolyte (and/or Li_2O_2). This leads to enhanced cyclic performance.

1.3.2.2 Aluminum–Air Battery

Oxygen reduction reactions (ORRs) occur at cathode in metal–air batteries. ORR catalysts like Pt, Pt–Ru alloys, and some metal oxides have been used in air cathodes

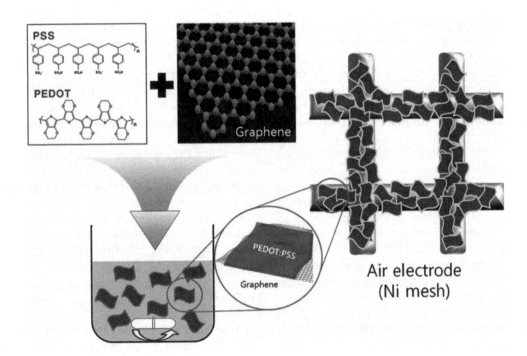

FIGURE 1.13 Schematic diagram showing graphene/PEDOT:PSS composite fabrication process (Image included in the article's Creative Commons license) [128].

[129,130]. Pt-based alloys suffer from high cost and carbon monoxide poisoning on the electrode surface, thus reducing the oxygen amount in the electrochemical reaction and shortening the time required for reaching a steady state. Therefore, attempts are made for finding an alternative to platinum [131]. Kuo et al. (2015) prepared a composite electrode for an aluminum–air battery that led to improved ORR of the air electrode due to matching of α- and β-MnO_2 with PEDOT [132]. The catalyst powders of α-MnO_2 and β-MnO_2 were prepared by the hydrothermal method with different precursors, whereas PEDOT was subsequently deposited on the screen-printed MnO_2/carbon paper electrodes by oxidative chemical vapor deposition. The half-cell polarization curve test strongly depends on crystalline phases of MnO_2. Density functional theory (DFT) and experimental studies showed that the conductivity of PEDOT/α-MnO_2 is higher than PEDOT/β-MnO_2 due to structural effect-mediated improvements in charge transfer. Therefore, PEDOT deposited on α-MnO_2/carbon paper was used as composite cathode in aluminum–air battery. The composition and electrochemical properties of PT-based air cathodes for different metal–air batteries have been presented in Table 1.10.

1.3.2.3 Zinc–Air Battery

The printability of PEDOT has been used for preparing efficient air-electrode material [131,135]. Conducting inks using noble metals such as silver are readily available as compared to galvanic metals such as Al, Mg, or Zn. Hilder et al. (2009) prepared a flexible battery on paper through screen-printing a zinc/carbon/polymer composite as anode on one side of the sheet [133]. The PEDOT cathode was prepared on the other side of the sheet through inkjet printing a pattern of iron(III)p-toluenesulfonate as a solution in butan-1-ol onto paper that was followed by vapor-phase polymerization of monomer. A solution of lithium chloride and lithium hydroxide was used as electrolyte between the two electrodes by inkjet printing on the paper, where it was absorbed into the sheet cross-section. Zinc/carbon–PEDOT/air battery on a polyethylene naphthalate substrate gave a discharge capacity up to 1.4×10^4 mAh m^{-2}. But, the performance of paper-based battery was found to be low having an open-circuit voltage of about 1.2 V and a discharge capacity of 5.0×10^3 mAh m^{-2}. This observation is due to the limited ability of paper/electrolyte combination to take up anode oxidation products, before suffering reduction during ionic mobility. They also discussed the effects of different zinc/carbon/binder combinations, differences in application method for the zinc/carbon composite and various electrolyte compositions.

Conjugated polymer nanosheets belong to new class of 2D soft materials but are rarely developed due to challenges in dimensionality control and lack of synthetic strategies. Su et al. (2016) prepared a sulfur-enriched conjugated polymer nanosheet (2DP-S) with a high aspect ratio [134]. It possesses the chemical identity of cruciform-fused polymeric backbone having quinoidal PT and poly(p-phenylenevinylene) along horizontal and vertical directions, respectively. Two alternating single–double carbon–carbon bonds were shared in each repeating unit. Carrier mobilities up to 0.1 ± 0.05 cm^2 V^{-1} s^{-1} were observed through Terahertz time domain spectroscopy.

1.3.3 PTs as Anodic Materials

Conductive polymers possess specific electrochemical, electronic and photochemical properties owing to their π-conjugated systems that facilitate their application in various fields [136–139]. The redox reactions of conductive polymers with anion doping and dedoping of cathodes have been utilized for charge storage [140–143]. Conductive polymers like polyacetylene (PA) [144], polypyrrol (PPy) [145], PT [146] and polyacene derivatives [147] showed redox reactions with n doping in the potential range of 2.0–0.5 V vs. Li/Li$^+$. Previous studies conclude that the n-doping reaction is inappropriate for anode-active material in LIBs due to the low specific capacity and structure instability [148]. However, the redox reactions of anode with cation doping and dedoping have not been fully studied for charge storage. The theoretical capacity determined by the doping level was 8.2×10^4 mAh kg^{-1} for PT with anion doping of 25 mol.% [141].

1.3.3.1 PTs as Active Materials for Anode

1.3.3.1.1 Li-Ion Batteries (LIBs)

Kuwabata et al. (1998) studied the electrochemical properties of a composite film made from synthetic graphite and poly(3-n-hexylthiophene) (graphite/PHT) as an anode active material in rechargeable LIBs [149]. Charge capacity of 4.35×10^4 mAh kg^{-1} and coulombic efficiency of 94.6% were obtained for graphite-free PHT film in a potential range between 2.0 and 0 V vs. Li/Li$^+$. A specific capacity of 3.12×10^5 mAh kg^{-1} was obtained for the deintercalation of Li$^+$. The irreversible capacity observed during the first cycle of the charge–discharge test of graphite was reduced upon using PHT in place of PVDF binder. The n-type-doped PHT worked as a good electrically conducting binder for graphite and did not require mechanical pressing during electrode preparation.

Kwon et al. (2018) used SWCNTs anchored with carboxylated PT links for high-capacity LIB anode materials [150]. Poly[3-(potassium-4-butanoate) thiophene (PPBT) was used to act as a bridge between SWNT networks and the anode materials, which included monodispersed Fe_3O_4 spheres (sFe_3O_4) and silicon nanoparticles (Si NPs). π-Conjugated backbone of PPBT and carboxylated (COO−) alkyl side chains attracted the π-electron surface of SWNT and interacted chemically with surface hydroxyl (−OH) species of the active material, respectively, to form a carboxylate bond. Electrode thickness changes were

TABLE 1.10

Composition and electrochemical properties of PT-based air cathodes for different metal-air batteries

S. no.	Working electrode (cathodic material)			Counter electrode	Electrolyte/ solvent	Method of preparation/Oxidant	Separator	Working voltage (V)	Current density (mA kg^{-1})	Charge capacity/ discharge capacity (mA m^{-2})	Stable cycle life	Energy density (Wh kg^{-1})/ power density (mW m^{-2})	Reference
	Active material	Conductive additive	Binder										
Li-air battery													
1	PEDOT:PSS-coated grapheme	–	–	Li	LiTFSI/ TEGDME	–	Whatman glass filter	2.35–4.35	4.0×10^5	–	> 100	–	Yoon et al. (2016) [128]
Aluminum–air battery													
2	PEDOT/MnO$_2$/carbon composite paper air electrode	Carbon black	PTFE	Al sheet	KOH	Oxidative chemical vapor deposition/FeCl$_3$	–	−0.7–0	$5.0 \times 10^{5\ a}$	–	–	–	Kuo et al. (2015) [132]
Zinc–air batteries													
3	Air cathode PEDOT printed on paper	–	–	Zinc/carbon/polymer composite	LiCl/LiOH electrolyte	Inkjet printing of (FepTS)$_3$ in butan-1-ol onto paper, followed by vapor-phase polymerization/40% butan-1-ol	–	1.2 (OCV)	–	–/1.4 × 10^4	–	–	Hilder et al. (2009) [133]
4	Air cathode KOH, N/S-2DPC-60	–	–	Zn foil	O$_2$ saturated 0.1 M KOH	Electrochemical method/H$_2$O$_2$	Microporous membrane (25 μm) polypropylene membrane, Celgard 5550	0.3–0.75 0.75 (OCV)	$2.1 \times 10^{4\ a}$	–	–	–/6.9 × 10^3	Su et al. (2016) [134]

PEDOT: poly(3,4-ethylene dioxythiophene).
PEDOT-PSS: poly(3,4-ethylene dioxythiophene): poly(styrenesulfonate).
PTFE: polytetrafluoroethylene.
LiTFSI: lithium bis(trifluorosulfonyl imide).
N/S-2DPC-60: sulfur and nitrogen co-doped porous carbon nanosheets with cruciform-fused polymeric backbone consisting of quinoidal polythiophene and poly(p-phenylenevinylene).
OCV: open circuit voltage.
TEGDME: tetraethylene glycol dimethyl ether.
(FepTS)$_3$: iron(III)ptoluenesulfonate.
n^a: The unit has been taken as mA m^{-2}.

substantially decreased as cracked/pulverized particles resulting from repeated active material volume changes occurring during charging and discharging processes were effectively captured by the above-mentioned architecture. Formation of stable solid–electrolyte interface (SEI) layers, reduction of electrode resistance and enhanced electrode kinetics were observed leading to excellent electrochemical performance. SWNT–Si NP-based electrode had better cycling performance at 0.5 C (1.250×10^6 mA kg^{-1}) as compared to pure Si NPs. n-Doped PT with reversible redox behavior can be used as anodic material for LIBs. But low redox activity and rate performance limit its further study. However, complete n-doping is hard to achieve for PTs; hence, low doping level of 0.25–0.3 has been reported for electro-polymerized PTs [151–153].

Si has been used as anode material for LIBs due to its high theoretical specific capacity of 4200 mA hg^{-1}, natural abundance and relatively low cost [154]. Si interaction with Li occurs through electrochemical alloying for obtaining high capacity in place of usual intercalation process in graphite-based anodes [155]. Continuous alteration of Si structure and surface area as a result of expansion/lithiation and contraction/de-lithiation during cycling leads to formation of a dynamic SEI layer. This leads to high irreversible capacity as a result of consumption of large quantities of electrolyte during the charge/discharge process, fracture and de-lamination of the electrode, low conductivity, short cycle life and high rate of capacity fading [156–158]. PT-coated silicon composite anode materials have been prepared through in-situ chemical oxidation polymerization mainly at α position [159]. PT offers better electric contact between silicon particles; therefore, Si/PT composite electrodes attain better cycling performance as compared to the bare Si anode. The specific capacity of this anode was measured to be 478 mAh g^{-1} after 50 cycles. McGraw et al. (2016) used one-step solid-state in-situ thermal polymerization approach for the synthesis of SiNP–PEDOT (SiNP–PEDOT) nanocomposites for LIB anodes [160]. The in-situ polymerized SiNP–PEDOT nanocomposite shows enhanced lithiation–de-lithiation kinetics, conductivity and rate capability as compared to the ex-situ SiNP–PEDOT nanocomposite.

An all-polymer battery system having polypyrrole and poly(3-styryl-4,4-didecyloxyterthiophene) (poly(OC$_{10}$DASTT)), a functionalized polyterthiophene as the cathode and anode material, respectively, was developed by Wang et al. (2006) [161]. The anode was prepared by casting undoped neutral poly(OC$_{10}$DASTT) from chloroform solution directly on carbon-fiber mat or Ni/Cu-coated nonwoven polyester. High discharge efficiency of over 94% was obtained with discharge capacity of 3.91×10^4 mAh kg^{-1} for battery having poly(OC$_{10}$DASTT) on Ni/Cu-coated nonwoven polyester anode. Wang et al. (2010b) used poly(OC$_{10}$DASTT) or poly(OC$_{10}$STT) polymer electrodeposited on stainless steel mesh and Ni/Cu-coated nonwoven polyester fabric as anode with a polypyrrole cathode in LIB [162]. Poly(OC$_{10}$DASTT)-Ni/Cu-coated fabric anode

displayed a discharge capacity of 4.52×10^4 mAh kg^{-1}. Electropolymerized poly(4,4-didecyloxyterthiophene) (poly(OC$_{10}$STT)) on Ni/Cu-coated fabric exhibited a maximum discharge capacity of 9.47×10^4 mAh kg^{-1}. Repeated charge/discharge cycling leads to decrease in capacity for both the polymers owing to mechanical degradation of the polymers.

All-organic rechargeable batteries possibly will be cheap, sustainable and environment friendly, hence suitable for use as large-scale electric energy-storage devices [163]. But lack of suitable organic anode materials has hindered the development of such a new generation of batteries. Zhu et al. (2013) used chemically polymerized n-dopable poly(3,4-dihexylthiophene)/carbon nanofiber composite as high-capacity anode materials for all-organic LIBs [163]. It offered very high reversible electrochemical capacity of 3.0×10^5 mAh kg^{-1} due to n-type redox reactions and superior capacity retention of 95% after hundred cycles. The theoretical capacity is calculated to be 1.06×10^5 mAh kg^{-1}.

ZnO has theoretical capacity of 9.78×10^5 mAh kg^{-1} which is almost thrice as compared to commercially used graphite anode (3.72×10^5 mAh kg^{-1}) [164,165]. It also possesses higher Li-ion diffusion coefficient than other transition metal oxides [166]. Xu et al. (2015) synthesized ZnO/C hierarchical porous nanorods through one-pot wet-chemical reaction followed by thermal calcinations for use as ultralong-life anode material for LIBs [167]. It contained numerous nanograins with hierarchical micro/nanostructure. In-situ synchrotron high-energy XRD study was performed which revealed that ZnO/C hierarchical porous nanorods involved a two-step reversible lithiation mechanism during charge/discharge. Part of ZnO and Zn remain at the end of the first discharge and charge process, respectively. This results into low coulombic efficiency in the initial few cycles. PEDOT–PSS coating greatly improved their reversible capacity and rate performance of ZnO/C hierarchical porous nanorod-based anode. Reversible capacity of 6.239×10^5 mAh kg^{-1} was obtained after 1500 cycles at 1 C with high-rate capability and long-cycle stability owing to high electronic conductivity of PEDOT–PSS coating and hierarchical structure of ZnO/C porous nanorods.

Structure-design strategy for preparing thiophene-containing conjugated microporous polymers (CMPs) has been adopted by Zhang et al. (2018) for use as anode materials in LIBs [168]. High redox-active thiophene content, highly crosslinked porous structure and improved surface area are responsible for high electrochemical performance of CMPs. Poly(3,3′-bithiophene) with crosslinked structure and surface area of 696 m^2 g^{-1} shows a discharge capacity of 1.215×10^6 mAh kg^{-1} at 4.5×10^4 mA kg^{-1}. High-rate capability and excellent cycling stability with capacity retention of 6.63×10^5 mAh kg^{-1} at 5.0×10^5 mA kg^{-1} after 1000 cycles were obtained. Structure–performance relationships were also studied that offers fundamental understanding of the rational design of CMP anode materials for high-performance LIBs.

Reversible redox reactions take place between cations and the nanostructures of conductive polymers, such as polypyrrole and PT derivatives [166]. Specific capacity of 4.44×10^4 mAh kg^{-1} was obtained for anodes made up of PT nanostructures at a current density of 2.0×10^4 mA kg^{-1}. Introduction of a carboxy group to the thiophene ring increased the specific capacity up to 9.63×10^5 mAh kg^{-1} [169]. But enhanced electrochemical properties were not obtained for bulk-size conductive polymers. They suggested that conductive-polymer nanostructures can be used for developing metal-free, high-performance charge-storage devices. The specific capacity was reported to vary within 20% during 1000 cycles (approximately). Levi et al. (2002) attempted to use PT polypyrrole as anode and cathode, respectively, for rechargeable batteries [170]. n-Doping of PT was done with Tetraethylammonium (TEA) cations. Almost complete deactivation of film was observed with respect to n-doping, even if Li salt was added in small amount to the electrolyte. They suggested the use of fluorosubstituted polyphenyl(thiophene) in order to overcome this drawback. Table 1.11 compares the composition and electrochemical properties of PT-based anodes for different LIBs.

1.3.3.1.2 Li–S Batteries

Formation of dendrite lithium on the negative electrode is a prime concern in case of rechargeable batteries. This leads to low cycling efficiency due to electrically isolated dendrite lithium, high corrosion rate due to their high surface area and internal short circuit due to their growth through the separator. This results into high currents that ultimately heat the battery to the point where highly reactive Li reacts explosively with the electrolyte and cathodic materials. Therefore, it is a prime safety hazard for rechargeable Li batteries. Ma et al. (2014) prepared a protective layer of conductive polymer, PEDOT-co-PEG, on the surface of Li anode for a LSB [171]. The protective layer prevents the formation of SEI between the ether-based electrolyte and Li anode. It inhibits the corrosion of Li anode due to lithium polysulfides efficiently and suppresses the growth of Li dendrites. The discharge capacity was found to be 8.15×10^5 mAh kg^{-1} after 300 cycles at 0.5 C having an average coulombic efficiency of 91.3%, with 2.5–3.0 kg m^{-2} sulfur loading (approximately) on the electrode and commercial electrolyte. Figure 1.14 presents the impedance value of batteries having Li anode with and without the protective conductive polymer layer. Before cycling the impedance value of both the batteries is very close, indicating that the PEDOT-co-PEG layer does not alter the fast ion transport behavior of Li anode. The rearrangement of insulating sulfur on the surface of cathode, separator and anode during the charging and discharging process is the prime cause of the increased resistance.

But, after 100 cycles, the LSB with protected Li anode had much smaller interfacial resistance as compared to the cell with pristine Li electrode. This indicated that the cell with the modified Li electrode had more stable and less resistive SEI layer which assisted more efficient Li-ion transfer at the interfaces during cycling resulting into suppressed rearrangement of insulating active material during the charge/discharge process due to suppression of shuttle effect.

Figure 1.15 represents the surface morphologies of the pristine and surface-protected Li anode after 100 cycles. Surface of the pristine Li anode is loosely packed after 100 cycles, symptomatic of severe dendrite growth and corrosion, whereas the protected Li anode had relatively smoother and denser surface morphology suggestive of corrosion reaction and dendrite growth suppression. Composition and electrochemical properties of PT-based anodes for different types of LSBs, NIBs, potassium-ion batteries (KIBs) and solar rechargeable batteries have been tabulated in Table 1.12.

1.3.3.1.3 Na-Ion Batteries

Sodium ion has larger radius as compared with Li^+ ion which hinders intercalation kinetics and makes the traditional graphite anodes unusable in NIBs. Zhang et al. (2017b) designed PEDOT-encapsulated β-FeOOH nanorods as NIB anodes through a two-step method, including surfactant-assisted hydrothermal synthesis of β-FeOOH nanorods and encapsulation of the obtained nanorods [172]. It delivers a high discharge capacity of 7.26×10^5 mAh kg^{-1} and a reversible capacity of 4.68×10^5 mAh kg^{-1} over 50 cycles. They proposed the formation of PEDOT shells as the causative factor for improvement of electrochemical properties.

1.3.3.1.4 Potassium-Ion Batteries (KIBs)

Despite of low price and ecofriendly nature, KIBs have not gained much attention [175,176]. Standard potential of KIBs (2.93 V) is lower than NIBs (2.71 V), that is, KIBs have higher theoretical working voltage and energy density [176]. KIBs may also have higher ionic conductivity and power in liquid electrolytes because of lower charge density due to large radius and smaller solvated cations [177,178]. Zeng et al. (2018) synthesized nano-sized, porous and amorphous PT as an organic anode for high-performance KIB [173]. A reversible capacity of 5.8×10^4 mAh kg^{-1} was observed at 3.0×10^4 mA kg^{-1} with a good capacity retention even after 80 cycles. The coulombic efficiency increases during the second cycle due to the formation of stable SEI films. A discharge capacity of 4.5×10^4 mAh kg^{-1} was achieved after 100 cycles with an average discharge capacity of 5.6×10^4 mAh kg^{-1}. The curves of charge–discharge were observed to be highly overlapped, signifying good cycling stability.

TABLE 1.11

Composition and electrochemical properties of PT-based anodes for different types of Li-ion batteries (LIBs)

S. no.	Active material	Conductive additive	Binder	Counter electrode/ reference electrode	Electrolyte/ solvent	Method of preparation/ oxidant	Separator	Working voltage (V)	Specific capacity (mAh kg^{-1})/ current density (mA kg^{-1})	Charge capacity/ discharge capacity (mAh kg^{-1})	Reversible capacity/ irreversible capacity (mAh kg^{-1})	Columbic efficiency (%)	Reference
1	Composite of synthetic Graphite and PHT on a Cu foil Graphite/5% PHT (A) Graphite/10% PHT (A) Graphite/15% PHT (C)	–	PVDF	Li foil/Li	LiClO$_4$/EC:DEC (50:50 v/v)	Electropolymerization method Thin film formation/–	Porous PC film	2.0–0	$3.12 \times 10^5/<$ $1.0 \times 10^{4\,a}$	43.5/46.0 $A = 3.27 \times 10^5/$ 4.07×10^5 $B = 3.14 \times 10^5/$ 3.85×10^5 $C = 3.09 \times 10^5/$ 3.52×10^5	$A = -8.0 \times 10^4$ $B = -7.1 \times 10^4$ $C = -4.3 \times 10^4$	94.6 $A = 80.3$ $B = 81.6$ $C = 87.8$	Kuwabata et al. (1998) [149]
2	70 wt.% PT-coated nano-Si Si/20% PT (A) Si/30% PT (B) Si/40% PT (C)	20 wt.% carbon black	10 wt.% NaCMC	Li/Li	LiPF$_6$/EC:EMC: DMC (1:1:1 v/v) and 3 vol.% VC	COP/FeCl$_3$	Celgard 2325 microporous film	0.02–1.5	$4.78 \times 10^5/5.0$ $\times 10^4$ (1st & 2nd cycles) 1.0×10^5 (rest of cycles)	$A = -/2.384$ $\times 10^6$ $B = -/2.245$ $\times 10^6$ $C = -/2.202$ $\times 10^6$	–	–	Wang et al. (2016) [159]
3	Si-PEDOT Nanocomposites on Cu foil	–	–	Li foil/Li	LiPF$_6$/EC:DMC (1:1 v/v)	In-situ and ex-situ thermal polymerization/–	Celgard polypropylene membrane	0.05–1.00	In-situ and ex-situ 1.55 × 10^6 and 1.355 10^6/1.0 × 10^6 (for 1000 cycle)	–	–	–	McGraw et al. (2016) [160]
4	Poly(OC$_{10}$ DASTT) coated on C-fiber mat (A)/ Ni/Cu-coated nonwoven polyester (B)	–	–	PPy electro-polymerized on stainless-steel mesh/Ag	LiPF$_6$/EC:DMC TBAPF$_6$/PC	Electrochemical method Drop casting technique/–	Celgard 2500 (MPPM)	0.2–1.65	$-5.0 \times 10^{2\,a}$	$A = 1.91 \times 10^4$ 1.80×10^4 $B = 4.11 \times 10^4$ 3.91×10^4	–	Cycle life = 50	Wang et al. (2006) [161]
5	Poly (OC$_{10}$DASTT)	–	–	PPy electro-polymerized on stainless-steel mesh/Ag	LiPF$_6$/EC:DMC (1:1 v/v)	Electrodeposition on stainless steel mesh and Ni/Cu-coated nonwoven polyester fabric/–	Celgard 2500 MPPM	0.2–1.65	$-5.0 \times 10^{2\,a}$	-9.47×10^4 -3.71×10^4 (50th cycle)	–	95	Wang et al. (2010) [162]

6	90 wt.% PDHT/C nanofibers	–	10 wt.% PTFE	PTPA film on Li/Li	LiPF$_6$/EC:DMC:EMC (1:1:1 v/v/v)	COP/-	–	3.0–0.02	$3.0 \times 10^5/4.0 \times 10^4$		2.90×10^5/–	99	Zhu et al. (2013) [163]
7	79 wt.% PEDOT–PSScoated ZnO/C hierarchical porous nanorods	14 wt.% carbon black	9.8 wt.% Polyacrylic acid	Li/Li	LiPF$_6$/EC:EMC (3:7 v/v %) and FC electrolyte	Ball milling followed by vacuum drying/–	Celgard 2325	3.0–4.0	$-/9.78 \times 10^5$ [a]	7.878×10^5	7.481×10^5 (after 500 cycles) 6.239×10^5 (after 1500 cycles)/–	59.6	Xu et al. (2015) [167]
8	50 wt.% P33DT	40 wt.% acetylene black	10 wt.% sodium alginate	Li/Li	LiPF$_6$/EC:EMC:DMC (1:1:1 v/v/v)	COP/FeCl$_3$	Celgard 2400 MPPM film	0.9–2.5	1.215×10^6		1.215×10^6/–	99	Zhang et al. (2018) [168]

SWCNTs: single-walled carbon nanotubes.
Poly(OC$_{10}$DASTT): poly(3-styryl-4,4-dideceyloxyterthiophene).
FC electrolyte: fluorinated carbonate-based electrolyte.
PT: polythiophene.
PHT: poly(3-n-hexylthiophene).
PPBT: poly[3-(potassium-4-butanoate) thiophene].
P33DT: poly(3,3′-bithiophene).
PDHT: poly(3,4-dihexylthiophene).
PVDF: poly(vinylidene) fluoride.
PTFE: polytetrafluoroethylene.
PPy: polypyrrol.
NaCMC: sodium carboxymethyl cellulose.
CMC: carboxymethyl cellulose.
TBAPF$_6$: tetrabutylammonium hexafluorophosphate.
PTPA: polytriphenylamine.
COP: chemical oxidative polymerization.
PC: propylene carbonate.
EC: ethylene carbonate.
DEC: diethyl carbonate.
EMC: ethylmethyl carbonate.
DMC: dimethyl carbonate.

n[a]: Unit conversion has been done by following the relation, 1 C = 9.78×10^5 mA kg^{-1} [167].

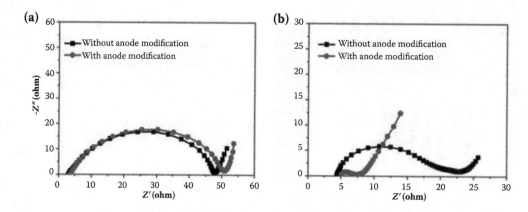

FIGURE 1.14 AC impedance spectra before (a) and after 100 cycles (b) of Li–S batteries assembled with different lithium anodes (Reprinted with permission from Royal Society of Chemistry, London, Great Britain) [171].

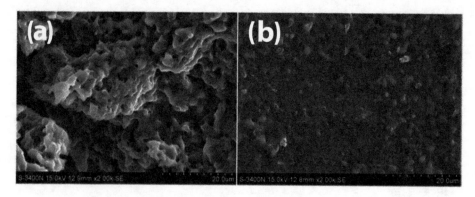

FIGURE 1.15 The surface morphologies of the Li anode after 100 cycles: (a) the pristine Li anode; (b) the surface-protected Li anode (Reprinted with permission from Royal Society of Chemistry, London, Great Britain) [171].

1.3.3.1.5 *Solar Rechargeable Batteries*

Electrochemical rechargeable batteries can convert electrical energy to chemical energy and vice versa, but usually not store solar energy. On the other hand, photovoltaic and photo-electrochemical cells can convert solar light to electricity, but are unsuitable for electric energy storage. Solar energy can be utilized more efficiently if conversion of solar energy to electric energy along with its storage can be carried out in situ in a photo-electrochemical cell. Liu et al. (2012) reported a solar rechargeable battery constructed from a hybrid TiO_2/PEDOT photo-anode and ClO_4^--doped polypyrrole counter-electrode [174]. Here, the dye-sensitized photoanode acts as positive charge storage, and p-doped PPy counter-electrode serves for electron storage in $LiClO_4$ electrolyte. Porous carbon sheet was used as current collector of PEDOT. An absorbent paper piece impregnated with electrolyte was used as the separator. It demonstrates rapid photo-charge at light illumination and stable electrochemical discharge in the dark, leading to in-situ solar-to-electric conversion and storage.

1.3.3.2 PTs as Binders

PEDOT:PSS has been used as a binder in mesocarbon microbead composite anode by Courtel et al. (2011) for LIBs. It results into significant capacity retention at higher charging rates [179]. Salem et al. (2016) used ionically functionalized PT conductive polymers as binders for silicon and graphite anodes for LIBs [180]. Si anode can result into about ten times higher capacity (theoretical capacity = 4.2×10^6 mA kg^{-1}) than the graphite one. However, poor chemical interaction and electrical conductivity between Si and the binder lead to volume changes during battery cycling of Si causing detrimental effects. Electrically conductive PTs having ionic alkyl carboxylate groups of various lengths—poly[3-(lithium acetate) thiophene-2,5-diyl] (PT-3-LiA), poly[3-(lithium-4-butanoate) thiophene-2,5-diyl] (PT-3-LiB), and poly[3-(lithium-6-hexanoate) thiophene-2,5-diyl] (PT-3-LiH)—act as effective multifunctional binders for Si and commercial graphite anodes in LIBs. The polymer with shorter side chain (PT-3-LiA) gave the highest reversible capacity upon pairing with graphite or

TABLE 1.12

Composition and electrochemical properties of PT-based anodes for different types of Li–S battery (LSB), Na-ion battery (NIB), K-ion battery (KIB), and solar rechargeable battery

S. no.	Working electrode (anodic material)			Electrolyte/ solvent	Counter-electrode/ reference electrode	Method of preparation/ oxidant	Separator	Working voltage (V)	Specific capacity (mAh kg⁻¹)/ current density (mA kg⁻¹)	Charge capacity/ discharge capacity (mAh kg⁻¹)	Reversible capacity/ irreversible capacity (mAh kg⁻¹)	Columbic efficiency (%)	Reference
	Active material	Conductive additive	Binder										
Li–S battery													
1	Li anode with 10-mm thick PEDOT-co-PEG coating			$LiN(CF_3 SO_2)_2$ (LiTFSI)/ DOL and DME	80% S–C, 10 wt.% AB, 5 wt.% CMC, 5 wt.% SBR/Li foil	Thin film formation/–	Celgard 2400	1.8–2.6	$-/8.375 \times 10^{5}$ [a]	$-/8.15 \times 10^{5}$ (after 300 cycles)	–	91.3	Ma et al. (2014) [171]
Sodium–ion battery													
3	80 wt.% PEDOT encapsulated β-FeOOH nanorods dissolved in NMP	10 wt.% AB	10 wt.% PVDF	$NaPF_6$/EC: DMC (1:1 v/v)	Metallic sodium/–	Surfactant-assisted hydrothermal synthesis/–	Celgard 2400	0–2.8	$-/2.0 \times 10^{4}$	5.93×10^{5}/ 7.26×10^{5}	$4.68 \times 10^{5}/-$	81.7	Zhang et al. (2017b) [172]
Potassium-ion battery													
4	Nanosized PT with a diameter of 10–30 nm	–		KFSI/DEC: EC (1:1 v/v)	K/K	$COP/FeCl_3$	Glassy-fiber	0–3	$-/3.0 \times 10^{4}$	$-/7.6 \times 10^{4}$ $-/4.5 \times 10^{4}$ (100th cycle)	$5.8 \times 10^{4}/-$	–	Zeng et al. (2018) [173]

(Continued)

TABLE 1.12 (Cont.)

S. no.	Working electrode (anodic material)			Counter-electrode/reference electrode	Electrolyte/solvent	Method of preparation/oxidant	Separator	Working voltage (V)	Specific capacity (mAh kg⁻¹)/current density (mA kg⁻¹)	Charge capacity/discharge capacity (mAh kg⁻¹)	Reversible capacity/irreversible capacity (mAh kg⁻¹)	Columbic efficiency (%)	Reference
	Active material	Conductive additive	Binder										
Solar rechargeable battery													
5	Dye-sensitized hybrid TiO_2/poly(3,4-ethylene-dioxythiophene, PEDOT) photo-anode	–		PPy/Ag	Photo-electrochemical polymerization	$LiClO_4$/–	–	0.25–0.76	–/8.0 × 10³	–/8.3 × 10³	–	~0.1	Liu et al. (2012) [174]

PEDOT: poly(3,4-ethylene dioxythiophene).

$TBAPF_6$: tetrabutylammonium hexafluorophosphate.

LiTFSI: lithium bis(trifluorosulfonyl imide).

PEDOT-co-PEG: poly(3,4-ethylenedioxythiophene)- co-poly(ethylene glycol).

PVDF: poly(vinylidene)fluoride.

PPy: polypyrrol.

NMP: N-methylpyrrolidone.

AB: acetylene Black.

EC: ethylene carbonate.

DMC: dimethyl carbonate.

DME: dimethyl ether.

DOL: 1,3-dioxolane.

CMC: carboxymethyl cellulose.

SBR: styrene–butadiene rubber.

DEC: diethyl carbonate.

COP: Chemical oxidative polymerization.

n^a.: Unit conversion has been done by following the relation, 1 C = 1.675 × 10⁶ mA kg⁻¹ [70].

silicon, reaching 3.0×10^6 mAh kg^{-1}, as compared to PEDOT:PSS binder and sodium carboxymethyl cellulose (NaCMC) binder. The better performance of these multifunctional binders is due to their ability to maintain their doping level and conductivity along with good interaction with Si surface during cycling. The multifunctional polymer acts as binder and a conductor. The PT backbone provides electrical conductivity, whereas the carboxylate ionic group facilitates good interaction with Si particles during expansion (lithiation) and contraction (de-lithiation) processes. Table 1.13 presents the electrochemical properties of PT-based binder and conduction-promoting agents for LIBs.

1.3.3.3 PTs as Conduction Promoting Agents (CPAs)

Shao et al. (2014) used a water-soluble composite PEDOT:PSS as a conduction-promoting agent and carboxymethyl cellulose (CMC) as binder for Si anode of LIBs [181]. Higher initial coulombic efficiencies enhanced cycling and rate performances. More favorable electrochemical kinetics has been observed as compared to the electrodes using CMC binder with acetylene black as a conducting agent. Addition of conductive PEDOT:PSS facilitates the formation of homogeneous and continuous conducting bridges all through the electrode and enhances the compaction density of electrode by decreasing the content of the common conducting agent, acetylene black.

1.3.4 PTs as Battery Separators

Separators are electronically insulating membranes, placed between the anode and cathode, in order to avoid any physical contact between these electrodes. They form an important part of battery and their ionic resistivity is altered by manipulating its thickness and porosity in order to meet the desired level of activity. It also acts as a channel for lithium-ion transportation while battery charging and discharging phenomenon, through the micro–nano pores present in its basic structure. Separators were generally used with the main aim of increasing safety and high-temperature performance of the batteries [182,183].

1.3.4.1 Li-Ion Batteries

The development of finely porous, microporous, and solvent swellable membranes has also not been able to entirely eliminate the internal shorting problem for Li secondary cells. D.L. Foster (1989) obtained a patent on separator for Li batteries and Li batteries including the separator with an object of eliminating the internal shorting by developing a separator material that would react with any Li dendrite penetrating the separator [184]. He developed a multilayer separator, including porous membrane and an electroactive polymeric

material contained within the separator layers. The proposed polymer could be polyvinylpyridine, poly-3 methylthiophene, PT, or polyaniline. Poly-3 methylthiophene can be electrodeposited by first sputtering a thin layer of metal on the separator surface followed by electrodeposition of a very thin layer of polymer on the metallized surface.

Different types of separators like cedar shingles, sausage casing, cellulosic papers, cellophane, nonwoven fabrics, foams, ion-exchange membranes, and microporous flat sheet membranes made from polymeric materials, etc., have been used in batteries over the years. Kinoshita et al. (1985) had surveyed the different types of membranes/separators used in electrochemical systems along with batteries [185]. Boehnstedt and Spotnitz presented a detailed review of battery separators used in lead acid and LIBs, respectively [186,187]. Monolayer and triple-layered polyolefin-based separators are the most commonly studied for LIBs. However, these separators have hydrophobic surface affecting their electrolyte retention ability. Therefore, focus of the scientific community has changed from the cathode modification toward the alteration of the separators. Arora and Zhang (2004) reviewed the various types of separators on the basis of applications in batteries along with chemical, mechanical, and electrochemical properties and prominently discussed the separators for LIBs [188]. Thomas et al. (2004) [189] reported a mathematical model explaining the use of electroactive polymers such as PT to provide overcharge protection for LIBs. As soon as the cell potential goes beyond the oxidation potential of polymer, cell transformation takes place from a battery to a resistor for over a timescale of few minutes, after which a steady-state overcharge condition is achieved.

Zhang et al. (2013) developed a redox-active PT-modified separator for safety control of LIBs [190]. Poly (3-butylthiophene) (P3BT) was incorporated into the micropores of a commercial porous polyolefin film. It was tested for overcharge protection of $LiFePO_4/Li_4Ti_5O_{12}$ LIBs. This separator can switch between electronically insulating and conductive state, reversibly in order to maintain the charge voltage of $LiFePO_4/Li_4Ti_5O_{12}$ cells at a safe value of 2.4 V. This results from reversible p-doping and dedoping of the redox-active conductive polymer P3BT embedded in the separator due to changes of the cathodic potential from an overcharge to a normal operating state, thus preventing the battery from voltage runaway. These separators can be used as a part of internal and self-protecting mechanism for commercial LIBs and other rechargeable batteries due to their reversible effect and non-detrimental nature on the cell performance. Reversible capacity of 5.7×10^4 mAh kg^{-1} at a current rate of 1.0×10^2 Ma kg^{-1} was obtained. Voltage runaway resulting from overcharge is a major concern regarding the unsafe behaviors of LIBs. A highly reversible and effective overcharge protection of

TABLE 1.13

Composition and electrochemical properties of PT-based binder and PT-based conduction-promoting agents for Li-ion batteries (LIBs)

S. no.	Working electrode (anodic material)			Counter-electrode/separator	Electrolyte	Method of preparation/oxidant	Working voltage (V)	Specific capacity (mAh kg⁻¹)/current density (mA kg⁻¹)	Charge capacity/discharge capacity (mAh kg⁻¹)	Reversible capacity/irreversible capacity (mAh kg⁻¹)	Columbic efficiency (%)	Reference
	Active material	Conductive additive	Binder									
Binder												
1	Silicon and graphite	–	PT-3-LiA PT-3-LiB PT-3-LiH	Li/Celgard 3501 separators	$LiPF_6$/EC:DMC (1:1 v/v) with 5% FEC as electrolyte	chemical oxidation method/$FeCl_3$	0.005–2	3.0×10^6/3.5×10^5 ᵃ	$-/4.66 \times 10^5$, 4.55×10^5 and 4.10×10^5, resp.	1.16×10^5, 1.30×10^5 and 9.0×10^4 resp./1.86×10^5	–	Salem et al. (2016) [180]
Conduction-promoting agent												
1	Si	Aqueous PEDOT/ PEDOT:PSS	CMC	Li foil/ Celgard 2400	$LiPF_6$/EC:DEC: DMC (1:1:1 v/v/v)	High-speed mixing/–	0.01–1.5	$-/2.0 \times 10^5$	$-/3.814x\ 10^6$	$1.834x\ 10^6$/–	98.5 (after 100 cycles)	Shao et al. (2014) [181]

PEDOT: poly(3,4-ethylene dioxythiophene).

PEDOT-PSS: poly(3,4-ethylene dioxythiophene): poly(styrenesulfonate).

PT-3-LiA: poly[3-(lithium acetate) thiophene-2,5-diyl].

PT-3-LiB: poly[3-(lithium-4-butanoate) thiophene-2,5-diyl].

PT-3- LiH: poly[3-(lithium-6-hexanoate) thiophene-2,5-diyl].

CMC: carboxymethyl cellulose.

COP: chemical oxidative polymerization.

FEC: fluoroethylene carbonate.

EC: ethylene carbonate.

DEC: diethyl carbonate.

DMC: dimethyl carbonate.

nᵃ: Unit conversion has been done by following the relation, 1 C = 4.20×10^6 mA kg⁻¹ [180].

the P3BT-modified separator was observed for LiFePO$_4$/Li$_4$Ti$_5$O$_{12}$ cells.

Shin et al. (2015) prepared surface-modified polyethylene (PE) separators by thin coating with poly(3,4-ethylenedioxythiophen)-co-poly(ethylene glycol) (PEDOT-co-PEG) conductive copolymer and aluminum fluoride for LIBs [191]. Significant reduction in thermal shrinkage and an improved electrolyte uptake was observed for the surface-modified separators. They used carbon-negative electrodes and LiNi$_{1/3}$Co$_{1/3}$Mn$_{1/3}$O$_2$-positive electrodes and evaluated the cycling performances. Superior cycling performance was obtained for the cells containing the surface-modified separators as compared to the cells containing pristine PE separator at ambient and elevated temperature.

1.3.4.2 Li–S Batteries

Rapid decays in capacity, insulating nature of active sulfur (5×10^{-27} S m^{-1} at 25°C), and shuttling phenomena are among the few drawbacks of LSBs. Different polymers have been tested for higher electrolyte intake [192,193]. However, the processing of required material generally involves painstaking chemical synthesis and uses very low loadings of sulfur active mass, thus restricting the probability of commercialization of LSBs. Park et al. (2014) reported a low-cost, thin, flexible, and mechanically robust alkali-ion electrolyte separator [194]. They worked on LIB using a redox flow-through cathode for LIB and NIB having an insertion host as a cathode. Polypropylene (PP) separator having thin films of PEDOT:PSS was used. The theoretical capacity of NIB was reported to be 1.20×10^5 mA kg^{-1}. They used a thin mixed Li$^+$/electronic-conducting film on the redox-flow cathode in order to block a soluble ferrocene redox molecule from crossing the cathode side to the anode in LIB. Osmosis issue was minimized by solute concentration balancing on both the sides of the separator where the cathodic side contained a soluble redox molecule.

Abbas et al. (2016) used a bifunctional separator for overcoming the shuttling process of lithium polysulfides responsible for the degradation of LSB capacity [195]. The shuttling process is observed due to high solubility of lithium polysulfides in organic electrolytes during electrochemical charge/discharge processes [196]. PEDOT:PSS was sprayed on pristine separator in order to obtain long-cycle LSBs. An electrostatic shield of negatively charged SO$_3^-$ groups present in PSS is formed around soluble lithium polysulfides due to mutual coulombic repulsion interactions. On the contrary, PEDOT shows chemical interactions with insoluble polysulfides (Li$_2$S, Li$_2$S$_2$). Effective protection from the shuttling phenomenon is achieved owing to the dual shielding effect of PEDOT:PSS by restricting lithium polysulfides to the cathodic side of the battery. Electrochemical performance of the battery is also improved because the separator surface changes from hydrophobic to hydrophilic due to PEDOT:PSS coating, resulting in greater wettability, improved

electrolyte intake, and electrochemical performance. Even after running the battery for 1000 cycles at 0.25 C, a very small ultralow decay of 0.0364% per cycle was observed. This is much better than the pristine separator and is also one of the lowest recorded values reported at low current density. Other advantages of this separator under consideration include its versatility, efficiency under stress conditions, flawless performance, and economical and simple nature. The cycling stability of LSBs was greatly enhanced with losses as low as 0.0364% per cycle for 1000 cycles at a low C-rate of 0.25. This LSB has the lowest-reported degradation rate per cycle at a low C-rate [195].

1.3.4.3 Li–O$_2$ Batteries

Li–O$_2$ batteries are expected to become capable next-generation energy-storage systems due to excellent theoretical capacities. But low conductivity of the discharge product leads to large overpotential, thus limiting its use. Different redox mediators (RMs) have been experimented to reduce the overpotential produced during the charging process, thus promoting the oxidation of Li$_2$O$_2$. But reaction between the RM and Li degrades the Li metal anode. Lee et al. (2017) proposed an effective method for checking migration of RM toward the anodic side of Li [197]. They used a commercial glass fiber separator (GF/C, Whatman) coated with a negatively charged polymer mixture of PEDOT:PSS. The migration of 5,10-dihydro-5,10-dimethylphenazine (DMPZ) RM toward counter-electrode of the Li metal anode was suppressed, which was ascertained through visual redox couple diffusion test, morphological investigation, and X-ray diffraction studies. As the unwanted reaction between RM and Li anode was reduced due to the presence of advanced separator, the catalytic activity of DMPZ was enhanced automatically. The round-trip efficiency of 90% was observed up to the 20th cycle.

Mirzaee and Pour (2018) fabricated a paper ultracapacitor based on the BaTiO$_3$/PEDOT:PSS separator film and polyvinyl alcohol (PVA) electrolyte [198]. A film of the graphite nanoparticles (GNPs) was coated on the piece of white commercial paper (3 cm)-substrate to work as an electrode. Typically, paper substrate has a porosity that allows the transfer of electrons between the two sides of the separator. A dielectric layer was coated on both sides of the separator along with fabrication of a blocking layer, in order to combat this defect. The dielectric layer was prepared by addition of PEDOT/PSS to BaTiO$_3$ and dispersing the mixture through ultrasonicator. Electrochemical performance of the ultracapacitor was measured through cyclic voltammetry and galvanostatic techniques. The internal resistance was reported to be 80 Ω, obtained from the Nyquist curve. The composition and electrochemical properties of PT separator-based batteries have been tabulated in Table 1.14.

TABLE 1.14

Composition and electrochemical properties of PT separator-based batteries

S. no.	Anode			Cathode			Separator	Electrolyte/ solvent	Method of preparation/ oxidant	Working potential (V)	Specific capacity (mAh kg⁻¹)/ current density (mA kg⁻¹)	Charge capacity/ discharge capacity (mAh kg⁻¹)	Reversible capacity/ irreversible capacity (mAh kg⁻¹)/ columbic efficiency (%)	Reference
	Anodic material	conductive additive	Binder	Cathodic material	Conductive additive	Binder								
Li-ion batteries														
1	80 wt.% $Li_4Ti_5O_{12}$	12 wt.% AB	8 wt.% PTFE	80 wt.% $LiFePO_4$	12 wt.% AB	8 wt.% PVDF	P3BT@ polyolefin film	$LiPF_6$/EC: DMC:EMC (1:1:1 = v/v/v)	electrochemical oxidation/ $FeCl_3$	2.5–4.2	$-/1.0 \times 10^2$	$-/1.40 \times 10^5$	$5.7 \times 10^4/$ $-/100$	Zhang et al. (2013) [190]
2	88 wt.% MCMB onto a copper foil	8 wt.% super-P carbon	4 wt.% PVDF	85.0 wt.% $LiNi_{1/3}Co_{1/3}Mn_{1/3}O_2$ onto an Al foil	7.5 wt.% super-P carbon	7.5 wt.% PVDF	PEDOT-co-PEG and AlF_3 particle surface-modified polyethylene	$LiPF_6$/EC: EMC:DEC (3:5:2 v/v) and 5 wt.% FEC	Thin film coating/–	3.0–4.5	$-/1.0 \times 10^{6\,a}$	$-/1.822 \times 10^5$	$-/-$	Shin et al. (2015) [191]
Li-S batteries														
3	Li foil mounted on a stainless-steel rod	–	–	Li_2O_2 @ GDL	–	–	PEDOT:PSS@ polypropylene separator (Celgard 2500)	LiTFSI/EC: DEC (1:1 v/v)	Spin coating/ –	2.5–4.0	$-/1.20 \times 10^5$	$-/-$	$-/-$	Park et al. (2014) [194]
4	Li	–	–	80 wt.% graphene/ sulfur	10 wt.% carbon black (super P)	10 wt.% PVDF	Porous Celgard 2500 PP/PE pristine and bifunctional @PEDOT:PSS	LiTFSI/ DME:DOL (1:1 v/v) with 1 wt.% $LiNO_3$ as an additive	Spray coating/–	1.5–2.8	$-/1.675 \times 10^{5\,b}$	$1.26 \times 10^6/$	$1.02 \times 10^6/$ $-/92$	Abbas et al. (2016) [195]
Na-ion batteries														
5	60% $Fe_2(CN)_6$ and $Na_2MnFe(CN)_6$	30% carbon black	10% PTFE	–	–	–	–	$NaClO_4$/EC: DEC (1:1 v/v)	–	3.10–2.78	$-/2.40 \times 10^{5\,c}$	$7.53 \times 10^4/$	$-/-$	Park et al. (2014) [194]

Li–O₂ batteries

6	Li	–	–	Li_2O_2 on the surface of the GDL	–	PEDOT:PSS coated onto the commercial GF/C separator	LiTFSI/DEGDME, DMPZ was used as the RM	Spin coating/–	2.4–4.2	–/5.2 × 10⁴ ᵃ	–/–	–/–	Lee et al. (2017) [197]

PEDOT-PSS: poly(3,4-ethylene dioxythiophene):poly(styrenesulfonate).

PVDF: poly(vinylidene)fluoride.

PTFE: polytetrafluoroethylene.

DEGDME: diethylene glycol dimethyl ether.

PEDOT-co-PEG: poly(3,4-ethylenedioxythiophene)-co-poly(ethylene glycol).

P3BT: poly(3-butylthiophene).

DMPZ: 5,10-dihydro-5,10-dimethylphenazine.

LiTFSI: lithium bis(trifluorosulfonyl imide).

MCMB: mesocarbon microbeads.

GDL: gas diffusion layer.

PE: polyethylene.

PP: polypropylene.

RMs: redox mediators.

EC: ethylene carbonate.

EMC: ethylmethyl carbonate.

DMC: dimethyl carbonate.

DME: dimethyl ether;

DOL: 1,3-dioxolane.

DEC: diethyl carbonate.

AB: acetylene black.

nᵃ. The unit has been taken as mA m⁻².

nᵇ. Unit conversion has been done by following the relation, 1 C = 1.675 × 10⁶ mA kg⁻¹ [70].

nᶜ. Unit conversion has been done by following the relation, 1 C = 1.20 × 10⁵ mA kg⁻¹ [194].

1.3.5 PTs as Electrolytes

PTs have been used as solid electrolytes in electrolyte capacitors, but their battery applications as electrolyte are limited and not much literature is available. The European Patent EP-A 340 512, however, explains the fabrication of a solid electrolyte from 3,4-ethylene-1,2-dioxythiophene along with the use of its cationic polymers, prepared through oxidative polymerization, as solid electrolyte in electrolyte capacitors [199]. Poly (3,4-ethylenedioxythiophene) possesses higher electrical conductivity; hence, it was used as a replacement for manganese dioxide or charge transfer complexes in solid electrolyte capacitors, since it decreases the equivalent series resistance of the capacitor and improves the frequency performance. Jonas et al. (1990) obtained a patent on solid electrolytes [200]. JP-A 2000–021687 stated that high-frequency performance of electrolyte capacitors can be enhanced by using poly(3,4-ethylene-dithiathiophene) as a solid electrolyte, but it is difficult to prepare [201]. Similarly, Reuter et al. (2008) obtained patent on PTs having alkyleneoxythia thiophene units in electrolyte capacitors [202]. However, current leakage in case of electrolyte capacitors having polymeric solid electrolytes is about ten times higher than that of electrolyte capacitors with manganese dioxide as solid electrolyte [203]. High current losses result in earlier flattening of

the battery, in case of mobile electronics. Cutler et al. (2002) prepared PEDOT polyelectrolyte-based electrochromic films via electrostatic adsorption [2].

1.3.6 PTs as Coin-cell Cases

Aluminum-ion battery (AIB) electrical energy-storage system suffers from mediocre volumetric capacity (~7.4 × 10^5 mAh m^{-3}) cathode materials, and it has rarely been applied in the widely used coin-cell configuration. Bare coin cells are also inappropriate for use in AIBs due to the corrosion of stainless steel cases in the presence of chloroaluminate electrolytes. Huang et al. (2017) presented the synthesis of free-standing monolithic nanoporous graphene foam with high density and stability [204]. Glass fiber membrane and commercial carbon felt were used as separator and spacer, respectively. They used PEDOT layer coating on the inner surface of positive and negative coin-cell cases in order to prevent any direct contact between stainless steel and electrolyte. Figure 1.16 shows the correlation between theoretical weight density and pore diameter of highly porous graphene foam at a fixed surface area of 7.62 × 10^5 m^2 kg^{-1}. It had volumetric and gravimetric capacities up to 1.22 × 10^7 mAh m^{-3} and 1.51 × 10^5 mAh kg^{-1}, respectively. Galvanostatic charge/discharge measurement of the blank cell

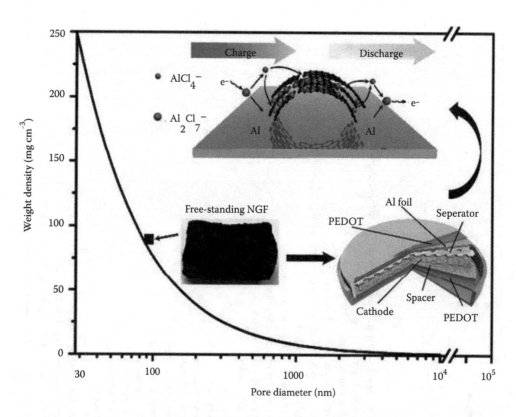

FIGURE 1.16 The correlation between theoretical weight density and pore diameter of highly porous graphene foam at a fixed surface area of 762 m^{2}g^{-1}. Inset is the illustration of the application of freestanding monolithic NGF cathode in a coin-cell-type AIB. (Reprinted with permission from Royal Society of Chemistry, London, Great Britain) [204].

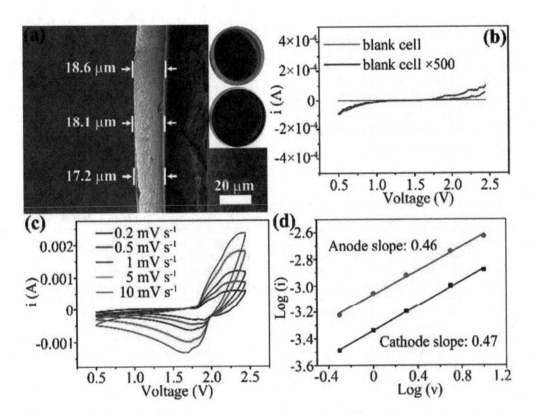

FIGURE 1.17 (a) SEM image of PEDOT coating. Inset is the photographs of PEDOT-coated coin cell. (b) CV analysis of the coin cell without NGF cathode at the scan rate of 5 mV s^{-1}. (c) CV curves of NGF cathode recorded at scan rates of 0.2, 0.5, 1, 5, and 10 mV s^{-1}. (d) The log(i)–log(v) plots of both anodic and cathodic peaks and their linear fitting slopes (Reprinted with permission from Royal Society of Chemistry, London, Great Britain) [204].

without nanoporous graphene foam cathode delivered negligible capacity (~1.0×10^2 mAh kg^{-1}). This verifies high reliability of the cell design. Even after 100 cycles typical peaks of Fe or Cr, the major stainless steel compositions were not identified form the XPS spectrum of the glass fiber separator after cycling. This proves that PEDOT coating provides effective protection by overcycling and is a very reliable battery prototype design. Dendrite formation was not observed on the Al anode surface after cycling, again proving outstanding safety of this AIB configuration. The SEM image of PEDOT coating and CV analysis of the coin cell with and without NGF cathode has been shown in Figure 1.17.

1.3.7 PTs as Li–O₂ Catalyst

Large charge–discharge voltage hysteresis is a barrier limiting practical applications of Li–O₂ batteries [205]. The discharge of Li–O₂ batteries occurs at about 2.7 V. Electrocatalytic systems, including precious and nonprecious metals, transition metal oxides, and high surface area carbon materials, have been worked upon to increase the electron transfer kinetics. But, organic compounds have not received much focus in this field. Nasybulin et al. (2013) studied the catalytic properties of

PEDOT during oxygen reduction/evolution reactions in Li–O₂ batteries [206]. PEDOT was synthesized through in-situ chemical polymerization of EDOT monomer in carbon matrix. PEDOT reduces the overvoltage of charging process and voltage by 0.7–0.8 V for the first cycle. This is due to the electrocatalytic effect of PEDOT attributed to its redox activity. PEDOT performs the role of a mediator in electron transfer during charging and discharging processes. Irreversible decomposition of electrolyte reduces the electrocatalytic effect of PEDOT.

1.4 CONCLUSION

The prime focus of this chapter has been the conductive polymers of thiophenes, a sulfur heterocycle, known as PTs. Synthesis of these materials has been discussed as molecular structure plays a major role in determining their properties. Both electrochemical and chemical routes are followed for PT synthesis, FeCl₃ method being the most common between them. PEDOT and PEDOT: PSS are among the most explored PTs either in bare form or in combined form with other molecules to give composites, nanocomposites, nanosheets, nanofibers, etc. PTs have been widely used in different fields as organic semiconductors, supercapacitor, fuel cells, batteries, etc.

They are p- and n-dopable due to their ability to transport either holes or electrons. PTs have been used in battery as active materials, binder, and conduction-promoting agents in cathodic and anodic materials in LIBs, LSBs, NIBs, and Zn- and Al-based batteries. PT-based textile batteries have also been discussed. PTs have also been used as air cathode, battery separators, electrolytes, coin-cell cases, Li–O$_2$ catalyst, etc. A lot of work on utilization of PT in battery technology is still going on.

REFERENCES

1. Kobayashi, M., J. Chen, T.C. Chung, F. Moraes, A. J. Heeger and F. Wudl. 1984. Synthesis and properties of chemically coupled poly(thiophene). *Synth. Met.* 9: 77–86.
2. Cutler, C.A., M. Bouguettaya and J.R. Reynolds. 2002. PEDOT polyelectrolyte based electrochromic films via electrostatic adsorption. *Adv. Mat.* 14: 684–688.
3. Said, S.M., S.M. Rahman, B.D. Long and M.F.M. Sabri. 2015. Preparation of TEG material based on conducting polymer PEDOT: PSS through treatment by nitric acid. *WIT Trans. Ecol. Environ.* 206: 251–258.
4. Sundfors, F., H. Gustafsson, A. Ivaska and C. Kvarnström. 2010. Characterisation of the aluminium- electropolymerised poly(3,4-ethylenedioxythiophene) system. *Solid State Electrochem.* 14: 1185–1195.
5. Hernandez-Labrado, G.R., R.E. Contreras-Donayre, J.E. Collazos-Castro and J.L. Polo. 2011. Subdiffusion behavior in poly(3,4-ethylenedioxythiophene): polystyrene sulfonate (PEDOT:PSS) evidenced by electrochemical impedance spectroscopy. *Electroanal. Chem.* 659: 201–204.
6. Lee, H.J., J. Lee and S.M. Park. 2010. Electrochemistry of conductive polymers. 45. Nanoscale conductivity of PEDOT and PEDOT:PSS composite films studied by current-sensing AFM. *Phys. Chem. B* 114: 2660–2666.
7. Kirchmeyer, S. and K. Reuter. 2005. Scientific importance, properties and growing applications of poly(3,4-ethylenedioxythiophene). *J. Mater. Chem.* 15: 2077–2088.
8. Schopf, G. and G. Koßmehl. 1997. Polythiophenes-electrically conducting polymers. *Adv. Polym. Sci.* 129: 1–166.
9. Tourillon, G. and F. Gamier. 1982. New electrochemically generated organic conducting polymers. *J. Electroanal. Chem. Interfacial Electrochem.* 135: 173–178.
10. Sato, M., S. Tanaka and K. Kaeriyama. 1985. Electrochemical preparation of highly conducting polythiophene films. *J. Chem. Soc. Chem. Commun.* 0: 713–714.
11. Roncali, J., A. Yassar and F. Gamier. 1989. Films minces de poly(thiophènes) hautement conducteurs. *J. Chim. Phys.* 86: 85–92.
12. Waltman, R.J., J. Bargon and A.F. Diaz. 1983. Electrochemical studies of some conducting polythiophene films. *J. Phys. Chem.* 87: 1459–1463.
13. Tourillon, G. and F. Gamier. 1983. Effect of dopant on the physicochemical and electrical properties of organic conducting polymers. *J. Phys. Chem.* 87: 2289–2292.
14. Roncali, J., A. Yassar and F. Gamier. 1988. Electrosynthesis of highly conducting poly(3-methylthiophene) thin films. *J. Chem. Soc. Chem. Commun.* 0: 581–582.
15. Yassar, A., J. Roncali and F. Gamier. 1989. Conductivity and conjugation length in poly(3-methylthiophene) thin films. *Macromolecules.* 22: 804–809.
16. Inganas, O., B. Liedberg, C.R. Wu and H. Wynberg. 1985. A new route to polythiophene and copolymers of thiophene and pyrrole. *Synth. Met.* 11: 239–249.
17. Yamamoto, T., K. Sanechika and A. Yamamoto. 1980. Preparation of thermostable and electric-conducting poly (2,5-thienylene). *J. Polym. Sci. Polym. Lett. Ed.* 18: 9–12.
18. Lin, J.W.P. and L.P. Dudek. 1980. Synthesis and properties of poly(2,5-thienylene). *J. Polym. Sci. Polym. Chem.* 18: 2869–2873.
19. Sugimoto, R., S. Takeda, H.B. Gu and K. Yoshino. 1986. Preparation of soluble polythiophene derivatives utilizing transition metal halides as catalysts and their property. *Chem. Express.* 1: 635–638.
20. Pomerantz, M., J.J. Tseng, H. Zhu, S.J. Sproull, J. R. Reynolds, R. Uitz, H.J. Arnott and H.I. Haider. 1991. Processable polymers and copolymers of 3-alkylthiophenes and their blends. *Synth. Met.* 41: 825–830.
21. Pomerantz, M., H. Yang and Y. Cheng. 1995. Poly(alkyl thiophene-3-carboxylates). Synthesis and characterization of polythiophenes with a carbonyl group directly attached to the ring. *Macromolecules.* 28: 5706–5708.
22. Yamamoto, T., A. Morita, Y. Miyazaki, T. Maruyama, H. Wakayama, Z.H. Zhou, Y. Nakamura, T. Kanbara, S. Sasaki and K. Kubota. 1992. Preparation of π-conjugated poly(thiophene-2,5-diyl), poly(p-phenylene), and related polymers using zerovalent nickel complexes. Linear structure and properties of the π-conjugated polymers. *Macromolecules.* 25: 1214–1223.
23. Souto Maior, R.M., K. Hinkelmann and F. Wudl. 1990. Synthesis and characterization of two regiochemically defined poly(dialkylbithiophenes): a comparative study. *Macromolecules.* 23: 1268–1279.
24. Wang, Z., Y. Wang, D. Xu, E.S.W. Kong and Y. Zhang. 2010. Facile synthesis of dispersible spherical polythiophene nanoparticles by copper(II) catalyzed oxidative polymerization in aqueous medium. *Synth. Met.* 160: 921–926.
25. Najar, M.H. and K. Majid. 2013. Synthesis, characterization, electrical and thermal properties of nanocomposite of polythiophene with nanophotoadduct: a potent composite for electronic use. *J. Mater. Sci.* 24: 4332–4339.
26. Uygun, A., O. Turkoglu, S. Sen, E. Ersoy, A.G. Yavuz and G.G. Batir. 2009. The electrical conductivity properties of polythiophene/TiO$_2$ nanocomposites prepared in the presence of surfactants. *Curr. Appl. Phys.* 9: 866–871.
27. Mehdinia, A., N. Khodaee and A. Jabbari. 2014. Fabrication of graphene/Fe$_3$O$_4$@polythiophene nanocomposite and its application in the magnetic solid-phase extraction of polycyclic aromatic hydrocarbons from environmental water sample. *Anal. Chim. Acta.* 868: 1–9.
28. Hammer, B.A.G., M.A. Reyes-Martinez, F.A. Bokel, F. Liu, T.P. Russell, R.C. Hayward, A.L. Briseno and T. Emrick. 2014. Reversible, self cross-linking nanowires from thiol-functionalized polythiophene diblock copolymers. *ACS Appl. Mater. Interfaces.* 6: 7705−7711.
29. Kim, H.J., M. Skinner, H. Yu, J.H. Oh, A.L. Briseno, T. Emrick, B.J. Kim and R.C. Hayward. 2015. Water processable polythiophene nanowires by photo-cross-linking and click-functionalization. *Nano Lett.* 15: 5689–5695.
30. Umeda, R., H. Awaji, T. Nakahodo and H. Fujihara. 2008. Nanotube composites consisting of metal nanoparticles and polythiophene from electropolymerization of terthiophene-functionalized metal (Au, Pd) nanoparticles. *J. Am. Chem. Soc.* 130: 3240–3241.
31. Samitsu, S., T. Shimomura, S. Heike, T. Hashizume and K. Ito. 2008. Effective production of poly(3-alkylthiophene)

nanofibers by means of whisker method using anisole solvent: structural, optical, and electrical properties. *Macromolecules.* 41: 8000–8010.

32. Balakrishnan, K., M. Kumar and A. Subramania. 2014. Synthesis of polythiophene and its carbonaceous nanofibers as electrode materials for asymmetric supercapacitors. *Adv. Mater. Res.* 938: 151–157.

33. Alabadi, A., S. Razzaque, Z. Dong, W. Wang and B. Tan. 2016. Graphene oxide-polythiophene derivative hybrid nanosheet for enhancing performance of supercapacitor. *J. Power Sources.* 306: 241–247.

34. Xiao, W., J.S. Chen, Q. Lu and X.W. Lou. 2010. Porous spheres assembled from polythiophene (PTh)-coated ultrathin MnO_2 nanosheets with enhanced lithium storage capabilities. *J. Phys. Chem. C* 114: 12048–12051.

35. Zhai, L. and R.D. McCullough. 2004. Regioregular polythiophene/gold nanoparticle hybrid materials. *J. Mater. Chem.* 14: 141–143.

36. Kwasi-Effah, C.C. and T. Rabczuk. 2018. Dimensional analysis and modelling of energy density of lithium-ion battery. *J. Energy Storage.* 18: 308–315.

37. Mauger, A. and C.M. Julien. 2017. Critical review on lithium-ion batteries: are they safe? Sustainable? *Ionics.* 23 (8): 1933–1947.

38. Lyu, H., J. Liu, S. Mahurin, S. Dai, Z. Guo and X.G. Sun. 2017. Polythiophene coated aromatic polyimide enabled ultrafast and sustainable lithium ion batteries. *J. Mater. Chem. A.* 5: 24083–24090.

39. Bai, Y.M., P. Qiu, Z.L. Wen and S.C. Han. 2010. Improvement of electrochemical performances of $LiFePO_4$ cathode materials by coating of polythiophene. *J. Alloys Compd.* 508: 1–4.

40. Trinh, N.D., M. Saulnier, D. Lepage and S.B. Schougaard. 2013. Conductive polymer film supporting $LiFePO_4$ as composite cathode for lithium ion batteries. *J. Power Sources.* 221: 284–289.

41. Xing, Y., Y.B. He, B. Li, X. Chu, H. Chen, J. Ma, H. Du and F. Kang. 2013. $LiFePO_4$/C composite with 3D carbon conductive network for rechargeable lithium ion batteries. *Electrochim. Acta.* 109: 512–518.

42. Kempaiah, D.M., D. Rangappa and I. Honma. 2012. Controlled synthesis of nanocrystalline Li_2MnSiO_4 particles for high capacity cathode application in lithium-ion batteries. *Chem. Commun.* 48: 2698–2700.

43. Xu, H.Y., H. Wang, Z.Q. Song, Y.W. Wang, H. Yan and M. Yoshimura. 2004. Novel chemical method for synthesis of LiV_3O_8 nanorods as cathode materials for lithium ion batteries. *Electrochim. Acta.* 49: 349–353.

44. Liu, L., L. Jiao, J. Sun, Y. Zhang, M. Zhao, H. Yuan and Y. Wang. 2008. Electrochemical performance of $LiV_{3-x}Ni_xO_8$ cathode materials synthesized by a novel low-temperature solid-state method. *Electrochim. Acta.* 53: 7321–7325.

45. Kubiak, P., M. Pfanzelt, J. Geserick, U. Hormann, N. Husing, U. Kaiser and M. Wohlfahrt-Mehrens. 2009. Electrochemical evaluation of rutile TiO_2 nanoparticles as negative electrode for Li-ion batteries. *J. Power Sources.* 194: 1099–1104.

46. Jouanneau, S., A. Salle, A. Verbaere and D. Guyomard. 2005. The origin of capacity fading upon lithium cycling in $Li_{1.1}V_3O_8$. *J. Electrochem. Soc.* 152: A1660–A1667.

47. Guo, H., L. Liu, H. Shu, X. Yang, Z. Yang, M. Zhou, J. Tan, Z. Yan, H. Hu and X. Wang. 2014. Synthesis and electrochemical performance of LiV_3O_8/polythiophene composite as cathode materials for lithium ion batteries. *J. Power Sources.* 247: 117–126.

48. Fan, X., C. Luo, J. Lamb, Y. Zhu, K. Xu and C. Wang. 2015. PEDOT encapsulated FeOF nanorod cathodes for high energy lithium-ion batteries. *Nano Lett* 15: 7650–7656.

49. Gomez-Romero, P. 2001. Hybrid organic-inorganic materials-in search of synergic activity. *Adv. Mater.* 13: 163–174.

50. Murugan, A.V., C.W. Kwon, G. Campet, B.B. Kale, T. Maddanimath and K. Vijayamohanan. 2002. Electrochemical lithium insertion into a poly(3,4-ethylenedioxythiophene)PEDOT/V_2O_5 nanocomposite. *J. Power Sources.* 105: 1–5.

51. Kaschak, D.M., S.A. Johnson, D.E. Hooks, H.-N. Kim, M.D. Ward and T.E. Mallouk. 1998. Chemistry on the edge: a microscopic analysis of the intercalation, exfoliation, edge functionalization, and monolayer surface tiling reactions of R-zirconium phosphate. *J. Am. Chem. Soc.* 120: 10887–10894.

52. Han, Y.S., I. Park and J.H. Choy. 2001. Exfoliation of layered perovskite, $KCa_2Nb_3O_{10}$, into colloidal nanosheets by a novel chemical process. *J. Mater. Chem.* 11: 1277–1282.

53. Dungey, K.E., M.D. Curtis and J.E. Penner-Hahn. 1998. Structural characterization and thermal stability of MoS_2 intercalation compounds. *Chem. Mater.* 10: 2152–2161.

54. Wang, L., J.L. Schindler, J.A. Tomas, C.R. Kannewurf and M.G. Kanatzidis. 1995. Entrapment of polypyrrole chains between MoS2 layers via an in situ oxidative polymerization encapsulation reaction. *Chem. Mater.* 7: 1753–1755.

55. Templeton, T.L., Y. Yoshida, X.Z. Li, A.S. Arrott, A. E. Curzon, F. Hamed, M.A. Gee, P.J. Schurer and J. L. LaCombe. 1993. Magnetic properties of particles formed by H_2 reduction of co-ferrite or γ-FeOOH particles wrapped with exfoliated MoS_2. *IEEE Trans. Magn.* 29: 2625–2627.

56. Murugan, A.V., M. Quintin, M.H. Delville, G. Campet, C. S. Gopinath and K. Vijayamohanan. 2006. Exfoliation-induced nanoribbon formation of poly(3,4-ethylene dioxythiophene) PEDOT between MoS_2 layers as cathode material for lithium batteries. *J. Power Sources.* 156: 615–619.

57. Häupler, B., A. Wild and U.S. Schubert. 2015. Carbonyls: powerful organic materials for secondary batteries. *Adv. Energy Mater.* 5: 1402000–1402034.

58. Han, X.Y., G.Y. Qing, J.T. Sun and T.L. Sun. 2012. How many lithium ions can be inserted onto fused C6 aromatic ring systems? *Angew. Chem. Int. Ed.* 51: 5147–5151.

59. Luo, W., M. Allen, V. Raju and X.L. Ji. 2014. An organic pigment as a high-performance cathode for sodium-ion batteries. *Adv. Energy Mater.* 4: 1400549–1400554.

60. Liang, Y., Z. Tao and J. Chen. 2012. Organic electrode materials for rechargeable lithium batteries. *Adv. Energy Mater.* 2: 742–769.

61. Wu, H., S.A. Shevlin, Q. Meng, W. Guo, Y. Meng, K. Lu, Z. Wei and Z. Guo. 2014. Flexible and binder-free organic cathode for high-performance lithium-ion batteries. *Adv. Mater.* 26: 3338–3343.

62. Huang, Y., K. Li, J. Liu, X. Zhong, X. Duan, I. Shakir and Y. Xu. 2017. Three-dimensional graphene/polyimide composite-derived flexible high-performance organic cathode for rechargeable lithium and sodium batteries. *J. Mater. Chem. A.* 5: 2710–2716.

63. Wu, H.P., Q. Yang, Q.H. Meng, A. Ahmad, M. Zhang, L. Y. Zhu, Y.G. Liu and Z.X. Wei. 2016. A polyimide derivative containing different carbonyl groups for flexible lithium ion batteries. *J. Mater. Chem. A.* 4: 2115–2121.

64. Deng, S.R., L.B. Kong, G.Q. Hu, T. Wu, D. Li, Y.H. Zhou and Z.Y. Li. 2006. Benzene-based polyorganodisulfide cathode materials for secondary lithium batteries. *Electrochim. Acta.* 51: 2589–2593.

65. Amaike, M. and T. Iihama. 2006. Chemical polymerization of pyrrole with disulfide structure and the application to lithium secondary batteries. *Synth. Met.* 156: 239–243.

66. Liu, M., S.J. Visco and L.C. De Jonghe. 1990. Electrode kinetics of organodisulfide cathodes for storage batteries. *J. Electrochem. Soc.* 137: 750–759.

67. Picart, S. and E. Genies. 1996. Electrochemical study of 2,5-dimercapto-1,3,4-thiadiazole in acetonitrile. *J. Electroanal. Chem.* 408: 53–60.

68. Kiya, Y., G.R. Hutchison, J.C. Henderson, T. Sarukawa, O. Hatozaki, N. Oyama and H.D. Abrun. 2006. Elucidation of the redox behavior of 2,5-dimercapto-1,3,4-thiadiazole (DMcT) at poly(3,4-ethylenedioxythiophene) (PEDOT)-modified electrodes and application of the DMcT-PEDOT composite cathodes to lithium/lithium ion batteries. *Langmuir.* 22: 10554–10563.

69. Zhu, L., H.J. Peng, J. Liang, J.Q. Huang, C.M. Chen, X. Guo, W. Zhu, P. Li and Q. Zhang. 2015. Interconnected carbon nanotube/graphene nanosphere scaffolds as freestanding paper electrode for high-rate and ultra-stable lithium–sulfur batteries. *Nano Energy.* 11: 746–755.

70. Yan, J., X. Liu, X. Wang and B. Li. 2015. Long-life, high-efficiency lithium/sulfur batteries from sulfurized carbon nanotube cathodes. *J. Mater. Chem. A.* 3: 10127–10133.

71. Li, W., Q. Zhang, G. Zheng, Z.W. Seh, H. Yao and Y. Cui. 2013. Understanding the role of different conductive polymers in improving the nanostructured sulfur cathode performance. *Nano Lett.* 13: 5534–5540.

72. Chen, H., W. Dong, J. Ge, C. Wang, X. Wu, W. Lu and L. Chen. 2013. Ultrafine sulfur nanoparticles in conducting polymer shell as cathode materials for high performance lithium/sulfur batteries. *Sci. Rep.* 3: 1910.

73. Li, H., M. Sun, T. Zhang, Y. Fang and G. Wang. 2014. Improving the performance of PEDOT-PSS coated sulfur@activated porous graphene composite cathodes for lithium–sulfur batteries. *J. Mater. Chem. A.* 2: 18345–18352.

74. Oschmann, B., J. Park, C. Kim, K. Char, Y.E. Sung and R. Zentel. 2015. Copolymerization of polythiophene and sulfur to improve the electrochemical performance in lithium–sulfur batteries. *Chem. Mater.* 27: 7011–7017.

75. Su, D., M. Cortie, H. Fan and G. Wang. 2017. Prussian blue nanocubes with an open framework structure coated with PEDOT as high-capacity cathodes for lithium–sulfur batteries. *Adv. Mater.* 29: 1700587.

76. Zhang, M., Q. Meng, A. Ahmad, L. Mao, W. Yan and Z. Wei. 2017. Poly (3, 4-ethylenedioxythiophene)-coated sulfur for flexible and binder-free cathode of lithium-sulfur batteries. *J. Mater. Chem. A.* 5: 17647–17652.

77. Yin, X., Y. Li, Y. Feng and W. Feng. 2016. Polythiophene/graphite fluoride composites cathode for high power and energy densities lithium primary batteries. *Synth. Met.* 220: 560–566.

78. Groult, H. and A. Tressaud. 2018. Use of inorganic fluorinated materials in lithium batteries and in energy conversion systems. *Chem. Commun.* 54: 11375–11382.

79. Lee, Y.S. 2008. Syntheses and properties of fluorinated carbon materials. *J. Fluor. Chem.* 126: 392–403.

80. Li, Y., Y.Y. Feng and W. Feng. 2013. Deeply fluorinated multi-wall carbon nanotubes for high energy and power densities lithium/carbon fluorides battery. *Electrochim. Acta.* 107: 343–349.

81. Li, Y., Y.F. Chen, W. Feng, F. Ding and X.J. Liu. 2011. The improved discharge performance of Li/CF$_x$ batteries by using multi-walled carbon nanotubes as conductive additive. *J. Power Sources.* 196: 2246–2250.

82. Meduri, P., H.H. Chen, X.L. Chen, J. Xiao, M. E. Gross, T.J. Carlson, J.G. Zhang and Z.D. Deng. 2011. Hybrid CFx–Ag$_2$V$_4$O$_{11}$ as a high-energy, power density cathode for application in an underwater acoustic microtransmitter. *Electrochem. Commun.* 13: 1344–1348.

83. Zhang, Q., A.D. Astorg, P. Xiao, X. Zhang and L. Liu. 2010. Carbon-coated fluorinated graphite for high energy and high power densities primary lithium batteries. *J. Power Sources.* 195: 2914–2917.

84. Groult, H., C.M. Julien, A. Bahloul, S. Leclerc, E. Briot and A. Mauger. 2011. Improvements of the electrochemical features of graphite fluorides in primary lithium battery by electrodeposition of polypyrrole. *Electrochem. Commun.* 13: 1074–1076.

85. Ong, S.P., V.L. Chevrier, G. Hautier, A. Jain, C. Moore, S. Kim, X. Ma and G. Ceder. 2011. Voltage, stability and diffusion barrier differences between sodium-ion and lithium-ion intercalation materials. *Energy Environ. Sci.* 4: 3680–3688.

86. Shang, C., S. Dong, S. Zhang, P. Hu, C. Zhang and G. Cui. 2015. A Ni$_3$S$_2$-PEDOT monolithic electrode for sodium batteries. *Electrochem. Commun.* 50: 24–27.

87. Ellis, B.L., W.R. Makahnouk, Y. Makimura, K. Toghill and L.F. Nazar. 2007. A multifunctional 3.5 V iron-based phosphate cathode for rechargeable batteries. *Nat. Mater.* 6: 749–753.

88. Prosini, P., M. Lisi, D. Zane and M. Pasquali. 2002. Determination of the chemical diffusion coefficient of lithium in LiFePO$_4$. *Solid State Ionics.* 148: 45–51.

89. Ali, G., J.H. Lee, D. Susanto, S.W. Choi, B.W. Cho, K. W. Nam and K.Y. Chung. 2016. Polythiophene-wrapped olivine NaFePO$_4$ as a cathode for Na-ion batteries. *ACS Appl. Mater. Interfaces.* 8: 15422–15429.

90. Casado, N., M. Hilder, C. Pozo-Gonzalo, M. Forsyth and D. Mecerreyes. 2017. Electrochemical behavior of PEDOT/lignin in ionic liquid electrolytes: suitable cathode/electrolyte system for sodium batteries. *ChemSusChem.* 10: 1783–1791.

91. Wang, S., Y. Wang, S. Chen, H. Hou and H. Li. 2016. Enhanced capacity and stability of K2FeO4 cathode with poly(3-hexylthiophene) coating for alkaline super-iron battery. *Electrochim. Acta.* 213: 132–139.

92. Kitani, A., M. Kaya and K. Sasaki. 1986. Performance study of aqueous polyaniline batteries. *J. Electrochem. Soc.* 133: 1069–1073.

93. Somasiri, N.L.D. and A.G. Macdiarmid. 1988. Polyaniline: characterization as a cathode active material in rechargeable batteries in aqueous electrolytes. *J. Appl. Electrochem.* 18: 92–95.

94. Shaolin, M., Y. Jinhai and W. Yuhua. 1993. A rechargeable Zn/ZnCl$_2$, NH$_4$Cl/polyaniline/carbon dry battery. *J. Power Sources.* 45: 153–159.

95. Simons, T.J., M. Salsamendi, P.C. Howlett, M. Forsyth, D. R. MacFarlane and C. Pozo-Gonzalo. 2015. Rechargeable

Zn/PEDOT battery with an imidazoliumbased ionic liquid as the electrolyte. *ChemElectroChem.* 2: 2071–2078.

96. Zeng, Y., X. Zhang, Y. Meng, M. Yu, J. Yi, Y. Wu, X. Lu and Y. Tong. 2017. Achieving ultrahigh energy density and long durability in a flexible rechargeable quasi-solid-state Zn–Mn$_2$ battery. *Adv. Mater.* 29: 1700274.

97. Wang, Z., N. Bramnik, S. Roy, G.D. Benedetto, J.L. Zunino, III and S. Mitra. 2013. Flexible zinc-carbon batteries with multiwalled carbon nanotube/conductive polymer cathode matrix. *J. Power Sources.* 237: 210–214.

98. Hudak, N.S. 2013. Rechargeable aluminum batteries with conducting polymers as positive electrodes. SANDIA REPORT SAND2014-0068.

99. Yu, X., Y. Fu, X. Cai, H. Kafafy, H. Wu, M. Peng, S. Hou, Z. Lv, S. Ye and D. Zou. 2013. Flexible fiber-type zinc–carbon battery based on carbon fiber electrodes. *Nano Energy.* 2: 1242–1248.

100. Pan, H., Y. Shao, P. Yan, Y. Cheng, K.S. Han, Z. Nie, C. Wang, J. Yang, X. Li, P. Bhattacharya, K.T. Mueller and J. Liu. 2016. Reversible aqueous zinc/manganese oxide energy storage from conversion reactions. *Nat. Energy.* 1: 16039.

101. Zhang, N., F. Cheng, Y. Liu, Q. Zhao, K. Lei, C. Chen, X. Liu and J. Chen. 2016. Cation-deficient spinel ZnMn$_2$O$_4$ cathode in Zn(CF$_3$SO$_3$)$_2$ electrolyte for rechargeable aqueous Zn-Ion battery. *J. Am. Chem. Soc.* 138: 12894–12901.

102. Huang, M., Y. Zhang, F. Li, L. Zhang, R.S. Ruoff, Z. Wen and Q. Liu. 2014. Self-assembly of mesoporous nanotubes assembled from interwoven ultrathin birnessite-type MnO$_2$ nanosheets for asymmetric supercapacitors. *Sci. Rep.* 4: 3878.

103. Gaikwad, A.M., G.L. Whiting, D.A. Steingart and A. C. Arias. 2011. Highly flexible, printed alkaline batteries based on mesh-embedded electrodes. *Adv. Mater.* 23: 3251–3255.

104. Hu, L., H. Wu, F.L. Mantia, Y. Yang and Y. Cu. 2010. Thin, flexible secondary Li-Ion paper batteries. *ACS Nano.* 4: 5843–5848.

105. Wadia, C., P. Albertus and V. Srinivasan. 2011. Resource constraints on the battery energy storage potential for grid and transportation applications. *J. Power Sources.* 196: 1593–1598.

106. Odhiambo, S.A., G.D. Mey, C. Hertleer, A. Schwarz and L.V. Langenhove. 2014. Discharge characteristics of poly (3,4-ethylene dioxythiophene): poly(styrenesulfonate) (PEDOT:PSS) textile batteries; comparison of silver coated yarn electrode devices and pure stainless steel filament yarn electrode devices. *Text. Res. J.* 84: 347–354.

107. Das, P.R., L. Komsiyska, O. Osters and G. Wittstock. 2015. PEDOT: PSS as a functional binder for cathodes in lithium ion batteries. *J. Electrochem. Soc.* 162: A674–A678.

108. Eliseeva, S.N., O.V. Levin, E.G. Tolstopjatova, E. V. Alekseeva, R.V. Apraksin and V.V. Kondratiev. 2015. New functional conducting poly-3,4-ethylenedioxythiopene: polystyrenesulfonate/carboxymethylcellulose binder for improvement of capacity of LiFePO$_4$-based cathode materials. *Mater. Lett.* 161: 117–119.

109. Levin, O.V., S.N. Eliseeva, E.V. Alekseeva, E.G. Tolstopjatova and V.V. Kondratiev. 2015. Composite LiFePO$_4$/poly-3,4-ethylenedioxythiophene cathode for lithium-ion batteries with low content of nonelectroactive components. *Int. J. Electrochem. Sci.* 10: 8175–8189.

110. Ma, T., Q. Zhao, J. Wang, Z. Pan and J. Chen. 2016. A sulfur heterocyclic quinone cathode and a multifunctional binder for a high-performance rechargeable lithium-ion battery. *Angew. Chem. Int. Ed.* 55: 6428–6432.

111. Wang, Z., Y. Chen, V. Battaglia and G. Li. 2014. Improving the performance of lithium–sulfur batteries using conductive polymer and micrometric sulfur powder. *J. Mater. Res.* 29: 1027–1033.

112. Pan, J., G. Xu, B. Ding, Z. Chang, A. Wang, H. Dou and X. Zhang. 2016. PAA/PEDOT: PSS as a multifunctional, water-soluble binder to improve the capacity and stability of lithium-sulfur batteries. *RSC Adv.* 6: 40650–40655.

113. Frysz, C.A., X.P. Shui and D.D.L. Chung. 1996. Carbon filaments and carbon black as a conductive additive to the manganese dioxide cathode of a lithium electrolytic cell. *J. Power Sources.* 58: 41–54.

114. Li, M.Q., M.Z. Qu, X.Y. He and Z.L. Yu. 2009. Effects of electrolytes on the electrochemical performance of Si/graphite/disordered carbon composite anode for lithium-ion batteries. *Electrochim. Acta.* 54: 4506–4513.

115. Rai, A.K., J. Gim, S.W. Kang, V. Mathew, A.L. Tuan, J. Kang, J. Song, B.J. Paul and J. Kim. 2012. Improved electrochemical performance of Li$_4$Ti$_5$O$_{12}$ with a variable amount of graphene as a conductive agent for rechargeable lithium-ion batteries by solvothermal method. *Mater. Chem. Phys.* 136: 1044–1051.

116. Lee, J.H., G.S. Kim, Y.M. Choi, W.I. Park, J.A. Rogers and U. Paik. 2008. Comparison of multiwalled carbon nanotubes and carbon black as percolative paths in aqueous-based natural graphite negative electrodes with high-rate capability for lithium-ion batteries. *J. Power Sources.* 184: 308–311.

117. Endo, M., Y.A. Kim, T. Hayashi, K. Nishimura, T. Matusita, K. Miyashita and M.S. Dresselhaus. 2001. Vapor-grown carbon fibers (VGCFs) basic properties and their battery applications. *Carbon.* 39: 1287–1297.

118. Wang, G., Q. Zhang, Z. Yu and M. Qu. 2008. The effect of different kinds of nano-carbon conductive additives in lithium ion batteries on the resistance and electrochemical behavior of the LiCoO$_2$ composite cathodes. *Solid State Ionics.* 179: 263–268.

119. Zhong, H., A. He, J. Lu, M. Sun, J. He and L. Zhang. 2016. Carboxymethyl chitosan/conducting polymer as water-soluble composite binder for LiFePO$_4$ cathode in lithium ion batteries. *J. Power Sources.* 336: 107–114.

120. Yan, L.L., X. Gao, J.P. Thomas, J.H.L. Ngai, H. Altounian, K.T. Leung, Y. Meng and Y. Li. 2018. Ionically cross-linked PEDOT: PSS as a multi-functional conductive binder for high-performance lithium-sulfur battery. *Sustain. Energy Fuels.* 2: 1574–1581.

121. Park, C.S., K.S. Kim and Y.J. Park. 2013. Carbon-sphere/Co$_3$O$_4$ nanocomposite catalysts for effective air electrode in Li/air batteries. *J. Power Sources.* 244: 72–79.

122. Yoon, T.H. and Y.J. Park. 2013. Polydopamine-assisted carbon nanotubes/Co$_3$O$_4$ composites for rechargeable Li-air batteries. *J. Power Sources.* 244: 344–353.

123. Adams, B.D., C. Radtke, R. Black, M.L. Trudeau, K. Zaghib and L.F. Nazar. 2013. Current density dependence of peroxide formation in the Li–O$_2$ battery and its effect on charge. *Energy Environ. Sci.* 6: 1772–1778.

124. Gallant, B.M., D.G. Kwabi, R.R. Mitchell, J. Zhou, C. V. Thompson and Y. Shao-Horn. 2013. Influence of Li$_2$O$_2$ morphology on oxygen reduction and evolution kinetics in Li–O$_2$ batteries. *Energy Environ. Sci.* 6: 2518–2528.

125. Lu, J., Y. Lei, K.C. Lau, X. Luo, P. Du, J. Wen, R. S. Assary, U. Das, D.J. Miller, J.W. Elam, H.M. Albishri,

D.A. El-Hady, Y.K. Sun, L.A. Curtiss and K. Amine. 2013. A nanostructured cathode architecture for low charge overpotential in lithium-oxygen batteries. *Nat. Commun.* 4: 2383.

126. Peng, Z., S.A. Freunberger, Y.H. Chen and P.G. Bruce. 2012. A reversible and higher-rate Li-O_2 battery. *Science.* 337: 563–566.

127. Riaz, A., K.N. Jung, W. Chang, S.B. Lee, T.H. Lim, S. J. Park, R.H. Song, S. Yoon, K.H. Shina and J.W. Lee. 2013. Carbon-free cobalt oxide cathodes with tunable nanoarchitectures for rechargeable lithium–oxygen batteries. *Chem. Commun.* 49: 5984–5986.

128. Yoon, D.H., S.H. Yoon, K.S. Ryu and Y.J. Park. 2016. PEDOT: PSS as multi-functional composite material for enhanced Li-air-battery air electrodes. *Sci. Rep.* 6: 19962.

129. Mojtahedi, M., M. Goodarzi, B. Sharifi and J.V. Khaki. 2011. Effect of electrolysis condition of zinc powder production on zinc–silver oxide battery operation. *Energy Convers. Manage.* 52: 1876–1880.

130. Huang, Y., Y. Lin and W. Li. 2013. Controllable syntheses of α- and δ-MnO_2 as cathode catalysts for zinc-air battery. *Electrochim. Acta.* 99: 161–165.

131. Winther-Jensen, B., O. Winther-Jensen, M. Forsyth and D. R. MacFarlane. 2008. High rates of oxygen reduction over a vapor phase–polymerized PEDOT electrode. *Science.* 321: 671–674.

132. Kuo, Y.L., C.C. Wu, W.S. Chang, C.R. Yang and H. L. Chou. 2015. Study of poly (3,4-ethylenedioxythiophene)/MnO_2 as composite cathode materials for aluminum-air battery. *Electrochim. Acta.* 176: 1324–1331.

133. Hilder, M., B. Winther-Jensenb and N.B. Clark. 2009. Paper-based, printed zinc–air battery. *J. Power Sources.* 194: 1135–1141.

134. Su, Y., Z. Yao, F. Zhang, H. Wang, Z. Mics, E. Cánovas, M. Bonn, X. Zhuang and X. Feng. 2016. Sulfur-enriched conjugated polymer nanosheet derived sulfur and nitrogen co-doped porous carbon nanosheets as electrocatalysts for oxygen reduction reaction and zinc–air battery. *Adv. Funct. Mater.* 26: 5893–5902.

135. Winther-Jensen, B. and N.B. Clark. 2008. Controlled release of dyes from chemically polymerised conducting polymers. *React. Funct. Polym.* 68: 742–750.

136. Heinze, J., B.A. Frontana-Uribe and S. Ludwings. 2010. Electrochemistry of conducting polymers-persistent models and new concepts. *Chem. Rev.* 110: 4724–4771.

137. Kim, F.S., G. Ren and S.A. Jenekhe. 2011. One-dimensional nanostructures of π conjugated molecular systems: assembly, properties, and applications from photovoltaics, sensors, and nanophotonics to nanoelectronics. *Chem. Mater.* 23: 682–732.

138. Vyas, V.S., V.W. Lau and B.V. Lotsch. 2016. Soft photocatalysis: organic polymers for solar fuel production. *Chem. Mater.* 28: 5191–5204.

139. Inagi, S. 2016. Fabrication of gradient polymer surfaces using bipolar electrochemistry. *Polym. J.* 48: 39–44.

140. Snook, G., P. Kao and A. Best. 2011. Conducting-polymer-based supercapacitor devices and electrodes. *J. Power Sources.* 196: 1–12.

141. Song, Z. and H. Zhou. 2013. Towards sustainable and versatile energy storage devices: an overview of organic electrode materials. *Energy Environ. Sci.* 6: 2280–2301.

142. Muench, S., A. Wild, C. Friebe, B. Häupler, T. Janoschka and U.S. Schubert. 2016. Polymer-based organic batteries. *Chem. Rev.* 116: 9438–9484.

143. Lukatskaya, M.R., B. Dunn and Y. Gogotsi. 2016. Multidimensional materials and device architectures for future hybrid energy storage. *Nat. Commun.* 7: 12647.

144. Čaja, J., R.B. Kaner and A.G. MacDiarmid. 1984. A rechargeable battery employing a reduced polyacetylene anode and a titanium disulfide cathode. *J. Electrochem. Soc.* 131: 2744–2750.

145. Mahammadi, A., O. Inganäs and I. Lundström. 1986. Properties of polypyrrole-electrolyte-polypyrrole cells. *J. Electrochem. Soc.* 133: 947–949.

146. Chowdhury, A.N., Y. Harima, K. Kunugi and K. Yamashita. 1996. p- and n-type conductance of electrochemically synthesized poly(3-methyl thiophene) films. *Electrochim. Acta.* 41: 1993–1997.

147. Yata, S., K. Sakurai, T. Osaki and Y. Inoue. 1990. Studies of porous polyacenic semiconductors toward application III. Characteristics of practical batteries employing polyacenic semiconductive materials as electrodes. *Synth. Met.* 38: 185–193.

148. Novák, P., K. Müller, K.S.V. Santhanam and O. Hass. 1997. Electrochemically active polymers for rechargeable batteries. *Chem. Rev.* 97: 207–282.

149. Kuwabata, B.S., N. Tsumura, S. Goda, C.R. Martin and H. Yoneyam. 1998. Charge-discharge properties of composite of synthetic graphite and poly(3-n-hexylthiophene) as an anode active material in rechargeable lithium-ion. *J. Electrochem. Soc.* 145: 1415–1420.

150. Kwon, Y.H., K. Minnici, J.J. Park, S.R. Lee, G. Zhang, E. S. Takeuchi, K.J. Takeuchi, A.C. Marschilok and E. Reichmanis. 2018. SWNT anchored with carboxylated polythiophene "links" on highcapacity li-ion battery anode materials. *J. Am. Chem. Soc.* 140: 5666−5669.

151. Kaneto, K., Y. Kohno, K. Yoshino and Y. Inuishi. 1983. Electrochemical preparation of a metallic polythiophene film. *J. Chem. Soc. Chem. Commun.* 7: 382–383.

152. Kaneto, K., K. Yoshino and Y. Inuishi. 1983. Characteristics of polythiophene battery. *J. Appl. Phys.* 22: L567–L568.

153. Kaufman, J.H., T.C. Chung, A.J. Heeger and F. Wudl. 1984. Poly(thiophene): a stable polymer cathode material. *J. Electrochem. Soc.* 131: 2092–2093.

154. Zhang, W.J. 2011. A review of the electrochemical performance of alloy anodes for lithium-ion batteries. *J. Power Sources.* 196: 13–24.

155. Chan, C.K., R. Ruffo, S.S. Hong, R.A. Huggins and Y. Cui. 2009. Structural and electrochemical study of the reaction of lithium with silicon nanowires. *J. Power Sources.* 189: 34–39.

156. Ryou, M.H., J. Kim, I. Lee, S. Kim, Y.K. Jeong, S. Hong, J.H. Ryu, T.S. Kim, J.K. Park, H. Lee and J.W. Choi. 2013. Mussel-inspired adhesive binders for high-performance silicon nanoparticle anodes in lithium-ion batteries. *Adv. Mater.* 25: 1571–1576.

157. Liu, W.R., M.H. Yang, H.C. Wu, S.M. Chiao and N. L. Wu. 2005. Enhanced cycle life of Si anode for Li-ion batteries by using modified elastomeric binder. *Electrochem. Solid-State Lett.* 8: A100–A103.

158. Kim, H., E.J. Lee and Y.K. Sun. 2014. Recent advances in the Si-based nanocomposite materials as high capacity anode materials for lithium ion batteries. *Mater. Today.* 17: 285–297.

159. Wang, Q.T., R.R. Li, X.Z. Zhou, J. Li and Z.Q. Lei. 2016. Polythiophene-coated nano-silicon composite anodes with enhanced performance for lithium-ion batteries. *J. Solid State Electrochem.* 20: 1331–1336.

160. McGraw, M., P. Kolla, R. Cook, B. Yao, Q. Qiao, J. Wu and A. Smirnova. 2016. One-step solid-state in-situ thermal polymerization of silicon-PEDOT nanocomposites for the application in lithium-ion battery anodes. *Polymer.* 99: 488–495.

161. Wang, C.Y., A.M. Ballantyne, S.B. Hall, C.O. Too, D.L. Officer and G.G. Wallace. 2006. Functionalized polythiophene-coated textile: a new anode material for a flexible battery. *J. Power Sources.* 156: 610–614.

162. Wang, C.Y., G. Tsekouras, P. Wagner, S. Gambhir, C.O. Too, D. Officer and G.G. Wallace. 2010. Functionalised polyterthiophenes as anode materials in polymer/polymer batteries. *Synth. Met.* 160: 76–82.

163. Zhu, L.M., W. Shi, R.R. Zhao, Y.L. Cao, X.P. Ai, A.W. Lei and H.X. Yang. 2013. n-Dopable polythiophenes as high capacity anode materials for all-organic Li-ion batteries. *J. Electroanal. Chem.* 688: 118–122.

164. Shen, X., D. Mu, S. Chen, R. Huang and F. Wu. 2014. Electrospun composite of ZnO/Cu nanocrystals implanted carbon fibers as an anode material with high rate capability for lithium ion batteries. *J. Mater. Chem. A.* 2: 4309–4315.

165. Guo, R., W. Yue, Y. An, Y. Ren and X. Yan. 2014. Graphene-encapsulated porous carbon-ZnO composites as high performance anode materials for Li-ion batteries. *Electrochim. Acta.* 135: 161–167.

166. Xie, J., N. Imanishi, A. Hirano, Y. Takeda, O. Yamamoto, X.B. Zhao and G.S. Cao. 2011. Determination of Li-ion diffusion coefficient in amorphous Zn and ZnO thin films prepared by radio frequency magnetron sputtering. *Thin Solid Films.* 519: 3373–3377.

167. Xu, G.L., Y. Li, T. Ma, Y. Ren, H.H. Wang, L. Wang, J. Wen, D. Miller, K. Amine and Z. Chen. 2015. PEDOT-PSS coated ZnO/C hierarchical porous nanorods as ultralong-life anode material for lithium ion batteries. *Nano Energy.* 18: 253–264.

168. Zhang, C., Y. He, P. Mu, X. Wang, Q. He, Y. Chen, J. Zeng, F. Wang, Y. Xu and J.X. Jiang. 2018. Toward high performance thiophene-containing conjugated microporous polymer anodes for lithium-ion batteries through structure design. *Adv. Funct. Mater.* 28: 1705432.

169. Numazawa, H., K. Sato, H. Imai and Y. Oak. 2018. Multistage redox reactions of conductive polymer nanostructures with lithium ions: potential for high-performance organic anodes. *NPG Asia Mater.* 10: 397–405.

170. Levi, M.D., Y. Gofer and D. Aurbach. 2002. A synopsis of recent attempts toward construction of rechargeable batteries utilizing conducting polymer cathodes and anodes. *Polym. Adv. Technol.* 13: 697–713.

171. Ma, G., Z. Wen, Q. Wang, C. Shen, J. Jin and X. Wu. 2014. Enhanced cycle performance of a Li–S battery based on a protected lithium anode. *J. Mater. Chem. A.* 2: 19355–19359.

172. Zhang, M., D. Han and P. Lu. 2017. PEDOT encapsulated b-FeOOH nanorods: synthesis, characterization and application for sodium-ion batteries. *Electrochim. Acta.* 238: 330–336.

173. Zeng, G., Y. An, H. Fei, T. Yuan, S. Qing, L. Ci, S. Xiong and J. Feng. 2018. Green and facile synthesis of nanosized polythiophene as an organic anode for high-performance potassium-ion battery. *Funct. Mater. Lett.* 11: 1840003.

174. Liu, P., H.X. Yang, X.P. Ai, G.R. Li and X.P. Gao. 2012. A solar rechargeable battery based on polymeric charge storage electrodes. *Electrochem. Commun.* 16: 69–72.

175. Zhao, J., X. Zou, Y. Zhu, Y. Xu and C. Wang. 2016. Electrochemical intercalation of potassium into graphite. *Adv. Funct. Mater.* 26: 8103–8110.

176. Ji, B., F. Zhang, X. Song and Y. Tang. 2017. A novel potassium-ion-based dual-ion battery. *Adv. Mater.* 29: 1700519.

177. Zou, X., P. Xiong, J. Zhao, J.H.Z. Liu and Y. Xu. 2017. Recent research progress in non-aqueous potassium-ion batteries. *Phys. Chem. Chem. Phys.* 19: 26495–26506.

178. Komaba, S., T. Hasegawa, M. Dahbi and K. Kubota. 2015. Potassium intercalation into graphite to realize high-voltage/high-power potassium-ion batteries and potassium-ion capacitors. *Electrochem. Commun.* 60: 172–175.

179. Courtel, F.M., S. Niketic, D. Duguay, Y. Abu-Lebdeh and I.J. Davidson. 2011. Water-soluble binders for MCMB carbon anodes for lithium-ion batteries. *J. Power Sources.* 196: 2128–2134.

180. Salem, N., M. Lavrisa and Y. Abu-Lebdeh. 2016. Ionically-functionalized poly(thiophene) conductive polymers as binders for silicon and graphite anodes for Li-ion batteries. *Energy Technol.* 4: 331–340.

181. Shao, D., H. Zhong and L. Zhang. 2014. Water-soluble conductive composite binder containing PEDOT:PSS as conduction promoting agent for Si anode of lithium-ion batteries. *ChemElectroChem.* 1: 1679–1687.

182. Woo, J.J., Z. Zhang and K. Amine. 2014. Separator/electrode assembly based on thermally stable polymer for safe lithium-ion batteries. *Adv. Energy Mater.* 4: 1301208.

183. Yan, L., Y.S. Li and C.B. Xiang. 2005. Preparation of poly (vinylidene fluoride)(pvdf) ultrafiltration membrane modified by nano-sized alumina (Al2O3) and its antifouling research. *Polymer.* 46: 7701–7706.

184. Foster, D.L. 1989. Separator for lithium batteries and lithium batteries including the separator. US Patent 4,812,375 (45).

185. Kinoshita, K. and R. Yeo. 1985. *Survey on Separators for Electrochemical Systems.* LBNL, Berkeley, CA.

186. Boehnstedt, W., ed. 1999. *Handbook of Battery Materials.* J.O. Besenhard, VCH Wiley, Amsterdam and New York.

187. Spotnitz, R., ed. 1999. *Handbook of Battery Materials.* J.O. Besenhard, VCH Wiley, Amsterdam and New York.

188. Arora, P. and Z.J. Zhang. 2004. Battery separators. *Chem. Rev.* 104: 4419–4462.

189. Thomas-Alyea, K.E., J. Newman, G. Chen and T.J. J. Richardson. 2004. Modeling the behavior of electroactive polymers for overcharge protection of lithium batteries. *Electrochem. Soc.* 151: A509–A521.

190. Zhang, H., Y. Cao, H. Yang, S. Lu and X. Ai. 2013. A redox-active polythiophene-modified separator for safety control of lithium-ion batteries. *J. Polym. Sci. Part B.* 51: 1487–1493.

191. Shin, W.K., J.H. Yoo and D.W. Kim. 2015. Surface-modified separators prepared with conductive polymer and aluminum fluoride for lithium-ion batteries. *J. Power Sources.* 279: 737–744.

192. Lee, Y., M.H. Ryou, M. Seo, J.W. Choi and Y.M. Lee. 2013. Effect of polydopamine surface coating on polyethylene separators as a function of their porosity for high-power li-ion batteries. *Electrochim. Acta.* 113: 433–438.

193. Cao, C., L. Tan, W. Liu, J. Ma and L. Li. 2014. Polydopamine coated electrospun poly(vinyldiene fluoride) nanofibrous membrane as separator for lithium-ion batteries. *J. Power Sources.* 248: 224–229.

194. Park, K., J.H. Cho, K. Shanmuganathan, J. Song, J. Peng, M. Gobet, S. Greenbaum, C.J. Ellison and J.B. Goodenough. 2014. New battery strategies with a polymer/Al2O3 separator. *J. Power Sources.* 263: 52–58.

195. Abbas, S.A., M.A. Ibrahem, L. Hu, C. Lin, J. Fang, K.M. Boopathi, P. Wang, L. Li and C.W. Chu. 2016. Bifunctional separator as a polysulfide mediator for highly stable Li-S batteries. *J. Mater. Chem. A.* 4: 9661–9669.

196. Huang, J.Q., X.F. Liu, Q. Zhang, C.M. Chen, M.Q. Zhao, S.M. Zhang, W. Zhu, W.Z. Qian and F. Wei. 2013. Entrapment of sulfur in hierarchical porous graphene for lithium–sulfur batteries with high rate performance from –40 to 60° C. *Nano Energy.* 2: 314–321.

197. Lee, S.H., J.B. Park, H.S. Lim and Y.K. Sun. 2017. An advanced separator for Li–O$_2$ batteries: maximizing the effect of redox mediators. *Adv. Energy Mater.* 7: 1602417-1-6.

198. Mirzaee, M. and G.B. Pour. 2018. Design and fabrication of ultracapacitor based on paper substrate and BaTiO$_3$/PEDOT: PSS separator film. *Recent Pat. Nanotechnol.* 12: 192–199.

199. Jonas, F., G. Heywang and W. Schmidtberg. 1988. European Patent EP 340 512 (Bayer AG).

200. Jonas, F., G. Heywang and W. Schmidtberg. 1990. Solid electrolytes, and electrolyte capacitors containing same. US Patent 4,910,645.

201. Wang, C., J.L. Schindler, C.R. Kannewurf and M.G. Kanatzidis. 1995. Poly(3,4-ethylenedithiathiophene). A new soluble conductive polythiophene derivative. *Chem. Mater.* 7: 58–68.

202. Reuter, K., U. Merker and F. Jonas. 2008. Polythophenes having alkyleneoxythia thophene units in electrolyte capacitors. US Patent 7,341,801 B2.

203. Horacek, I., T. Zednicek, M. Komarek, J. Tomasko and S. Zednicek. 2001. Improved ESR on MnO$_2$ tantalum capacitors at wide voltage range. Proceedings of the 15th European Passive Components Symposium CARTS-Europe, Copenhagen, Denmark, pp. 24–29.

204. Huang, X., Y. Liu, H. Zhang, J. Zhang, O. Noonan and C. Yu. 2017. Free-standing monolithic nanoporous graphene foam as high performance aluminum-ion battery cathode. *J. Mater. Chem. A.* 5: 19416–19421.

205. Lu, Y.C., H.A. Gasteiger, M.C. Parent, V. Chiloyan and Y. Shao-Horn. 2010. The influence of catalysts on discharge and charge voltages of rechargeable Li–oxygen batteries. *Electrochem. Solid State Lett.* 13: A69–A72.

206. Nasybulin, E., W. Xu, M.H. Engelhard, X.S. Li, M. Gu, D. Hu and J.G. Zhang. 2013. Electrocatalytic properties of poly(3,4-ethylenedioxythiophene) (PEDOT) in Li-O$_2$ battery. *Electrochem. Commun.* 29: 63–66.

2 Synthetic Strategies and Significant Issues for Pristine Conducting Polymers

Mriganka Sekhar Manna and Kalyan Gayen
Department of Chemical Engineering
NIT Agartala
West Tripura, India

CONTENTS

2.1 INTRODUCTION

Polymers are customarily used in a variety of applications due to their insulating property. However, in the last two decades, researchers have shown that polymers with π-conjugation in the backbone could have exhibited conductivity characteristics [1]. The incorporation of oxidizing or reducing agents (dopants) into this polymer matrix dramatically increases the conductivity and other properties of such polymers so as to be applicable in various fields of science and engineering, especially in the field of energy storage of various kinds. The techniques of doping of appropriate dopants into the matrix of conducting polymers matrix lead to a further dramatic increase in the conductivity of such conjugated polymers to a value as high as 10^5 S cm^{-1} comparable to the conductivity (10^6 S cm^{-1}) of copper [2] (Table 2.1). The induced conductivity depends on various factors— viz. nature of dopant, synthesis technique, processing technique of synthesized polymers, the crystallinity achieved thereby, and molecular structure of fabricated polymers. However, the dopants being a kind of smaller molecule, the macromolecular characteristics, and properties of polymers are consequently deteriorated in terms of mechanical, chemical, and thermal stabilities and processibilities, especially while higher proportion (30–50 wt.%) of dopants are used. The higher conductivity of such polymers provides a remarkable scope for their applications. The comparatively lower weight and ease of fabrication of various types of conducting polymers (CPs) provide the opportunity for their utilization in various fields. The conventional polymers such as polyethylene and polypropylene are made up of essentially σ-bonds between monomers. The charge (electrons) of polymeric atoms connected by σ-bonds along the chain of polymer is not characteristically portable. On the other hand, conjugated π-electrons in alternative single and double bonds along the chain backbone or ring structure of conducting polymers (polyaniline, polypyrrole, etc.) are characteristically mobile. The π-conjugation electrons that are injected by doping or other techniques in polymeric atoms of backbone chain lead to the modification of polymers and they, in turn, become electrically more conductive. The doping of dopants in the polymer matrix facilitates the orientation of these π-electrons and results in an electron imbalance for charge mobility. The extended π-conjugated structure of CPs allows the new electron propulsion to move along the backbone chain on the application of electric potential across the material.

A number of polymers such as polyacetylene, polyaniline (PAn), polypyrrole (PPy), poly(p-)phenylene, and polythiophene (PTh) are electrically conductive due to the presence of these π-electrons. Conductive polymers exhibit highly reversible redox behavior and the unusual combination of properties of metals and common plastics. They have unique conduction mechanism and good environmental stability even in the presence of oxygen and water. These polymers could be highly promising for much technological utilization due to their chemical versatility, stability, ease of processability, and low cost. Conducting polymers can be classified into three categories, namely electron conducting, proton conducting, and ion conducting polymers. Each of the type has been used mostly in optical and electronic applications such as batteries, display units, wires, optical signal processing, materials for storage of information, and conversion and storage of solar energy. The discovery of light-emitting polymers in 1990 in the Cavendish Laboratory at Cambridge University has been considered the turning point to the development of conducting polymers. Thereafter, new developments both on a fundamental level and from the manufacturing aspect have been carried out in this field of science. Consequently, the employment of metals and metal alloys in electronic devices has been replaced by conducting polymeric materials. The critical problems associated with this replacement are less oxidative stability of CPs and lifetime of the devices made up with them. However, the very low weight of polymeric devices employed in LED display, solar power system, televisions, etc., can be utilized if oxidation stability can be improved.

The conduction of electrons in a conducting polymer (CP) is explained by the energy level of the electrons in the outer shell of an atom. The bandgap is defined as the energy level required for promoting an electron from outer shell or the highest occupied energy level or valence band to the empty band or conduction band immediately above it. Metals are electrically conductive as they have zero bandgap. On the other hand, insulators like nonmetals and many of the common polymers have large bandgaps (1.5–4 eV) that electrons are unable to flow from valence band to conduction band. The bandgaps in polymers are minimized by altering the chemical structure of polymer backbone by suitable processing methods like heating, reactions, and mostly by doping appropriate dopants into the polymer matrix. Some nonconductive polymers can be made electrically conductive by addition of respective dopants; those are necessarily either electron

TABLE 2.1

Comparison of conductivity and specific gravity of metals and polymer (Joel R. Fried, 2nd ed., 2003)

Material	Conductivity[a] (S cm^{-1})	Specific gravity
Silver	10^6	10.5
Copper	6×10^5	8.9
Aluminum	4×10^5	2.7
Platinum	10×10^4	21.4
Mercury	1×10^4	13.5
Carbon fiber	0.05×10^4	1.7–2
Carbon black-filled polyethylene	0.001×10^4	1
PTFE (Teflon)	10^{-8}	2.1–2.3
Polyacetylene (non-doped)	1×10^4	1
Polyacetylene (doped)	15×10^4	1
Polythiophene (doped)	1×10^4	1

[a] Unit of siemens (S) per cm.

donors (e.g., alkali metals (Na, K, Li)) or electron acceptors (e.g., I_2, Br_2, arsenic pentafluoride AsF_5, borate (BF_4^-), and chlorate (ClO_4^-)) (Table 2.2). Therefore, the dopants are of two types, namely electron donors (such as alkali metal ions, e.g., Na, K, Li) and electron acceptors (such as AsF_5, I_2, Br_2, BF_4^-, ClO_4^-, and p-methylphenylsulfonate). Polysulfurnitride $(SN)_n$ with small conductivity (1000 S cm^{-1}) usable purposefully as conducting material was found as a polymeric material before 1973. But it was not further investigated due to the explosive nature of its precursor (S_2N_2). Later, natural rubber doped with carbon black or acetylene black was observed to possess conducting property. Natural rubber doped with acetylene black was used as an antistatic device in large establishments like hospitals for safety purpose [1]. Naarmann et al. reported that the polycyclic aromatic compounds

reacted with halogens could be transformed to charge transferring complex salts with moderate conductivity [3]. The research for the development of conducting charge transfer salts is the beginning of the discovery of conducting polymers [4]. In the early 1970s, researchers demonstrating salts of tetrathiafulvalene showed the almost metallic range of conductivity. This indicated that organic compounds like polymers can also transport current. Diaz and Logan examined the use of polyaniline film as electrodes [5]. In 1977, Alan J. Heeger, Alan MacDiarmid, and Hideki Shirakawa reported similar high conductivity in oxidized iodine-doped polyacetylene [6]. They were awarded the 2000 Nobel Prize in Chemistry for the discovery and development of conductive polymers. An important application of conducting polymers thereafter started as organic light-emitting diodes (OLEDs) [7].

TABLE 2.2

Various conducting polymers (Joel R. Fried, 2nd ed., 2003)

Conducting polymer	Repeating unit	Dopant	Conductivity (S cm^{-1})
Polyaniline		HCl	200
Polyisothionaphthalene		BF_4^-, ClO_4^-	50
Poly(3-alkyltiophene)		BF_4^-, ClO_4^- $FeCl_4^-$	10^3–10^4
Poly(p-phenylene)		AsF_5, Li, K	10^3
Poly(p-phenylene vinylene)		AsF_5	10^4
Polypyrrole		BF_4^-, ClO_4^-	500–700
Polythiophene		BF_4^-, ClO_4^-	10^3

2.2 CONDUCTION MECHANISM

Chemically synthesized conductive polymers have very low conductivities as the electrons of π-bond are mostly localized and their movement is restricted within two adjoining double-bonded atoms. On the other hand, doped polymers are highly conductive due to the delocalization of π-electrons through the large number of double-bonded atoms. An electron is removed from the valence band in the case of p-doping, and in another case (i.e., for n-doping) an electron is added to the conduction band. The positive charge carriers (holes) and negative charge carriers (electrons) produced by both kinds of doping (p or n) move to the opposite directions to the cathode and anode, respectively, when an electric field is applied across the doped polymer. This movement of charge carriers is responsible for electrical conductivity. The radical cations or polarons (holes) are generated by doping oxidative dopants, such as AsF_5, I_2, Br_2, BF_4^-, and ClO_4^-, into the π-backbone of the polyacetylene chain by removing one electron [8]. The other electron (i.e., the lone electron of π-bond) travels easily along the molecule. Further oxidation produces a bipolaron that condenses pairwise into the so-called solitons. Polyaniline is easily synthesized in polymerization oxidation reaction in protonic acid aqueous solution. But, the conjugated π-bonds of polyaniline are highly susceptible to chemical and electrochemical oxidation or reduction and the conductive and optical properties are deteriorated. However, such properties can be enhanced with the controlling of the extent of oxidation and reduction. The extent of oxidation results in three different polymers in varied oxidation states. Pernigraniline is fully oxidized, emeraldine or quinoid imine structure is partially oxidized, and leuco-emeraldine is a fully reduced or only benzenoid amine structure. Only the emeraldine form of polyaniline which is both oxidized and reduced partially exhibits the electrical conductivity, and it is stable too [8]. The leuco-emeraldine is easily oxidized in the presence of air, and the pernigraniline is very much degradable.

2.3 SYNTHESIS OF CONDUCTING POLYMERS

The processes for the synthesis of conducting polymers are primarily of two types, viz. chemical synthesis and electrochemical copolymerization. The chemical synthesis involves the connectivity of the C–C bond among the monomers by the application of heat, pressure, exposure of light, and the presence of a catalyst. The advantage of this process is a higher yield of conducting polymers with the disadvantage of lower purity. On the other hand, the electrochemical copolymerization involves the electrochemical process where the conduction polymers film is deposited on the electrodes from the electrolyte of monomers in an electrochemical cell. The conducting polymer obtained is pure in comparison, but the process is very

specific to limited conducting polymers. Electrochemical copolymerization is of two types based on the mode of application of electric potential in the electrochemical process. Applied potential is cyclic in cyclic voltammetric process, whereas it can be both cyclic and constant in potentiostatic method [9]. Photochemical synthesis and doping into the conducting polymers are other two methods available for the synthesis of conducting polymers. Conductive polymers are successfully being synthesized in recent time for the reduction of the problems associated with the stability and solubility. The problems are minimized using chemical reactions and experimentations on the doping of appropriate dopants with characteristics of electron donor or acceptor. A wide range of various techniques are discussed for the purpose. The conduction mechanism of conducting polymers was initially demonstrated with an example of polyacetylene as it is with the simpler molecular framework. However, the poor solubility, high melting point, and very low stability to the oxygen and moisture have resulted in the appropriateness for the technological applications of it. There are other four classes of conducting polymers—viz. PAn, PPy, PTh, and polyphenylene vinylene (PPV)—based on the technology used for their preparations. Polyaniline is the unique polymer that is prepared by doping with a strong acid. It has various forms based on the pH level of the medium in which it has been fabricated. Unlike other conducting polymers, neutral emeraldine form of polyaniline is soluble in solvents like N-methyl pyrrolidone (NMP) that is polar but does not provide H^+. However, laterally substituted derivatives of the other forms are substantially soluble in conventional solvent; hence, fabrication is processable. Poly(3-hexylthiophene) has been extensively investigated as a laterally substituted conjugated polymer. A wide range of various techniques is discussed for the purpose.

2.3.1 SYNTHESIS THROUGH POLYMERIZATION

Conducting polymers are prepared with chain-growth or step-growth polymerization.

2.3.1.1 Chain-Growth Polymerization

Polyaniline is a polymer of aniline suitably synthesized in chain growth polymerization reaction. Polymerization takes place in an aqueous acidic medium. Various grades of polyaniline are obtained by chemical or electrochemical oxidation depending on the extent of oxidation. The conductivity of polymers so obtained ranges from 10^{-11} to 100 S cm^{-1} depending on both the oxidants: ammonium peroxydisulfate and extent of oxidation. Another example of chain-growth polymerization is the synthesis of poly(p-phenylene) (PPP). PPP with excellent thermal and oxidative stability is synthesized by Freidel–Crafts polymerization from the benzene as the monomer:

$$\text{(1)}$$

2.3.1.2 Step-Growth Polymerization

Polyphenylene sulfide can be prepared by condensation of *p*-dichlorobenzene with sodium sulfide [10]. Others conducting polymers that are prepared by this method are polyvinylidene sulfide, polythiophene sulfide with appropriate dichlorosulfide, and anhydrous sodium sulfide [11,12]. Polyphenylene oxide and polyparaphenylene can also be synthesized in the step-growth polymerization process:

$$\text{(2)}$$

Polyphenylene oxide and polyparaphenylene can also be synthesized in the step-growth polymerization process. The reaction between magnesium in ether and *p*-dibromobenzene is catalyzed by nickel chloride bipyridyl, and a sodium salt of *p*-bromophenol is condensed by Ullmann condensation for the preparation of polyparaphenylene [13] and polyphenylene oxide, respectively:

$$\text{(3)}$$

2.3.2 Synthesis by Doping with Compatible Dopants

The conductive property of polymer is enhanced by the doping process. In the doping, negative or positive charge on the polymer backbone is created by either oxidation or reduction. The polymers with C = C conjugation at the backbones exhibit properties like optical transmission, ionization capacity, and substantial affinity to electrons. The oxidation and reduction are easier and they are more reversible compared to those in pristine non-conducting polymers. The conductivity in polymers is induced with the exposure of the polymers in either gas phase or solution to a chemical agent that can transfer the charge. Excellent thermally stable conductive polyacetylene is obtained by doping with arsenic pentafluoride into acetylene matrix. Lewis acid is reacted with the polymers in this process. Polyacetylene is reacted with Lewis acid and Lewis base to have p-doped polyacetylene and n-doped polyacetylene, respectively:

$$(CH)_n \quad + \quad A \quad = \quad (CH)_x^+ A^- \quad \text{(4)}$$

Polyacetylene   Lewis acid   p-doped polyacetylene

$$(CH)_n \quad + \quad B \quad = \quad (CH)_x^- B^+ \quad \text{(5)}$$

Polyacetylene   Lewis base   n-doped polyacetylene

Polyacetylene is used in solar cells and batteries. It is critical to oxidation by air and moisture. Two other common conductive polymers, viz. polypyrrole and polythiophene, are also synthesized by doping. Polypyrrole is synthesized from pyrrole by electropolymerization. On the other hand, polythiophene is synthesized by anodic oxidation of thiophene monomer. Both the polymers are very stable to oxidation but with lower conductivity as compared to polyacetylene. Doped conducting polymers, especially polypyrrole and polyaniline, have shown substantial attention because of their better environmental stability, good electrical conductivity, and economic importance:

$$\text{(6)}$$

Dispersion of dopant in the polymer solution with proper agitation facilitates the aggregation of dopant on the surface of entangled chains of polymers. The dopant concentration is higher up to 50%. The dopant molecules are incorporated in the quasi-one-dimensional polymer backbone that alters the order of chain, and polymers are reorganized for the conductivity. The induced conductivity depends on the factors such as ionic nature and concentration of dopants, homogeneity of the dopants in the polymer matrix, the mobility of the generated carrier, crystallinity, and morphology of the polymers. The dopants in the polymers result in the formation of defects in the conjugated double-bonded carbon atoms to form carriers such as solitons, polarons, and bipolarons in the backbone [14]. Lewis et al. describe the transport of charge by tunneling transitions between localized states. Theoretical formulation is provided for the calculation of transient AC and DC conductivities depending on temperature, time, and frequency with a good agreement with experimental conductivities for a number of polymers. The change in the C–C bond length of the polyacetylene chain has been confirmed by X-ray diffraction. The length increases in the polyacetylene by doping with donor and decreases by doping with the acceptor.

2.3.2.1 Types of Doping Agents

Doping agents may be classified as oxidizing and reducing neutral organic molecules or easily ionizable inorganic salts (Table 2.3). Ionic dopants are either oxidized or reduced by transfer of an electron with the polymer, and the counter-ion remains with the polymer so as to maintain the process neutral. On the other hand, neutral organic dopant compounds are ionized into positive or negative ions with or without chemical modification during the doping process. In the electrochemical doping process, dopant molecules are dissociated to anion that neutralizes the positive charge of the polymer. The

TABLE 2.3
Various classes of dopants

Dopant	Ions that work	Doped polymer[a]
Neutral dopant		
I_2	I_2^-	PA, PPS, PPP
Br_2	Br_2^-	PA(*trans*)
AsF_5	AsF_6^-	PA, PPS, PPP
$FeCl_3$	$FeCl_4^-$	PA
$AlCl_3$	$AlCl_4^-$	PP
$SnCl_4$	$SnCl_2^-$	PA
Na	Na^+	PA
K	K^+	PPP
Ionic dopant		
$LiClO_4$	ClO_4^-	PPY, PTh
$FeClO_4$	ClO_4^-	PA, PPy, P3MT
$LiBF_4$	BF_4^-	PPY, PTh
$LiAsF_6$	AsF_6^-	PA
$(CH_3)NPF_6$	PF_6^-	PTy, PTh
	$(CH_3)_4N^+$	PTh
Na-naphthalide	Na^+	PA(*trans*)
Organic dopant		
CF_3COOH	COO^-	PPy
CF_3SO_3Na	SO_3^-	PPy
$CH_3C_6H_4SO_3H$	SO_3^-	PPy
Polymeric dopant		
PVS	SO_3^-	PPy, PAn
PPS	SO_3^-	PPy, PAn
PS-Co-MA	COO^-	PPy

[a] PA: polyacetylene, PPS: polyphenylene sulfide, PPP: polyparaphenylene, PPy: polypyrrole, PTh: polythiophene, PAn: polyaniline, P3MT: poly(3-methyl thiophene), PVS: polyvinyl sulfide.

anionic organic dopants are generally incorporated into polymers from aqueous electrolyte during the anodic film deposition of the polymer on the anode. Some of the polymeric dopants are functionalized polymer electrolytes containing amphiphilic anions.

2.3.2.2 Doping Techniques
Various convenient and low-cost techniques of doping—viz. gaseous doping, solution doping, and electrochemical doping are widely used. Polymers are exposed to the dopant in the vapor phase and under vacuum in the gaseous doping process. The concentration of dopant in the doped polymer can easily be controlled by the parameters such as temperature, the extent of vacuum, and the time of exposure of vapor to the polymer. In case of solution doping, dopants, polymer, and other chemicals should be soluble in the appropriate polar organic solvents such as nitromethane, toluene, and tetrahydrofuran. Doping into the polymer is generally accomplished simultaneously with

the polymerization reaction in the electrochemical process of doping. However, only ionic dopants are doped into the otherwise conducting polymer that is soluble in the polar solvents such as nitromethane, dichloroethane, acetonitrile, and tetrahydrofuran. The conducting polymers are used as the electrolyte in the electrochemical cell. The other methods of doping include self-doping, doping induced by irradiation, and ion-exchange doping. The ionizable functional group such as sulfonate group of poly [3(2-ethane sulfonate)thiophene] acts as an internal dopant in a self-doping process and no external dopant is required. High-energy irradiations (e.g., gamma-ray, neutron radiation, and electron beam) are employed to incorporate the neutral dopants into the polymers. The neutral molecules SF_6 and I_2 decompose to active dopant species under high energy before they (dopant species) are doped into the conducting polymers. The doped polythiophene is obtained in the presence of either SF_6 decomposed by gamma-ray or I_2 energized by neutron irradiation [15].

2.3.2.3 Mechanism of Doping
Either positive or negative charge carriers are generated in the polymer as doping agents are either strong oxidizing or reducing agents, and the general scheme of the reaction process is represented as follows:

$$Polymer + Dopant \xrightarrow{Oxidation} (Doped\ polymer)^+ + Dopant^-$$
(7)

$$Polymer + Dopant \xrightarrow{Reduction-} (Doped\ polymer)^- + Dopant^+$$
(8)

Apart from oxidation and reduction reactions, doping leads to the rearrangement of charges in the atoms attached in the backbone of polymer chains, and as a result, the new structure is formed with changes in the bond length of C–C bond of a polymer such as a polyacetylene (Figure 2.1). This change in the bond length depends on the acceptor or donor used for the doping. The charged species known as solitons are charge defects with no spin. A reducing or donor-type dopant introduces an electron to the polymer chain that couples with the neutral defects resulting in a negative soliton with zero spins. On the other hand, an oxidizing or acceptor-type dopant abstracts as an electron from the polymer chain and a positive spinless soliton are formed. The doping of polyacetylene may explain the mechanism. A soliton is formed in the polymer chain at a very low doping concentration, whereas polaron is formed in case of higher level of doping. With the increasing extent of doping, polarons interact with each other to form bipolaron. The existence of bipolarons has been reported in different highly doped conducting polymers such as polyparaphenylene, polypyrrole,

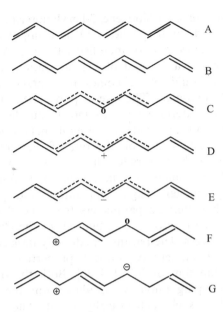

FIGURE 2.1 Various forms of charge carriers in doping conducting polymers: A and B: *trans*-polyacetylene in two different phases; C: neutral soliton; D: positive soliton; E: negative soliton; F: polaron; G: bipolaron.

polythiophene, poly(3-methyl thiophene), and polyacetylene. The bipolarons are considered as the best carrier species that respond to the induction of the highest level of conductivity into the doped conducting polymers.

2.3.2.4 Influence of Doping on Conductivity

The extent of induced conductivity of polymer by doping depends mostly on the chemical reactivity of the dopant. The selection of dopant for a polymer matrix and the compatible conditions for the doping are also important factors. The other factors that contribute to this enhancement of conductivity include methods of synthesis and the structures available, processing of polymers' degree of crystallinity, and the temperature. The *cis*-polyacetylene is more conductive as compared to the *trans*-polyacetylene, whereas both are doped with the AsF$_5$ dopant. However, the same dopant is not effective for different polymers. Polyacetylene doped with iodine has the conductivity 10–12 times more than the undoped polyacetylene. But, same iodine of comparatively weaker oxidizing ability is unable to increase the electrical conductivity while it is doped into the polyphenylene sulfide and polyparaphenylene. On the other hand, stronger oxidizing dopant ASF$_5$ can successfully enhance the conductivity of polyphenylene sulfide and polyparaphenylene while doped with it. The conductivity of polymers depends on the extent of doping. For vapor-phase doping, the doping level increases with the exposure time of dopant vapor to the polymers. The conductivity increases linearly with the concentration of dopant. But, in a very few cases, a sharp increase in the conductivity is observed with a very small increment of dopant concentration. This sharp

increase is assumed to be the higher mobility of the charged carrier at that concentration due to interchain interaction.

2.3.3 ELECTROCHEMICAL POLYMERIZATION

Conducting polymers such as polypyrrole, polythiophene, and other polymers (e.g., aniline, benzene, and phenol) are prepared by electrochemical polymerization. The principle of the polymerization is similar to the metal plating on the electrode from the polyelectrolyte solution. Pyrrole in aqueous acetonitrile solvent containing tetraethylammonium tetrafluoroborate is electropolymerized in an electrochemical cell [16] for the preparation of conducting polypyrrole containing BF$_4^-$ as a dopant. A film of such conducting polymer is deposited on the surface of the platinum cathode. A thin but tough violet film of a copolymer of polypyrrole, polythiophene, and poly(thienylpyrrole) can be prepared on the surface of the cathode using the polyelectrolyte of 2-2'-thienylpyrrole [17].

2.3.4 PHOTOCHEMICAL SYNTHESIS

The photochemical process is reported to be a quick and inexpensive method. Moreover, it is less destructive to the environment. Thus, the method is advantageous for the fabrication of conducting polymers such as polypyrrole, polythiophene, and its derivatives [18]. Polypyrrole film and polybenzothiophene are prepared on the surface of a compatible solid surface by the photosensitized polymerization. The metal complexes such as ruthenium (Ru(II)) and copper(Cu(II)) complexes are oxidized to Ru(III) and Cu(III) by photo-irradiation for the generation of active electron, and the photopolymerization is conducted through oxidation by transfer of an active electron for the preparation of polypyrrole [19]. On the other hand, polybenzothiophene is prepared in acetonitrile using carbon tetrachloride (CCl$_4$) and tetrabutylammonium bromide [20]. Polyaniline has been recently prepared in the oxidative free radical-induced reactions via horseradish peroxide. The photochemical synthesis of polymerization can be accomplished in mild environmental condition as compared to that for chemical and electrochemical techniques.

2.4 VARIOUS ISSUES FOR SYNTHESIS

Polymerization reactions, in general, are complicated, and synthesis of conducting polymers is in particular of complex nature. High purity of all the chemicals such as solvent, catalyst, initiator, and particularly the monomers is required for the synthesis of conducting polymers. The conditions and the atmosphere of synthesis must be appropriately controlled for obtaining high-molecular-weight polymers. Almost every type of catalysts such as Ziegler–Natta catalysts employed for the synthesis of conducting polymers is susceptible to either moisture or oxygen and other polar solvents [21]. Consequently, dry and inert environments are prerequisites for the preparation of conducting polymers. A very little variation in the conditions or purity of the chemicals or inappropriate catalysts significantly changes the properties of the polymers prepared. Many of the conducting polymers synthesized are susceptible to the heat and light, and may undergo isomerization to become polymers of lesser conductivity or deterioration of other properties required for their employment as conducting polymers. The extent of oxidation and the oxidative state of the synthesized polymer render the conductivity and other properties. Oxidative doping is preferential, and the resulting oxidative state of conducting polymers is the most contributing factor for their properties. Such polymers are basically macro-salts; consequently, their processing is found to be difficult owing to the diminished solubility in organic solvents and water. The π-conjugated polymeric chains are rigid and possess strong interactions of the atoms along the polymer backbone. Consequently, conventional polymers are insoluble and infusible materials lacking processability. The significant change in the structure of conjugated pristine polymer enables the modification of electronic properties, and this development has a significant contribution in electronic industries. Moreover, the charge induced to the organic backbone by doping is often unstable in the presence of atmospheric moisture. The solubilizing agents are added to the processing to enhance the processability during the synthesis of many polymers and that further complicates the synthesis and also contributes to the deterioration of properties of conducting polymers. It is evident from the theoretical thermodynamics that some of the conductive polymers may even be completely insoluble to any solvent and their processing can only be accomplished by dispersion [22]. However, it is recently reported that lateral substitution of suitable functional groups in the conjugated conducting polymers renders solubility to ease the processability without significant loss in their conductivity. The inherent instability is the other issue for the utility of polymers, especially in the doped form at ambient conditions. However, in recent time, conducting polymers with substantial stability have been synthesized with advanced techniques. Some of the techniques for the purpose have been demonstrated with some specific examples of such systems.

2.4.1 Vapor-Phase Polymerization

Inherently conducting polymers (ICPs) have limited uptake due to insolubility, and methods to manufacture them in a usable form have been problematic for the application in consumer devices. Vapor-phase polymerization (VPP) provides a convenient route for the production of thin films of both soluble and insoluble ICPs, and thereby nano-composites. The parameters that affect the growth of polymers for a range of different monomers (thiophene, 3-hexylthiophene, pyrrole, 3,4-ethylene dioxythiophene, etc.) have importance for design and fabrication of modified CPs using VPP [23]. VPP process is a method of polymerization where an oxidant in liquid phase comes in contact with the vaporized monomers. The rate of polymerization at the liquid–vapor interface depends mostly on the concentrations of both the oxidant and monomers available at the interface, and also on the rate of their replenishments. Hence, the factors for efficient mixing of the oxidant and monomers are necessarily important. The physical parameters (e.g., temperature and pressure) and the characteristic parameters (e.g., the shape and the size of the reactor) are optimized for the efficacy of the VPP process. The parameters directly impact on the properties (conductivity, optical properties, etc.) of resulting thin film of ICP [24–26]. Both the oxidizing agent and doping anions are available from the selected oxidant that is characteristically a salt. The oxidants have varied oxidation strength. Hence, an individual oxidant incorporated into polymers through VPP provides different oxidation state of the fabricated polymers. The standard electrode potential of the cation of the oxidizing agent is the measure of its strength. Fe^{3+} with a standard electrode potential of 0.77 V is found as the most commonly used cation to dope the ICPs. The cation (Fe^{3+}) and any of the anions from Cl^-, Tos^- [27], and sulfonates [24,28] are paired. These anions can stabilize the polarons and bipolarons which are mobilized along the conjugated C–C double bond at the backbone of the ICPs that result in the enhancement of conductivity of ICPs. Subramanian et al. employed sulfonate-based anions for doping polypyrrole (PPy) via VPP. The length of the alkyl chain attached to the parent sulfonic acid anion is varied during the doping. Benzene sulfonic acid, p-ethylbenzenesulfonic acid, and p-dodecylbenzenesulfonic acid are the different sources of sulfonates. Nair et al. proposed 'top-down' approach for polymerization at both the surface and bulk of the oxidant [29]. The monomers diffuse through the polymer layer, and rate of diffusion is found as the limiting step in the polymerization reaction. The nanofibers of polystyrene complexes with $Fe(Tos)_3$ and the complexes are employed as the substrate. Fabretto et al. have explicitly described both 'top-down' and 'bottom-up' mechanisms of VPP [23]. The polymer layer is formed and the oxidant diffuses up through the formed polymer layer in the 'top-down' mechanism and through the forming polymer layer to access additional oxidant. The results of the X-ray photoelectron spectroscopy (XPS) and time-of-flight secondary ion mass spectroscopy (ToF-SIMS) analyses provided the depth profile of polymer layer doped with oxidant, and the 'bottom-up' mechanism has been proved to be the most effective. The concentration of Fe^{3+} species increases with the depth of polymer layer. The trace amount of Fe^{3+} is found on the top surface after a substantial amount of time. The ICP fibers exposed to the vapor-phase monomer are coated with the oxidants. Such coating on the thread has the applications as a sensor for hydrogen gas [30], bio-wires [31], and electrochromic devices [32]. The faster rate of polymerization of pyrrole in case of complex substrate like three-dimensional (3D) shape is more attractive compared to poly(3,4-ethylene dioxythiophene) (PEDOT) in spite of relatively lower conductivity of PPy. Substrate of complex 3D shape can even be coated with an ICP film through VPP along with normal type of substrate.

2.4.2 HYBRID CONDUCTING POLYMERS

The recent trends for the development of efficient CPs include the approaches for the preparation of multifunctional nanomaterials/nanocomposites, that is, hybrid conducting polymers (HCPs) with efficacy and certain benefits. Zhang et al. have investigated the means of conductivity of HCPs and enunciated mechanism to regulate their other properties and also challenges in developing better nanomaterial-based CPs [33]. The HCPs possess characteristics of CPs and at the same time those of the organic/inorganic nanoparticles. The combination of both types of properties in a single material has been attractive as the research in the field of conducting polymers [34]. The development of HCPs is expected to overcome the inherent problems with CPs such as poor processability, very low solubility, very less stability, and insufficient yield. The primary technology for the synthesis of HCPs and the mechanism of conduction have been reported [34,35]. Various applications of HCPs include manufacture of devices for the storage of various forms of energy, anti-corrosive coatings, sensors, EMI shielding, and also in biomedical applications. Nanostructure CPs have larger surface area per unit mass owing to the smaller particle size. They also possess a higher electrical conductivity and a superior electrochemical activity. Accordingly, they are appropriate for various applications. The composites of a CP and a nanostructured material are promising to have improved performance in different areas of applications such as energy storage and sensors.

2.4.3 NANOSTRUCTURE CONDUCTING POLYMERS

One of the recent trends for tuning the intrinsic properties of the polymers is the synthesis of nano-dimensional conducting polymers for the introduction of biocompatibility and solution processability. Owing to the larger surface area per unit mass, nano-structured conducting polymers have the certain advantages. Higher conductivity of charge due to the shorter path for mobilization of ion and better electrochemical activity are also available in nanostructured CPs. Hence, they are suitable in the applications for energy storage and conversion. An understanding of the electronic behavior based on the size and shape of nanostructured polymer has been investigated to know the effectiveness and possible scope for the storage of energy in solar cells, fuel cells, supercapacitors, and rechargeable battery

applications. Conducting polymer nanostructures (CPNs) have been fabricated with the help of templates in the process of polymerization. The hard template process employing membranes of polycarbonate, polyester, and anodic aluminum oxide (AAO) is a widely used process for the synthesis of CPNs [36]. The nanorods, nanofibers, and nanotubes of CPs such as PEDOT, PAn, PPy, poly(3-hexylthiophene) (P3HT), and poly(p-phenylene vinylene) (PPV) are synthesized by using the polycarbonate PTM (PC-PTM) and the AAO membranes. Liu et al. have fabricated coaxial nanowires of pore diameter up to 200 nm through AAO hard template. The simultaneous growth of core MnO_2 and shell PEDOT in an electrochemical deposition has been observed [37]. On the contrary, the nanowires or nanotubes from mesoporous silica have been found to be interconnected through the micro- or mesopores. Conducting polymeric filaments of PAn have been fabricated by Bein et al. in the form of the 3-nm-wide hexagonal channel of the aluminosilicate MCM-41 [38]. PPy and PAn nanostructures are used as a counter-electrode in dye-sensitized solar cell (DSSC) applications and can be prepared by the chemical and electrochemical processes. On the other hand, double-layered PAn consisting of a nanoparticle layer and a nanofiber layer has been prepared by an electro-deposition route. Significantly increased active sites of double-layered PAn are used to increase the performance of the solar cell. The efficiency of conversion of solar energy to electrical energy by using double-layered PAn-based DSSC has been reported to be higher compared to that by using chemically deposited PAn based DSSCs. The higher efficiency of conversion is due to the double-layered nanostructures that results in an enhancement of kinetics for transfer of charge and higher electrocatalytic activity in the redox reactions [39].

2.4.4 NARROW BANDGAP CONDUCTING POLYMERS

The copolymerization of aromatic monomers and o-quinoid hetero-cycles or donor–acceptor monomers is the effective strategy for the synthesis of low-bandgap conducting polymers [40]. The bandgap depends on the geometry of aromatic rings in conjugated polymers such as polyparaphenylene, polypyrrole, and polythiophene. The more the quinoid characteristics of the backbone of the polymer chain, the lesser is the bandgap. The bandgap is thus not minimum for the alternation of bond length of zero which is the case for compounds like polyacetylene. This finding is consequently employed for the design of new conjugated organic polymers with minimum bandgaps, enhanced solubility, and novel optical properties as well [41]. Lamba et al. synthesized imine-bridged poly(p-phenylene) derivatives for the maximization of extended π-conjugation to enhance the solubility of highly insoluble poly(p-phenylene). A different class of polymers known as narrow bandgap polymers is therefore developed to cope up with the disadvantages of doped conducting polymers. External dopants are not incorporated into the polymers to make them conductive. Instead, such conducting polymers are obtained by the reduction of aromaticity of polymers, that is, by an increase of quinoid form and by the introduction of empty p-orbitals in the atoms of polymeric backbone chain [21,35]. The problems of stability and processability inherent to the doped polymers can be abridged by the introduction of narrow bandgap conducting polymers.

2.4.5 Synthesis in Supercritical CO_2

The electrochemical synthesis of CPs, specifically PPy and PAn, in supercritical carbon dioxide ($scCO_2$) has been reported by Anderson et al. [42]. They have synthesized the PPy and PAn by the electropolymerization process by employing the $scCO_2$ as a solvent for polymer dissolution. The $scCO_2$ contained 0.011 mol of acetonitrile as a modifier and 7.2×10^4 mol of tetrabutylammonium hexafluorophosphate (TBAPF6) as the electrolyte. PAn has also been successfully electrochemically synthesized in $scCO_2$. The better electrical conductivity of synthesized PPy and PAn in $scCO_2$ as compared to those films prepared electrochemically in aqueous solution has been confirmed by cyclic voltammetry (CV). The films produced in $scCO_2$ are visibly smoother and flatter and exhibit distinctive surface characteristics that could be proved to be advantageous in optical, dielectric, and anti-corrosion applications. The $scCO_2$ is considered as the green solvent and is an alternative to environmentally hazardous solvents such as sulfuric acid and acetonitrile traditionally used in the synthesis of conducting polymer [43]. PPy and PAn synthesized in $scCO_2$ have recently received much attention for their use as corrosion inhibitor free-standing conductive membranes and in microelectronic devices [44].

2.4.6 Biodegradability and Biocompatibility of Conducting Polymers

The development of biocompatible CPs with electrical conductivity similar to that of metals and at the same time with the property of biodegradability is a recent trend for varied applications [45]. Biodegradable CPs are used as suitable matrices of biomolecules to enhance the stability, speed, and sensitivity of various biomedical devices [46]. They are versatile as their properties can be readily modulated by the technique of surface functionalization, and many possible dopants can be incorporated into the biomolecules matrices. Various CPs can be functionalized with certain biomolecules so as to be applicable as biological sensing elements in different signaling pathways that are required for cellular processes. The modification of surface of the CPs can be used in order to provide physicochemical and biological guidance for the promotion of cell proliferation at the polymer–tissue interface to be employed to induce various cellular mechanisms in medical applications and bioengineering. The application includes for tissue engineering, regenerative medicine, and biomedical imaging as well as biomedical implants, bioelectronics devices, consumer electronics, neuroprosthetic devices, cardiovascular applications, drug delivery, biosensors, and actuators. Modifications of the surface through functionalization priory for incorporation of biomolecules into the CPs can be accomplished by both physical and chemical modifications. Biomolecules are doped into intrinsic conducting polymers, or chemical modification has been extensively investigated using biomolecules as dopants [47], or active moieties are immobilized on the surface of the CPs [48]. On the other hand, the surface roughness is enhanced through physical modification by different techniques. The techniques include generation of microporous films using polystyrene sphere templates, fabrication of composites of nanoparticles and polylactide [49] and the formation of composites of biomolecules to avail the 'fuzzy' structures [47]. Laminin peptide sequences and polysaccharides are widely used biological dopants (Collier et al. 2000; Finkenstadt 2005). Electro-polymerized PEDOT upon the surface of gold electrodes with glutamate (Glu) as the dopant has potential neural applications [50]. The PEDOT/Glu coating exhibits lower impedance over a substantial range of frequencies than pure gold with no such coating. The electrical signal of good quality is available with these coated CPs of lower impedance.

2.5 APPLICATIONS

Various applications of conductive polymers include electrodes in lightweight batteries such as lithium-ion batteries; variable transmission windows, electrochromic display in television, computer and mobile phones, etc; and biosensors and nonlinear optical devices. Polymeric lithium batteries such as lithium-poly(p-phenylene) and lithium-polyacetylene are widely used because of lighter, flexible, easier-to-fabricate materials compared to other materials. Doping of lithium in the polymer matrix provides the electrons to be mobilized in conductive polymers. Conducting polymers are used as both electrodes and electrolytes. Polymeric lead-acid batteries are used in automotive and other small energy supply. The other advantages of lithium-doped polymeric batteries are high reliability, lightweight, non-leakage of the electrolyte solution, ultrathin film form, flexibility, and high energy density (100 W dm^{-3}). The conductive polymer can be used as the cathode. Various conductive polymers that are used as cathode include polyacetylene, poly(p-phenylene), polypyrrole, and polyaniline. The anode can be made of Li dissolved in aluminum foil that provides an energy density of 320 Wh kg^{-1}. Others uses of conductive polymers are as solid electrolytes. Two examples are poly(alkyl sulfide)s and poly phosphazenes. Various lithium salts such as lithium tetrafluoroborate ($LiBF_4$), lithium hexafluoroarsenate ($LiAsF_6$), lithium perchlorate ($LiClO_4$), and lithium trifluoromethanesulfonate ($LiCF_3SO_3$) are added to solid polymer electrolytes. Li salts have low crystal lattice energies that donate the electrons very easily to the conducting polymers. Other possible applications include sensors, conductive paints, semiconductor circuits, low current wires, and electromechanical actuators for electromagnetic shielding. Conducting polymers are useful in antistatic application, that is, removing a large amount of static electricity in electronic industries. This can be accomplished by coating the conducting polymer on an insulating surface. Repeated oxidation and reduction of the polymeric backbone provides the opportunity of polymer rechargeable batteries. For example, PPy lithium cell is a useful rechargeable battery in comparison with Ni–Cd cell. Polymeric batteries are nontoxic and environmentally safe. A solar cell made of conducting polymers is a device for utilizing solar energy and replacing inorganic cells. PPy, PAn, PTh, and their derivatives are used in the active layer of gas sensors. Such conducting polymers are used as the structural material and to serve as discriminating layers in electronic chemical sensors. A chemiresistor is formed by two electrodes at the contact point. The conducting polymer is applied over an insulating substrate. The difference in current represents the output signal. The output of the interaction between a neutral gas and an organic semiconductor provides principle of transduction in field effect transistor (FET) sensors. PAn and its derivatives are most frequently used for protection against corrosion when they are electrodeposited onto oxidizable metals, stainless, and iron sheets. Artificial intelligent materials or smart polymers can remember numerical configuration and can confirm when exactly the same stimulus is given. This property is utilized in generating a password for high-security applications. Polymer sheets coated with conducting polymers are used as an inexpensive and better adhesive for the fabrication of printed circuit boards in comparison with conventional copper-coated epoxy resin.

2.6 FUTURE SCOPE FOR APPLICATIONS

Fabrication of composites, especially nanocomposites, of intrinsic conducting polymers is a new trend to have special structures and properties with improved electric conductivity. Various kinds of sensors and biosensor devices—viz. conductometric, impedimetric, potentiometric, amperometric, and voltammetric devices—are being developed with the use of conducting polymers as active, sensing, or catalytic layers. Enzymes or other biologically active compounds are entrapped into matrices of intrinsic CPs. The biocompatibility of several conducting polymers can be applicable in medicine as artificial muscles, limbs, and nerves with further development. Scanning microscopy and electrochemical quartz crystal nano-balance may be developed with a sophisticated technique for the synthesis of biocompatible conducting polymers being developed in the laboratories worldwide. The copolymerization of monomer derivatives is potential for the fabrication of CPs with fine-tuning of flexibility,

crystallinity, and optimized chemical and mechanical stabilities for better processibility so that they can be used in heterojunction solar cells and other applications. The syntheses of smart materials through functionalization of conducting polymers that respond well to the environment have a great opportunity for applications in the biosensor. The self-doped polymers also have great opportunity to address the challenges in the transport of ionic charge during redox switching and other limitations of the CPs. The improved nanocomposites and hybrid conducting polymeric materials will be important materials in the future. Electro-conducting nanomaterials—viz. nanofibers, nanorods, and other nanostructures—based on the supramolecular self-assembly of CPs have the potential in the enhancement of the photoluminescence efficiency in surface resonance coupling due to their capacity for charge transfer. Microstructure CPs are supposed to improve the efficiency of the polymer-based transistors and electrochemical cells. The nanoparticle and organometallic inks prepared from CPs are expected to be used in solar cells, light emitting devices, display devices, sensors etc.

2.7 CONCLUSIONS

The latest advances in the synthesis of CPs and their composites are presented in this chapter. The recent achievements concerning the synthesis issues are reviewed. The promising applications in the areas of supercapacitors and batteries, sensors and biosensors, photoelectrochemistry, electrocatalysis, and corrosion inhibition are emphasized, and several other utilizations are mentioned. Several CPs—including polyaniline, polypyrrole, polythiophene, polyvinylpyrrolidone, poly(3,4-ethylene dioxythiophene), poly(m-phenylenediamine), polynaphthylamine, poly(p-phenylene sulfide), and their carbon nanotube reinforced nanocomposites—are discussed in this chapter. The physical, electrical, structural, and thermal properties of polymers based on the method of synthesis are discussed. A brief discussion on carbon nanotubes regarding their purification, functionalization, properties, and production is reported. The conducting polymer/carbon nanotube and nanocomposites are frequently used in storage devices for energy, supercapacitors, solar cells, diodes, coatings sensors, and biosensors.

ABBREVIATIONS

AAO	Anodic aluminum oxide
AFM	Atomic force microscopy
CPs	Conducting polymers
CPNs	Conducting polymers nanostructures
EMI	Electromagnetic interface
FET	Field effect transistor
DSSC	Dye-sensitized solar cell

HCPs	Hybrid conducting polymers
ICPs	Intrinsic conducting polymers
LHNH	Light harvesting nanohetero
LiClO$_4$	Lithium perchlorate
LiBF$_4$	Lithium tetrafluoroborate
LiCF$_3$SO$_3$	Lithium trifluoromethanesulfonate
PAn	Polyaniline
PPy	Polypyrrole
PC-PTM	Polycarbonate poly(tetramethylene)
PDPB	Polydiphenylbutadiyne
P3HT	Poly(3-hexylthiophene)
PMT	Poly(3-methylthiophene)
PPV	Poly(p-phenylenevinylene)
PTM	Particle track-etched membranes
PEDOT	Poly(3,4-ethylenedioxythiophene)
PSCs	Polymer solar cells
PTh	Polythiophene
scCO$_2$	Supercritical carbon dioxide
ToF-SIMS	Time-of-flight secondary ion mass spectroscopy
XPS	X-ray photoelectron spectroscopy

REFERENCES

1. Flory, P.J., Statistical thermodynamics of semi-flexible chain molecules. *Proceedings of the Royal Society of London. Series A*, 1956. **234**: pp. 606–673. doi: 10.1098/rspa.1956.0015.

2. Negi, Y.S. and P.V. Adhyapak, Development in polyaniline conducting polymers. *Journal of Macromolecular Science, Polymer Reviews*, 2002. **C42**: pp. 35–53.

3. Naarmann, H., *Polymers, electrically conducting*. 2000: Ullmann's Encyclopedia of Industrial Chemistry, Copyright © 2011 Wiley-VCH Verlag GmbH & Co. KGaA. doi: 10.1002/14356007.a21_429.

4. Ferraris, J.S., D.O. Cowan, W. Verlstein, and J.H. Perlstein, Electron transfer in a new highly conducting donor-acceptor complex. *Journal of the American Chemical Society*, 1973. **95**(3): pp. 948–949.

5. Diaz, A.F. and A.J. Logan, Electroactive polyaniline films. *Journal of Electroanalytical Chemistry*, 1980. **111**: pp. 111–114.

6. Shirakawa, H., E.J. Louis, A.G. MacDiarmid, C.K. Chiang, and A.J. Heeger, Synthesis of electrically conducting organic polymers: Halogen derivatives of polyacetylene, (CH) x. *Journal of the Chemical Society, Chemical Communications*, 1977. **16**: p. 578.

7. Burroughs, J.H., D.D.C. Bradley, A.R. Brown, R.N. Marks, K. MacKay, R.H. Friend, P.L. Burns, and A.B. Holmes, Light-emitting diodes based on conjugated polymers. *Nature*, 1990. **347**(6293): pp. 539–541.

8. Kumar, S., Conducting polymers and their characterization. *International Research Journal of Engineering and Technology*, 2016. **3**(5): pp. 479–482.

9. Kesik, M., et al., Synthesis and characterization of conducting polymers containing polypeptide and ferrocene side chains as ethanol biosensors. *Polymer Chemistry*, 2014. **5**(21): pp. 6295–6306.

10. Hill, H.W. and D.J. Brady, *Encyclopedia of chemical technology*. Vol. 18. 1982: Wiley, New York.

11. Ikeda, Y., M. Ozaki, and T. Arakawa, Synthesis of poly (vinylene sulfide) and its electrical properties. *Journal of the Chemical Society, Chemical Communications*, 1983. (24): pp. 1518–1519. doi: 10.1039/C39830001518.

12. Mohammad, F., Compensation behavior of electrically conductive polythiophene and polypyrrole. *Journal of Physics D : Applied Physics*, 1998. **31**(8): pp. 951–959.

13. Yamamoto, T., K. Sanechika, and A. Yamamoto, Preparation of thermostable and electric-conducting poly(2,5-thienylene). *Journal of Polymer Science: Polymer Letters Edition*, 1980. **18**(1): pp. 9–12.

14. Lewis, T.J., A simple general model for charge transfer in polymers. *Faraday Discussions of the Chemical Society*, 1989. **88**(0): pp. 189–201.

15. Yoshino, K., et al., *Molecular crystals and liquid crystals*, 1985. **121**: p. 255.

16. Diaz, A.F., K.K. Kanazawa, and G.P. Giardini, Electrochemical polymerization of pyrrole. *Journal of the Chemical Society, Chemical Communications*, 1979. (14): pp. 635–636. doi: 10.1039/C39790000635.

17. Naitoh, S., K. Sanui, and N. Ogata, Electrochemical synthesis of a copolymer of thiophene and pyrrole-poly(thienylpyrrole). *Journal of the Chemical Society, Chemical Communications*, 1986. (17): pp. 1348–1350. doi: 10.1039/C39860001348.

18. Deronzier, A. and J.-C. Moutet, Polypyrrole films containing metal complexes: syntheses and applications. *Coordination Chemistry Reviews*, 1996. **147**: pp. 339–371.

19. Kern, J.-M. and J.-P. Sauvage, Photochemical deposition of electrically conducting polypyrrole. *Journal of the Chemical Society, Chemical Communications*, 1989. (10): pp. 657–658.

20. Iyoda, T., M. Kitano, and T. Shimidzu, New method for preparing poly(benzo[c]thiophene) thin films by photopolymerization. *Journal of the Chemical Society, Chemical Communications*, 1991. (22): pp. 1618–1619.

21. Maiti, S., Recent trends in conducting polymers: problems and promises. *India Journal of Chemistry*, 1994. **33A**: pp. 524–539.

22. Nalwa, H.S., *Handbook of nanostructured materials and nanotechnology*. 4th edition. Vol. 5. 2000: USA Academic Press, New York.

23. Brooke, R., et al., Recent advances in the synthesis of conducting polymers from the vapor phase. *Progress in Materials Science*, 2017. **86**: pp. 127–146.

24. Winther-Jensen, B., et al., Vapor phase polymerization of pyrrole and thiophene using iron(III) sulfonates as oxidizing agents. *Macromolecules*, 2004. **37**(16): pp. 5930–5935.

25. Wu, D., et al., Temperature-dependent conductivity of vapor-phase polymerized PEDOT films. *Synthetic Metals*, 2013. **176**: pp. 86–91.

26. Ali, M.A., et al., Effects of solvents on poly(3,4-ethylene dioxythiophene) (PEDOT) thin films deposited on a (3-aminopropyl)trimethoxysilane (APS) monolayer by vapor phase polymerization. *Electronic Materials Letters*, 2010. **6**(1): pp. 17–22.

27. Ali, M., et al., Effects of the FeCl 3 concentration on the polymerization of conductive poly(3,4-ethylene dioxythiophene) thin films on (3-aminopropyl) trimethoxysilane monolayer-coated SiO 2 surfaces. *Metals and Materials International*, 2009. **15**: pp. 977–981.

28. Subramanian, P., et al., Vapour-phase polymerization of pyrrole and 3,4-ethylenedioxythiophene using iron(III) 2,4,6-trimethylbenzenesulfonate. *Australian Journal of Chemistry*, 2009. **62**: pp. 133–139.

29. Nair, S., E. Hsiao, and S.H. Kim, Melt-welding and improved electrical conductivity of nonwoven porous nanofiber mats of poly(3,4-ethylene dioxythiophene) grown on electrospun polystyrene fiber template. *Chemistry of Materials*, 2009. **21**(1): pp. 115–121.

30. Al-Mashat, L., et al., Electropolymerized polypyrrole nanowires for hydrogen gas sensing. *The Journal of Physical Chemistry C*, 2012. **116**(24): pp. 13388–13394.

31. Choi, S.J. and S.M. Park, Electrochemical growth of nanosized conducting polymer wires on gold using molecular templates. *Advanced Materials*, 2000. **12**(20): pp. 1547–1549.

32. Invernale, M., Y. Ding, and G. Sotzing, The effects of colored base fabric on electrochromic textile. *Coloration Technology*, 2011. **127**: pp. 167–172.

33. Zhang, L., et al., Recent progress on nanostructured conducting polymers and composites: synthesis, application and future aspects. *Science China Materials*, 2018. **61**(3): pp. 303–352.

34. Iqbal, S. and S. Ahmad, Recent development in hybrid conducting polymers: synthesis, applications and future prospects. *Journal of Industrial and Engineering Chemistry*, 2018. **60**: pp. 53–84.

35. Meer, S., A. Kausar, and T. Iqbal, Trends in conducting polymer and hybrids of conducting polymers/carbon nanotube: a review. *Polymer-Plastic Technology and Engineering*, 2016. **55**(13): pp. 1416–1440.

36. Lee, W. and S.-J. Park, Porous anodic aluminum oxide: anodization and templated synthesis of functional nanostructures. *Chemical Reviews*, 2014. **114**(15): pp. 7487–7556.

37. Liu, R., J. Duay, and S.B. Lee, Electrochemical formation mechanism for the controlled synthesis of heterogeneous MnO2/poly(3,4-ethylene dioxythiophene) nanowires. *ACS Nano*, 2011. **5**(7): pp. 5608–5619.

38. Wu, C.-G. and T. Bein, Conducting polyaniline filaments in a mesoporous channel host. *Science*, 1994. **264**(5166): p. 1757.

39. Tang, Q., et al., Counter electrodes from double-layered polyaniline nanostructures for dye-sensitized solar cell applications. *Journal of Materials Chemistry A*, 2013. **1**(2): pp. 317–323.

40. Brédas, J.L., Relationship between band gap and bond length alternation in organic conjugated polymers. *The Journal of Chemical Physics*, 1985. **82**(8): pp. 3808–3811.

41. Lamba, J.J.S. and J.M. Tour, Imine-bridged planar poly(p-phenylene) derivatives for maximization of extended.pi.-conjugation. The common intermediate approach. *Journal of the American Chemical Society*, 1994. **116**(26): pp. 11723–11736.

42. Anderson, P.E., et al., Electrochemical synthesis and characterization of conducting polymers in supercritical carbon dioxide. *Journal of the American Chemical Society*, 2002. **124**(35): pp. 10284–10285.

43. Cooper, A.I., Polymer synthesis and processing using supercritical carbon dioxide. *Journal of Materials Chemistry*, 2000. **10**(2): pp. 207–234.

44. Li, W. and L. Calle, Micro-encapsulation for corrosion detection and control. 2019.

45. Ravichandran, R., et al., Application of conducting polymers and their issues in biomedical engineering. *Journal of the Royal Society Interface*, 2010. **7**: pp. S559–S579.

46. Hackett, A.J., J. Malmström, and J. Travas-Sejdic, Functionalization of conducting polymers for biointerface applications. *Progress in Polymer Science*, 2017. **70**: pp. 18–33.

47. Cui, X. and D.C. Martin, Electrochemical deposition and characterization of poly(3,4-ethylene dioxythiophene) on neural microelectrode arrays. *Sensors and Actuators B: Chemical*, 2003. **89**(1): pp. 92–102.

48. Zhong, Y., et al., Stabilizing electrode-host interfaces: a tissue engineering approach. *Journal of Rehabilitation Research & Development*, 2001. **38**(6): pp. 627–632.

49. Yang, J. and D.C. Martin, Microporous conducting polymers on neural microelectrode arrays: I Electrochemical deposition. *Sensors and Actuators B: Chemical*, 2004. **101**(1): pp. 133–142.

50. Peramo, A., et al., In situ polymerization of a conductive polymer in acellular muscle tissue constructs. *Tissue Engineering Part A*, 2008. **14**(3): pp. 423–432.

3 Conducting Polymer-Derived Materials for Batteries

U. Naresh and R. Jeevan Kumar
Department of Physics
Sri Krishnadevaraya University
Anantapuramu, India

S. Ramesh and K. Chandra Babu Naidu
Department of Physics
GITAM Deemed to be University
Bangalore, India

D. Baba Basha
Department of Physics
College of Computer and Information Sciences, Majmaah University
Al'Majmaah, KSA

Prasun Banerjee and K. Srinivas
Department of Physics
GITAM Deemed to be University
Bangalore, India

CONTENTS

3.1 INTRODUCTION

Nowadays, the modern life due to urgency needs the usage of electronic gadgets and devices. This approach is increased enormously in daily life style. The electronic world depends on the energy-storage devices commonly named as batteries. Hence, the general requirements of batteries are charge-storage capacity, energy density, electrochemical properties, thermal stability (safety), and so on needed to improve requirements of electronic world. The statistics and literature suggest that most of the electronic devices are being operated with help of the Li-ion batteries. However, the development in modern technology needs the high energy storage, mechanical strength, as well as renewable, recycled, and cost-effective energy-storage devices. The researchers have focused on the new inventions and research on battery devices by overcoming the disadvantages in case of batteries such as Li-ion, lithium–sulfur, and sodium-ion batteries, as well as some conventional batteries [1].

The scientist John Goodenough fabricated Li-ion battery (LIB) first in 1991 and it was marketed by Sony Company. Eventually, it had great importance due to its energy-storage capacity improvement and compatibility in availability of their resources [2]. In the later days, the usage of LIBs is increased in every field which was connected to electronic gadgets such as smart phones, iPods, computers, UPS systems, and satellites. The fundamental principle behind the LIB is conversion of chemical energy into electrical energy. The basic parts of the LIB are positive electrode and negative electrode that are generally called as cathode and anode, respectively. The electrolyte acts as the medium in between anode and cathode to transfer charges through connected circuits. However, the working of battery depends primarily on the electrolyte and secondarily on anode and cathode. The capacity of the LIB battery depends on the rate of lithium-ion transformation via electrolyte, so that an electrolyte plays a major role for the working of battery effectively with better capacity and storage of energy.

An electrolyte is generally divided into two types: liquid-based electrolyte (organic) and solid polymer electrolyte (SPE). Even the liquid-based electrolyte strikes some disadvantages such as leakage of charge, flammable, and stability and has less volumetric energy density. These materials need more flexibility because some of the electronic devices need to be carried from one place to the other place. But, it is very uncomfortable with liquid-based electrolyte. On the other hand, solid-based electrolyte overcomes disassociation of liquid electrolytes; therefore, it can be considered as a promising material for various applications and advantageous over the liquid-based electrolyte. However, solid polymer electrolyte has little drawbacks such as low ionic conduction, low mechanical strength, flexible battery size, fire resistance, and chemically inert. But researchers have asserted strongly in this group of materials for electronic devices [7], solar cell [8], fabrication of LEDs [4], sensing devices, etc.

To prepare environment-friendly materials and those having high ionic conduction, good electrochemical properties, and thermal stability (safety) are the main challenges for the researchers to go through the alternate novel materials. Moreover, the application of electrolyte or electrode put together is more significant. For instance, a small operational voltage induces less electrochemical stability. Meanwhile, the more oxygen content increases chemical tolerance. Nevertheless, the requirement and applications reinforced the development of solid polymer electrolyte (SPES) with the new class of materials like gel-type polymer composites, conjugated or conducting polymer derivative composites as electrolyte. In this chapter, the author's attention is to focus on the conducting polymer electrolyte and electrode materials for the battery devices.

Since last few years, the researchers have considerable interest in order to develop new modernizations with conducting polymers as well as their composites in bulk and nano was started toward the battery applications. Conducting polymers are the class of the polymers that have conducting property and it was first discovered by Shirakawa in 1997 [9]. The researchers found and induced a list of conducting polymers based on factors such as structure, chemical formula, and strength of conductivity as well as their physical properties and stabilities. Apart from this list, we picked the major conducting polymers such as polyaniline (PANI), polypyrrole (PPy), and poly(3,4-ethylenedioxythiophene) (PEDOT). Due to the contribution of these polymers in the field of solid polymer electrolyte and supercapacitors, the innovation of new electronic devices was grown rapidly [10].

The synthesis of conducting polymers (CPs) is a crucial step because of the synthesis conditions such as doping method; temperature range may strongly influence the morphological, physicochemical properties, conductivity of the ions, mechanical properties, and thermal stability of the conducting polymers. Generally, this type of polymer can be prepared by common polymerization technique [1]. Table 1 presented in reference [1] showed a few important common synthesis methods. Each synthesis technique interpreted different morphological behavior such as rods, fibers, rings, wires, and particles in nano-state and bulk state. [1, 11] The polymer morphology can be changed by dissolution of actual pattern [12]. For example, Chen et al. prepared the CP composite electrode for the purpose of electro-catalysis by rearranging pattern of the PANI conducting polymer [3]. Likewise, several researchers study various polymer-derived materials using many synthesis techniques. The very common polymerization techniques for altering and doping are listed below. Generally, the conducting polymer synthesis techniques are [10]: chemical polymerization, electrochemical polymerization, photochemical polymerization, concentrated polymerization, emulsion polymerization, inclusion polymerization, solid-state polymerization, plasma polymerization, pyrolysis, soluble precursor, etc.

Changes in morphology of the electrode can also be responsible to enhance charge transfer efficiency of the battery. Zedda et al. [13] used various techniques to improve charge efficiency of electrode by altering the electrode conducting part with conducting polymers [13]. Hence, a conducting polymer has the capability to improve the electrode compatibility by alteration of surface morphology by incorporating respective dopants. In a recent study, PPy, PANI-based multifunctional-derived materials such as capacitors and electrodes came into the existence by alternation of polymers using several transition metals, ceramic polymer composites, and magneto-polymer composites [14, 15].

Various cooperative special effects were noted by the inclusion of electrodes or electrolyte cells with the coating or incorporation of conducting polymer systems, and their derived composites [16] toward the electronic storage or device applications. Moreover, conducting polymers also exhibit different types of outstanding properties in addition to the electron conduction such as optical, mechanical, magnetic properties, and shielding of electromagnetic wave. Martin et al. [17] studied that the conducting polymer nanofibers of first and second order exhibited the high electrical conductivity rather than the metallic nanostructures. Naturally, the addition of polymers can decrease the electrical conductivity due to the fact that the polymer has its own insulating nature. Nevertheless, the researchers strongly suggested that the addition of conducting polymers can increase the transformation electrons (conductivity) greater than or equal to the conductivity obtained using the transitions element.

3.2   THEORY

Figure 1 of reference [17] shows the basic structure of LIB, before going to brief discussion about the conducting polymer mechanism of energy-storage device. It is worthwhile that one has to define the terms that are the parts of the battery. Battery is a device that works for the energy transformation device from chemical energy to electrical energy using two electrodes, that is, anode and cathode, respectively, with a barrier located at the middle of the electrodes.

Cathode is a positive electrode that accepts electrons through the circuit and can control the chemical reaction (reduction), whereas anode is the negative electrode that gives the electrons through the circuit by chemical reaction (oxidation). Meanwhile, the important part of the battery is a barrier (separator) to avoid the short-circuit problem in battery. This is generally treated as electrolyte and made up of liquid or solid and polymer gel type of matter. The function of the separator is migration of ions between electrodes. In the external circuit the transport of the electrons depends upon the potential difference between anode and cathode. More generally, the potential difference is divided as open circuit voltage (apart from external current in circuit) and closed circuit voltage (with external current in circuit). While the increase of charge is referred to as charging of battery, discharge is defined as release of charges. The basic properties of battery are energy density (maximum energy per unit volume in cell), coulombic efficiency (discharge time over the charging time), transition temperature (the temperature at which phase transition changes), and charging rate (quantity of charging rate and discharge rate of current in cell).

The utility of polymer electrolyte determined by various properties may be likely based on safety (unflammable), plasticity, environment-friendly nature, disintegration reaction, etc. These types of flexible properties are noticed using solid polymer electrolyte (SPE) but not the liquid electrolyte or gel type electrolyte [17]. So far, researchers found liquid ionic conduction greater than the SPE [18].

Feasibly, the SPEs have got huge challenges over liquid electrolyte as it allows diffusion of ions into the highly porous composition. The previous reports presented the projection of conducting polymers on the surface of the electrode that is small solution of this type of the problem [19]. But, it is practically very difficult in nature due to exchange of ions as well as synthesis.

The ionic conductivity of battery linearly increases with the increase of free charges and their numbers. In the synthesis of polymer composites, the dielectric constant of medium may depend on the kind of dopant, the organic solvent, and ion pairing between each other [20, 21]. In case of polymer materials, the dielectric constant is quite low that leads to the disadvantage for the SPE application [22]. The polymer dissolution in solvent is also related to ions' incorporation of polymer templates in the structure and its molecular system. Moreover, an ideal electrolyte nature is observed by the substitution of required element into the conducting polymer for the application of SPEs [23]. The experimental ionic conductivity of a polymer electrolyte is most probably measured by AC-impedance analyzer of thin electrolyte located between electrodes [24].

Bruce and Gray [26] found that the ion exchange between electrodes can be divided into two types: transport number and transference number. The LIB transference number is a measure of the mobility of the cation relative to the anion in a single-salt SPE which is usually denoted as the number of moles of Li transferred by migration per Faraday of charge. This is different from the transport number that is defined as the fraction of the current carried by the specific species. However, the transport number and transference numbers are almost equal in reduction limit due to the absence of ionic association. The transference number is associated to large conductivity of Li$^+$-ion mobility. Ideally, transference number is equal to 1 in case of Li battery [4, 27].

The condition for stability of a polymer electrolyte or electrodes is explained in terms of mechanical and electrochemical stabilities. However, the stability of battery truly depends upon the thermal stability of the material. Therefore, this can be a key to achieve electrochemical and mechanical stabilities. This is a hard-hitting command for several substances over the liquid electrolyte, because the electrodes are prone to either electrochemical oxidation (at the cathode), reduction (at the anode), or both [28]. Instead of thermal stability, the LIB somewhat constitutes kinetically stable nature by protecting electrolyte through formation of stable conducting polymer layer at the surface of the electrolyte and electrodes. The layer controls the additives of the electrode reaction [29]. The polymer electrolytes normally form the polymer layers, although their chemical and morphological properties are being characterized to very limited resources [30, 31]. And also, some difficulties associated with the X-ray photoelectron spectroscopy (XPS) or Fourier transform infrared spectroscopy (FTIR) studies naturally used for cycling investigations. Once the mechanical stability of the electrolyte's shear modulus exceeds 6 GPa, the Li transformation can be concealed and Li metal can be used for a possible anode material [32].

The morphology of polymers and their composites determines the accurate electrical conductivity due to the fact that the carrier density is varied based on the size of the grains and particles. If it is microscopic, it can possess less charge density when compared to nanosized particles. This fact can also be applicable to conducting polymer properties, such as surface current density and their electrochemical behavior. Previous reports confirmed that the conducting polymers in nanoscale perform better electrochemical activity than those in bulk scale. This fact can provide interest to researchers to synthesize conducting polymer composites as nanowires, nanorods, polymer-decorated nanoparticles, nanobowls, and nanosized thin films for the battery electrolyte, catalytic, and electrode application [33]. For example, Winther-Jensen et al. used vapor polymerization technique and prepared a 350–399-nm PEDOT solid film on the gore-tex membrane surface as shown in figure 2 of their publication [34]. Further, it is observed that the prepared system exhibited similar catalytic nature of Pt/C catalyst [34]. In addition, the system reinforced to change the electrical conductivity of film-coated membrane. Interestingly, the synthesized PEDOT electrode showed an easiness greater than 1500 h by acquiring this life-time stability. This nature indicates that the morphology of PEDOT conducting polymer showed considerable impact on the stability of the electrode.

The dimension of the conducting polymer system can also influence the physical and chemical properties. Indeed, it is proven by many researchers. For instance, Zhu et al. [35] formed 3D-PEDOT as double-layered hollow microspheres via identity assemblage polymerization technique as shown in figure 2a-c of reference. [35]. They found the enhancement of inflated surface area in the case of redox reaction. It summarizes the enlarged catalytic nature (figure 2d & 2e) [35]. In addition, another important conducting polymer candidate is PANI that occupied a position to amplify the conduction and electrochemical properties. The Li and their group of materials are used to fabricate PANI-coated Pt/c shell-type shape in the range of 5 nm (see figure 2f) [35]. The new structure allowed electron momentum on the surface of shell that works as a separator among carbon description to acidic surroundings as shown in figure 2f [35].

3.3 DISCUSSION ON CONDUCTING POLYMER-DERIVED MATERIALS

3.3.1 PEDOT DERIVATIVES

3.3.1.1 Structural Properties

Poly(3, 4-ethylenedioxythiophene) (see Figure 3.1) [32] or polystyrene sulfonate (PSS) is the full form of PEDOT. The poly(4-styrene sulfonate) (PSS) is the conducting polymer and it is the combination of monomers. They are: (i) odium polystyrene sulfonate (made up of sodium polystyrene solfonate) and (ii) poly(3, 4-ethylenedioxythiophene) generally shortened as PEDOT. The PEDOT carries positive charge associated with polythiophene conducting polymer. This mixer forms macromolecular salt and its chemical structure as shown in figure 1 of reference [32].

Naturally, it appears as transparent and conductive with its chemical structure. Hence, it may be used for several applications. The important application with this virtual and visible PEDOT:PSS structure leads to the preparation of thin films for optical application, whereas it is used as electrostatic electrode and electrolyte application due to the electron conducting nature [11, 36–38].

Zheng et al. [39] fabricated an electrode using carbon cloth associated with Bi_2O_3 coated by PEDOT electrode using solvothermal technique. Previous scientists first prepared carbon cloths and then the Bi_2O_3 particles are injected via the solvothermal method using electro-

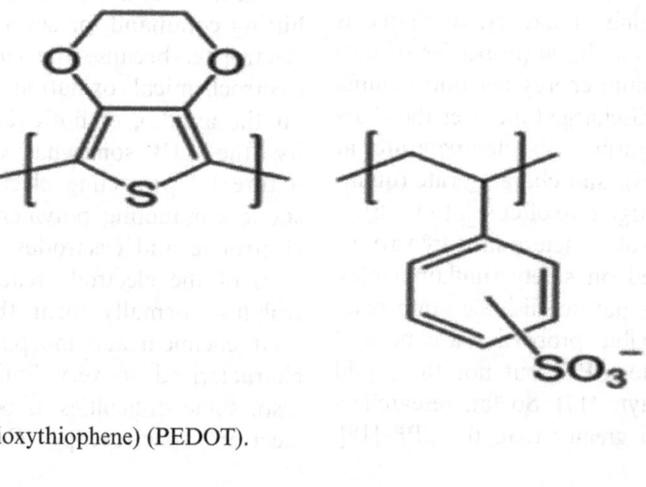

FIGURE 3.1 Poly (3,4-ethylenedioxythiophene) (PEDOT).

deposition technique. The layer of PEDOT deposition is clearly shown in figure 2 of the published work [39]. The shape of the system is converted to CC/Bi_2O_3 decorated by PEDOT nanoarrays (NAs). Interestingly, it is found that the convenient flexibility of PEDOT and the size of the formed nanoarrays in the range of 4 μm (diameter). This illustrates the fact that the solvothermal treatment shows the effect of the carbon fibers on color which are changed at end. It is clear from scanning electron microscopy (SEM) pictures of figure 2 in reference [39]. Furthermore, the X-ray powder diffraction (XRD) information of this material expressed the distance between two nanofibers of 0.30 and 0.25 μm with the Bi_2O_3 phases [39–41]. Finally, they concluded that this nanofiber sheet may be used as electrode and electrolyte [41–43].

Kirahira et al. [44] fabricated PEDOT:PSS conducting polymer film and found specific resistance with an electrode by transmission line technique on adding the different solvents to PEDOT:PSS film. It led to a unique reduction that increases the electrical conductivity. This polymer binder can eliminate the need of an interactive material in between the electrode and additional additive conductor [44]. Qingshuo Wei et al. [45] reported that the PEDOT combinations are considered organic thermoelectric modules. These materials are suitable for battery devices by overcoming the flame oxidation in the devices. Additionally, the thermal energy raised in this material can be converted to electric charge [45–48]. Yamashita et al. [49] studied that the solvent of ethylene glycol increased carrier density in PEDOT. They also used the combination of terahertz (THz) time-domain transmission spectroscopy (0.1–3 THz) and broadband reflectance spectroscopy (4–800 THz) for obtaining conduction spectra. Further, the spectrum is fitted and concluded that the increase in EG is responsible for the increase in carrier density [49, 50]. Lin et al. [51] analyzed the carrier transport process of ZnO particle-doped poly(3,4-ethylenedioxythiophene) with poly(4-styrene sulfonate) (PEDOT:PSS). It is noted that the dopant enhances the high electron carrier density and it was attributed due to incorporation of ZnO particles into the PEDOT:PSS system. It also revealed that the increase of conduction in the conducting polymer composite provides the potential application in field of battery electrode fabrication [51].

Scott B et al. [52] reported a strong agreement in their work, that is, 7 types of gold electrodes with 13 cells are taken for checking the charge density and surface area current transition. Meanwhile, they synthesized PEDOT polymer nanowires [52]. In their comparison, the gold electrode does not comprise the efficiency in modulation and cellular voltage. Then the PEDOT nanowires showed the experimental evidence for the charge-storage density, current transformation with respect to the surface area as depicted in figure 2 of reference [52]. In their experiment, the surface area modification is controlled by using phosphate buffered saline (PBS). Nanowires of PEDOT:PSS do not encourage the acted potentials [52, 53].

3.3.1.2 Electrochemical Studies of PEDOT and Its Derivatives

Zhang et al. [54] reported the influence of PEDOT on Li_2NiF_4–PEDOT electrode composite for LIB application. Their electrochemical study as shown in figure 6a [54] revealed that the discharge capacity of Li_2NiF_4–PEDOT is higher than the earliest one. This fact is attributed to the Li_2NiF_4-storage capacity that is greater than the Li_2NiF_4–PEDOT composite. Hence, the discharge capacity becomes very low; this may have happened by the high storage capacity of Li_2NiF_4.

Meanwhile, the second discharge capacity for Li_2NiF_4 obtained a very less value around 350 μA g^{-1}. In fact, this type of observation is caused by interfacial interaction between Li and NI/LiF [55]. Moreover, the charge and discharge capacities of Li_2NiF_4–PEDOT cathode material are observed to be 400 mAh g^{-1} and 7%, respectively. These are very near to standard value; therefore, the determined columbic efficiency is 72%, and this value is greater than the translation electrodes [56].

The greater specific capacity of Li_2NiF_4–PEDOT is responsible for the reputable energy density; even it exhibits the usual power of ~1.6 V. On the other hand, the discharge curve is quite different throughout the second cycle rather than the first one. But the capacity variation appears to be very small. Herein, one more important aspect is that throughout the charge process the voltage variation might have occurred due to the nickel oxide (NiO) and the change of Ni^{2+} ions.

3.3.1.3 Magnetic Properties

In this chapter, we discussed about the magnetic properties of the PEDOT-based composite electrodes. Basically, the polymer is amorphous in nature; however, the production of conducting polymers leads to the existence of polymer composites as electrodes, electrolytes, and various derived materials for the battery application [58–61]. Recently, some of the researchers studied magnetic properties for the electromagnetic shielding applications, electrodes, and electrolyte properties in order to deduce the relations between electrical properties. The authors focused on energy storage, coulomb efficiency, and magnetic properties, such as superpara magnetism, and types of magnetic and general magnetic properties of coercivity, saturation magnetization, and M–H loop studies.

In order to study the magnetic behavior of PEDOT and its composites for electrolyte application, Amithabha et al. [62] prepared the PEDOT–DBSA–Fe_3O_4 (DBSA, dodecylbenzene sulfonicacid) conducting polymer by using the oxidative polymerization method. Further, Fe_3O_4 nanoparticles were injected via colloidal dispersion technique [62]. They recorded the hysteresis curve (M–H loop) at room temperature by the variation of concentration in composites as shown in figure 7 of published work. [62]. With the help of M–H loop and the properties such as saturation magnetization (M_s), coercivity (H_c), and remanence ratio (M_r/M_s), the existence of superparamagnetism is studied. In these studies, it is found that the magnetization of pure ferrite nanoparticles is acquired at an elevated magnetic field of 90–95 emu g^{-1} for bulk samples [63–66]. However, ferromagnetic nanoparticles attribute low saturation magnetization in the case of particles which contains core-shell-like morphology [67]. The samples P50 and P500 (different concentration) showed smaller values than the pure iron oxide. It was due to increases of spin disorder at the surface of the film. The magnetic moment of film is deviated due to high anisotropy and broken exchange interaction. Moreover, the susceptibility of sample at different frequencies is carried out as shown in figure 6 of reference [67].

3.3.2 PPY FOR THE ENERGY-STORAGE DEVICES

3.3.2.1 Structural Property of PPy

The PPy (Figure 3.2) [68] has important applications in optoelectronics due to its light responsive nature. The PPy-based systems such as films, coated nanoparticles, electrode materials, membranes, nanofibers, NAs, and any flexible electronic devices are prepared [68–71]. Hence, these types of materials acquired a wide range of applications in the field of optoelectronics [37, 68–77]. However, the downside of PPy electrodes is poor stability in the process of ion doping and diffusing. the PPy-based materials may not be able to control mechanical force attributed to the enlargement and attenuation of PPy chain-like structure. This is the base for searching the structurally stable PPy-based materials. The researchers took these challenges and made efforts to synthesize novel PPy-based systems such as PPy/PSS, PPy/Fe$_3$O$_4$, polymer graphene combination, and many derivatives that came into the existence [78–82].

PPy-based (reduced graphene oxide) RGO ferrite platinum (Fe$_3$O$_4$@pt@RGO@PPY) material was prepared [82] for the photocurrent electrode via hydrothermal-assisted ultra-sonic oxidative polymerization technique. PPy is one of the most studied conducting polymers due to its interesting electrical and mechanical stabilities. The PPy-based composite film [82] for electrolyte application is obtained by the electrochemical addition of carbozole. It was found that PPy-based composite seemed to be stable and electro-active for the battery electrolyte application. The higher concentration of dopant creates bi-polarons; hence, they provided the possibility of existence for strength of ions [82].

The surface morphology of organic polymers related to electrochemical devices such as sensors and battery devices is studied [83]. It is noticed that the polymers exhibited higher current density with enhanced specific capacitance. In addition, the PPy nanosheets are synthesized that can reflect considerable performance of electrochemical components (see figure 1a) [83]. Hence, the surface morphology of the conjugated polymers seemed to be like a constant conductive path without any boundaries. These will help in increasing the conductivity due to the ions. Practically, the kind of application for electrode contains a specified morphology. In view of this, the conductive polymer-based nanoparticles such as PPy derivatives are synthesized for the electrode of LIB applications [83]. The obtained size of the nanoparticles is in the range of 50–100 nm and conduction mechanism is found to be increased.

On the other hand, thin PPy sheet can be possible to act as the separator in case of LIB. This type of PPy paper separator is not only used to enhance the thermal and chemical stabilities but it also acts as flexible electrolyte. Moreover, the ionic loss and conduction process may have happened due to excess oxidation [84]. This indicates that the electronically conducting polymer (i.e., PPy) is converted into an electronically insulating mesoporous material. This is suitable for the usage in systematic study of the influence of the separator structure on the performance of lithium-based batteries. It also showed that the usage of over-oxidized PPy separator can yield an improved electrochemical performance of lithium metal-based cells in terms of enhanced cycling stability and rate performance [85–87]. The enriched surface of these materials offered the functionalization possibilities and the straightforward variation of the pore structure of the over-oxidized material. These results

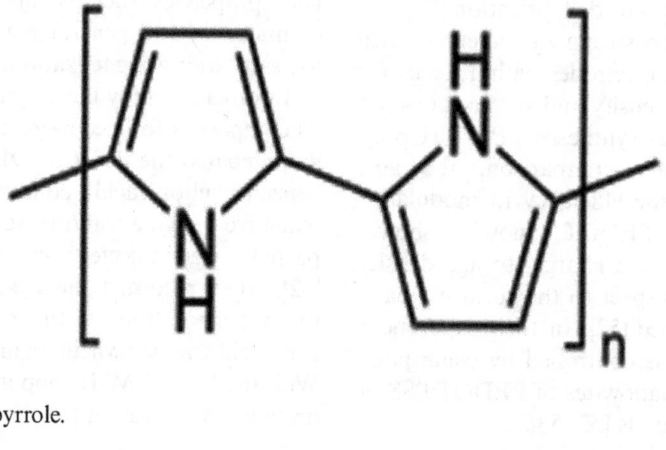

FIGURE 3.2 Structure of polypyrrole.

showed the use of new generation of separators that can be rationally tailored to increase electrochemical performance of lithium metal batteries [47].

3.3.2.2 Electrochemical Properties of Polypyrrol

Han et al. [88] prepared a PSS/PPy conductive polymer for the application of Li sulfur cathode material. The material is found to be well suitable for cathode material. In this view, PPy makes easy conduction, whereas the PSS is neutralized. In the whole synthesis, the PSS worked to control the particle size by blocking polarons. Hence, the decrease of particle size is responsible for the increase of electron conductivity. Moreover, synthesized polymer nanoparticles create better chances to the sulfur cathode as an active material. The specific capacity, current density, and discharge capacity are studied, and the obtained results are shown in figure 3 of reference [88]. The results revealed the combination of PSS/PPy with high discharge capacity which can be possible to recycle. As shown in figure 3 of reference [88], the relative performance is made among the PSS, PPy, and PPy/PSS cells. It is analyzed that the PSS/PPy cell is extremely reduced and reutilized. Usually, the PPy shows good capacity due to its intrinsic electric conductivity.

As shown in figure 3 [88], the relative performance is made between the PSS, PPy, and PPy/PSS cells. Furthermore, it is noted that PSS/PPy cell showed extremely reduced reutilization. On the other hand, PSS cell with better ionic conductivity offers good discharge capacity. However, the sulfonates (negatively charged entities) create the repulsion between Li and PS systems. This induces the loss of electrochemical contact. In addition, the obtained high reversible capacity leads to an active material protection in batteries. The copolymer neutralizes the sulfur cathode mechanism chemically and physically [88–91].

Manthiram et al. [92] prepared PPy/S material aided by poly(2-acrylamido-2-methyl-1-propanesulfonic acid) (PAAM/PSA). The formation of the prepared samples and electrochemical activity is shown in figure 3 (cycling performance vs. C rate) [93]. This composite delivered a capacity of 390 Ah g^{-1} by offering 50 cycles with a unit current density. This type of nature is obtained in the previous reports [48, 93–102]. Me et al. [11] synthesized sulfur cathode with carbon as the layer of array of PPy and Li-ion sulfur battery. This PPy/CMK-8/S showed good electrochemical activity. The warping structure is stable even after 40 charge/discharge cycles which further demonstrated that the volume expansion is well buffered by this three-dimensional (3D) structural design [103]. The structure of this novel material made us first to eliminate oxygen groups from graphene oxide. This is termed RGO (reduced graphene oxide). Later, the ferrite and Pt ferrite nanoparticle are prepared through the hydrothermal method. Then, the induced oxidative polymerization technique is used to prepare the PPy-coated Pt ferrite nanoparticles. Afterwards, this composite dispersed in ultra-sonicator and graphene

oxide solution into the ultra-sonicator. Then, the whole system is irradiated under the ultraviolet (UV) radiation after 8 hours. Finally, Fe_3O_4@pt@RGO@PPY is formed as microspheres. This virtual appearance is clearly shown in schema 2 of reference [104]. RGO is an attractive material to improve the electronic conductivity, response movement, and stability of structure for the PPy-based materials [105–109]. On the other hand, strong π–π stacking interaction between RGO and PPy can yield more delocalized electrons in them [110–113]. In addition, it can make the chain structure of PPy more ordered [114] and lower down hopping barriers of electrons [114–117]. Thus, it results in easier way for electron movement in PPy and RGO; consequently, it increases the electrical conductivity. Furthermore, strong π–π stacking can enhance structural stability of PPy-based materials by creating distortion stress of RGO induced by sp^3 conjugation. This makes the PPy-based materials with higher tolerance to ensure mechanical stress caused by volume change that is accompanied with charging and discharging cycling during electrochemical tests [118].

3.3.2.3 Magnetic Properties

Magnetic conducting polymer (PPy) composites with morphology of nano- and microspheres are considered multifunctional materials. These materials attained wide applications in the field of biomedicine and in the field of bioengineering (cell analysis, as drug carriers, and irradiation therapy). Researchers found a relation between polymer electrode and magnetic material due to its peculiar properties [119, 120]. Luo et al. [121] predicted the PPy/Fe_3O_4 spheres synthesized via common ion effect (ultrasonication-assisted polymerization and doping treatment) and co-precipitation technique. They noticed the morphology as spheres with the size equal to 30–40 nm. The architecture is formed as the ferrite nanoparticles covered by the PPy polymers. This whole structure seemed to be core shell as shown in figure 5 of reference [121]. The hysteresis loop is carried out at room temperature, and it revealed the information about the magnetization of the pure Fe_3O_4 nanoparticles greater than the PPy-doped Fe_3O_4 composites. In fact, this happens due to the non-magnetic polymer that can be responsible for the decreased magnetization of the PPy with ferrite nanoparticles. Significantly, PPy/Fe_3O_4 system oriented as superparamagnetic nature since it has smaller value than the hypothetical value of superparamagnetic particle [121–127].

3.3.3 PANI FOR BATTERY APPLICATION

3.3.3.1 Structural Properties

Although PANI (Figure 3.3) [128] is treated as homopolymer, the molecular structure has bensenoid state or quinonoid state or may exist in both states. The PANI oxidization states are: (i) complete lecoemeraldine state (hydrogen atoms are attached with nitrogen) that contains benzenoid

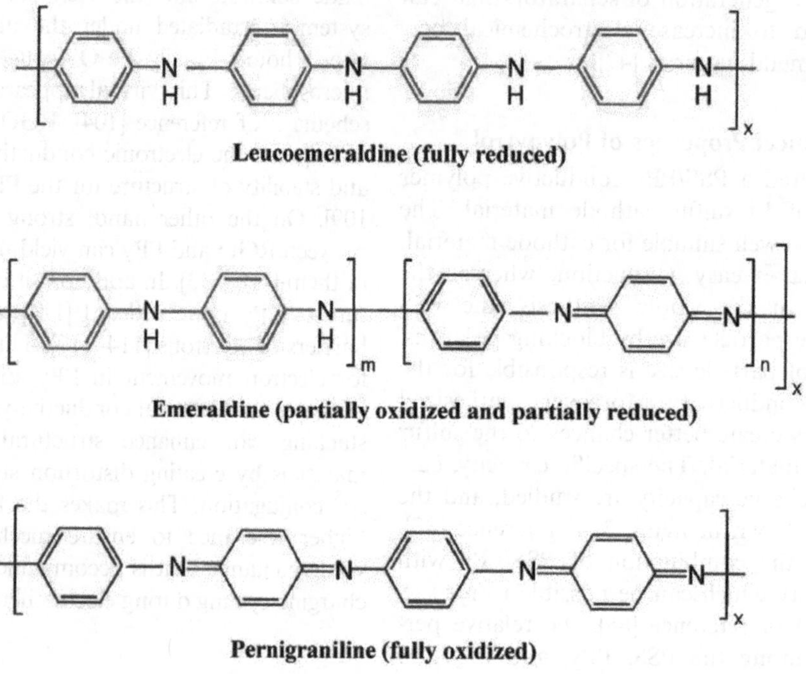

Leucoemeraldine (fully reduced)

Emeraldine (partially oxidized and partially reduced)

Pernigraniline (fully oxidized)

FIGURE 3.3   Structure of polyaniline (PANI).

and (ii) pernigraniline state (no hydrogen atoms are attached with nitrogen) that contains quinonoid. If it becomes emaraldine salt (electrically active), the benzenoid and quinonoid groups are to be balanced. Depending on the type of the dopants and their polarity, the bandgap of PANI is reduced which in turn enhances electrical conductivity.

The recent reports for the manufacturing of PANI and its composites are considered particular aspects for the electronic devices, battery electrolytes, and sensor applications [128–131]. However, the researchers reported PANI synthesis and interpreted the kind of structures in the form of nanoparticles and nanowires, and nanofilms that helped the existence commercial polymers for electrolyte application in batteries [132]. Moreover, PANI-doped derivatives or intrinsic PANI and coated PANI materials induced the new ionic conducting solid polymer electrolyte [133–137]. There were many reports available on PANI and its derivatives. Yeng et al. [139] studied the PANI characteristics toward Li-ion cathode applications. It is prepared via absorption polymerization method and then its film was prepared using the solution-casting method. It is used as modified polyelectrolyte film and achieved good energy density and outstanding recharge [130, 139–143].

3.3.3.2   Electrochemical Properties of PANI for Battery Electrode

PANI conducting polymer can be used not only in Li-ion battery but also in sodium ion battery. Sodium-ion battery (SIB) is replaced by LIB due to the similar properties and low-cost element [145]. As electrodes, most of the organic

systems are taken for SIB electrodes [146–148]. The polymer material is the one of the sources for overcoming some shortages. Xu et al. [149] synthesized HCl-doped PANI (substrate)-sputtered $Ni_{80}Fe_{20}$ film for the formation of $Ni_{80}Fe_{20}$/HCl-PANI film composite. The electrical properties are measured as temperature-dependent resistance curve (R-T) of the front to back surface in proper order. From the measurements, it is evident that the synthesized film exhibits semiconducting nature [140, 149].

Han et al. [150] prepared PANI nanofibers through the in-situ polymerization method. This nanofiber shows an excellent utility as cathode material for the SIB due to reflection of high reversible capacity and rate of capability. figure 5 of reference [150] shows excellent evidence for PANI hollow nanofibers with stable columbic efficiency [150]. Chen et al. [147] made an attempt to prepare polymer cathode materials for solid-state electrochemical batteries. Electrolyte anions are computed using PANI-induced cyno group and obtained voltage variation using density functional tight binding (DFTB). These observations concluded that these types of organic polymers are reasonable for the cathode application [147, 151, 152].

However, except some drawbacks like insolubility, diffusion issue, and stability, PANI has several advantages. All the interactions and solubility may become the key phenomena in charge-storage devices [153, 154]. Common demerits may have overcome through step by step for the polymer instead of complete rejection. Chang et al. [155] prepared energy-storage device of W18O49/PANI-EC using assembled PANI and it is found that PANI has capability to add charges and to conduct electron transport naturally. They measured that W18O49/PANI-EC discharge capacity is higher than the $W_{18}O_{49}$/EC battery. Liu et al. [146] synthesized nitrogen-doped graphene oxide with assembling PANI via general in-situ polymerization technique. The systems exists nanowires that lead to increase the current density and capacity of energy-storage device [146]. From all the investigations of PANI mixtures in the form of films, nanowires, and nanoparticles showed effective uses of PANI in electrochemical devices which increase the columbic efficiency and charge transport.

3.3.3.3   Magnetic Properties of PANI

Bahadur et al. [156] prepared battery anode with a novel materiel that is Gr–Fe₃O₄–PANI (graphite–magnetite–polyaniline) for the promising thermal stability, charge density, energy capacity, better cyclic performance, and reversible capacity. The hydrothermally prepared magnetite hallow rods are polymerized through in-situ polymerization technique. Then, the particles are enveloped with a sheet of graphene by ultrasonication. Finally, the synthesized polymer samples are analyzed with different techniques. In the electrochemical analysis, it is found that the morphology is good and it can be used as anode in the battery. The measurements such as cyclic voltammetry, galvanostatic charge–discharge curves, and life cycle test offered impressive appearance [156]. PANI was successfully decorated on the Fe_3O_4 hallow rods with graphene. The anode has large surface area with high thermal stability (around 620°C). Moreover, the high current density is achieved up to 5 Ag^{-1}, good cyclability, and coulomb efficiency (99.25). Besides, charge and discharge tests are carried out successfully in the range of voltage (up to 3 V) adjacent to Li^+/Li at various current densities. Significantly, the phase transformation is recognized that communicates ferric and ferrous ions into iron. Therefore, it can be concluded that manufacturing the GrFe₃O₄–PANI-based material for LIB electrode is a huge step for developing energy-storage devices [156–164].

3.4   SUMMARY AND CONCLUSIONS

In this chapter, we discussed the recent progress about conducting polymer-based battery devices. The availability of conducting polymers and their multipliers is increased rapidly. However, the present discussion clears that among all conducting polymers, PANI, PPy and PEDOT nanostructures are being given more importance due to their surface morphology that can be varied for different synthesis techniques. Hence, the possibility of increasing electrolyte properties such as surface current density, coulomb charge density, and charge-storage capacity achieved successfully. These conducting polymer composite synthesis techniques with the existence of nanofibers, nanospheres, and nanowires are described. Moreover, it is noticed that the electrochemical properties for electrode and electrolyte depend on the surface morphological properties such as size, shape, and orientation of the conducting polymer composite. We also described the electrolyte and electrode with their requirement of thermal stability, mechanical, and flexibility for battery devices. Thus, we concluded that better battery requirements can be possible with the conducting polymers such as PEDOT, PPy, and PANI.

REFERENCES

[1] Han Y, Dai L. 2019. Conducting polymers for flexible super capacitors. *Macromol Chem Phys* 220: 1800355.

[2] Wang JG, Wei BQ, Kang F. 2014. Bioimaging based on fluorescent carbon dots. *RSC Adv* 4: 199.

[3] Chen W, Rakhi RB, Alshareef HN. 2013. Morphology-dependent enhancement of the pseudo-capacitance of template-guided tunable polyaniline nanostructures. *J Phys Chem C* 117: 15009.

[4] Mindemark J, Lacey MJ, Bowden T, Brandell D. 2018. Beyond PEO alternative host materials for Li+-conducting solid polymer electrolytes. *Prog Polym Sci* 81: 114–43.

[5] Roy D, Cambre JN, Sumerlin BS. 2010. Future perspectives and recent advances in stimuli-responsive materials. *Prog Polym Sci* 35: 278–301.

[6] Mindemark J, Edman L. 2016. Illuminating the electrolyte in light-emitting electrochemical cells. *J Mater Chem* 4: 420–32.

[7] Barbosa PC, Silva MM, Smith MJ, Gonc,alves A, Fortunato E. 2008. Solid-state electro-chromic devices based on poly (trimethylene carbonate) and lithium salts. *Thin Solid Films* 516: 1480–83.

[8] Wang C, Gu P, Hu B, Zhang Q. 2015. Recent progress in organic resistance memory with small molecules and inorganic–organic hybrid polymers as active elements. *J Mater Chem C* 3: 10055–65.

[9] Song Z, Zhou H. 2013. Towards sustainable and versatile energy storage devices: an overview of organic electrode materials. *Energy Environ Sci* 6: 2280–2301.

[10] Wojnarowska Z, Paluch KJ, Shoifet E, Schick C, Tajber L, Knapik J, Wlodarczyk P, Grzybowska K, Hensel-Bielowka S, Verevkin SP, Paluch M. 2015. Molecular origin of enhanced proton conductivity in anhydrous ionic systems. *J Am Chem Soc* 13: 1157–64.

[11] Peled E, Golodnitsky D, Ardel G. 1997. Advanced model for solid electrolyte interphase electrodes in liquid and polymer electrolytes. *J Electrochem Soc* 144: L208–10.

[12] Wang Y, Agapov AL, Fan F, Hong K, Yu X, Mays J, Sokolov AP. 2012. Decoupling of ionic transport from segmental relaxation in polymer electrolytes. *Phys Rev Lett* 108 (8): 088303.

[13] Wojnarowska Z, Feng H, Fu Y, Cheng S, Carroll B, Kumar R, Novikov VN, Kisliuk AM, Saito T, Kang N-G, Mays JW, Sokolov AP, Bocharova V. 2017. Effect of chain rigidity on the decoupling of ion motion from segmental relaxation in polymerized ionic liquids: ambient and elevated pressure studies. *Macromolecules* 50 (17): 6710e6721.

[14] Kisieliute A, Popov A, Apetrei R-M, Carac G, Morkvenaite-Vilkonciene I, Ramanaviciene A, Ramanavicius A. 2019. Towards microbial biofuel cells: improvement of charge transfer by self-modification of micro organisms with conducting polymer – polypyrrole. *Chem Eng J* 356: 1014–21.

[15] Zebda A, Tingry S, Innocent C, Cosnier S, Forano C, Mousty C. 2011. Hybrid layered double hydroxides-polypyrrole composites for construction of glucose/O_2 biofuel cell. *Electrochim Acta* 56: 10378–84.

[16] Kashyap HK, Annapareddy HVR, Raineri FO, Margulis CJ. 2011. How is charge transport different in ionic liquids and electrolyte solutions? *J Phys Chem B* 115 (45): 13212–21.

[17] Richardson-Burns SM, Hendricks JL, Martin DC. 2007. Electrochemical polymerization of conducting polymers in living neural tissue. *J Neural Eng* 4: L6–L13.

[18] Lee H, Yanilmaz M, Toprakci O, Fu K, Zhang X. 2014. A review of recent developments in membrane separators for rechargeable lithium-ion batteries. *Energy Environ Sci* 7: 3857–86.

[19] Song JY, Wang YY, Wan CC. 1999. Review of gel-type polymer electrolytes for lithium-ion batteries. *J Power Sources* 77: 183–197.

[20] Cowie JMG, Cree SH. 1989. Electrolytes dissolved in polymers. *Annu Rev Phys Chem* 40: 85–113.

[21] Mindemark J, Sun B, Brandell D. 2015. Hydroxyl-functionalized poly (trimethy-lenecarbonate) electrolytes for 3D-electrode configurations. *Polym Chem* 6: 4766–74.

[22] Mindemark J, Törmä E, Sun B, Brandell D. 2015. Copolymers of trimethylene carbonate and epsilon-caprolactone as electrolytes for lithium-ion batteries. *Polymer* 63: 91–98.

[23] Pesko DM, Jung Y, Hasan AL, Webb MA, Coates GW, Miller T, III. 2016. Effect of monomer structure on ionic conductivity in a systematic set of polyester electrolytes. *Solid State Ionics* 289: 118–24.

[24] Angell CA. 2017. Polymer electrolytes–some principles, cautions, and new practices. *Electrochim Acta* 250: 368–75.

[25] Sun B, Tehrani P, Robinson ND, Brandell D. 2013. Tailoring the conductivity of PEO-based electrolytes for temperature-sensitive printed electronics. *J Mater Sci* 48: 5756–67.

[26] Bruce PG, Gray FM. 1995. Polymer electrolytes II: physical principles. In: PG Bruce, editor. *Solid State Electrochemistry*. Cambridge: Cambridge University Press, pp. 119–62.

[27] Armand M. 1987. Current state of PEO-based electrolyte. In: JR MacCallum, CA Vincent, editors. *Polymer Electrolyte Reviews*, vol. 1. London: Elsevier, pp. 1–22.

[28] Doyle M, Fuller TF, Newman J. 1994. The importance of the lithium ion transference number in lithium/polymer cells. *Electro Chim Acta* 39: 2073–81.

[29] Klett M, Giesecke M, Nyman A, Hallberg F, Lindström RW, Lindbergh G, Furó I. 2012. Quantifying mass transport during polarization in a Li ion battery electrolyte by in situ 7Li NMR imaging. *J Am Chem Soc* 134: 14654–57.

[30] Edström K, Herstedt M, Abraham DP. 2006. A new look at the solid electrolyte interphase on graphite anodes in Li-ion batteries. *J Power Sources* 153: 380–84.

[31] Tehrani Z, Korochkina T, Govindarajan S, Thomas DJ, O'Mahony J, Kettle J, Claypole TC, Gethin DT. 2015. Ultra-thin flexible screen printed rechargeable polymer battery for wearable electronic applications. *Org Electron* 26: 386–494.

[32] Guo Z, Qiao Y, Liu H, Ding C, Zhu Y, Wan M, Jiang L. 2012. Self-assembled hierarchical micro/nano-structured PEDOT as an efficient oxygen reduction catalyst over a wide pH range. *J Mater Chem* 22: 17153.

[33] Zhang SS. 2006. A review on electrolyte additives for lithium-ion batteries. *J Power Sources* 162: 1379–94.

[34] Winther-Jensen B, Winther-Jensen O, Forsyth M, MacFarlane DR. 2008. High rates of oxygen reduction over a vapor phase-polymerized PEDOT electrode. *Science* 321: 671–74.

[35] Wang J, Wang J, Kong Z, Lv K, Teng C, Zhu Y. 2017. Conducting-polymer-based materials for electrochemical energy conversion and storage. *Adv Mater* 29: 1703044.

[36] Xu C, Sun B, Gustafsson T, Edström K, Brandell D, Hahlin M. 2014. Interface layer formation in solid polymer electrolyte lithium batteries: an XPS study. *J Mater Chem A* 2: 7256–64.

[37] Monroe C, Newman J. 2005. The impact of elastic deformation on deposition kinetics at lithium/polymer interfaces. *J Electrochem Soc* 152: A396–A404.

[38] Cheng T, Zhang YZ, Zhang JD, Lai WY, Huang W. 2016. High-performance freestanding PEDOT: PSS electrodes for flexible and transparent all-solid state super capacitors. *J Mater Chem A* 4: 10493–99.

[39] Zheng S, Fu Y, Zheng L, Zhu Z, Chen J, Niu Z, Yang D. 2019. PEDOT engineered Bi_2O_3 nano sheet arrays for flexible asymmetric supercapacitors with boosted energy density. *J Mater Chem A* 7: 5530–38.

[40] Park H, Lee SH, Kim FS, Choi HH, Cheong IW, Kim JH. 2014. Enhanced thermoelectric properties of PEDOT:PSS nanofilms by a chemical dedoping process. *J Mater Chem A* 2: 6532–39.

[41] Pan J, Xu G, Ding B, Chang Z, Wang A, Dou H, Zhang X. 2016. PAA/PEDOT:PSS as a multifunctional, water-soluble binder to improve the capacity and stability of lithium-sulfur batteries. *RSC Adv* 6: 40650–55.

[42] de Kok MM, Buechel M, Vulto SIE, van de Weijer P, Meulenkamp EA, de Winter SHPM, Mank AJG, Vorstenbosch HJM, Weijtens CHL, van Elsbergen V. 2004. Modification of PEDOT:PSS as hole injection layer in polymer LEDs. *Phys Status Solidi A* 201: 1342–59.

[43] Kim H, Nam S, Lee H, Woo S, Ha CS, Ree M, Kim Y. 2011. Influence of controlled acidity of hole-collecting buffer layers on the performance and lifetime of polymer: fullerene solar cells. *J Phys Chem C* 115: 13502–10.

[44] Kirihara K, Wei Q, Mukaida M, Ishida T. 2018. Reduction of specific contact resistance between the conducting polymer PEDOT: PSS and a metal electrode by addition of a second solvent during film formation and a post-surface treatment. *Synth Met* 246: 289–96.

[45] Wei Q, Mukaida M, Kirihara K, Naitoh Y, Ishida T. 2015. Recent progress on PEDOT based thermoelectric materials. *Materials* 8: 732–50.

[46] Zhao X, Ahn HJ, Kim KW, Cho KK, Ahn JH. 2015. Polyaniline-coated mesoporous carbon/sulfur composites for advanced lithium sulfur batteries. *J Phys Chem C* 119: 7996–8003.

[47] Lindell L, Burquel A, Jakobsson FLE, Lemaur V, Berggren M, Lazzaroni R, Cornil J, Salaneck WR, Crispin X. 2006. Transparent, plastic low-work function poly (3,4- ethylenedioxythiophene) electrodes. *Chem Mater* 18: 4246.

[48] Jakobsson FLE, Crispin X, Lindell L, Kanciurzewska A, Fahlman M, Salaneck WR, Berggren M. 2006. Toward all-plastic flexible light emitting diodes. *Chem Phys Lett* 433: 110–114.

[49] Wusten J, Potje-Kamloth K. 2008. Organic thermogenerators for energy autarkic systems on flexible substrates. *J Phys D Appl Phys* 41: 41135113.

[50] Khodagholy D, Doublet T, Gurfinkel M, Quilichini P, Ismailova E, Leleux P, Herve T, Sanaur S, Bernard C,

Malliaras GG. 2011. Highly conformable conducting polymer electrodes for in vivo recordings. *Adv Mater* 23: H268–H272.

[51] Lin Y-J, Tsai C-L, Su Y-C, Liu D-S. 2012. Carrier transport mechanism of poly(3,4-ethylenedioxythiophene) doped with poly(4-styrenesulfonate) films by incorporating ZnO nanoparticles. *Appl Phys Lett* 100: 253302. doi:10.1063/1.4730391.

[52] Scott B. Thourson1,2 & Christine K. Payne3,4 2017 Modulation of action potentials using PEDOT: PSS conducting polymer microwires. *Sci Rep* 7: 10402. doi:10.1038/s41598-017-11032-3.

[53] Seong B, Lee H, Lee J, Lin L, Jang HS, Byun D. 2018. Biomimetic, flexible, and self- healable printed silver electrode by spontaneous self-layering phenomenon of a gelatin scaffold. *ACS Appl Mater Interfaces* 10: 25666–72.

[54] Zhang M. 2017. Fabrication of Li2NiF4-PEDOT nanocomposites as conversion cathodes for lithium-ion batteries. *J Alloys Comp* 723: 139–45.

[55] Sovizi MR, Fahimi Z. 2018. Honeycomb polyaniline-dodecyl benzene sulfonic acid(hPANI-DBSA)/sulfur as a new cathode for high performance Li-S batteries. *J Taiwan Inst Chem E* 86: 270–80.

[56] Sen P, Xiong Y, Zhang Q, Park S, You W, Ade H, Kudenov MW, O'Connor BT. 2018. Shear-enhanced transfer printing of conducting polymer thin films. *ACS Appl Mater Interfaces* 10: 31560–67.

[57] Bubnova O, Crispin X. 2018. Towards polymer-based organic thermoelectric generators. *Energy Environ Sci* 5: 9345–62.

[58] Burgt Y, Lubberman E, Fuller EJ, Keene ST, Faria GC, Agarwal S, Marinella MJ, Talin AA, Salleo A. 2017. A non-volatile organic electrochemical device as a low-voltage artificial synapse for neuro morphic computing. *Nat Mater* 16: 414–19.

[59] Lieser G, Biasi L, Scheuermann M, Winkler V, Eisenhardt S, Glatthaar S, Indris S, Geßwein H, Hoffmann MJ, Ehrenberg H, Bindera JR. 2015. Sol-gel processing and electrochemical conversion of inverse spinel-type Li2NiF4. *J Electrochem Soc* 162 (4): A679–A686.

[60] Li H, Richter G, Maier J. 2015. Reversible formation and decomposition of LiF clusters using transition metal fluorides as precursors and their application in rechargeable Li batteries. *Adv Mater* 15 (9): 736–39.

[61] Xuan Y, Sandberg M, Berggren M, Crispin X. 2012. An all-polymer-air PEDOT battery. *Org Electron* 13: 632–37.

[62] De A, Sen P, Poddara A, Dasa A. 2009. Synthesis, characterization, electrical transport and magnetic properties of PEDOT–DBSA–Fe_3O_4 conducting nanocomposite. *Synth Met* 159 (2009): 1002–1007.

[63] Wu F, Yushin G, Wu F, Yushin G. 2017. Conversion cathodes for rechargeable lithium and lithium-ion batteries. *Energy Environ Sci* 10 (2): 435–59.

[64] Wang Y, Zhu C, Pfattner R, Yan H, Jin L, Chen S, Molina-Lopez F, Lissel F, Liu J, Rabiah NI, Chen Z, Chung JW, Linder C, Toney MF, Murmann B, Bao Z. 2017. Highly stretchable, transparent, and conductive polymer. *Sci Adv* 3: e1602076.

[65] Xiao YH. 2014. Electrochemical polymerization of poly (hydroxymethylated-3,4-ethylenedioxythiophene) (PEDOT-MeOH) on multichannel neural probes. *Sensor Actuat B-Chem* 99: 437–43.

[66] Kim G-H, Shao L, Zhang K, Pipe KP. 2013. Engineered doping of organic semiconductors for enhanced thermoelectric efficiency. *Nat Mater* 12: 719–23.

[67] Higgins MJ, Molino PJ, Yue ZL, Wallace GG. 2012. Organic conducting polymer-protein interactions. *Chem Mat* 24: 828–39.

[68] Bubnova O, Khan ZU, Malti A, Braun S, Fahlman M, Berggren M, Crispin X. 2011. Optimization of the thermoelectric figure of merit in the conducting polymer poly (3,4-ethylenedioxythiophene). *Nat Mater* 10 (2011): 429–33.

[69] Kommeren S, Coenen MJJ, Eggenhuisen TM, Slaats T, Gorter H, Groen P. 2018. Combining solvents and surfactants for inkjet printing PEDOT:PSS on P3HT/PCBM in organic solar cells. *Org Electro* 61: 282–88.

[70] Groenendaal BL, Jonas F, Freitag D, Pielartzik H, Reynolds JR. 2000. Poly(3,4-ethylenedioxythiophene) and its derivatives: past, present, and future. *Adv Mater* 12: 481–94.

[71] Richardson-Burns SM. 2007. Polymerization of the conducting polymer poly(3,4-ethylenedioxythiophene) (PEDOT) around living neural cells. *Biomaterials* 28: 1539–52.

[72] Thapa PS, Ackerson BJ, Grischkowsky DR, Flanders BN. 2009. Directional growth of metallic and polymeric nanowires. *Nanotechnology* 20: 235307.

[73] Thapa PS. 2009. Directional growth of polypyrrole and polythiophene wires. *App Phys Lett* 94: #033104.

[74] Morris JD, Thourson SB, Panta K, Flanders BN, Payne CK. 2017. Conducting polymer nanowires for control of local protein concentration in solution. *J Phys D Appl Phys* 50: #174003.

[75] Wong JY, Langer R, Ingber DE. 1994. Electrically conducting polymers can noninvasively control the shape and growth of mammalian-cells. *Proc Natl Acad Sci USA* 91: 3201–04.

[76] Okuzaki H, Ishihara M. 2003. Spinning and characterization of conducting microfibers. *Macromol Rapid Commun* 24: 261–64.

[77] Jayaram DT, Luo Q, Thourson SB, Finlay A, Payne CK. 2017. Controlling the resting membrane potential of cells with conducting polymer microwires. *Small* 13(27): #1700789.

[78] Kang M. 2014. Subcellular neural probes from single-crystal gold nanowires. *ACS Nano* 8: 8182–89.

[79] Robinson JT. 2012. Vertical nanowire electrode arrays as a scalable platform for intracellular interfacing to neuronal circuits. *Nat Nanotechnol* 7: 180–84.

[80] Robinson JT, Jorgolli M, Park H. 2013. Nanowire electrodes for high-density stimulation and measurement of neural circuits. *Front Neural Circ* 7: #38.

[81] Xie C, Lin ZL, Hanson L, Cui Y, Cui BX. 2012. Intracellular recording of action potentials by nano pillar electroporation. *Nat Nanotechnol* 7: 185–90.

[82] Lin ZLC, Xie C, Osakada Y, Cui Y, Cui BX. 2014. Iridium oxide nanotube electrodes for sensitive and prolonged intracellular measurement of action potentials. *Nat Commun* 5: #3206.

[83] Khodagholy D. 2011. Highly conformable conducting polymer electrodes for in vivo recordings. *Adv Mater* 23: H268–H272.

[84] Wei Q, Mukaida M, Kirihara K, Naitoh Y, Ishida T. 2014. Thermoelectric power enhancement of PEDOT:PSS in high-humidity conditions. *Appl Phys Express* 7: 031601.

[85] Lee SH, Park H, Son W, Choi HH, Kim JH. 2014. Novel solution-process able, doped semiconductors for application in thermoelectric devices. *J Mater Chem A* 2: 13380–87.

[86] Frischmann PD, Hwa Y, Cairns EJ, Helms BA. 2016. Redox-active super molecular polymer binders for lithium-sulfur batteries that adapt their transport properties in operando. *Chem Mater* 28: 7414–21.

[87] Yan L, Gao X, Thomas JP, Jenner N, Altounian H, Leung KT. 2018. Ionically cross-linked PEDOT:PSS as a multi-functional conductive binder for high-performance lithium-sulfur battery. *Sustain Energy Fuels* 2: 1574–81.

[88] Han P, Chung S-H, Manthiram A. 2019. Designing a high-loading sulfur cathode with a mixed ionic-electronic conducting polymer for electrochemically stable lithium-sulfur batteries. *Energy Storage Mater* 17: 317–24.

[89] Li G, Pickup PG. 1999. Ion transport in a chemically prepared polypyrrole/poly (styrene-4-sulfonate) composite. *J Phys Chem B* 103: 10143–48.

[90] Mangold K-M, Weidlich C, Schuster J, Jüttner K. 2005. Ion exchange properties and selectivity of PSS in an electrochemically switchable PPy matrix. *J Appl Electrochem* 35: 1293–301.

[91] Atanasoska L, Naoi K, Smyrl WH. 1992. XPS studies on conducting polymers: polypyrrole films doped with perchlorate and polymeric anions. *Chem Mater* 4: 988–94.

[92] Manthiram A, Milroy C. 2016. An elastic, conductive, electroactive nano composite binder for flexible sulfur cathodes in lithium-sulfur batteries. *Adv Mater* 28: 9744–51.

[93] Zhang Q, Sun Y, Xu W, Zhu D. 2014. Organic thermoelectric materials: emerging green energy materials converting heat to electricity directly and efficiently. *Adv Mater* 26: 6829–51.

[94] Hernández G, Lago N, Shanmukaraj D, Armand M, Mecerreyes D. 2017. Polyimide-polyether binders-diminishing the carbon content in lithium-sulfur batteries. *Mater Today Energy* 6: 264–70.

[95] Chen H, Wang C, Dai Y, Ge J, Lu W, Yang J. 2016. *In situ* activated polycation as a multifunctional additive for Li-S batteries. *Nano Energy* 26: 43–49.

[96] Zhong YJ, Liu Z, Zheng X, Luo SL, Yuan NY, Ding JN. 2016. Rate performance enhanced Li/S batteries with a Li ion conductive gel-binder. *Solid State Ion* 289: 23–27.

[97] Gao H, Lu Q, Yao Y, Wang X, Wang F. 2017. Significantly raising the cell performance of lithium sulfur battery via the multifunctional polyaniline binder. *Electrochim Acta* 232: 414–21.

[98] Tsao Y, Chen Z, Rondeau-Gagne´ S, Zhang Q, Yao H, Chen S. 2017. Enhanced cycling stability of sulfur electrodes through effective binding of pyridine-functionalized polymer. *ACS Energy Lett* 2: 2454–62.

[99] Ai G, Dai Y, Ye YF, Mao W, Wang Z, Zhao H. 2015. Investigation of surface effects through the application of the functional binders in lithium sulfur batteries. *Nano Energy* 16: 28–37.

[100] Nakazawa T, Ikoma A, Kido R, Ueno K, Dokko K, Watanabe M. 2016. Effects of compatibility of polymer binders with solvate ionic liquid electrolytes on discharge and charge reactions of lithium-sulfur batteries. *J Power Sources* 307: 746–52.

[101] Hwa Y, Cairns EJ. 2018. Polymeric binders for the sulfur electrode compatible with ionic liquid containing electrolytes. *Electrochim Acta* 271: 103–109.

[102] Kim HM, Hwang JY, Aurbach D, Sun YK. 2017. Electrochemical properties of sulfurized-polyacrylonitrile cathode for lithium-sulfur batteries: effect of polyacrylic acid binder and fluoro ethylene carbonate additive. *J Phys Chem Lett* 8: 5331–37.

[103] Ma G, Wen Z, Jin J, Lu Y, Rui K, Wu X, Wu M, Zhang J. 2013. Enhanced performance of lithium sulfur battery with polypyrrole warped mesoporous carbon/sulfur composite. *J Power Sources* 254: 353–59.

[104] Zhang W, Shen C, Lu G, Ni Y, Lu C, Xu Z. 2018. Synthesis of PPy/RGO-based hierarchical material with super-paramagnetic behavior and understanding its robust photo current driven by visible light. *Synth Met* 241: 17–25.

[105] Wang W, Li G, Wang Q, Li G, Ye S, Gao X. 2013. Sulfur-polypyrrole/grapheme multi-composites as cathode for lithium-sulfur battery. *J Electrochem Soc* 160: A805–10.

[106] Ma G, Wen Z, Jin J, Lu Y, Rui K, Wu X. 2014. Enhanced performance of lithium sulfur battery with polypyrrole warped mesoporous carbon/sulfur composite. *J Power Sources* 254: 353–9.

[107] Zhang Y, Zhao Y, Bakenov Z, Tuiyebayeva M, Konarov A, Chen P. 2014. Synthesis of hierarchical porous sulfur/polypyrrole/multi-walled carbon nano tube composite cathode for lithium batteries. *Electrochim Acta* 143: 49–55.

[108] Zhang Y, Zhao Y, Konarov A, Gosselink D, Li Z, Ghaznavi M. 2007. One-pot approach to synthesize PPy@S core–shell nano-composite cathode for Li/S batteries. *J Nanopart Res* 15: 1–7.

[109] Wang C, Wan W, Chen JT, Zhou HH, Zhang XX, Yuan LX. 2013. Dual core–shell structured sulfur cathode composite synthesized by a one-pot route for lithium sulfur batteries. *J Mater Chem A Mater Energy Sustain* 1: 1716–23.

[110] Zhang Y, Zhao Y, Konarov A, Gosselink D, Soboleski HG, Chen P. 2013. A novel sulfur/polypyrrole/multi-walled carbon nanotube nanocomposite cathode with core–shell tubular structure for lithium rechargeable batteries. *Solid State Ion* 238: 30–35.

[111] Zhang Y, Zhao Y, Konarov A, Gosselink D, Soboleski HG, Chen P. 2013. A novel nano-sulfur/polypyrrole/graphene nanocomposite cathode with dual-layered structure for lithium rechargeable batteries. *J Power Sources* 241: 517–21.

[112] Wang J, Lu L, Shi D, Tandiono R, Wang Z, Konstantinov K. 2013. Sulfur– composite for use in lithium-sulfur batteries. *Chem Plus Chem* 78: 318–24.

[113] Liang X, Kwok CY, Lodi-Marzano F, Pang Q, Cuisinier M, Huang H. 2016. Tuning transition metal oxide–sulfur interactions for long life lithium sulfur batteries: the "Goldilocks" principle. *Adv Energy Mater* 6: 1–9, 1501636.

[114] Wei W, Du P, Liu D, Wang Q, Liu P. 2018. Facile one-pot synthesis of well-defined coaxial sulfur/polypyrrole tubular nano composites as cathodes for long-cycling lithium sulfur battery. *Nanoscale* 10: 13037–44.

[115] Liu Y, Yan W, An X, Du X, Wang Z, Fan H. 2018. A polypyrrole hollow nanosphere with ultra-thin wrinkled shell: synergistic trapping of sulfur in lithium-sulfur batteries with excellent elasticity and buffer capability. *Electrochim Acta* 271: 67–76.

[116] Xin P, Jin B, Li H, Lang X, Yang C, Gao W, Zhu Y, Zhang W, Dou S, Jiang Q. 2014. Facile synthesis of sulfur-polypyrrole as cathodes for lithium-sulfur batteries. *Chem Electro Chem* 4: 115–21.

[117] Qian W, Gao Q, Zhang H, Tian W, Li Z, Tan Y. 2017. Cross linked polypyrrole grafted reduced graphene

oxide-sulfur nanocomposite cathode for high perform-ance Li-S battery. *Electrochim Acta* 235: 32–41.

[118] Yin F, Liu X, Zhang Y, Zhao Y, Menbayeva A, Bakenov Z. 2017. Well-dispersed sulfur anchored on interconnected polypyrrole nanofiber network as high performance cathode for lithium-sulfur batteries. *Solid State Sci* 66: 44–9.

[119] Kazazi M. 2016. Synthesis and elevated temperature per-formance of a polypyrrole-sulfur-multi-walled carbon nanotube composite cathode for lithium sulfur batteries. *Ionics* 22: 1103–12.

[120] Xie Y, Zhao H, Cheng H, Hu C, Fang W, Fang J, Xu J, Chen Z. 2016. Facile large-scale synthesis of core–shell structured sulfur@polypyrrole composite and its applica-tion in lithium-sulfur batteries with high energy density. *ACS Appl Energy Mater* 175: 522–28.

[121] Luo Y-L, Fan L-H, Xu F, Chen Y-S, Zhang C-H, Wei Q-B. 2010. Synthesis and characterization of Fe_3O_4/PPy/P(MAA-co-AAm) trilayered composite micro-spheres with electric, magnetic and pH response characteristics. *Mater Chem Phys* 120: 590–97.

[122] Zhao Y, Zhu W, Chen GZ, Cairns EJ. 2016. Polypyrrole/TiO_2nanotube arrays with coaxial heteroge-neous structure as sulfur hosts for lithium sulfur batteries. *J Power Sources* 327: 447–56.

[123] Zhang J, Shi Y, Ding Y, Zhang W, Yu G. 2016. In situ reactive synthesis of polypyrrole-MnO_2 coaxial nanotubes as sulfur hosts for high-performance lithium-sulfur battery. *Nano Lett* 16: 7276–81.

[124] Dong Y, Liu S, Wang Z, Liu Y, Zhao Z, Qiu J. 2015. Sulfur-infiltrated graphene-backboned mesoporous carbon nanosheets with a conductive polymer coating for long-life lithium-sulfur batteries. *Nanoscale* 7: 7569–73.

[125] Liu C. 2014. Enhancement of long stability of Li–S bat-tery by thin wall hollow spherical structured polypyrrole based sulfur cathode. *RSC Adv* 4: 21612–18.

[126] Li W, Zheng G, Yang Y, Seh ZW, Liu N, Cui Y. 2013. High-performance hollow sulfur nano-structured battery cathode through a scalable, room temperature, one-step, bottom-up approach. *Proc Natl Acad Sci USA* 110 (18): 7148–53.

[127] Jianga H, Zhaoa L, Gaia L, Wanga Y, Houc Y, Liub H. 2015. Conjugation of methotrexate onto dedoped Fe_3O_4/PPy nano spheres to produce magnetic targeting drug with controlled drug release and targeting specificity for He La cells. *Synth Met* 207: 18–25.

[128] Negi YS, Adhyapak PV. 2002. Development in polyaniline conducting polymers. *J Macromol Sci Part C: Polym Rev* 42: 35–53.

[129] Srinives S, Sarkar T, Mulchandani A. 2013. Nano thin polyaniline film for highly sensitive chemiresistive gas sensing. *Electroanalysis* 25: 1439–45.

[130] Carquigny S, Sanchez J-B, Berger F, Lakard B, Lallemand F. 2009. Ammonia gas sensor based on elec-trosynthesized polypyrrole films. *Talanta* 78: 199–206.

[131] Al-Mashat L, Debiemme-Chouvy C, Borensztajn S, Wlodarski W. 2012. Electro-polymerized polypyrrole nanowires for hydrogen gas sensing. *J Phys Chem C* 116: 13388–94.

[132] Zainal MF, Mohd Y. 2015. Characterization of PEDOT films for electrochromic applications. *Polym Plast Technol Eng* 54: 276–81.

[133] Patra S, Barai K, Muni Chandraiah N. 2008. Scanning electron microscopy studies of PEDOT prepared by vari-ous electrochemical routes. *Synth Met* 158: 430–35.

[134] Wang Z, Pan R, Xu C, Ruan C, Edström K, Strømme M, Nyholm L. 2018. Conducting polymer paper derived sep-arators for lithium metal batteries. *Energy Storage Mater* 13: 283–92.

[135] Tuan CV, Tuan MA, Hieu NV, Trung T. 2012. Electro-chemical synthesis of polyaniline nanowires on Pt inter-digitated microelectrode for room temperature NH3 gas sensor application. *Curr Appl Phys* 12: 1011–16.

[136] Liang X, Zhang M, Kaiser MR, Gao X, Konstantinov K, Tandiono R. 2015. Split-half-tubular polypyrrole@sul-fur@polypyrrole composite with a novelthree-layer-3D structure as cathode for lithium/sulfur batteries. *NanoE-nergy* 11: 587–99.

[137] Wang Z, Li X, Cui Y, Yang Y, Pan H, Wang Z. 2014. Improving the performance of lithium-sulfur battery by blocking sulfur diffusing paths on the host materials. *J Electrochem Soc* 161: A1231–5.

[138] Tuan CV, Tuan MA, Hieu NV, Trung, T. 2012. Electro-chemical synthesis of polyaniline nano wires on Pt inter-digitated microelectrode for room temperature NH3 gas sensor application. *Curr Appl Phys* 12: 1011–16.

[139] Yang L, Qiub W, Liub Q. 1996. Polyaniline cathode material for lithium batteries. *Solid State Ionics* 8: 819–24.

[140] Han H, Lu H, Jiang X, Zhong F, Ai X, Yang H, Cao Y. 2019. Polyaniline hollow nano fibers prepared by control-lable sacrifice-template route as high-performance cathode materials for sodium-ion batteries. *Electrochim Acta* 301: 352–58.

[141] Chen Y, Luder J, Ng M-F, Sullivan M., Manzhosa S. 2017. Polyaniline and CN-functionalized polyaniline as organic cathodes for lithium and sodium ion batteries: a combined molecular dynamics and density functional tight binding study in solid state. *Phys Chem Chem Phys* 20: 232–37, C7CP06279F.

[142] Groenendaal L, Jonas F, Freitag D, Pielartzik H, Reynolds JR. 2000. Poly (3,4-ethylenedioxythiophene) and its derivatives: past, present, and future. *Adv Mater* 12: 481–94.

[143] Partridge AC, Harris P, Andrews MK. 1996. High sensi-tivity conducting polymer sensors. *Analyst* 121: 1349–53.

[144] Zhou X, Chen F, Yang J. 2015. Core@shell sulfur@poly-pyrrole nanoparticles sandwiched in graphene sheets as cathode for lithium-sulfur batteries. *J Energy Chem* 24: 448–55.

[145] Yuan G, Wang H. 2014. Facile synthesis and performance of polypyrrole-coated sulfur nanocomposite as cathode materials for lithium/sulfur batteries. *J Energy Chem* 23: 657–61.

[146] Liu Z, Li D, Li Z, Liu Z, Zhang Z. 2017. Nitrogen-doped 3D reduced graphene oxide/polyaniline composite as active material for supercapacitor electrodes. *Appl Surf Sci* 422: 339–47.

[147] Chen Y, Luder J, Ng M-F, Sullivan M, Manzhos S. 2017. Polyaniline and CN-functionalized polyaniline as organic cathodes for lithium and sodium ion batteries: a combined molecular dynamics and density functional tight binding study in solid state. *Phys Chem Chem Phys* 20 (1): 232–37. doi:10.1039/C7CP06279F.

[148] Shanshan X, Yan J, Feng Z, Qiu S, Shao F, Li M, Yu M, Qiu H. 2016. Electrical transport properties of Ni80Fe20/HCl-PANI composites. *Thin Solid Films* 608: 44–49.

[149] Yang Y, Yu G, Cha JJ, Wu H, Vosgueritchian M, Yao Y. 2011. Improving the performance of lithium-sulfur batter-ies by conductive polymer coating. *ACS Nano* 5: 9187–93.

[150] Wang X, Zhang Z, Yan X, Qu Y, Lai Y, Li J. 2015. Interface polymerization synthesis of conductive polymer/graphite oxide@sulfur composites for high-rate lithium-sulfur batteries. *Electrochim Acta* 155: 54–60.

[151] Li W, Zhang Q, Zheng G, She ZW, Yao H, Cui Y. 2013. Understanding the role of different conductive polymers in improving the nanostructured sulfur cathode performance. *Nano Lett* 13: 5534–40.

[152] Wu F, Chen J, Chen R, Wu S, Li L, Chen S. 2011. Sulfur/polythiophene with acore/shell structure: synthesis and electrochemical properties of the cathode for rechargeable lithium batteries. *J PhysChem C* 115: 6057–63.

[153] Wu F, Wu S, Chen R, Chen J, Chen S. 2010. Sulfur–polythiophene composite cathode materials for rechargeable lithium batteries. *Electrochem Solid-State Lett* 13: A29–A31.

[154] Ma G, Wen Z, Jin J, Lu Y, Wu X, Wu M, Chen C. 2014. Hollow polyaniline sphere@sulfur composites for prolonged cycling stability of lithium-sulfur batteries. *J Mater Chem A Mater Energy Sustain* 2: 10350–54.

[155] Chang X, Hu R, Sun S, Liu J, Lei Y, Liu T, Dong L, Yin Y. 2018. Sunlight-charged electrochromic battery based on hybrid film of tungsten oxide and polyaniline. *Appl Surf Sci* 441: 105–112.

[156] Bahadur A, Iqbal S, Shoaib M, Saeed A. 2018. Electrochemical study of specially designed graphene-Fe$_3$O$_4$-polyaniline nanocomposite as a high performance anode for lithium-ion battery. *Dalton Trans* 47 (42): 15031–37. doi:10.1039/C8DT03107J.

[157] Li X, Rao M, Lin H, Chen D, Liu Y, Liu S, Liao Y, Xing L, Xu M, Li W. 2015. Sulfur loaded in curved graphene and coated with conductive polyaniline: preparation and performance as a cathode for lithium-sulfur batteries. *J Mater Chem A MaterEnergy Sustain* 3: 18098–104.

[158] Li X, Rao M, Chen D, Lin H, Liu Y, Liao Y. 2015. Sulfur supported by carbon nanotubes and coated with polyaniline: preparation and performance as cathode of lithium-sulfur cell. *Electrochim Acta* 166: 93–9.

[159] Ding K, Bu Y, Liu Q, Li T, Meng K, Wang Y. 2015. Ternary-layered nitrogen-doped graphene/sulfur/polyaniline nano architecture for the high-performance of lithium-sulfur batteries. *J Mater Chem A Mater Energy Sustain* 3: 8022–27.

[160] Moon S, Jung YH, Kim DK. 2015. Enhanced electrochemical performance of a crosslinked polyaniline-coated graphene oxide-sulfur composite for rechargeable lithium-sulfur batteries. *J Power Sources* 294: 386–92.

[161] Wu F, Chen J, Li L, Zhao T, Chen R. 2011. Improvement of rate and cycle performance by rapid polyaniline coating of a MWCNT/sulfur cathode. *J Phys Chem C* 115: 24411–17.

[162] Qiu Y, Li W, Li G, Hou Y, Zhou L, Li H. 2014. Polyaniline-modified cetyltrimethyl ammonium bromide-graphene oxide-sulfur nano composites with enhanced performance for lithium-sulfur batteries. *Nano Res* 7: 1355–63.

[163] Liu Y, Zhang J, Liu X, Guo J, Pan L, Wang H, Su Q, Du G. 2014. Nanosulfur/polyaniline/graphene composites for high-performance lithium-sulfur batteries: one pot in-situ synthesis. *Mater Lett* 133: 193–96.

[164] Zhu P, Zhu J, Yan C, Dirican M, Zang J, Jia H. 2018. In situ polymerization of nanostructured conductive polymer on 3D sulfur/carbon nanofiber composite network as cathode for high-performance lithium-sulfur batteries. *Adv Mater Interfaces* 5: 1–10, 1701598.

4 An Overview on Conducting Polymer-Based Materials for Battery Application

Satyabadi Martha

Government (SSD) +2 Science College (HSS), Badabharandi
Umerkote, India

Dilip Kumar Mishra

Department of Physics, Faculty of Engineering and Technology (ITER)
Siksha 'O' Anusandhan Deemed to be University
Bhubaneswar, India

CONTENTS

4.1 INTRODUCTION

Owing to the high population growth, energy consumption in the world has been continuously increasing. Owing to the technological advances in portable electronic industries, the demand toward the fabrication of high power density energy-storage devices such as secondary batteries and supercapacitors is constantly increasing [1]. Mostly, the energy-storage materials and devices are based on lithium (Li)-ion battery [2]. Nowadays, energy-storage materials based on conducting polymers (CPs) have gained much

attention for improving the electrochemical performance of device. Li-ion batteries were introduced way back in 1991 by the Sony Corporation, since then many researches have been carried out on them till date [3]. The detailed chemistry of the material is reported by Reddy *et al.* [4] Owing to their small radius ions and easy insertion of foreign atoms into electrode resources, the energy density of a Li-ion battery is superior to the corresponding supercapacitor and Na-ion battery. Li-ion batteries are extensively used in electronic items such as cellular phones and laptops. The major drawbacks that restrict the usage of Li-ion battery in various fields are its high cost and safety issues [5–8]. The use of volatile organic electrolytes in Li-ion battery has raised safety concerns, as the lithium metal electrodes may react with the electrolyte and might lead to dendrites formation, which will directly affect the long-term stability of the battery [8].

In recent years, the development of solid polymer electrolyte systems that can inhibit dendrite growth and improve its practical specific energy capacity, energy density, and power efficiency is of prime interest for the scientific community [8]. The synthesis and characterization of hybrid composites for the development of efficient battery are now of great concern [9,10]. Inorganic compounds have been extensively used in Li-ion-based batteries [11] due to easier transportation of Li$^+$ ion and electron for superior accommodation of ions. Owing to their high crystallinity, outstanding conductivity, and redox properties, inorganic materials are mostly used as "electrodes" in energy-storage applications [12,13].

Owing to low weight, flexibility, toughness, low cost, and easy tailoring of the shape and size, polymeric materials have gained much more importance and found as a competitive material in electrochemical applications [8]. Conducting polymeric materials can be easily incorporated into electronic paper, textiles, and structural panels because of their ease of processing into different shapes [14–16]. The π-conjugated system in organic semiconductors (polyacetylene) is shown in Figure 4.1.

These materials are not only used to hoard energy but also to do so independent of form factor [8]. It is also used to directly combine with polymer solar cells and light emitting diodes to produce inexpensive devices [17,18]. The properties of CPs can be changed by varying its fabrication procedure [19]. The ingredients and methods for the preparation of CPs are often cost-effective. Researchers have adopted various synthesis tools such as screen printing, inkjet printing, or doctor blading for improved activity [8,20,21].

It has been reported that mainly two types, i.e., metal or doped semiconductors (p-type or n-type) can be fabricated by various chemical routes [22–27]. Electrochemical method is one of the most advanced, efficient, and economical procedures for the preparation of organic semiconductor, which can conduct electricity even at room temperature. Because of this unique advantage, organic semiconductors are efficiently used as electrode materials for the fabrication of light batteries [28].

In this chapter, we are focusing on the possible uses of various CPs for fabrication of battery, its characterization, and performance.

4.2 PRINCIPLE OF CONDUCTING POLYMER BATTERY

Generally, p-type and n-type conducting polymeric materials can be prepared by the interstitial or substitutional doping of anions or cations to pristine semiconductor [29]. In electrochemical reaction, the p-type-doped polymers are easily reduced, whereas the n-type-doped polymers are oxidized, so the electrochemical reactions involved in the p-type- and n-type-doped polymers are reversible [8,29]. The reaction can be understood from the following mechanism [29]:

$$[W]_n + anU^+ \rightleftharpoons [W^{(-a)}(U^+)_a]_n$$

where "W" symbolizes the monomer unit of the CP, "a" stands for doping level, "U" for the types of pre-owned cation used, whereas "n" is the number of units repeated. The above process is similar for undoped and "n-type" doped polymers.

$$[W^{(+a)}(Y^-)_a]_n + anU \rightleftharpoons W_n + anUY$$

The above mechanism shows electrochemical undoping and redoping of p-type-doped polymer.

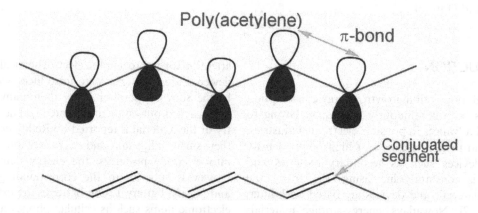

Poly(acetylene)

π-bond

Conjugated segment

FIGURE 4.1 π-conjugated system of organic conjugated polymer [reproduced from Reference 15].

4.3 ASSORTMENT OF CONDUCTING POLYMER ELECTRODES FOR BATTERY APPLICATION

Secondary batteries can be made up by using conjugated polymers as the electrode-active material (dopable/doped), and the doping and undoping reactions during the electrochemical process are helpful for charging and discharging processes [28,29]. The electrolyte plays an important role in secondary battery. Generally, the compounds that can be easily ionized to ionic dopant species are considered as good electrolytes. It is not an important issue that whether the batteries in their primary state are uncharged or charged, because both the states are interconvertible through electrochemical doping procedure. Thus, a large diversity of rechargeable battery can be fabricated. The secondary batteries are mainly classified as per the following types. In Type I secondary battery, the anode of the secondary battery consists of an n-type CP material as anode. The n-type CP material is used in combination with various companionable electrolytes and cathodes. In such secondary battery, the discharging process is mainly due to the electrochemical undoping of the cation-doped conjugated polymer anode and the releasing of the cation dopants from the polymer to the electrolytic system. During uncharged state, the anode material is in undoped form.

In Type II secondary battery, the cathode is in the charged condition and consists of a p-type conducting cathodic polymer. Herein, the cathode is used with a variety of well-suited anodes and electrolytes. The discharging mechanism is same as that of Type I, which releases anionic dopant species to the electrolytic system.

In Type III secondary battery, n-type and p-type CPs are used as anode-active material and cathode-active material. The discharging process involves simultaneous electrochemical undoping. In this process, the dopant species are released to the electrolytic system simultaneously.

4.4 MECHANISM OF CONDUCTING POLYMERS IN RECHARGEABLE BATTERIES

The fundamental operating principle for electrochemical devices mainly involves the following steps (1) separation (or ionization) of charge carrier, (2) transport of charged species, and (3) recombining of charge. The above working steps are observed in various electrochemical devices, such as batteries, supercapacitors fuel cells, photovoltaic, and photoelectrochemical devices [8,28–30]. The operating principle of the different electrochemical devices involves different mechanisms. In each case, the interface plays a critical role as through which the charge transport occurs. For fabrication of an energy-storage device, such as battery, two specialized bodies of electrodes, namely cathode-active and anode-active material are required, which are separated by an ionically conducting but electronically insulating "electrolyte". The selected electrodes must exhibit good electronic conductivity, good stability, and high catalytic activity. In a simple energy-storage device, the reduction reaction takes place at cathode, whereas the oxidation reaction takes place at anode [8,28–30].

Recently, ample researches have been carried out on the synthesis of secondary Li-ion battery. In secondary rechargeable batteries, the inorganic metal ions migrate from anode to cathode, as electrons flow from the anode to cathode through an external circuit, which results in producing energy during discharge. The electrolytes play an important role as the metal ions can move freely from anode to cathode through the electrolytic medium. During the charging process, the metal ions and electrons move in opposite direction by external voltage so electrical energy can be stored in the form of chemical energy. Figure 4.2 shows the principle that occurs in the secondary battery [8].

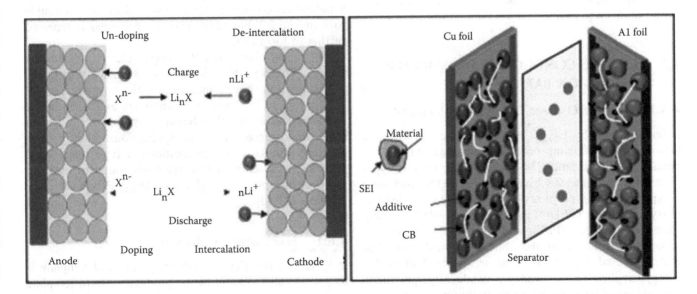

FIGURE 4.2 Schematic representation of the principle of lithium-ion battery [Reproduced from Reference 8; Permission License Id: 4547880775716].

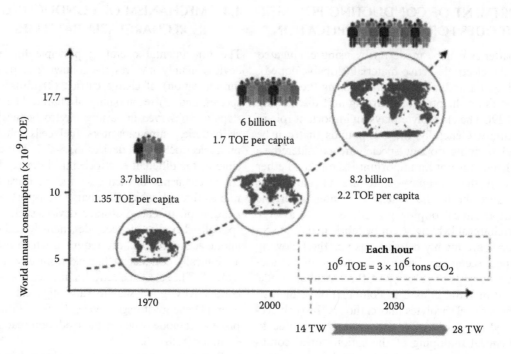

FIGURE 4.3 The scenario for past, present, and future, up to 2050, of the world's energy needs is shown. TOE = ton of oil equivalent [Reproduced from reference 3].

For fabrication of a typical battery, parameters such as gravimetric capacity, energy density, volumetric capacity, cycle ability, capability, and self-discharge properties are mainly focused. The function of anode-active materials is to monitor three different mechanisms, namely redox reaction, intercalation–deintercalation mechanism, and alloying–dealloying reactions while classic shrinking core model, anisotropic shrinking core model, and domino-cascade model are monitored by cathode-active materials [31–33]. Figure 4.3 shows energy scenario in past, present, and future needs up to 2050.

4.5 ORGANIC CONDUCTING POLYMER FOR LITHIUM-ION BATTERY

4.5.1 TYPES OF ORGANIC CONDUCTING POLYMERS

After the work by Heeger *et al.*, the research work on finding new conducting polymeric materials has gained much more importance [34–36]. To date, various types of conducting materials have been discovered, including polypyrrole (PPy), polyacetylene (PA), poly (3,4-ethylenedioxythiophene) (PEDOT), polyaniline (PANI), poly (phenylenevinylene) (PPV), and so on [37]. Figure 4.4 shows the structure of the abovementioned polymeric materials used tremendously in secondary battery for energy application. Nowadays, conducting polymeric nanomaterials have gained much attention in both basic and applied research.

4.6 SYNTHESIS OF CONDUCTING POLYMER

CPs have been synthesized by most of the researchers via chemical or electrochemical oxidation technique. The electrochemical oxidation of monomers and succeeding coupling reaction of the charged monomers generally produce giant polymeric molecules. The preparation of PPy is shown in Figure 4.5 [38]. It can be synthesized by simple oxidative polymerization of pyrrole units. One electron oxidation of pyrrole will lead to generate a radical cation, subsequently two radical cations couple to form the 2,2′-bipyrrole. The above process is repeated further to form PPy.

But the synthesis of CP nanoparticles is bit complex than the simple CPs. The use of templates during the polymerization process may lead to the formation of CP nanoparticles. To fabricate morphology and size-tunable nanostructures, template-based synthesis has gained much importance for device applications. Mainly three different routes, namely conventional soft template, hard template, and template-free synthesis route are used to fabricate morphology and size-tailored CPs and nanoparticles [39–42].

4.6.1 HARD-TEMPLATE METHOD

It is well known that the advantage of hard template is to have a better control of the product, so hard template route is recognized as the general technique for the synthesis of size- and shape-oriented CP nanoparticles.

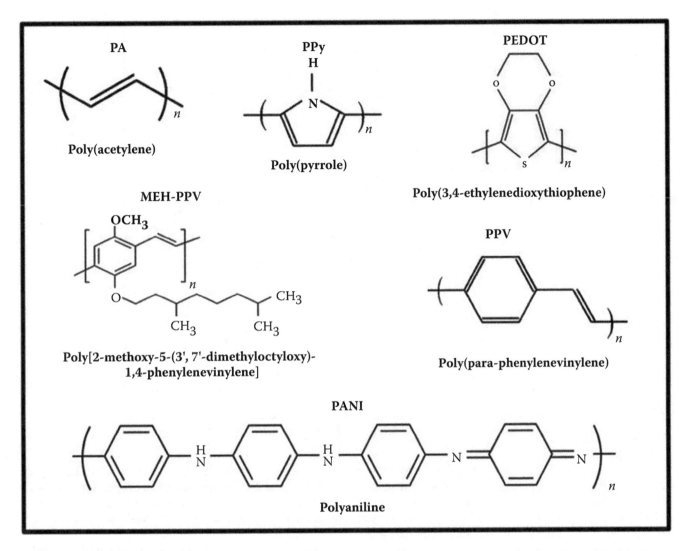

FIGURE 4.4 Molecular structure of some representative conducting polymers. [Reproduced from reference 37; Permission License Id: 4547731016904].

The fabrication of structure-specific organic semiconductors was carried out by using the porous templates. The porous template is the direct and successful technique to generate nanostructures similar to the size of the pores in the templates [43,44]. The hollow spaces of the template have been effectively filled by the desired materials/precursors using various synthesis techniques such as physical and chemical vapor deposition. Once the incorporation of precursors was completed, then the template was removed and exposed to the required chemical environment to produce the desired materials with the replicated nanostructures. This advantage of the technique is to produce nanostructures of one-dimensional (1D) to three-dimensional (3D) ordered or disordered materials [45].

Polyester membranes and anodic aluminum oxide (AAO) are used as common hard templates for the synthesis of CP [46–48]. In addition to these hard templates, other commonly used templates, namely zeolites [49], silica [50], oxides [51], polyoxometallates [52] are used for

the synthesis of various types of controllable 1D nanostructure of various CPs by chemical technique [53–58]. Martin *et al.* first developed this method and since then it has been widely used to produce more uniform nanotube- and nanowire-like structure. The morphology of nanostructures can be optimized by modifying the reaction condition and carefully regulating the template pores. Using nonconnected AAO porous membranes as template, CPs can be synthesized in the isolated pores of AAO. Hence the structure of the polymer is almost same as that of the void space in templates after removing the template scaffold, and consequently, aligned 1D nanotube and nanowire structures can be prepared. By using electrochemical deposition technique, the formation of nanowires by using AAO template is reported by Liu *et al.* [59] It has been reported that the mesoporous silica also plays an important role in preparing nanotube- or nanowire-based materials. Conducting filaments of PANI (3 nm) by using MCM-41 has been reported by Wu and Bein [60].

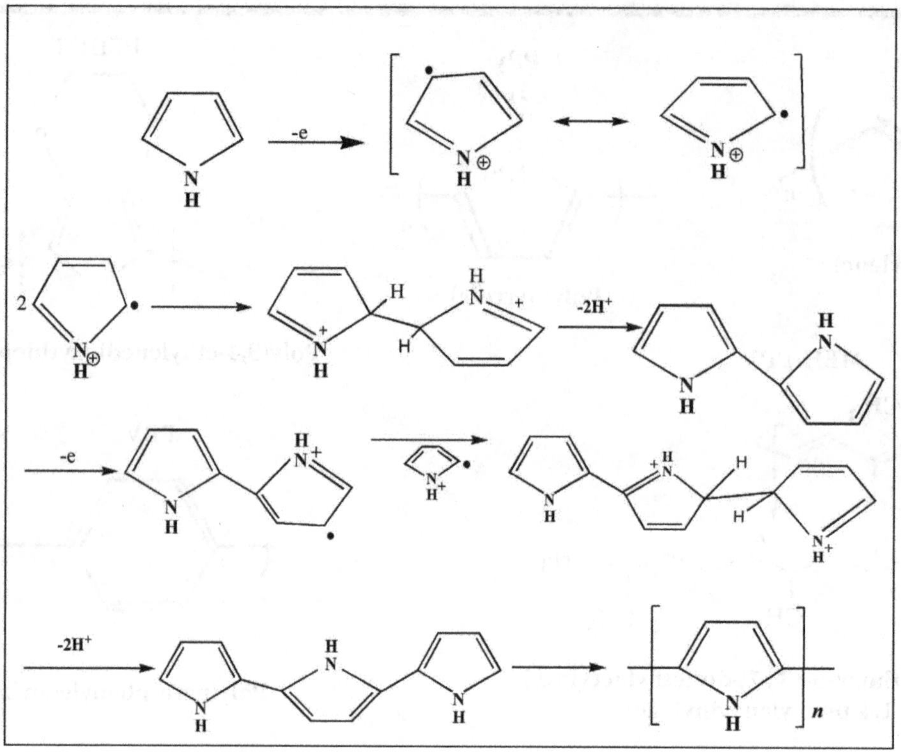

FIGURE 4.5   Schematic representation of conduction mechanism of PPY [Reproduced from reference 38].

Manohar *et al.* described a very simple way to produce structure-directing CPs by chemical oxidative synthesis of bulk quantities 100–180 nm diameter nanofibers of PEDOT in powder form by using V_2O_5 nanofiber seed templates [61]. In another work, Ppy nanotubes are synthesized by using $FeCl_3$ and methyl orange [62]. One-dimensional V_2O_5/PANI core-shell nanobelts by using V_2O_5 nanobelt template is reported by Li *et al.* [63] The biggest drawback of using the hard template route is its removal from the final product, which requires high temperature. During heating at high temperature, structural deformation of the polymer also takes place, so the overall preparation procedure is monotonous and is unsuitable for the large-scale production.

4.6.2   Soft-template Method

Soft-template method is gaining much importance for its simple fabrication process and also its removal process is quite easier [64–69]. This method is mainly based on the self-assembly of surfactants, which is helpful for shape-controlled conducting [70]. In emulsion polymerization soft-template route, the morphology of the synthesized CPs depends on the isolated or aggregated micellar state. The effect of concentration of surfactant toward synthesis of PPy nanosphere is reported by Jayakannan *et al.* He has obtained Coral-Ppy nanospheres at higher surfactant concentration while the well-defined Ppy nanospheres

were obtained at lower surfactant concentration [71]. β-naphthalene sulfonic acid is efficient toward the synthesis of PANI nanowires as reported by Wan *et al.* Zhang *et al* reported Ppy nanostructures with various shapes by soft template route [72].

Morin *et al.* reported conjugated nanowires by polymerization of butadiynes. Recently, porphyrin-based 1D materials have been synthesized by linking the monomers of porphyrin units by polymerization of butadiyne in the gel state [73,74]. PPy, PANI, and PEDOT with clip-like structure have been prepared by Manohar *et al.* [75]

One-step synthesis of PEDOT nanostructures with spindle-like or vesicle-like shapes by using $FeCl_3$ as a reductant is reported elsewhere [76]. Hulvat *et al.* reported that hexagonal fibrillary PEDOT nanostructures can be converted to hexagonal liquid crystals through electropolymerization method [77,78]. Komiyama *et al.* developed block co-polymer-templated electropolymerization technique to prepare PPy- and thiophene-based CP nanoparticles [79]. DNA is also used as a template for the preparation of uniform conductive Ppy nanowires as reported by Moon *et al.* Si base with DNA templates is helpful for the synthesis of PANI nanowires [80–82]. Enzyme also played a great role in synthesizing PEDOT nanostructures as reported by Pomposo *et al.* [83] Among other templates, ionic liquids serve as soft templates in synthesizing various conductive polymers. Perfluorooctane sulfonic acid as soft template helps in the synthesis of hollow spheres of PANI. In this work, perfluorooctane sulfonic acid acts as a dopant, which induces superhydrophobicity [84]. Park *et al.* successfully controlled the anisotropic growth of PANI nanostructures (nanofibers nanorods and nanospheres) by utilizing a polymeric stabilizer, poly(*N*-vinylpyrrolidone). They reported that even a slight variation in the polymerization condition can result in a variety of morphology, which in turn greatly affects the performance of the material.

4.6.3　Template-free Technique

Template-free technique is regarded as one of the environmentally friendly, simple, and a low-cost technique to fabricate CPs [85,86]. It does not require a template or post treatment for template removal, which is the main advantage of this technique. Moreover, the nanostructures formed are uniform, so they are easily scalable and reproducible. Template-free synthetic strategies include interfacial (self-assembly) polymerization, electrospinning, and electropolymerization. CPs of various forms such as nanotubes, hollow spheres, and nanofibers are fabricated without using any template. Typically, 1D PANI morphology is prepared without using any template [87]. Guo *et al.* prepared PANI hollow nanospheres by using controllable incontinuous nanocavities without surfactant [88,89]. Tseng *et al.* reported the fabrication of individually addressable PPy, PANI, etc. nanowires on microelectrode junctions by electrochemical method [90]. To date, ample work has been carried out for the synthesis of variety of PANI nanostructures such as nanotubes, nanowires or fibers by the template-free method [91,92].

4.6.4　Self-Assembly or Interfacial

Self-assembly is a method by which a disordered system is converted to an organized structure by local interactions among the components without any external force. Interfacial polymerization (IP) technique is mainly useful to synthesize conductive polymer by oxidative coupling processes at low temperature. The main advantage of this method is that it has limited side reactions. In this process, two reactive monomers or agents are first allowed to dissolve in two immiscible liquids, respectively, and the chemical reaction is carried out at the interface of those two liquids. The distribution of mass and charge through the liquid–liquid interface control the main fundamental properties (crystallinity, size, and shape) of the CPs [93–96]. Nanotubes of PANI have been synthesized without using any templates, as reported by Du and co-workers [97].PANI nanofibers with diameters of 30–50 nm have been synthesized by interfacial polymerization method, as per Huang *et al* [98].

4.6.5　Electrospinning

It is one of the most competent techniques for the preparation of uniform and aligned CP nanofibers and composites [37,99]. When the surface tension of the solution gets higher than the repulsion forces of a charged solution, the spun fibers are deposited on a collector as a nonwoven web [37]. Chronakis *et al.* [100] reported Ppy/PEO nanofibers in nanometer range with improved electrical conductivity. In another work, PEDOT: poly(styrenesulfonate), (PSS)/PVP nanofibers were reported to have fabricated by electrospun method [37,101–103]. Well-aligned 1D fibers of PEDOT have been synthesized by electrospinning and oxidative chemical polymerization [37,101–104]. The main advantage of the electrospinning process is that it can produce continuous long nanofibers. But the drawback is that to assist fiber formation, nonconducting polymers or sometimes supports are added, which has adverse effect by decreasing the conductivity of electrospun fibers [37].

4.7　CHARACTERIZATION

Researchers have analyzed the CPs by various advanced microscopic, optoelectronic techniques in order to study their fundamental properties and potential applications in energy devices. Scanning electron microscopy is generally used to study the surface texture of the polymeric materials. In addition, there are various other techniques, such as atomic force microscopy (AFM), transmission electron microscopy (TEM), and cryo-TEM, which are used to study the surface and bulk behavior of the polymeric materials [22,23]. Infrared (IR) absorption and AFMIR techniques are used to study the bonding between CP and their macroscopic behavior. Cyclic voltammetry (CV) is used to study the electronic states of the CP and impedance spectroscopy is used to study charge transfer resistance and capacitance behavior of CP [105].

4.7.1　Surface Characterization by AFM and AFMIR

AFM is comprehensively used to investigate the morphology, phase segregation, surface roughness, and connectivity of the polymeric chains. By using AFM, Pruneanu *et al.* [106] observed PPy nanowires by taking DNA as a template. The formation of cross-linked PPy has been observed by Taranekar *et al.* from AFM image [102]. The surface energy, charge distribution, and the concentration of the doped material can be obtained by electrical force microscopy [103,107]. Double-tip scanning tunneling microscope (STM) is also helpful for measuring the conductivity. Takami and co-workers measured the conductivity of polydiacetylene thin films by double-tip STM [108]. The scanning Kelvin probe microscope helps to identify the local surface potential and hence can distinguish regions with different chemical behavior or its composition [109]. The evidence of local structural inhomogeneity and irregular distribution of dopant in conducting polybithiophene has been reported by Semenikhin research group [110]. The stability of polyaniline nanofiber/gold nanoparticle composites in electric field has been studied over conductive atomic force microscope [111]. In AFMIR instrument, the sample absorbs the IR laser pulse hence it gets heated, which results in thermal expansion of the absorbing region of the sample. Owing to the immediate thermal expansion pulse, the AFM cantilever oscillates. The intensity of oscillations is directly related to the absorption phenomena. So, it is helpful to trace IR absorption spectra. So at a given wavelength, the instrument helps us for chemical mapping of the surface. Ghosh and co-workers have observed the polymer regions via AFMIR technique by using nano-IR instrument with IR source as the tunable pulsed laser [112].

4.7.2 TRANSMISSION ELECTRON MICROSCOPY

TEM is generally used to study the internal structure, crystallinity of the materials [37]. It helps to study the mechanism of formation of CPs [37]. By using an electrochemical liquid flow cell, Liu and co-workers have studied *in situ* electrochemical deposition of PEDOT by TEM [113].

Energy-dispersive X-ray spectroscopy is generally used to confirm the elemental composition of the material. Cryo-TEM is used to study the morphology in aqueous media. Remita and co-workers have reported various works and characterized the materials by utilizing cryo-TEM [114–116]. The *in situ* nucleation of PANI nanoparticles from micelles can also be studied through nuclear magnetic resonance spectroscopy [117].

4.7.3 ELECTROCHEMICAL CHARACTERIZATION

Electrochemical characterization is used to study the electronic behavior, the energy levels of valence band and conduction band, including the band gap of the CPs. The reversibility, cycle study, stability, and arrangements of the polymer films on substrate can be studied through electrochemical characterization. CV measurements help to know the redox behavior (both oxidizing and reducing property) of the electrode along with the oxidation and reduction potentials of the polymer [117,118]. In CV, a potential is continuously applied for current generation. The differential pulse voltammetry and squarewave voltammetry techniques can be very helpful when the output signals from the cathode or anode are very weak and not predicted accurately. In these techniques, after each potential step change the current can be determined.

Electrochemical impedance spectroscopy (EIS) along with CV helps to understand the electrochemical characterics of CPs including charge transfer resistance, the rate of charge transfer, charge transport process, double-layer capacitance, diffusion impedance, and solution resistance [119–121]. For scrutinizing the CPs for battery application, CV is very helpful for determining charge transfer process and the integrated surface area of CV.

The EIS plot generally exhibits two frequency-related regions, namely a semi-circle at high-frequency region and a sloping straight line in the low-frequency part. The semi-circle in the high-frequency region mainly arises due to the charge-transfer resistance (Rct), while the sloping straight line in the low-frequency region associated with the mass transfer [122]. The charge transfer properties of the samples are reflected by the radius of the semi-circle, the smaller the radius of the semicircle, faster is the charge transfer process and lower is the recombination of charge carriers [123]. Sometimes, one-half of the semi-circle is partially overlapped with the other half region depending on the property of the interfacial processes [124]. It has been reported that PPy film and PPy nanotube showed high impedance and low charge capacity density than PEDOT film and PEDOT nanotube [125,126].

4.8 APPLICATIONS OF VARIOUS CONDUCTING POLYMERS IN BATTERY

4.8.1 POLYACETYLENE BATTERY

Polyacetylene is used in battery as electrode material. It can be used both as an anodic and a cathodic material in battery. In the traditional batteries, it can be used as electrode in combination with conventional metals. The use of polyacetylene has offered excellent battery performance. It has been reported that the configuration n-(CH),–/LiClO$_4$ + propylene carbonate/p-(CH) showed an open circuit potential of 3.5 V and a current density of 28 mA/cm^2 [127]. Experimentally, the energy density has been calculated to be 424 W h/kg. [29,127] Cycle study confirmed that a polyacetylene battery can be used for 2000 successive charge/discharge cycles and is useful in a variety of temperature range (−20°C to 50°C). This configuration claims high energy density, flexibility, light weight, high power, and portability. The use of polyacetylene CP in the battery greatly improves the surface characteristics such as the uniform movement of Li ions in the film and a weaving of (CH) network and the prevention of precipitation of Li ion on the cathode surface. The use of dehydrated LiClO$_4$, propylene carbonate in the polyacetylene helps the battery to achieve long-term stability.

4.8.2 Polyaniline Batteries

Like polyacetylene, polyaniline is also another interesting CP that is extensively studied for the improvement of battery performance. It is also stable in air and humidity. Among its various forms, the following compound shows good electrochemical sensitivity, that is, benzenoid–quinonoid diiminium salt [29].

It can be synthesized through chemical or electrochemical method. Unlike polyacetylene batteries, the electrochemical cells are fabricated by using polyaniline electrodes and also metal electrodes. Owing to their simple synthesis, cycling stability, and energy density, they are significantly used in batteries. Polyaniline is a p-doped-state polymer, which is generally prepared by either electrochemical [128] or chemical [129] oxidation polymerization [130].

It has been reported that polyaniline-based CPs have a theoretical capacity of 295 mA h/g because of its high doping level (a full-oxidation state ($x_{max} = 1$). But the practical capacity of these CPs is subordinate than the theoretical capacity [organic molecules-polymer]. Practically, it is difficult to exceed the capacity than 150 mA h/g based on the total mass of P and A^- in the pristine state [130].

The polyaniline cell which was first reported in ref [131]. has a capacity of 13 A h/kg. The maximum capacity that is obtainable over $Zn/ZnSO_4$/polyaniline configuration is 108 A h/kg [132,133]. The energy density is observed to be 111 W h/kg with the coulombic efficiency -100% and more than 2000 charge/discharge cycles. It is important to note that both aqueous [134] and nonaqueous [135] batteries can be formulated by using polyaniline. Polyaniline can be used as anode material by using Li-doped aluminum [136] as cathode material.

Researchers have tried to prepare polyaniline/graphene composite paper or thin film [137–140], which also generates high electrochemical performance and high flexibility [130].

Polyaniline with Li–Al alloy [141] is now used in liquid crystal display (LCD) and memory back up of power source of random access memory (RAM). This packaged cell offers a common capacity of 4 mA h at a potential range of 3.3–2.0 V having 2000 charge/discharge cycles. It can be used in a wide temperature range, that is, −20°C to 60°C.

4.8.3 Poly (p-phenylene) Batteries

Poly (p-phenylene) (PPP) batteries are electroconducting polymers that have a high redox potential for electrochemical doping and undoping of anions in nonaqueous media [142]. PPP is used both as anode and cathode materials in the battery due to its extraordinary stability and processability. It can be selectively used along with other metals. While employing Li in the system, it shows an open circuit potential of 4.4 V, which is more than polyacetylene/Li battery due to elevated ionization potential of pristine PPP (5.6 eV) than polyacetylene (4.7 eV).

PPP has lower specific capacity so its energy density is not comparable to the inorganic electrodes. The major drawback is that at higher potential, PPP is very unstable due to irreversible oxidation reaction. The oxidation reaction at higher circuit potential makes the PPP units inactive during charge/discharge process. Moreover, the lower efficiency because of the self-oxidation process is a major disadvantage for the practical utility of CPs in rechargeable lithium batteries [130]. It is also not stable in conventional nonaqueous solvents such as propylene carbonate [143].

4.8.4 Heterocyclic Polymer Batteries

Nowadays, heterocyclic CP systems have gained much attention because of the good stability of heterocyclic compounds in air. The synthesis procedure of heterocyclic CP is very cheap and does not require any advanced reaction set up, thus becoming a competitive material in organic CP. Conjugated organic polymers have also been extensively used for multivalent metal-ion battery cathodes. It has been reported that PPy [144] and polythiophene [145] were employed in aluminum-ion battery cathodes [144–146]. Owing to electrochemical reversibility of polythiophene, it is also used as an electroactive material such as p-phenylene and PPP. A bio-derived melanin polymer is used as cathode-active material in magnesium-ion batteries [147]. These batteries show a 60 mA h/g Csp at 0.1 A/g and 15 mA h/g at 5.0 A/g and also are very stable [146]. Polythiophene is one of the interesting heterocyclic classes of electroactive material and can be used both as anode and cathode for battery application. In some cases, it has been proven that polythiophene is used for the solubility of the redox-active species [148]. Researches found that polymers containing thioethers and quinones can be used as energy-storage materials [149–151]. A polypyrrole-based battery fabricated by using Li–Al alloy showed a charge capacity of 130 A h/kg [152]. It has been reported that polythiophene cell [153] offers an open circuit potential of 2.8 V and a current of 5 rnA/cm^2 in a nonaqueous medium. It has also been reported a variety of carbonyl-based electrode-active compounds which contains various heteroaromatic structures [154]. The above-developed compounds were advanced as cathode-active materials for rechargeable Li-ion battery with respect to their benzene-fused analog (anthraquinone). [154] Hence, to investigate organic electrode materials for the enhancement of battery performance is still an active part of research.

It has been reported that a polythiophene-based battery offers a power density and an energy density of 2.5×10^4 W/kg and 140 W h/kg, respectively [155]. The assessment of such organic heterocyclic polymer electrodes gained much attention, and a clear understanding of their electrochemical behavior, optoelectronic, and structural properties is helpful for fabricating a useful system. The major advantage of the heterocyclic polymer battery is its stability at room temperature and that it can be easily modified to fabricate a suitable electrochemical system. For fabrication of secondary battery, poly dienothiophene has also been used as an effective electroactive material [156,157]. It is also stable in air. So, further research should be focused on the heterocyclic CP with respect to surface characteristics, long-term stability, flexibility, and so on, so that the future energy-related problems might be solved.

4.9 SUMMARY AND OUTLOOK

In this chapter, we concluded that the organic semi-conducting polymeric materials are promising candidates for secondary battery because of their low cost, stable operation, high capacity, high voltages, reversible electrochemistry, flexibility, and high performance. Organic electrodes also explore the use of multivalent ions for battery application. The use of CP electrodes for battery application is gaining importance day by day. Herein, we summarize the fabrication techniques of the secondary battery and characterization of CP for energy-storage applications. However, for industrial application of organic electrode materials the major disadvantage is their solubility in the electrolytic medium and low conductivity, which result in low capacities and low cycling stability. Successful functionalization with organic motifs can enhance the solubility and adjust the redox potentials in the desired electrolyte.

It is proposed that the modification of organic molecules by donating or accepting the electrons can refrain the redox potential. This process can be carried out by substituting carbon with other heteroatoms that can tune the oxidation and reduction potential. Using the most advanced characterization techniques such as STM, high-resolution TEM, and AFM may assist to recognize the advanced polymeric materials for better cycle activity and electrochemistry. The characterization technique is also helpful for the preparation of various kinds of nanostructure with variable shapes and sizes as different shapes of the organic polymer results in different electrochemical activities. It has also been highlighted that template method could be able to produce various nanostructures with unique shape, structure, and electrochemical properties. In some way, the chemical reduction process of organic molecules also enhances the conductivity of the molecules, which results in fast electro kinetics and high rate performance of the battery. For scalability and feasibility for the human being, the cost of the cell matters a lot. So combining organic electrode materials with other less expensive materials for modification in order to enhance the performance is highly necessary. Mostly, carbon-related sources such as graphene, carbon nitride might solve the issue for the generation of high-performance secondary battery. The development of more accurate computational methods such as DFT might be helpful for detailed electronic behavior, redox ability of the electrode materials for fabricating materials, and rationalizing behavior.

REFERENCES

1 S. H. Lee, C. Park, J. W. Park, S. J. Kim, S. S. Im and H. Ahn, Synthesis of conducting polymer-intercalated vanadate nanofiber composites using a sonochemical method for high performance pseudocapacitor applications, *J. Power Sources*, 2019, 414, 460–469.
2 D. Ferguson, D. J. Searles and M. Hankel, Biphenylene and phagraphene as lithium ion battery anode materials, *ACS Appl. Mater. Interfaces*, 2017, 9 (24), 20577–20584.
3 S. Isah, Advanced materials for energy storage devices (review), *Asian J. Nanosci. Mater.*, 2018, 1 (2), 90–103.
4 M. V. Reddy, G. V. Subba Rao and B. V. R. Chowdari, Metal oxides and oxysalts as anode materials for Li ion batteries, *Chem. Rev.*, 2013, 113, 5364–5457.
5 Focus on the US Batteries Market, US Reuters, Feb 5, 2009.
6 M. S. Islam and C. A. J. Fisher, Lithium and sodium battery cathode materials: computational insights into voltage, diffusion and nanostructural properties, *Chem. Soc. Rev.*, 2014, 43, 185.
7 B. Dunn, H. Kamath and J. M. Tarascon, Electrical energy storage for the grid: a battery of choices, *Science*, 2011, 334, 928.
8 P. Sengodua and A. Deshmukh, Conducting polymers and their inorganic composites for advanced Li-ion batteries: a review, *RSC Adv.*, 2015, 5, 42109–42130.
9 X. Gao, W. Luo, C. Zhong, D. Wexler, S. L. Chou, H. K. Liu, Z. Shi, G. Chen, K. Ozawa and J. Z. Wang, Novel germanium/polypyrrole composite for high power lithium-ion batteries. *Sci. Rep.*, 2014, 4, 6095.
10 K. C. White, High energy composite cathodes for lithiumion batteries, US Patent no. US20100124702 A1, US 12/272, 157, 2010.
11 P. G. Bruce, B. Scrosati and J. M. Tarascon, Nanomaterials for rechargeable lithium batteries, *Angew. Chem. Inter.*, 2008, 47, 2930.
12 H. Wu, G. Yu, L. Pan, N. Liu, M. T. McDowell, Z. Bao and Y. Cui, Stable Li-ion battery anodes by in-situ polymerization of conducting hydrogel to conformally coat silicon nanoparticles, *Nat. Commun.*, 2013, 4, 1943.
13 B. Liu, P. Soares, C. Checkles, Y. Zhao and G. Yu, Three-dimensional hierarchical ternary nanostructures for high-performance Li-ion battery anodes, *Nano Lett.*, 2013, 13, 3414.
14 D. Wei, D. Cotton and T. Ryhanen, All-solid-state textile batteries made from nano-emulsion conducting polymer inks for wearable electronics, *Nanomaterials*, 2012, 2, 268.
15 A. U. Agobil, H. Louis, T. O. Magu and P. M. Dass, A review on conducting polymers-based composites for energy storage application, *J. Chem. Rev.*, 2019, 1, 19–34.
16 S. Trohalaki, Energy focus: Li-ion batteries fabricated by spray painting, *MRS Bull.*, 2012, 37, 883.
17 H. Wu, G. Zheng, N. Liu, T. J. Carney, Y. Yang and Y. Cui, Engineering empty space between Si nanoparticles for lithium-ion battery anodes, *Nano Lett.*, 2012, 12, 904.
18 K. Bock, Polymer electronics systems-polytronics, *Proc. IEEE*, 2005, 93, 1400.
19 G. A. Snook, P. Kao and A. S. Best, Conducting-polymer-based supercapacitor devices and electrodes, *J. Power Sources*, 2011, 196, 1.
20 S. R. Forrest, The path to ubiquitous and low-cost organic electronic appliances on plastic, *Nature*, 2004, 428, 911.
21 T. Tsutsui and K. Fujita, The shift from "hard" to "soft" electronics, *Adv. Mater.*, 2002, 14, 949.

22 D. Sahoo, S. Nayak, K. H. Reddy, S. Martha and K. Parida, Fabrication of a Co(OH)2/ZnCr LDH "p–n" heterojunction photocatalyst with enhanced separation of charge carriers for efficient visible-light-driven H2 and O2 evolution, *Inorg. Chem.*, 2018, 57 (7), 3840–3854.

23 D. Kandi, S. Martha and K. M. Parida, Quantum dots as enhancer in photocatalytic hydrogen evolution: a review, *Int. J. Hydrogen Energy*, 2017, 42, 9467–9481.

24 S. Martha, P. C. Sahoo and K. M. Parida, An overview on visible light responsive metal oxide based photocatalysts for hydrogen energy production, *RSC Adv.*, 2015, 5, 61535–61553.

25 A. Nashim, S. Martha and K. M. Parida, Heterojunction conception of n-La2Ti2O7/p-CuO in the limelight of photocatalytic formation of hydrogen under visible light, *RSC Adv.*, 2014, 4, 14633–14643.

26 U. Routray, J. Mohapatra, V. V. Srinivasu and D. K. Mishra, Origin of magnetism probed through electron spin resonance studies in Zn1-xMnxO synthesized by chemical route, *Mater. Chem. Phys.*, 2018, 213, 52–55.

27 D. K. Mishra, J. Mohapatra, M. K. Sharma, R. Chattarjee, S. K. Singh, S. Varma, S. N. Behera, S. K. Nayak, and P. Entel, Carbon doped ZnO: synthesis, characterization and interpretation, *J. Magn. Magn. Mater.*, 2013, 329, 146–152.

28 M. E. Bhosale, S. Chae, J. M. Kim and J.-Y. Choi, Organic small molecules and polymers as an electrode material for rechargeable lithium ion batteries, *J. Mater. Chem. A*, 2018, 6, 19885–19911.

29 S. Pitchumani and V. Krishnan, Conducting polymer batteries – an assessment, *Electrochem.*, 1987, 3 (2), 117–121.

30 T. M. Gür, Review of electrical energy storage technologies, materials and systems: challenges and prospects for large-scale grid storage, *Energ. Environ. Sci.*, 2018, 11, 2696–2767.

31 P. Novak, O. Haas, K. S. V. Santhanam and K. Muller, Electrochemically active polymers for rechargeable batteries, *Chem. Rev.*, 1997, 97, 207.

32 S. B. Chikkannanavar, D. M. Bernardi and L. Liu, A review of blended cathode materials for use in Li-ion batteries, *J. Power Sources*, 2014, 248, 91–100.

33 V. Aravindan, J. Gnanaraj, Y. S. Lee and S. Madhavi, A review of blended cathode materials for use in Li-ion batteries, *J. Mater. Chem. A.*, 2013, 1, 3518.

34 T. Ito, H. Shirakawa and S. Ikeda, Simultaneous polymerization and formation of polyacetylene film on the surface of concentrated soluble Ziegler-type catalyst solution, *J. Polym. Sci.*, 1974, 12, 11–20.

35 H. Shirakawa, E. Louis, A. MacDiarmid, C. Chiang and A. Heeger, Synthesis of electrically conducting organic polymers: halogen derivatives of polyacetylene, (CH)x, *J. Chem. Soc. Chem. Commun.*, 1977, 16, 578.

36 A. J. Heeger, Semiconducting and metallic polymers: the fourth generation of polymeric materials (nobel lecture), *Angew, Chem. Int. Ed.*, 2001, 40, 2591–2611.

37 S. Ghosh, T. Maiyalagan and R. N. Basu, Nanostructured conducting polymers for energy applications: towards a sustainable platform, *Nanoscale*, 2016, 8, 6921–6947.

38 R. Kumar, S. Singh and B. C. Yadav, Conducting polymers: synthesis, properties and applications, *Int. Adv. Res. J. Sci. Eng. Tech.*, 2015, 2 (11), 105–109.

39 Y. Jin, J. Xu, L. Wang and Q. Lu, Template-free synthesis of nanorod-assembled hierarchical Zn1−xMnxS hollow nanostructures with enhanced pseudocapacitive properties, *Chem. Eur. J.*, 2016, 22, 18859–18864.

40 L. Pan, H. Qiu, C. Dou, Y. Li, L. Pu, J. Xu and Y. Shi, Conducting polymer nanostructures: template synthesis and applications in energy storage, *Int. J. Mol. Sci.*, 2010, 11, 2636–2657.

41 C. R. Martin, Template synthesis of electronically conductive polymer nanostructures, *Acc. Chem. Res.*, 1995, 28, 61–68.

42 S. Il Cho and S. B. Lee, Fast electrochemistry of conductive polymer nanotubes: synthesis, mechanism, and application, *Acc. Chem. Res.*, 2008, 41, 699–707.

43 G. Cao and D. Liu, Template-based synthesis of nanorod, nanowire, and nanotube arrays, *Adv. Colloid Interface Sci.*, 2008, 136, 45–64.

44 Z. Yao, C. Wang, Y. Li and N.-Y. Kim, AAO-assisted synthesis of highly ordered, large-scale TiO2 nanowire arrays via sputtering and atomic layer deposition, *Nanoscale Res. Lett.*, 2015, 10, 166.

45 Z. Tong, S. Liu, X. Li, J. Zhao and Y. Li, Self-supported one-dimensional materials for enhanced electrochromism, *Nanoscale Horiz.*, 2018, 3, 261–292.

46 C. R. Martin, Nanomaterials: a membrane-based synthetic approach, *Science*, 1994, 266, 1961–1966.

47 C. R. Martin, R. Parthasarathy and V. Menon, Template synthesis of electronically conductive polymers – preparation of thin films, *Electrochim. Acta*, 1994, 39, 1309–1313.

48 W. Lee and S. J. Park, Porous anodic aluminum oxide: anodization and templated synthesis of functional nanostructures, *Chem. Rev.*, 2014, 114, 7487–7556.

49 M. Treuba, M. A. L. Montero and J. Rieumont, Pyrrole nanoscaled electropolymerization: effect of the proton, *Electrochim. Acta.*, 2004, 49, 4341–4349.

50 M. S. Cho, H. J. Choi and W. S. Ahn, Enhanced electrorheology of conducting polyaniline confined in MCM-41 channels, *Langmuir.*, 2004, 20, 202–207.

51 Z. Zhang, J. Sui, L. Zhang, M. Wan, Y. Wei and L. Yu, Synthesis of polyaniline with a hollow, octahedral morphology by using a cuprous oxide template, *Adv. Mater.*, 2005, 17, 2854–2857.

52 P. J. Kulesza, M. Chojak, K. Miecznikowski, A. Lewera, M. M. Malik and A. Kuhn, Polyoxometallates as inorganic templates for monolayers and multilayers of ultrathin polyaniline, *Electrochem. Commun.*, 2002, 4, 510–526.

53 M. G. Han and S. H. Foulger, 1-Dimensional structures of poly(3,4-ethylenedioxythiophene) (PEDOT): a chemical route to tubes, rods, thimbles, and belts, *Chem. Commun.*, 2005, 3092–3094.

54 J. Fei, Y. Cui, X. Yan, Y. Yang, K. Wang and J. Li, Controlled fabrication of polyaniline spherical and cubic shells with hierarchical nanostructures, *ACS Nano.*, 2009, 3, 3714–3718.

55 S. De Vito and C. R. Martin, Toward colloidal dispersions of template-synthesized polypyrrole nanotubules, *Chem. Mat.*, 1998, 10, 1738–1741.

56 A. Rahman and M. K. Sanyal, Novel switching transition of resistance observed in conducting polymer nanowires, *Adv. Mater.*, 2007, 19, 3956–3960.

57 G. A. O'Brien, A. J. Quinn, D. Iacopino, N. Pauget and G. Redmond, polythiophene mesowires: synthesis by template wetting and local electrical characterisation of single wires, *J. Mater, Chem.*, 2006, 16, 3237–3241.

58 D. Bearden, J. P. Cannon and S. A. Gold, Solvent effects on template wetting nanofabrication of MEH-PPV nanotubules, *Macromolecules*, 2011, 44, 2200–2205.

59 R. Liu, J. Duay and S. B. Lee, Electrochemical formation mechanism for the controlled synthesis of heterogeneous

MnO2/Poly(3,4-ethylenedioxythiophene) nanowires, *ACS Nano*, 2011, 5, 5608–55619.

60 C. G. Wu and T. Bein, Conducting polyaniline filaments in a mesoporous channel host, *Science*, 1994, 264, 1757–1759.

61 X. Zhang, A. G. MacDiarmid and S. K. Manohar, Chemical synthesis of PEDOT nanofibers, *Chem. Commun.*, 2005, 0, 5328–5330.

62 L. J. Pan, L. Pu, Y. Shi, S. Y. Song, Z. Xu, R. Zhang and Y. D. Zheng, Synthesis of polyaniline nanotubes with a reactive template of manganese oxide, *Adv. Mater.*, 2007, 19, 461–464.

63 T. Y. Dai and Y. Lu, Water-soluble methyl orange fibrils as versatile templates for the fabrication of conducting polymer microtubules, *Macromol. Rapid Commun.*, 2007, 28, 629–633.

64 J. Jang and H. Yoon, Formation mechanism of conducting polypyrrole nanotubes in reverse micelle systems, *Langmuir*, 2005, 21, 11484–11489.

65 J. Jang, J. Bae and E. Park, Selective fabrication of poly (3,4-ethylenedioxythiophene) nanocapsules and mesocellular foams using surfactant-mediated interfacial polymerization, *Adv. Mater.*, 2006, 18, 354–358.

66 H. Yoon, M. Choi, K. J. Lee and J. Jang, Versatile strategies for fabricating polymer nanomaterials with controlled size and morphology, *Macromol. Res.*, 2008, 16, 85–102.

67 P. Anilkumar and M. Jayakannan, Divergent nanostructures from identical ingredients: unique amphiphilic micelle template for polyaniline nanofibers, tubes, rods, and spheres, *Macromolecules*, 2008, 41, 7706–7715.

68 B. H. Jones, K. Y. Cheng, R. J. Holmes and T. P. Lodge, Nanoporous poly(3,4-ethylenedioxythiophene) derived from polymeric bicontinuous microemulsion templates, *Macromolecules*, 2012, 45, 599–601.

69 K. J. Ahn, Y. Lee, H. Choi, M. S. Kim, K. Im, S. Noh and H. Yoon, Surfactant-templated synthesis of polypyrrole nanocages as redox mediators for efficient energy storage, *Sci. Rep.*, 2015, 5, 14097.

70 L. Xia, Z. Wei and M. Wan, Conducting polymer nanostructures and their application in biosensors, *J. Colloid. Interface. Sci.*, 2010, 341, 1–11.

71 M. J. Antony and M. Jayakannan, Amphiphilic azobenzenesulfonic acid anionic surfactant for water-soluble, ordered, and luminescent polypyrrole nanospheres, *J. Phys. Chem. B.*, 2007, 111, 12772–12780.

72 X. Zhang, J. Zhang, W. Song and Z. Kiu, Controllable synthesis of conducting polypyrrole nanostructures, *J. Phys. Chem. B.*, 2006, 110, 1158–1165.

73 M. Shirakawa, N. Fujita and S. A. Shinkai, A stable single piece of unimolecularly π-stacked porphyrin aggregate in a thixotropic low molecular weight gel: a one-dimensional molecular template for polydiacetylene wiring up to several tens of micrometers in length, *J. Am. Chem. Soc.*, 2005, 127, 4164–4165.

74 J. R. Néabo, S. Rondeau-Gagné, C. Vigier-Carrière and J. F. Morin, Soluble conjugated one-dimensional nanowires prepared by topochemical polymerization of a butadiynes-containing star-shaped molecule in the Xerogel state, *Langmuir*, 2013, 29, 3446–3452.

75 Z. Liu, X. Y. Zhang, S. Poyraz, S. P. Surwad and S. K. Manohar, Oxidative template for conducting polymer nanoclips, *J. Am. Chem. Soc.*, 2010, 132, 13158–13159.

76 S. Ghosh, H. Remita, L. Ramos, A. Dazzi, A. DenisetBesseau, P. Beaunier, F. Goubard, P.-H. Aubert and S. Remita, PEDOT nanostructures synthesized in hexagonal mesophases, *New J. Chem.*, 2014, 38, 1106–1115.

77 J. F. Hulvat and S. I. Stupp, Liquid-crystal templating of conducting polymers, *Angew Chem. Int. Ed.*, 2003, 42, 778–781.

78 J. F. Hulvat and S. I. Stupp, Anisotropic properties of conducting polymers prepared by liquid crystal templating, *Adv. Mater.*, 2004, 16, 589–592.

79 H. Komiyama, M. Komura, Y. Akimoto, K. Kamata and T. Iyoda, Longitudinal and lateral integration of conducting polymer nanowire arrays via block-copolymer-templated electropolymerization, *Chem. Mater.*, 2015, 27, 4972–4982.

80 R. Hassanien, M. Al-Hinai, S. A. F. Al-Said, R. Little, L. Ŝiller, N. G. Wright, A. Houlton and B. R. Horrocks, Preparation and characterization of conductive and photoluminescent DNA-templated polyindole nanowires, *ACS Nano.*, 2010, 4, 2149–2159.

81 H. K. Moon, H. J. Kim, N. H. Kim and Y. J. Roh, Fabrication of highly uniform conductive polypyrrole nanowires with DNA template, *Nanosci. Nanotechnol.*, 2010, 10, 3180–3184.

82 Y. Ma, J. Zhang, G. Zhang and H. He, Polyaniline nanowires on Si surfaces fabricated with DNA templates, *J. Am. Chem. Soc.*, 2004, 126, 7097–7101.

83 V. Rumbau, J. A. Pomposo, A. Eleta, J. Rodriguez, H. Grande, D. Mecerreyes and E. Ochoteco, First enzymatic synthesis of water-soluble conducting poly(3,4-ethylenedioxythiophene), *Biomacromolecules*, 2007, 2, 315–317.

84 Y. Zhu, D. Hu, M. Wan, L. Jiang and Y. Wei, Conducting and superhydrophobic rambutan-like hollow spheres of polyaniline, *Adv. Mater.*, 2007, 19, 2092–2096.

85 M. Wan, A template-free method towards conducting polymer nanostructures, *Adv. Mater.*, 2008, 20, 2926–2932.

86 Q. Zhao, R. Jamal, L. Zhang, M. C. Wang and T. Abdiryim, The structure and properties of PEDOT synthesized by template-free solution method, *Nanoscale Res. Lett.*, 2014, 9, 557–565.

87 H. D. Tran, J. M. D'Arcy, Y. Wang, P. J. Beltramo, V. A. Strong and R. B. Kaner, The oxidation of aniline to produce "polyaniline": a process yielding many different nanoscale structures, *J. Mater. Chem.*, 2011, 21, 3534–3550.

88 H. J. Ding, J. Y. Shen, M. X. Wan and Z. J. Chen, Formation mechanism of polyaniline nanotubes by a simplified template-free method, *Macromol. Chem. Phys.*, 2008, 209, 864–871.

89 H. W. Park, T. Kim, J. Huh, M. Kang, J. E. Lee and H. Yoon, Anisotropic growth control of polyaniline nanostructures and their morphology-dependent electrochemical characteristics, *ACS Nano*, 2012, 6, 7624–7633.

90 J. Wang, S. Chan, R. R. Carlson, Y. Luo, G. L. Ge, R. S. Ries, J. R. Heath and H. R. Tseng, Electrochemically fabricated polyaniline nanoframework electrode junctions that function as resistive sensors, *Nano Lett.*, 2004, 4, 1693–1697.

91 H. J. Ding, M. X. Wan and Y. Wei, Controlling the diameter of polyaniline nanofibers by adjusting the oxidant redox potential, *Adv. Mater.*, 2007, 19, 465–469.

92 H. J. Ding, Y. Z. Long, J. Y. Shen and M. X. Wan, Fe2 (SO4)3 as a binary oxidant and dopant to thin polyaniline nanowires with high conductivity, *J. Phys. Chem. B.*, 2010, 114, 115–119.

93 P. Dallasa and V. Georgakilas, Interfacial polymerization of conductive polymers: generation of polymeric nanostructures in a 2-D space, *Adv. Colloid Interface Sci.*, 2015, 224, 46–61.

94 C. Yuan, L. Zhang, L. Hou, J. Lin and G. Pang, Green interfacial synthesis of two-dimensional poly(2,5-dimethoxyaniline) nanosheets as a promising electrode for high

performance electrochemical capacitors, *RSC Adv.*, 2014, 4, 24773–24776.

95 H. Ma, Y. Luo, S. Yang, Y. Li, F. Cao and J. Gong, Synthesis of aligned polyaniline belts by interfacial control approach, *J. Phys. Chem. C.*, 2011, 115, 12048–12053.

96 T. Li, Z. Qin, B. Liang, F. Tian, J. Zhao, N. Liu and M. Zhu, Morphology-dependent capacitive properties of three nanostructured polyanilines through interfacial polymerization in various acidic media, *Electrochim. Acta*, 2015, 177, 343–351.

97 J. M. Du, J. L. Zhang, B. X. Han, Z. M. Liu and M. X. Wan, Polyaniline microtubes synthesized via supercritical CO2 and aqueous interfacial polymerization, *Synth. Met.*, 2005, 155, 523–526.

98 J. Huang, S. Virji, B. H. Weiller and R. B. Kaner, Polyaniline nanofibers: facile synthesis and chemical sensors, *J. Am. Chem. Soc.*, 2003, 125, 314–315.

99 Z. M. Huang, Y. Z. Zhang, M. Kotaki and S. Ramakrishna, A review on polymer nanofibers by electrospinning and their applications in nanocomposites, *Compos. Sci. Technol.*, 2003, 63, 2223–2253.

100 I. S. Chronakis, S. Grapenson and A. Jakob, Conductive polypyrrole nanofibers via electrospinning: electrical and morphological properties, *Polymer*, 2006, 47, 1597–1603.

101 J. Choi, J. Lee, J. Choi, D. Jung and S. E. Shim, Electrospun PEDOT:PSS/PVP nanofibers as the chemiresistor in chemical vapour sensing, *Synth. Met.*, 2010, 160, 1415–1421.

102 P. Taranekar, X. Fan and R. Advincula, Distinct surface morphologies of electropolymerized polymethylsiloxane network polypyrrole and comonomer films, *Langmuir*, 2002, 18, 7943–7952.

103 L. S. C. Pingree, O. G. Reid and D. S. Ginger, Electrical scanning probe microscopy on active organic electronic devices, *Adv. Mater.*, 2009, 21, 19–28.

104 Z. Q. Feng, J. Wu, W. Cho, M. K. Leach, E. W. Franzd, Y. I. Naimd, Z.-Z. Gug, J. M. Corey and D. C. Martin, Highly aligned poly (3,4-ethylene dioxythiophene) (PEDOT) nano- and microscale fibers and tubes, *Polymer*, 2013, 54, 702–708.

105 M. Ates, Review study of electrochemical impedance spectroscopy and equivalent electrical circuits of conducting polymers on carbon surfaces, *Progr. Org. Coating.*, 2011, 71, 1–10.

106 S. Pruneanu, S. A. F. Al-Said, L. Dong, T. A. Hollis, M. A. Galindo, N. G. Wright, A. Houton and B. R. Horrocks, Self-assembly of DNA-templated polypyrrole nanowires: spontaneous formation of conductive nanoropes, *Adv. Func. Mater.*, 2008, 18, 2444–2454.

107 J. N. Barisci, R. Stella, G. M. Spinks and G. G. Wallace, Study of the surface potential and photovoltage of conducting polymers using electric force microscopy, *Synth. Met.*, 2001, 124, 407–414.

108 K. Takami, J. Mizuno, M. Akai-kasaya, A. Saito, M. Aono and Y. Kuwahara, Conductivity measurement of polydiacetylene thin films by double-tip scanning tunneling microscopy, *J. Phys. Chem. B.*, 2004, 108, 16353–16356.

109 Y. Martin, D. W. Abraham, and H. L. Wickramasinghe, High-resolution capacitance measurement and potentiometry by force microscopy, *App. Phys. Lett.*, 1998, 52, 1103–1105.

110 O. A. Semenikhin, L. Jiang, T. Iyoda, K. Hashimoto, and A. Fujishima, Atomic force microscopy and kelvin probe force microscopy evidence of local structural inhomogeneity

and nonuniform dopant distribution in conducting polybithiophene, *J. Phys. Chem.*, 1996, 100, 18603–18606.

111 R. J. Tseng, J. Huang, J. Ouyang, R. B. Kaner, and Y. Yang, Polyaniline nanofiber/gold nanoparticle nonvolatile memory, *Nano Lett.*, 2005, 5, 1077–1080.

112 S. Ghosh, L. Ramos, A. Dazzi, A. Deniset-Besseau, S. Remita, P. Beaunier and H. Remita, Conducting polymer nanofibers with controlled diameters synthesized in hexagonal mesophases, *New J. Chem.*, 2015, 39, 8311–8320.

113 J. Liu, B. Wei, J. D. Sloppy, L. Ouyang, C. Ni and D. C. Martin, Direct imaging of the electrochemical deposition of Poly(3,4-ethylenedioxythiophene) by transmission electron microscopy, *ACS Macro Lett.*, 2015, 4, 897–900.

114 Z. Cui, C. Coletta, A. Dazzi, P. Lefrancois, M. Gervais, S. Neron and S. Remita, Radiolytic method as a novel approach for the synthesis of nanostructured conducting polypyrrole, *Langmuir*, 2014, 30, 14086–14094.

115 Z. Cui, C. Coletta, R. Rebois, S. Baiz, M. Gervais, F. Goubard, P.-H. Aubert, A. Dazzi and S. Remita, Radiation-induced reduction -polymerization route for the synthesis of PEDOT conducting polymers, *Rad. Phys. Chem.*, 2016, 119, 157–166.

116 S. Admassie, O. Inganäs, W. Mammob, E. Perzon and M. R. Andersson, Electrochemical and optical studies of the band gaps of alternating polyfluorene copolymers, *Synthetic Met.*, 2006, 156, 614–623.

117 X. Wu, K. Liu, L. Lu, Q. Han, F. Bei, X. Yang, X. Wang, Q. Wu and W. Zhu, In situ monitoring of the nucleation of polyaniline nanoparticles from sodium dodecyl sulfate micelles: a nuclear magnetic resonance study, *J. Phys. Chem. C.*, 2013, 117, 9477–9484.

118 S. J. Higgins, K. V. Lovell, R. M. Gamini Rajapaksea and N. M. Wals, Grafting and electrochemical characterisation of poly-(3,4-ethylenedioxythiophene) films, on Nafion and on radiation-grafted polystyrenesulfonate -polyvinylidene fluoride composite surfaces, *J. Mater. Chem.*, 2003, 13, 2485–2489.

119 E. Barsoukov and J. R. Macdonald, *Impedance Spectroscopy: Theory, Experiment, and Applications*, 2nd Edition, Wiley, Interscience, Hoboken, NJ, 2005.

120 M. D. Levi, Y. Gofer, D. Aurbach and A. Berlin, EIS evidence for charge trapping in n-doped poly-3-(3,4,5-trifluorophenyl) thiophene, *Electrochim. Acta*, 2004, 49, 433–444.

121 M. R. Abidian and D. C. Martin, Multifunctional nanobiomaterials for neural interfaces, *Adv. Funct. Mater.*, 2009, 19, 573–585.

122 D. Jun, J. Xia, S. Yin, H. Xu, M. He, H. Li, L. Xu and Y. Jiang, A g-C3N4/BiOBr visible-light-driven composite: synthesis via a reactable ionic liquid and improved photocatalytic activity, *RSC Adv.*, 2013, 3, 19624–19631.

123 J. Low, J. Yu, Q. Li and B. Cheng, Enhanced visible-light photocatalytic activity of plasmonic Ag and graphene co-modified Bi 2 WO 6 nanosheets, *Phy. Chem. Chem. Phys.*, 2014, 16, 1111–1120.

124 J. R. Macdonald, Impedance spectroscopy: old problems and new developments, *Electrochim. Acta*, 1990, 35, 1483–1492.

125 M. R. Abidian, J. M. Corey, D. R. Kipke and D. C. Martin, Conducting-polymer nanotubes improve electrical properties, mechanical adhesion, neural attachment, and neurite outgrowth of neural electrodes, *Small*, 2010, 6, 421–429.

126 M. R. Abidian and D. C. Martin, Experimental and theoretical characterization of implantable neural microelectrodes modified with conducting polymer nanotubes, *Biomaterials*, 2008, 29, 1273–1283.

127 T. Nakatomo, T. Homma, C. Yamamoto, K. Negishi and O. Omoto, A Long-lasting polyacetylene battery with high energy density, *Jpn. J. Appl. Physics.*, 1983, 22, 275.

128 K. Gurunathan, A. Vadivel Murugan, R. Marimuthu, U. P. Mulik and D. P. Amalnerkar, Electrochemically synthesised conducting polymeric materials for applications towards technology in electronics, optoelectronics and energy storage devices, *Mater. Chem. Phys*, 1999, 61, 173–191.

129 N. Gospodinova and L. Terlemezyan, Conducting polymers prepared by oxidative polymerization: polyaniline, *Prog. Polym. Sci.*,1998, 23, 1443.

130 Z. Song and H. Zhou, Towards sustainable and versatile energy storage devices: an overview of organic electrode materials, *Energy Environ. Sci.*, 2013, 6, 2280–2301.

131 R. desurville, M. Josefowicz, L. T. Yu, J. Perichon and R. Buvet, Electrochemical chains using protolytic organic semiconductors, *Electrochim. Acta*, 1968, 13, 1451.

132 A. Kitani, M. Kaya and K. Sasaki, Performance study of aqueous polyaniline batteries, *J. Electrochem. Soc.*, 1986, 133, 1069.

133 A. G. MacDiamid, S. L. Mu, N. L. D. Somasiri and W. Wu, Electrochemical characteristics of "polyaniline" cathodes and anodes in aqueous electrolytes, *Mol. Cryst. Liq. Cryst.*, 1985, 121, 187.

134 A. Kitani, M. Kaya, Y. Hiromoto and K. Sasaki, Performance study of LI-Polyaniline storage batteries, *Denki Kagaku*, 1985, 53, 592.

135 M. Kaya, A. Kitani and K. Sasaki, EPR studies of the charging process of polyaniline electrodes, *Chem. Lett.*, 1986, 15, 147–150.

136 E. M. Genies, A. A. Syed and C. Tsintavis, Electrochemical study of polyaniline in aqueous and organic medium. Redox and kinetic properties, *Mol. Cryst. Liq. Cryst.*, 1985, 121, 181–186.

137 D.-W. Wang, F. Li, J. Zhao, W. Ren, Z.-G. Chen, J. Tan, Z.-S. Wu, I. Gentle, G. Q. Lu and H.-M. Cheng, Fabrication of graphene/polyaniline composite paper via in situ anodic electropolymerization for high-performance flexible electrode, *ACS Nano*, 2009, 3, 1745.

138 Q. Wu, Y. Xu, Z. Yao, A. Liu and G. Shi, Supercapacitors based on flexible graphene/polyaniline nanofiber composite films, *ACS Nano*, 2010, 4, 1963–1970.

139 K. Zhang, L. L. Zhang, X. S. Zhao and J. Wu, Graphene/polyaniline nanofiber composites as supercapacitor electrodes, *Chem. Mater.*, 2010, 22, 1392–1401.

140 H.-P. Cong, X.-C. Ren, P. Wang and S.-H. Yu, Flexible graphene -polyaniline composite paper for high-performance supercapacitor, *Energy Environ. Sci.*, 2013, 6, 1185–1191.

141 T. Kita, M. Ogawa, Y. Masuda, T. Fuse, H. Daifuka, R. Fujio, T. Kawagoe and T. Matsunaga, Extended abstracts of the electrochemical society (USA), *Electrochem. Soc.*, 1986, 86, 37.

142 M. Morita, K. Komaguchi, H. Tsutsumi and Y. Matsuda, Electrosynthesis of poly(p-phenylene) films and their application to the electrodes of rechargeable batteries, *Elctrochem. Acta.*, 1992, 37, 1093–1099.

143 L. W. Shaklette, R. L. Elsenbanmer, R. R. Chance, J. M. Sowa, D. M. Jvory, G. G. Miller and R. H. Baughman, Electrochemical doping of poly-(p-phenylene) with application to organic batteries, *J. Chem, Soc. Chem. Commun.*, 1982, 0, 361–362.

144 M. Zhou, W. Li, T. Gu, K. Wang, S. Cheng and K. Jiang, A sulfonated polyaniline with high density and high rate Na-storage performances as a flexible organic cathode for sodium ion batteries, *Chem. Commun.*, 2015, 51, 14354–14356.

145 N. S. Hudak, Chloroaluminate-doped conducting polymers as positive electrodes in rechargeable aluminum batteries, *J. Phys. Chem. C.*, 2014, 118, 5203–5215.

146 T. B. Schon, B. T. McAllister, P.-F. Li and D. S. Seferos, The rise of organic electrode materials for energy storage, *Chem. Soc. Rev.*, 2016, 45, 6345–6404.

147 Y. J. Kim, W. Wu, S.-E. Chun, J. F. Whitacre and C. J. Bettinger, Catechol-mediated reversible binding of multivalent cations in eumelanin half-cells, *Adv. Mater.*, 2014, 26, 6572–6579.

148 S. H. Oh, C. W. Lee, D. H. Chun, J. D. Jeon, J. Shim, K. H. Shin and J. H. Yang, A metal-free and all-organic redox flow battery with polythiophene as the electroactive species, *J. Mater. Chem. A.*, 2014, 2, 19994–19998.

149 B. K. Pirnat, T. Bancic, M. Gaberscek, B. Genorio, A. Randon-Vitanova and R. Dominko, Anthraquinone-based polymer as cathode in rechargeable magnesium batteries, *Chem. Sus. Chem.*, 2015, 8, 4128–4132.

150 Z. Song, Y. Qian, T. Zhang, M. Otani and H. Zhou, Poly (benzoquinonyl sulfide) as a high-energy organic cathode for rechargeable Li and Na batteries, *Adv. Sci.*, 2015, 2, 1500124.

151 Y. Liang, Z. Chen, Y. Jing, Y. Rong, A. Facchetti and Y. Yao, Heavily n-dopable π-conjugated redox polymers with ultrafast energy storage capability, *J. Am. Chem. Soc.*, 2015, 137, 4956–4959.

152 N. Mermilliod, J. Tanguy and F. Petiot, A study of chemically synthesized polypyrrole as electrode material for battery applications, *J. EIectrochem. Soc.*, 1986, 133, 1073.

153 R. J. Waltman, A. F. Diaz and J. Bargon, Electroactive properties of polyaromatic molecules, *J. Electrochem. Soc.*, 1984, 131, 1452.

154 Y. Liang, P. Zhang, S. Yang, Z. Tao and J. Chen, Fused heteroaromatic organic compounds for high-power electrodes of rechargeable lithium batteries, *Adv. Energy Mater.*, 2013, 3, 600–605.

155 J. H. Haufman, T. C. Chung, A. J. Heeger and F. Wadl, Poly(Thiophene): a stable polymer cathode material, *J. Electrochem. Soc.*, 1984, 131, 2092–2093.

156 M. Biserni, A. Marinangeli and M. Mastragostino, Doped polydithienothiophene: a new cathode-active material, *J. EIectrochem. Soc.*, 1985, 132, 1597–1601.

157 P. Buttol, M. Mastragostino, S. Panero and B. Scrosati, The electrochemical characteristics of a polydithienothiophene electrode in lithium cells, *Electrochim. Acta*, 1986, 31, 783–788.

5 Polymer-Based Binary Nanocomposites

Pazhanivel Thangavelu and Priyadharshini Matheswaran
Smart Materials Interface Laboratory, Department of Physics
Periyar University
Salem, India

Navaneethan Duraisamy
Department of Chemistry
Periyar University
Salem, India

CONTENTS

5.1 INTRODUCTION

Owing to the rising energy crisis and environmental pollution, the scientific community was compelled to seek clean and renewable energy resources. Since the availability of renewable energy resources is region specific, various efforts are made to develop efficient and cost-effective energy storage technologies. Motivated from this idea, intensive research over the past decade has led to the development of nonconventional energy devices such as batteries, supercapacitors, and fuel cell. Of these, supercapacitors with ultrahigh power density and prolonged cycle life were considered as potential candidates in energy storage systems [1]. Figure 5.1 shows the Ragone plot of various energy storage systems [2]. Although notable development has been achieved in supercapacitors, there is a significant drawback related to their energy density due to shortage of suitable source materials (electrodes) with optimized electrochemical activity [3,4]. There are two different types of energy storage mechanism: electrostatic (electric double-layer capacitor, EDLC) and electrochemical (pseudocapacitors). The EDLC-type capacitors store charge via electrostatic interaction in space charge on the surface of the electrode and diffusive ions of the electrolyte, forming the compact Helmholtz layer (double layer) of thickness ~1 nm. The archetype of EDLC electrode materials is composed of carbonaceous materials such as activated carbon and carbon nanotubes (CNTs), which exhibit rapid charge–discharge process over 100,000 cycles due to high specific surface area, excellent electrical conductivity, controllable and interconnected porosity, and so on [5,6]. The type of charge storage mechanism in pseudocapacitors is rapid, reversible redox reaction over a broad potential, leading to exhibit relatively high energy density than EDLC. In addition, they also provide good lifetime and power density in comparison to batteries. However, there are limitations, for example, low conductivity forbids its use as an electrode material in commercial scale [7]. Substantial efforts such as various designs and facile synthesis procedure are made to achieve high-energy-density devices without significant loss of power density. Arguably, the possible way of enhancing the physiochemical properties is to incorporate transition metals into a matrix of high electrical conductivity [8].

Unlike conventional polymers, conducting polymers (CPs) are a unique class of organic polymers that exhibit

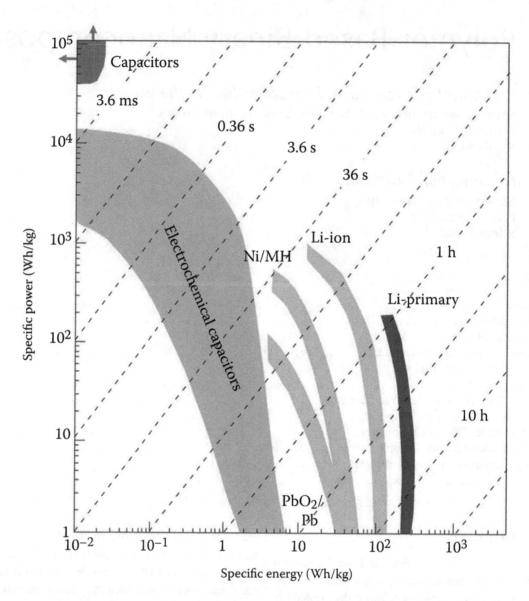

FIGURE 5.1 Ragone plot of various energy storage systems [2]. Reproduced with permission.

high delocalized π-conjugated polymer chain, excellent electrical conductivity, and possess high flexibility. However, segregation between HOMO and LUMO part of the sp² band due to Peierls distortion restricts their property of metallic conductivity [9,10]. To overcome the shortcomings of transition metals and to improve the conducting property of polymers, composite designs were chosen. The choice of this technique surpasses inherent disadvantages and combines with the best properties of individual components in the composite material [11].

5.2 BINARY COMPOSITES

A binary composite is a structural material composed of two phases, namely, a reinforcing phase and a matrix phase. Owing to the synergistic effect of their constituents, composite materials possess more significant advantages,

including high mechanical stability, improved electrochemical behavior, and so on. A CP's nanocomposite is a solid-phase material (metal or ceramic) with a conductive polymer, either structurally or compositionally falling within nanometer range (1–100 nm) at least in one of its phases. For the past two decades, researches have focused on merging nanomaterials with polymers because of their efficient electronic behavior through electronic interactions or structural modifications in the binary components. Also the physiochemical properties of binary composites are greatly influenced by molar ratio, nature of the constituent phase, and the geometrical dimension of the dispersed phase [12].

5.3 NANOSTRUCTURED CPs

The feasibility of CPs as a pseudocapacitive electrode material is much better when compared with the transition metals which facilitate large-scale production, rich theoretic specific capacitance, flexibility, good electrical conductivity, and are relatively cheaper. In contrast to transition metals, CPs store charge via reversible faradic reaction occurring throughout the material instead of only on the surface. To date, the major CPs are (i) polyaniline (PANI), (ii) polypyrrole (PPy), (iii) polythiophene (PTh), (iv) poly(3,4-ethylene dioxythiophene) (PEDOT), and their derivatives [10,12]. The corresponding chemical structures are shown in Figure 5.2.

PANI, the most promising electrode material, exhibits three different states of faradic reactions, namely (i) leucoemeraldine (LE) (fully reduced state), (ii) pernigraniline (PE) (fully oxidized state), and (iii) emeraldine base (EB) (half-oxidized state). Eventually, after protonation, PANI-EB possesses high capacitance, stability, and conductance, whereas both LE and PE are still insulators. In practical, the active material, that is, pure PANI is usually composed of a mixture of these three states. However, to achieve high electrochemical activity, this mixture should have significant contribution of emeraldine state [13]. Usually, PANI is prepared through oxidation of aniline monomer via electrochemical or chemical methods. Electrochemical polymerization, which is an oxidant- and additive-free process, is a rapid and eco-friendly method as compared to the chemical method. This process of synthesis yields binder-free electrodes. The structure of PANI incurred through electrochemical deposit is greatly influenced by the properties of substrates. In comparison with electrochemical polymerization, chemical method of polymerization includes the addition of oxidants and additives. With proper control over the addition of additives and oxidants, various

morphologies such as nanoflakes, nanotubes, and nanoflowers can be obtained [14]. A new design of PANI nanowires through thin polymerization over Cr/Au slices wrapped in polyethylene terephthalate thin layers opens a way to the construction of mini-supercapacitors. Wang et al. synthesized PANI nanowire array and assessed its redox performance in the KOH-based asymmetric capacitor and found out that the device exhibits a productive capacity of 950 F/g using the entire volume for storing charges. [15]. Probably, the one-dimensional nanowire array provides simple ion diffusion, resulting in high utilization of multi-redox sites throughout the volume and rapid doping and de-doping process. Similarly, Sivakumar and co-workers prepared PANI nanowires through interfacial polymerization, which exhibits a specific capacitance of 554 F/g at a constant current density of 1 A/g in aqueous supercapacitor [16]. Thus, the electrochemical operation of PANI is greatly influenced by morphological features.

On the other hand, researchers also analyzed the performance through computational and observational values in acidic medium. The experimental values calculated by various approaches were relatively smaller than the maximum theoretical specific capacitance of PANI (i.e., 2000 F/g) [17]. The lower electrochemical specific capacitance of PANI indicates that the contribution of effective PANI to capacitance was very less. The degree of effective PANI depends on both solvated counter ions and the conductivity of PANI. Hence, the electrochemical properties, especially the cycling performance of pure PANI, are greatly affected, leading to rapid capacitance fading.

PPy, electronic CP, is considered as a promising pseudocapacitive material for supercapacitor electrodes due to its

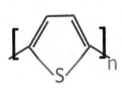

Polyaniline (PANI)

Polypyrrole (PPy)

Polythiophene (PTh)

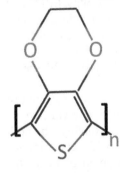

Poly (3,4 ethylenedioxythiophene) (PEDOT)

FIGURE 5.2 Chemical structures of conductive polymers.

ease of synthesis procedure, better conductivity, relatively high cycling stability, low cost, and eco-friendly nature. Further to enhance the electrochemical performance, a vast area of research has been carried out by varying several parameters such as synthesis procedure, template, and substrate [18,19]. PPy nanowire developed by template-free electrochemical polymerization strategy demonstrates that polymerization time plays a crucial role in deciding the geometrical structures and shows that nanowires in the range of 80–100 nm exhibit a maximal capacitance of 566 F/g and retain initial capacitance of 70% after 100 cycles [20]. Above phenomena conveys the fact that by merely optimizing the experimental conditions such as the amount of dopant, template, and substrate it is possible to yield efficient PPy-based electrodes to meet real-world application requirements. Yu et al. successfully applied PPy hydrogels for highly flexible devices, where the active material is synthesized via organic/aqueous biphasic interfacial procedure [21]. The as-fabricated symmetric device demonstrates not only high rate cycling performance but also negligible capacitance (380 F/g) loss at high bending state. Using phytic acid as a dopant, Rajesh et al. synthesized PPy thin film by the electropolymerization method [22]. The as-prepared PPy film exhibits a remarkable capacitance of 343 F/g at a scan rate of 5 mV/s, as well as the initial capacity reaction of 91% after around 4000 cycles at a current density of 10 A/g.

Similarly, interconnected porous mats of PPy nanofibers doped with two-dimensional copper phthalocyanine-$3,4^{1},4^{11},4^{111}$ Tera sulfonic acid tetra-sodium salt-(CuPcTs) [23] reduces the charge transfer resistance by inter-chain charge transport and chain alignment, and exhibits capacitance of 400 F/g at 0.2 A/g versus granular PPy of 232 F/g. The film-shaped intrinsical PPy electrode prepared through oxidation has reactive self-degradable template (methyl orange-$FeCl_3$) with nanotubes of length ranging from 5 to 6 μm with a diameter of about 50–60 nm [24], which also exhibits superior capacitance of 576 F/g with 82% retention over constant charging and discharging for 1000 charge–discharge cycles at 3 A/g.

From the above discussion, it is evident that nanostructured PPy is a promising electrode. Owing to its inherent advantages and also benefiting from its flexibility, PPy-based supercapacitors can be made wearable. As a complementary strategy, it should also possess excellent cycling stability, but severe structural pulverization and counter ion drain effect declined its stability, thereby limiting the use of PPy-based electrodes for practical applications.

PTh and its derivatives have attracted a huge interest in the area of supercapacitor application owing to its simple geometric modification, high stability in both states (doped/undoped), and good electrical conductivity. Various synthesis procedures were adopted to mend the electrochemical operation of PTh-based active material. Some of them are as follows. The film-shaped PTh synthesized via chemical bath deposition (CBD) exhibits

maximal efficiency of about 300 F/g for 5 mV/s in 0.0001 mm of $LiClO_4$/propylene carbonate [25]. Similarly, by using amorphous $FeCl_3$ as an oxidation agent, amorphous thin films of PTh were prepared via successive ionic layer adsorption and recorded an output of 252 F/g in 0.0001 mm $LiClO_4$ [26]. Ambadc et al. demonstrate solid-state symmetric flexible pseudocapacitor built by electropolymerization of PTh on to Titania wire [27]. The fabricated device shows a high efficiency of 1.35731 F/g with excellent cyclic retention for over 3000 constant charging and discharging processes retaining 97% of its initial capacitance. Pristine PTh as the active electrode, prepared via the chemical method by Lafargue et al., in a three-electrode system shows a capacitance of 40 mAh/g [28]. Using the as-prepared PTh electrode as a cathode, the constructed device exhibits a specific capacity of 260 F/g in comparison with a bare carbon electrode. Electrochemical analysis of an un-substituted PTh ultrathin film over active carbon electrode coated via oxidative chemical vapor deposition method shows a 50% of high specific capacitance and maintains 90% of capacity retention even after 5000 cycles. Gnanakan et al. analyzed the effect of the dopant through the synthesis of PTh and dihydroxy butanedioic acid-doped PTh through a cation-based surface active agent aided dilute polymerization [29]. The pure PTh shows 134 F/g, whereas the value of supercapacitor-doped PTh reached up to 156 F/g.

PEDOT, a derivative of PTh, has also gained attention due to its intrinsic characteristics such as rapid charge–discharge, wide potential window and its chemical stability, which is higher than PPy and PANI. Pandey et al. fabricated a solid-state symmetric supercapacitor using PEDOT films prepared through the pulse electrodeposited onto a flexible graphite sheet, which shows a specific capacitance of 57.5 F/g [30]. Similarly, PEDOT paper-based flexible supercapacitors were designed, in which active electrodes were prepared by polymerizing PEDOT on a flexible cellulose paper by interfacial polymerization technique. The device delivered an enhanced specific capacitance in a wide potential range from −1.0 to 1.0 V (versus SCE), with promising potential as cathode and anode [31]. Thus, the morphological and physiochemical attributes of PTh-dependent cells dramatically depend on several factors such as PTh's morphology, substrate, synthetic conditions, and dopant. However, recent studies regarding optimization of these parameters notably improved its specific capacitance and cycling stability. However, still PTh-based active electrodes could not compete PANI- and PPy-based electrodes because of the limitations in their characteristics, such as low specific capacitance and a rapid loss in rate performance.

Overall, pure CPs suffer swelling and shrinking by intercalation and de-intercalation of ions leading to degradation of mechanical stability, thereby inducing rapid fading of electrochemical performance. Generally, electrochemical supercapacitors based on conductive polymers mostly degrade within a thousand cycles of cycling. For instance, Sivaraman *et al.* showed that the cycling stability loss of 16.2% was observed for the first 200 cycles for a device designed from poly(3-methylthiophene) CPs [32]. Fonsca *et al.* prepared PMet with a template by electrochemical polymerization of 3-methylthiophene. The report showed that electrochemical supercapacitor fabricated through PMeT/PVDF/PC/ECL and PMet/PVDF as anode and cathode, respectively, in 1 M LiClO₄, shows a decrease of 16% specific capacitance over 150 cycles [33]. PPy could also suffer from a similar issue due to the volumetric variation observed during the doping/depositing mechanism. The loss of initial capacitance of electrochemical supercapacitors device formed with PPy-active electrodes could be as high as ~50% within thousand charge–discharge cycles carried out by providing ~2 mA/cm^2 [34]. Symmetric supercapacitor formed with PANI nanorods synthesized via *in situ* oxidative polymerization in perchloric acid shows an excellent specific capacitance but degrades to 29.5% after 1000 cycles. The major drawback that hinders CPs from the real-world application is cycling stability [35].

5.4 STRATEGIES TO IMPROVE PERFORMANCE

To overcome the disadvantage of CPs (i.e., cycling stability), following ways were adopted.

5.4.1 LOW-DIMENSIONAL CAPACITORS

By rendering short diffusion length, nanostructured capacitors (nanorods, nanofibers, and nanowire) can overcome volumetric changes due to intercalation/deintercalation mechanism, thereby reducing the cycling degradation problems. For instance, ordered nanometric PANI whiskers at 1 M H₂SO₄ characterized in a three-electrode cell shows a capacitance loss of only 5% over 3000 charge–discharge cycles in the applied current of 5 A/g, also yielding high specific capacitance of 900 F/g due to rapid penetration of electrolyte, short diffusion pathway, and minimized power loss [36].

5.4.2 HYBRID CAPACITORS

In case of conductive polymers, the p-doped state shows enhanced stability than the n-doped state. Usually, n-doped materials are used as anodes of asymmetric capacitors, which can be replaced by carbon-based electrodes. For example, a hybrid capacitor with activated carbon as anode, P-doped PTh derivative as cathode showed a stable performance over 10,000 cycles [37]. Also, the obtained results were higher than EDLC-based symmetric capacitors in terms of power density.

5.4.2.1 Hybrid Electrode Material

To realize the typical cases such as high-power density, prolonged cycling stability, and greater energy density, the choice was hybrid electrode materials, the third kind of electrode, as they are efficient to enhance chain structure, improve cycling stability, promote conductivity, and mitigate the mechanical stress. For example, Omar *et al.* show that pure PANI and PANI/copper cobaltite nanocomposite, respectively, show specific capacitance of 240 and 403 F/g at a current density of 40 mA/g. In fact, the fabricated device with activated carbon as an anode and PANI/CuCo₂O₄ as cathode also shows a stable performance improved to 113% in the initial 200 cycles. Then a slight decay to 94% after 3000 cycles was observed [38]. Regarding several methodologies, fabrication of composite electrodes is believed to be an efficient solution for enhancing the physiochemical properties of active electrodes such as metal oxides, metal sulfides, metal nitrides, and carbonaceous materials, with the polymer matrix. The enhanced performance of composite materials due to the integrated and anchored inorganic species in the polymer host matrix synergistically merges the best properties of each component. The optimized porosity, efficient cycling stability, and enhanced electrical conductivity of composites lead to a wide distribution of capacitance values through improved charge kinetics behavior and high charge mobility. Finally, the nature of active electrode based on pure CPs for supercapacitors is not ideal because of its instability and limited capacitance. Thus, the better property is achieved via the synergistic effect of composite materials comprising CPs with another type of materials. In the following sections, the detailed study on the advancement of various sorts of CP- related active materials is studied.

5.5 CP/CARBON-BASED BINARY COMPOSITE

Carbon materials have attracted great attention as a promising active electrode for supercapacitor application due to various outstanding advantages, such as good cycling, thermal stability, high electron conductivity, significant accessible surface area, and low resistivity, which are comparatively better than others; moreover, they have superior e$^-$ mobility. Employing pure carbon for the electrode yields a very inferior efficiency, which forbids its application on various fields. Several studies have shown that composites of carbon with other materials ideally exhibit better electrochemical activity. Researchers have paid great attention in CP/carbon composites, as it would show superior supercapacitor performance with improved mechanical properties. In CP-based supercapacitors, carbon materials are used as fillers. As carbon has several allotropes when composited with CPs, it exhibits different physiochemical properties. Hence carbon allotropes were studied, for example, CP with N-GO (nitrogen-doped graphene oxides), CNTs, carbon nanofibers, reduced graphene oxide, and carbon spheres and particles [39]. PTh/multiwalled CNT nanocomposite fabricated through electron-assisted chemical polymerization in solvents shows better capacitance than pure PTh and MWCNT individually [40]. The specific capacitance reaches up to 110 F/g at a scan rate of 60 mV/s and a supercapacitor based on such composite possesses 90% of its initial capacitance after 1000 cycles. Some other illustrations were shown by Wang and co-workers [41], who looked into the capacitance characteristics of conductive polymers by coating them on soft and rounded carbon (SRc) via electrochemical deposition for developing CPs/SRc composites. Such composite electrode (PTh/SRc) prepared achieves a specific capacitance of ≈719 F/g at current density 500 mA/g because of the high, accessible amount of area of a material per unit of mass of SRc. Also, after 1000 consecutive cycles, it retains 79% of its starting value for the given input of 10^3 mA/g. Alabadi et al. synthesized graphite [(thiophene-2,5-diyl)-co-para-chloro benzylidene] (GTCB) and graphene oxide-(thio-phene-2,5-diyl)-co-(benzylidene) (GOTB). The GOTB composite electrode offers less equivalent series resistance and improved efficiency of 296 F/g [42]. PTh/ multi-walled CNTs prepared by electron-assisted chemical polymerization exhibits the synergistic effect of composites in exhibiting a 216 F/g capacitance at an input current of 1 A/g with prominent ionic conductivity. Moreover, the electrochemical performance of CPs, especially cycling stability can be improved by utilizing CNT as additives or templates [43]. Also, CP/carbon-based composites combined rapid charge transfer stability of carbon with redox nature of CPs; thus, through this synergistic effect improved power and energy density of supercapacitors were achieved. During the synthesis of PPy/CNT composite, PPy was uniformly coated on air-plasma-activated CNT. A greater electrochemical activity for PPy/plasma-activated CNT in comparison with PPy/ CNT composite was observed. It was because of the conjugated structures formed from the active sites of PPy nitrogen groups, which actively enhance the interaction between PPy and CNT [44]. In another case, PPy/SWCNT-founded hybrid capacitor electrode shows capacitance of 256 F/g, which is comparatively higher than SWCNT and pure PPy individually. The enhancement of capacitance in a composite electrode is attributed to the uniform coating of PPy on SWCNT through which active sites of PPy were improved [45]. For the first time, Zhou et al. analyzed the long and short CNTs at PSS CNTs/PPy nanocomposite. They explained that the PPy/PSS composite with long CNT exhibits relatively superior capacitance and cycling behavior because of core-shell architecture, superior conducting network through the interconnection of abundant CNT, hierarchical porous structure, and its high surface area achieved via nanometer size [46]. Through one pot frigidity procedure, tubular PANI/multi-walled CNT nanocomposites were synthesized by Imani and Farzi. They explained that the tubular structure of composite on a large scale was possible by adopting a low-temperature oxidative polymerization procedure. When the content of MWCNT was 10%, the

sample showed a greater capacitance of 552.11 F/g higher than pristine PANI (411.52 F/g) [47]. In addition to CNT, graphene-based carbon also had a great role in the field of supercapacitors due to high thermal stability and excellent electrical conductivity. Xu et al. grown PANI on GO sheets and explained the dependency of growth on the concentration of monomer (aniline). The scheme was illustrated in Figure 5.3. As the concentration of aniline monomer is high (>0.06 M), aniline tends to nucleate in solution, whereas in low concentration it was dispersedly grown on layered GO sheets [48]. Using a growth mechanism, the as-prepared samples were optimized. Enhanced electrochemical performance of high capacitance (555 F/g) and better cyclability of 92% over 2000 cycles were observed for flocculent PANI/GO composite because of the interactive result within layered GO and PANI. Similarly, the optimized PANI/N-GO composites prepared via chemical precipitation method by Gomez et al. also achieved good pseudocapacitive characteristics [49]. Cong et al. initially synthesized a free-standing nitrogen-doped graphene paper followed by electropolymerization of PANI nanorods on the paper. As the prepared paper exhibits fast electrical pathways, good flexibility with low weight, it was directly used as working electrode. On account of interactive interaction among N-GO and PANI, an enhancement in capacitance (763 F/g) is attained without a noticeable loss in initial capacitance of 82% over 1000 cycles.[50]. On the other hand, apart from CNT and graphene, other carbon microstructures were also used to form composites with CPs; freestanding active electrodes were fabricated by Chau et al. using PANI and porous carbon nanofibers. Benefiting from the pseudocapacitive property of PANI, an excellent capacitance of 366 F/g at a scan rate of 0.1 v/s was attained. Nowadays, researchers also consider hollow spheres and carbon particles as ideal fillers of CPs [51]. Similar to the previous discussion, freestanding PANI along with acid-treated carbon particles was prepared by the results obtained from Khosrozadeh et al. The electrochemical analysis shows a maximal capacitance of 272.6 F/g at a given input of 0.63 A/g [52]. Shen and co-workers synthesized nano-PANI/hollow carbon spheres through one-pot polymerization and analyzed the pseudocapacitance properties. Tests resulted in exhibiting a rich capacitance of 435 F/g and had capacity retention of 60% from initial capacitance after 2000 charge–discharge cycles [53].

In summary, carbon materials could be promising candidates to be used as modifiers for CPs, especially to improve conductivity and cycling stability. Among various microstructures, freestanding 3D structures possess unique benefits, but they are difficult to prepare through the facile method. In addition, phenomenon like effectively depositing CPs on carbonaceous elements also plays a major role in deciding the electrochemical attributes of the prepared products. Hence optimization of several parameters has been carried out, which resulted in significant improvements, such as high specific capacitance and cycling. Solely

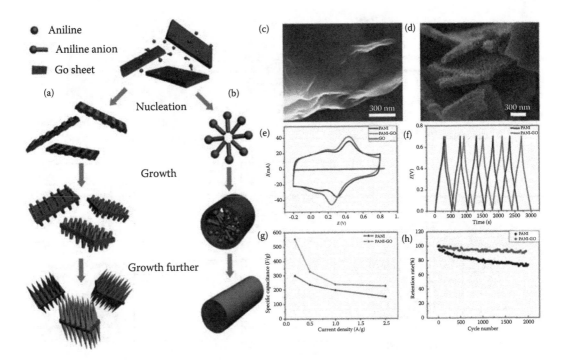

FIGURE 5.3 Illustration shows growth mechanism of low concentrated (a) and high concentrated (b) PANI in solution. SEM images of (c) GO and (d) PANI/GO reacted for 24 h. The electrochemical performance of PANI and PANI/GO (e–h) [48]. Reproduced with permission.

the energy and power density of such materials still lag behind lithium-ion-based batteries due to the limited accumulation of charges. This forbids it to meet real-world applications. To further promote the properties of CPs, it is suggested to fabricate CPs/metal oxides composites, since they possess high theoretical capacitance.

5.6   CP/METAL OXIDES BINARY COMPOSITES

Owing to the electrochemical faradic reactions (within the appropriate potential window), metal oxides, especially transition metal oxides, provide high energy density for electrochemical capacitors when compared to the carbonaceous stuff [54]. These ideal pseudoactive electrodes possess a rich theoretical capacitance, for example, 3000 F/g is achieved for Co_3O_4. The capacitance obtained for cobalt oxide through electrochemical tests is lower than its theoretical value, which may be ascribed to the low electron mobility and limited ion diffusion caused by poor conductivity. Nowadays, researchers investigate composites comprising two electroactive materials, for instance, CPs and metal oxides, which efficiently improves physiochemical performance of active electrodes. Roles of polymers toward enhancing physiochemical properties of metal oxides are (1) by steric and electrostatic stabilization mechanisms, where the polymers prevent aggregation of nanoparticles; (2) metal oxide nanoparticles were homogeneously distributed on the polymer matrix; (3) the accessible specific

surface area of metal oxides was improved because of the polymers; (4) to deliver the ions from electrolyte to oxide metals, a facile pathway was provided; and (5) adhesion of metal oxide nanoparticles with current collector was enhanced because of polymers. Chan *et al.* synthesized silane-coupled PANI nanorods/MnO_2 and PANI/MnO_2 nanocomposites. To improve reciprocity of manganese oxide and PANI, surface of metal oxide was altered through silane-matching reagent, compared to MnO_2/PANI nanorods, the silane-coupled MnO_2/PANI nanocomposites found a maximal capacitance of 415 F/g [55]. Through interchange process, Zhang *et al.* [56] prepared PANI with *n*-octadecyltrimethylammonium-intercalated manganese oxide and examined it in a three-electrode system. Although PANI was inserted in the layered MnO_2, it still shows good conductivity. The as-prepared complex presents 330 F/g at an input current of 1 A/g. The higher efficiency compared to pure PANI (187 F/g) may be because of synergetic consequence within polyaniline and MnO_2. Electrochemically prepared film-shaped conductive polymer composite (PPy, PANI, PEDOT with copper oxide) on GCE in sulfuric acid solution was studied in a three-electrode system to understand its electrochemical performance. In comparison with PEDOT/CuO (198.89 F/g) and PPy/CuO (20.78 F/g), the capacitance of PANI/CuO was higher (198.89 F/g) and was proved to be an ideal electrode of supercapacitor application [57]. A nanocomposite of PANI/nickel cobalt oxide prepared

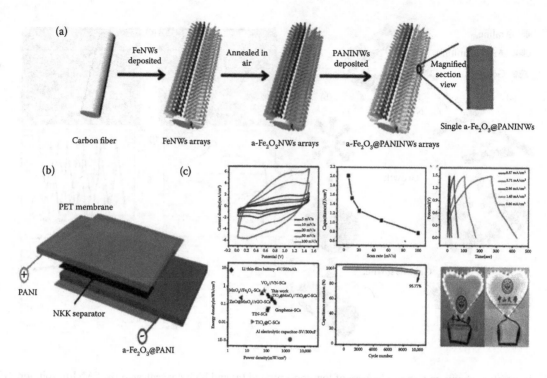

FIGURE 5.4 Schematic diagram of (a) α-Fe$_2$O$_3$@PANI core-shell nanowire structure; (b) flexible supercapacitor separated by NKK membrane, (c) summary plot of electrochemical activity of asymmetric supercapacitor device [59]. Reproduced with permission.

via one-pot chemical-assisted polymerization by Xu *et al.* showed an efficiency of 439.4 F/g also retained approximately 66% of its beginning capacitance after 1000 cycles at a given input of 5 mA/cm^2 [58]. Usually, core-shell structure nanowire array is synthesized through a two-step process, for example, consider Fe$_2$O$_3$/PANI. Initially, Fe$_2$O$_3$ nanowire array was developed on carbon fabric maintained at 500°C for 2 hours, which was followed by galvanostatic polymerization of PANI carried out on the surface of the α-Fe$_2$O$_3$ array on cloth. From Figure 5.4, it is ascribed that as-synthesized composite achieves a uniform coating of PANI on α-Fe$_2$O$_3$ nanowire array, thus paves rapid ion/electron mobility, large surface area, and high capacity retention [59]. Ji *et al.* developed MnO$_2$/PPy nanotube composite and depicted a prominent operation. The capacitance reaches up to 402 F/g at charge–discharge and retains the efficiency well of 88% of its beginning capacitance over 800 continuous charge–discharge cycles. Authors claim that the long-run cycling retention is possible because of the coated MnO$_2$ on PPy, which restricted its structural change of PPy while cycling [60]. Some other typical illustration was contributed by Qian and co-workers [61]. The as-prepared CuO core was covered by a shell of PPy via a direct electrochemical deposition on the conductive substrate. Finally, fabricated supercapacitor using CuO/PPy as active material exhibits outstanding performance of 1274 F/cm^3 and cycling stability of almost 100% at a given current of 250 mA/cm^3 over 3000 charge–

discharge cycles. Also, it has a desirable energy density of 28.35 m Wh/cm^3. A novel 3D PPy/CuO hybrid composite of nanowire array described by Zhou *et al.* was made of nickel foam, the binder-free active material shows excellent pseudocapacitive functioning such as rich capacitance of 2223 F/g at 1 mA/cm^2, good rate performance, and high capacity retention [62]. Aqueous asymmetric supercapacitor fabricated using the composite as cathode delivered high energy and power density of 43.5 Wh/kg and 5500 W/kg at 11.8 Wh/kg, respectively. Zhou *et al.* claimed that smart synergistic effect between CuO nanowires with PPy CP was the reason for such high performance of the materials. A composite of CP with metal oxide need not to be a cathode, but it could also be an active anode material. For example, Wang *et al.* prepared a core/shell-shaped CuO$_3$/PPy composite and characterized its electrochemical performance. As a negative electrode of hybrid supercapacitor, it operates in the potential window of −1.0 to 0.00 V and renders a maximum capacitance of 253 mF/cm^3. In this work, the author fabricated the device using CuO$_3$/PPy and Co(OH)$_2$ grown on carbon fibers as anode and cathode, respectively [63]. Owing to the prominent electrochemical performance of individual electrodes, the device showed an energy density and volumetric capacitance of about 1.02 m Wh/cm^3 and 2.865 F/cm^3, respectively. Sun *et al.* prepared vanadium pentoxide/PPy nanocomposite by an articulated two-pace procedure. The obtained results revealed that the composite comprehends the individual property of its constituents along with the synergistic effect in the hybrid active material [64]. Moreover, RuO_2/PEDOT nanotube composite exhibited a specific capacitance of 1217 F/g. When examined into a two-electrode system, the composite shows power and energy density of 20 kW/kg and 28 Wh/kg, respectively [65].

5.7 CP/METAL SULFIDES BINARY COMPLEXES

In recent decade, metal sulfides, multitudinous pseudocapacitive materials analogous to graphene have been extensively studied for a wide range of applications due to their intrinsic physiochemical properties. When compared with other electrode materials such as metal oxides and various carbon derivatives, metal sulfides are cheap and are found in abundance in nature. Besides, metal sulfides are better metallic conductors and they also undergo faradic redox reaction among various reduction/oxidation states of the metal ion. Owing to novel structural architectures and outstanding theoretical capacitance, few metal sulfides such as MoS_2, NiS, SnS, and CoS are being studied extensively. Despite their great advantages due to their peculiar layered architecture, they suffer short cycling stability, in which the capacitance fades rapidly. Therefore, it is of significance to formulate a potential way of raising their stability without sacrificing their capacitance. Hierarchically structured material is usually composited with CPs, as it buffers the volume changes while charging and discharging mechanisms. Thus, as an effective strategy, metal sulfides were hybridized with CPs. The raised redox attributes of metal sulfide/CPs complexes may be ascribed to the robust structure and synergistic effect between their constituents. For example, Long *et al.* prepared Ni_3S_2/PPy composite supported on Ni foam by a two-step eloquent synthesis procedure, such as hydrothermal, followed by simple electrochemical deposition [66]. In comparison with Ni_3S_2/NF (1.26 F/cm^2) electrode, Ni_3S_2@PPy/NF electrode shows a slight decrement in areal capacitance, yet it possesses a great power density and cycling stability. Moreover, a fabricated device using Ni_3S_2@PPy/NF as anode shows a 17.54 Wh/kg A and 179.33W kg/A energy and power density, respectively, at a constant current of 2.5 mA/cm^2 with a stable capacitance over 3000 charges–discharge cycles. With PANI, it was reported by Huang *et al.* that a PANI/MoS_2 composite synthesized via facile *in situ* polymerization method yielded 575 F/g at 1 A/g. The fabricated device exhibits an energy and power density of 265 Wh kg/A and 18.0 kW/kgA, respectively [67]. Authors claim that the high electrochemical performance of the supercapacitor was due to the positive synergistic effect of PANI along MoS_2. Sturdy and strong independent filmy MoS_2/PEDOT:PSS was developed by Ge *et al.* via a two-step procedure achieves efficiency of 141.4 F/cm^3, which is more prominent to its constituents MoS_2 (59.9 F/cm^3) and PEDOT:PSS (44.8 F/cm^3) and cycling stability of 98.6%

over 5000 cycles [68]. Authors explain that including PEDOT:PSS midway, the layered MoS_2 sheets improve charge mobility with low transfer resistance. An advanced electrode of PPy/molybdenum disulfide (MoS_2) nanocomposite was synthesized via facile one-pot oxidation polymerization through Ma and co-workers [69]. The electrochemical investigation of PPy-embedded MoS_2 nanosheet shows rich capacitance of 553.7 F/g, whose capacitance retention ranges 90% over 500 charge–discharge cycles at a current input of 1 A/g. This may be due to the narrowed thickness of PPy on MoS_2 by void space among sheet-like subunits and novel structure of MoS_2, which buffer the volumetric changes during galvanostatic cyclic process. Similarly, Hui Peng constructed a high-performance supercapacitor using CuS microspheres @PPy developed by *in situ* oxidative polymerization [70]. This experiment demonstrates a high redox performance say excellent specific capacitance of 427 F/g and long-run cycling retention of 88% throughout 1000 charge–discharge cycles. This is because of rapid electronic mobility of CuS due to its shortened ion-diffusive channel in spherical porous morphology. Using *in situ* oxidative polymerization, Ren *et al.* prepared a 10–20-nm PANI nanowire array on tubular MoS_2. The schematic illustration of the obtained tubular structure is shown in Figure 5(e). The capacitive performance of the active materials was optimized by controlled growth of nanowire by varying the amount of PANI in the nanocomposite. PANI, with a loading amount of 60%, exhibits a superior electrochemical performance (55 F/g and 82% of capacity retention at 0.5 A/g current density). Moreover, the fabricated symmetric device also shows an eminent capacitance of 124 F/g at 1 A/g plus cycling stability of 79% over 6000 cycles. The author claims that the 3D structure not only provides facile ion diffusivity through controlled porosity, but also maximizes the active surface area. Also, the synergistic effect between nanocomposites plays a significant role in charge mobility. Storage and the design withstand volumetric changes during charging and discharging [71]. A high conductive PANI/MoS_2 nanocomposite was synthesized by Wang et al. [72] through direct intercalation of aniline monomer and also by the addition of dodecyl sulfonic acid as an impurity. The researchers showed that in comparison with pure PANI electrode (131 F/g with 42% after 600 cycles), the composite MoS_2/PANI-38 gives enhanced electrochemical performance (390 F/g with 86% after 1000 cycles). Authors explained that high electrochemical activity of MoS_2/PANI-38 composite when compared with MoS_2/PANI-8, MoS_2/PANI-24 wt% was due to the high content of MoS_2 redox material. They also claimed that a better effect on the electrochemical behavior of the well-defined composite was because of the synergistic action between the CP and MoS_2-layered compound. Fu *et al.* prepared a three-dimensional spongia S-MoS_2/PANI complex by letting in polyvinyl pyrrolidone via facile hydrothermal method. When compared with C-MoS_2/PANI (composite without the addition of PVP) S-MoS_2

/PANI reaches a capacitance of 605 F/g at a constant current density of 1 A/g [73]. Hence, the obtained energy and power densities are 53.78 Wh/kg and 0.4 kW/kg, respectively. The remarkable electrochemical property of the composite was attributed toward the following phenomenon: (i) the structural degradation such as collapse and agglomeration while continuous cycle was restrained by the introduction of A-MoS$_2$ nanospheres, (ii) accessibility of electrolyte ions to the active site was enhanced by A-MoS$_2$. Due to the mobility of electrons and ions through short channels and adequate spaces, an intense electro conductibility between S-PANI and H$_2$SO$_4$ electrolyte was observed, and (iii) rapid electrical conductivity and extended interfacial areas provided by spongia-shaped loose morphology of the material benefit high redox performance. Electrochemical performance of petal-like NiS along PEDOT:PSS for high-performance hybrid capacitor was acquired by Rao et al.

via simple CBD method. The unique behavior of NiS/PEDOT:PSS with diethylene glycol facilitates fast electron and ion transferability, higher active sites, and good morphological stability. The specific capacitance reaches up to 750.6 F/g at a current density of 1.11 A/g. It also possesses good energy (24.52 Wh/kg) and power (138.88 W/kg) density [74]. Similar to the layer-structured graphene, a layered SnS$_2$ facilitates facile lithium-ion intercalation and deintercalation. Its composition with PANI nanoplates renders the shortest ion-diffusive pathway, high conducting channels with its adjacent nanoplates, which restrict volumetric contraction and expansion during the galvanostatic cyclic process. In some other cases, it also acts as a positive electrode of alkaline battery. In such cases, it gives reversible capacitance of 968.7 mAh/g along with cycling stability of 75.4% for 80 cycles. A facile strategy for the mass production of PANI/MoS$_2$ was proposed by Zhu et al. The

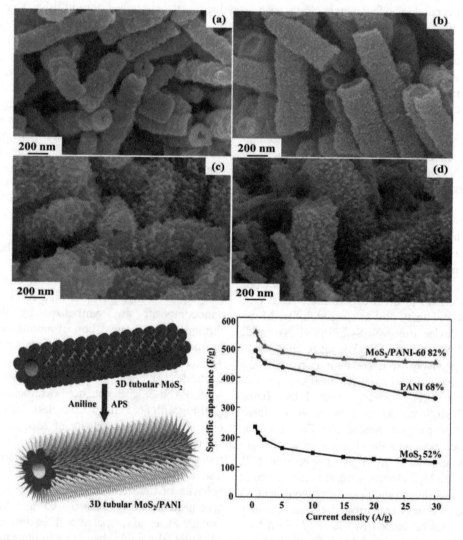

FIGURE 5.5   FESEM images of MoS$_2$/PANI hybrid materials with different amounts of PANI: (a) 35%, (b) 46%, (c) 60%, and (d) 73%. Schematic illustration of formation mechanism and the electrochemical performance of the fabricated device (e and f) [71]. Reproduced with permission.

as-prepared active material as an electrode exhibits an enhanced specific performance of 669 F/g in the potential window of ±0.6 V at 1 A/g. When the potential window varies to ±0.8 and ±1.0 V, the corresponding capacitance variations of 821 and 853 F/g, respectively, are obtained [75]. Double electrochemically active materials because of unique construction favor high interaction among electrode/electrolyte interface and shortened diffusive pathway. The high initial capacity retention of 91% was retained over 4000 cycles at a fixed charge–discharge cycles carried out by a constant supply of 10 A/g in the potential window of ±0.6 V.

5.8 OTHER CP-SUPPORTED BINARY COMPLEXES

Besides CP/C, CP/metal oxide, and CP/metal sulfide materials, there are several other CP-related active materials, which were reported recently that demonstrate considerably good redox reactions. Some of them were discussed here, Lee *et al.* [76] reported that $Mn_3(PO_4)_2$ nanoparticle-adorned PANI was prepared via a two-step procedure (i.e., sonochemical method followed by calcination). By optimizing the size of particles by varying the sonication time, the prepared metal phosphate is combined with pre-synthesized PANI to form the PANI–metal phosphate complex (schematically represented in Figure 5.6). Electrochemical investigation in aqueous electrolyte shows notably its mended electrochemical behavior. Moreover, the fabricated hybrid capacitor with activated charcoal as anode attained a maximum energy density of nearly 14 Wh/kg and a power density of 378 W/kg. It demonstrates the high electrical properties such as high electrochemical performance that attributes excellent specific capacity of 427 F/g plus long-term cycling stability of 88% over 1000 cycles. Transition metal nitrates were usually synthesized by the

nitridation process, i.e., the process of calcination was carried out under ammonia gas environment. In comparison with TiO_2, the prepared TiN exhibits enhanced electrochemical performance via efficient conductivity of TiN nanowire with TiO_2. Successive surfacing of carbon and PANI over titanium nitride paves core-shell-nanosized wire-shaped morphology. (i.e., PANI/carbon/TiN). The as-synthesized PANI/carbon/TiN, an array of nanowires, shows a rich capacitance of 1093 F/g and also retains 98% of its initial capacity over 2000 charging and discharging cycles [77]. Significant contribution of nanocomposite constituents given here was (i) nanowire array morphology rendered failures in electron and ion transfer pathways. (ii) Because of rapid doping and resulting faradic reaction, a large pseudocapacitance was contributed by PANI. (iii) Although the obtained specific capacitance from PANI/TiN is comparable, PANI/C/TiN composites without the center carbon shell PANI/TiN degrades rapidly. Enhanced stability of PANI/C/TiN was achieved via carbon shell, which effectively protects electrode material (TiN) from electrolyte corrosion. Recently, MOF, one of the major electrode materials, indicated the extraordinary potential of its application in energy storage devices. Therefore, numerous MOF-based electrochemical capacitors have been studied. To overcome the limiting factors of MOF-based electrodes, such as its less conductivity incorporation of CPs, would be an effective strategy. CPs along with MOF crystals provide additional redox capacitance through interconnected electron transportation channels that link isolated MOF. The fabrication of UIO-66/PPy hybrid involves two steps, initially electropolymerization followed by the incorporation of polymer in the potential compound [78]. The combined behavior of EDLC via porous UIO-66 and electronic conductivity of PPy improves electrochemical performance with high capacity retention over its rapid life. By *in situ* polymerization strategy, a cabbage-like

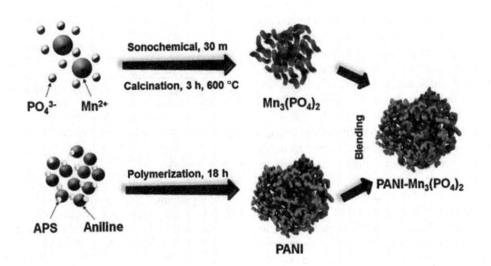

FIGURE 5.6 Schematic illustration for the formation of PANI Mn₃ (PO₄)₂ composite.

nanocomposite (PANI-hydroquinone) shows a good efficiency of 126.0 F/g at a scan rate of 0.005 V/s also its capacity retention was 85.1% over 500 cycles at 1 A/g. It is claimed that as a redox compound, hydroquinone was used and the conductive PANI provided facile electron conductive pathways. Zhan *et al.* synthesized a one-dimensional fiber, that is, PANI-decorated $Ni(OH)_2$ through a two-step facile hydrothermal method, which reveals phase change of $Ni(OH)_2$ from beta to gamma crystalline phase; this was further confirmed by the morphological transformation from nanoplate to 3D nanoflower. Owing to high surface area, flower-like nanocomposite showed good electrochemical properties, namely specific capacity of 55.50 C/g and capacity holding of 79.49% after 2500 charging and discharging cycles at an input of 1.5 mA/cm^2 [79]. Compact and high-whippy supercapacitors based on PPy/modified nanocellulose fibers display a gravimetric capacitance that reaches up to 127 F/g at 300 mA/cm^2. Also, the designed hybrid capacitor bridged the gap between conventional batteries and capacitors by nearly rendering 3 W/cm^3 (power density) and 3.1 mWh/cm^3 (energy density), respectively [80].

5.9 CONCLUSION

In summary, a study on CPs and its hybrids provide basic ideas of morphology and composition versus property relationship for its potential application in various fields. Here, we summarized various key findings that lead future research on pure and hybrid CP composites for energy storage systems. To date, a vast count of findings and advances has been adopted for the development of CP-based promising candidate for supercapacitors, because CPs offer scalable, cost-effective, highly stable, and excellent flexibility electrodes with unique pseudocapacitance process. However, desirable properties and functions of CPs were still limited for real-world applications. To enhance the physiochemical properties of CPs such as processability, stability, and metallic conductivity, it is important to improve their crystallinity, control over porosity, and morphology via optimized parameters, which include the amount of dopant and surfactant, oxidation level, type of surfactant, a method of polymerization and so on. To address this concern, recently, CP-based nanocomposites (along with carbon/inorganic transition metals) were considered, which comprise interesting properties of their original materials and ensure enhanced energy density and stability over charge–discharge cycles. Furthermore, to achieve improved behavior in hybrid materials, it is significant to disperse the organic/inorganic nanomaterials homogeneously in the given polymer matrix, which favors better interfacial pathways. Also, to decrease the resistance due to the combined area of various phases, interfacial polymerization method would exist as a promising channel to design homogeneous dispersion of materials in CP matrix with a regular arrangement of polymer chains. Also, better performance is possible by optimizing the proportion between constituents of composites.

However, microstructure and method of synthesis also have a vital part in influencing the operation of the active materials. On the whole, optimization of experimental parameters is a significant factor that serves as a significant parameter in enhancing the functioning of supercapacitors.

REFERENCES

[1] Yu A, Davies A, Chen Z. Electrochemical supercapacitors. *Electrochem Technol Energy Storage Convers* 2012; 1:317–82. doi:10.1002/9783527639496.ch8.

[2] Gogotsi Y, Simon P. Materials for electrochemical capacitors. *Nat Mater* 2008; 7:845–54.

[3] Simon P, Gogotsi Y, Dunn B. Where do batteries end and supercapacitors begin? *Science* 2014; 343:1210–1. doi:10.1126/science.1249625.

[4] Conway EB. Electrochemical supercapacitors scientific fundamentals and technological applications. *Adv Lithium-Ion Batter* 1999; 481–505. doi:10.1007/0-306-47508-1_17.

[5] Zhang LL, Zhao XS. Carbon-based materials as supercapacitor electrodes. *Chem Soc Rev* 2009; 38:2520–31. doi:10.1039/b813846j.

[6] Zeng S, Chen H, Cai F, Kang Y, Chen M, Li Q. Electrochemical fabrication of carbon nanotube/polyaniline hydrogel film for all-solid-state flexible supercapacitor with high areal capacitance. *J Mater Chem A* 2015; 3:23864–70. doi:10.1039/c5ta05937b.

[7] Conway BE, Birss V, Wojtowicz J. The role and utilization of pseudocapacitance for energy storage by supercapacitors. *J Power Sources* 1997; 66:1–14. doi:10.1016/S0378-7753(96)02474-3.

[8] Augustyn V, Simon P, Dunn B. Pseudocapacitive oxide materials for high-rate electrochemical energy storage. *Energy Environ Sci* 2014; 7:1597–614. doi:10.1039/c3ee44164d.

[9] Adjizian, J-J, Briddon P, Humbert B, Duvail J-L, Wagner P, Adda C, Ewels C. Dirac Cones in two-dimensional conjugated polymer networks. *Nat Commun* 2014; 5:5842. doi:10.1038/ncomms6842.

[10] Snook GA, Kao P, Best AS. Conducting-polymer-based supercapacitor devices and electrodes. *J Power Sources* 2011; 196:1–12. doi:10.1016/j.jpowsour.2010.06.084.

[11] Kalaji M, Murphy PJ, Williams GO. The study of conducting polymers for use as redox supercapacitors. *Synth Met* 1999; 102:1360–1. doi:10.1016/S0379-6779(98)01334-4.

[12] Yang J, Liu Y, Liu S, Li L, Zhang C, Liu T. Conducting polymer composites: Material synthesis and applications in electrochemical capacitive energy storage. *Mater Chem Front* 2017; 1:251–68. doi:10.1039/c6qm00150e.

[13] Bhadra S, Khastgir D, Singha NK, Lee JH. Progress in preparation, processing and applications of polyaniline. *Prog Polym Sci* 2009; 34:783–810. doi:10.1016/j.progpolymsci.2009.04.003.

[14] Silva CHB, Galiote NA, Huguenin F, Teixeira-Neto É, Constantino VRL, Temperini MLA. Spectroscopic, morphological and electrochromic characterization of layer-by-layer hybrid films of polyaniline and hexaniobate nanoscrolls. *J Mater Chem* 2012; 22:14052. doi:10.1039/c2jm31531a.

[15] Wang K, Huang J, Wei Z. Conducting polyaniline nanowire arrays for high performance supercapacitors. *J Phys Chem C* 2010; 114:8062–7. doi:10.1021/jp9113255.

[16] Sivakkumar SR, Kim WJ, Choi JA, MacFarlane DR, Forsyth M, Kim DW. Electrochemical performance of polyaniline nanofibres and polyaniline/multi-walled carbon nanotube composite as an electrode material for aqueous redox supercapacitors. *J Power Sources* 2007; 171:1062–8. doi:10.1016/j.jpowsour.2007.05.103.

[17] Li H, Wang J, Chu Q, Wang Z, Zhang F, Wang S. Theoretical and experimental specific capacitance of polyaniline in sulfuric acid. *J Power Sources* 2009; 190:578–86. doi:10.1016/j.jpowsour.2009.01.052.

[18] Sabouraud G, Sadki S, Brodie N. The mechanisms of pyrrole electropolymerization. *Chem Soc Rev* 2000; 29:283–93. doi:10.1039/a807124a.

[19] Zang J, Bao SJ, Li CM, Bian H, Cui X, Bao Q, et al. Well-aligned cone-shaped nanostructure of polypyrrole/RuO_2 and its electrochemical supercapacitor. *J Phys Chem C* 2008; 112:14843–7. doi:10.1021/jp8049558.

[20] Huang J, Wang K, Wei Z. Conducting polymer nanowire arrays with enhanced electrochemical performance. *J Mater Chem* 2010; 20:1117–21. doi:10.1039/b919928d.

[21] Yu P, Zhao X, Huang Z, Li Y, Zhang Q. Free-standing three-dimensional graphene and polyaniline nanowire arrays hybrid foams for high-performance flexible and lightweight supercapacitors. *J Mater Chem A* 2014; 2:14413–20. doi:10.1039/c4ta02721c.

[22] Rajesh M, Raj CJ, Kim BC, Cho BB, Ko JM, Yu KH. Supercapacitive studies on electropolymerized natural organic phosphate doped polypyrrole thin films. *Electrochim Acta* 2016; 220:373–83. doi:10.1016/j.electacta.2016.10.118.

[23] Nair S, Hsiao E, Kim SH. Fabrication of electrically-conducting nonwoven porous mats of polystyrene-polypyrrole core-shell nanofibers via electrospinning and vapor phase polymerization. *J Mater Chem* 2008; 18:5155–61. doi:10.1039/b807007e.

[24] Li M, Yang L. Intrinsic flexible polypyrrole film with excellent electrochemical performance. *J Mater Sci Mater Electron* 2015; 26:4875–9. doi:10.1007/s10854-015-2996-1.

[25] Patil BH, Patil SJ, Lokhande CD. Electrochemical characterization of chemically synthesized polythiophene thin films: performance of asymmetric supercapacitor device. *Electroanalysis* 2014; 26:2023–32. doi:10.1002/elan.201400284.

[26] Patil BH, Jagadale AD, Lokhande CD. Synthesis of polythiophene thin films by simple successive ionic layer adsorption and reaction (SILAR) method for supercapacitor application. *Synth Met* 2012; 162:1400–5. doi:10.1016/j.synthmet.2012.05.023.

[27] Ambade RB, Ambade SB, Salunkhe RR, Malgras V, Jin SH, Yamauchi Y, et al. Flexible-wire shaped all-solid-state supercapacitors based on facile electropolymerization of polythiophene with ultra-high energy density. *J Mater Chem A* 2016; 4:7406–15. doi:10.1039/c6ta00683c.

[28] Laforgue A, Simon P, Sarrazin C, Fauvarque JF. Polythiophene-based supercapacitors. *J Power Sources* 1999; 80:142–8. doi:10.1016/S0378-7753(98)00258-4.

[29] Gnanakan SRP, Murugananthem N, Subramania A. Organic acid doped polythiophene nanoparticles as electrode material for redox supercapacitors. *Polym Adv Technol* 2011; 22:788–93. doi:10.1002/pat.1578.

[30] Pandey GP, Rastogi AC. Poly(3,4-Ethylenedioxythiophene)-graphene composite electrodes for solid-state supercapacitors with ionic liquid gel polymer electrolyte. *ECS Trans* 2013; 45:173–81. doi:10.1149/04529.0173ecst.

[31] Anothumakkool B, Soni R, Bhange SN, Kurungot S. Novel scalable synthesis of highly conducting and robust PEDOT paper for a high performance flexible solid supercapacitor. *Energy Environ Sci* 2015; 8:1339–47. doi:10.1039/c5ee00142k.

[32] Sivaraman P, Thakur A, Kushwaha RK, Ratna D, Samui AB. Poly(3-methyl thiophene)-activated carbon hybrid supercapacitor based on gel polymer electrolyte. *Electrochem Solid-State Lett* 2006; 9:A435. doi:10.1149/1.2213357.

[33] Fonseca CP, Benedetti JE, Neves S. Poly(3-methyl thiophene)/PVDF composite as an electrode for supercapacitors. *J Power Sources* 2006; 158:789–94. doi:10.1016/j.jpowsour.2005.08.050.

[34] Sharma RK, Rastogi AC, Desu SB. Manganese oxide embedded polypyrrole nanocomposites for electrochemical supercapacitor. *Electrochim Acta* 2008; 53:7690–5. doi:10.1016/j.electacta.2008.04.028.

[35] Zhu ZZ, Wang GC, Sun MQ, Li XW, Li CZ. Fabrication and electrochemical characterization of polyaniline nanorods modified with sulfonated carbon nanotubes for supercapacitor applications. *Electrochim Acta* 2011; 56:1366–72. doi:10.1016/j.electacta.2010.10.070.

[36] Wang YG, Li HQ, Xia YY. Ordered whiskerlike polyaniline grown on the surface of mesoporous carbon and its electrochemical capacitance performance. *Adv Mater* 2006; 18:2619–23. doi:10.1002/adma.200600445.

[37] Di Fabio A, Giorgi A, Mastragostino M, Soavi F. Carbon-poly(3-methylthiophene) hybrid supercapacitors. *J Electrochem Soc* 2001; 148:A845. doi:10.1149/1.1380254.

[38] Omar FS, Numan A, Duraisamy N, Bashir S, Ramesh K, Ramesh S. A promising binary nanocomposite of zinc cobaltite intercalated with polyaniline for supercapacitor and hydrazine sensor. *J Alloys Compd* 2017; 716:96–105. doi:10.1016/j.jallcom.2017.05.039.

[39] Frackowiak E. Carbon materials for supercapacitor application. *Phys Chem Chem Phys* 2007; 9:1774. doi:10.1039/b618139m.

[40] Fu C, Zhou H, Liu R, Huang Z, Chen J, Kuang Y. Supercapacitor based on electropolymerized polythiophene and multi-walled carbon nanotubes composites. *Mater Chem Phys* 2012; 132:596–600. doi:10.1016/j.matchemphys.2011.11.074.

[41] Wang Y, Tao S, An Y, Wu S, Meng C. Bio-inspired high-performance electrochemical supercapacitors based on conducting polymer modified coral-like monolithic carbon. *J Mater Chem A* 2013; 1:8876–87. doi:10.1039/c3ta11348e.

[42] Alabadi A, Razzaque S, Dong Z, Wang W, Tan B. Graphene oxide-polythiophene derivative hybrid nanosheet for enhancing the performance of supercapacitor. *J Power Sources* 2016; 306:241–7. doi:10.1016/j.jpowsour.2015.12.028.

[43] Zhang H, Hu Z, Li M, Hu L, Jiao S. A high-performance supercapacitor based on a polythiophene/multiwalled carbon nanotube composite by electropolymerization in an ionic liquid microemulsion. *J Mater Chem A* 2014; 2:17024–30. doi:10.1039/c4ta03369h.

[44] Yang L, Shi Z, Yang W. Polypyrrole directly bonded to air-plasma activated carbon nanotube as electrode materials for high-performance supercapacitor. *Electrochim Acta* 2015; 153:76–82. doi:10.1016/j.electacta.2014.11.146.

[45] An Hyeok K, Jeon KK, Heo KJ, Lim Chu S, Bae Jae D, Lee Hee Y. High-capacitance supercapacitor using a nanocomposite electrode of single-walled carbon nanotube and polypyrrole. *J Electrochem Soc* 2002; 149. doi:10.1149/1.1491235.

[46] Zhou H, Han G, Xiao Y, Chang Y, Zhai HJ. A comparative study on long and short carbon nanotubes-incorporated polypyrrole/poly(sodium 4-styrenesulfonate) nanocomposites as high-performance supercapacitor electrodes. *Synth Met* 2015; 209:405–11. doi:10.1016/j.synthmet.2015.08.014.

[47] Imani A, Farzi G. Facile route for multi-walled carbon nanotube coating with polyaniline: Tubular morphology nanocomposites for supercapacitor applications. *J Mater Sci Mater Electron* 2015; 26:7438–44. doi:10.1007/s10854-015-3377-5.

[48] Xu J, Wang K, Zu SZ, Han BH, Wei Z. Hierarchical nanocomposites of polyaniline nanowire arrays on graphene oxide sheets with synergistic effect for energy storage. *ACS Nano* 2010; 4:5019–26. doi:10.1021/nn1006539.

[49] Gómez H, Ram MK, Alvi F, Villalba P, Stefanakos E, Kumar A. Graphene-conducting polymer nanocomposite as novel electrode for supercapacitors. *J Power Sources* 2011; 196:4102–8. doi:10.1016/j.jpowsour.2010.11.002.

[50] Cong HP, Ren XC, Wang P, Yu SH. Flexible graphene-polyaniline composite paper for high-performance supercapacitor. *Energy Environ Sci* 2013; 6:1185–91. doi:10.1039/c2ee24203f.

[51] Tran C, Singhal R, Lawrence D, Kalra V. Polyaniline-coated freestanding porous carbon nanofibers as efficient hybrid electrodes for supercapacitors. *J Power Sources* 2015; 293:373–9. doi:10.1016/j.jpowsour.2015.05.054.

[52] Khosrozadeh A, Xing M, Wang Q. A high-capacitance solid-state supercapacitor based on the free-standing film of polyaniline and carbon particles. *Appl Energy* 2015; 153:87–93. doi:10.1016/j.apenergy.2014.08.046.

[53] Shen K, Ran F, Zhang X, Liu C, Wang N, Niu X, et al. Supercapacitor electrodes based on nano-polyaniline deposited on hollow carbon spheres derived from cross-linked co-polymers. *Synth Met* 2015; 209:369–76. doi:10.1016/j.synthmet.2015.08.012.

[54] Holze R. Metal oxide/conducting polymer hybrids for application in supercapacitors. *Met Oxides Supercapacitors* 2017; 219–45. doi:10.1016/B978-0-12-810464-4.00009-7.

[55] Chen L, Sun LJ, Luan F, Liang Y, Li Y, Liu XX. Synthesis and pseudocapacitive studies of composite films of polyaniline and manganese oxide nanoparticles. *J Power Sources* 2010; 195:3742–7. doi:10.1016/j.jpowsour.2009.12.036.

[56] Zhang X, Ji L, Zhang S, Yang W. Synthesis of a novel polyaniline-intercalated layered manganese oxide nanocomposite as electrode material for electrochemical capacitor. *J Power Sources* 2007; 173:1017–23. doi:10.1016/j.jpowsour.2007.08.083.

[57] Ates M, Serin MA, Ekmen I, Ertas YN. Supercapacitor behaviors of polyaniline/CuO, polypyrrole/CuO and PEDOT/CuO nanocomposites. *Polym Bull* 2015; 72:2573–89. doi:10.1007/s00289-015-1422-4.

[58] Xu H, Wu JX, Chen Y, Zhang JL, Zhang BQ. Facile synthesis of polyaniline/NiCo2O4 nanocomposites with enhanced electrochemical properties for supercapacitors. *Ionics (Kiel)* 2015; 21:2615–22. doi:10.1007/s11581-015-1441-z.

[59] Li GR, Lu X-F, Chen X-Y, Zhou W, Tong Y-X. α-Fe2O3@PANI core-shell nanowire arrays as negative electrodes for asymmetric supercapacitors. *ACS Appl Mater Interfaces* 2015; 7:14843–50. doi:10.1021/acsami.5b03126.

[60] Ji J, Zhang X, Liu J, Peng L, Chen C, Huang Z, et al. Assembly of polypyrrole nanotube@MnO2 composites with an improved electrochemical capacitance. *Mater Sci Eng B* 2015; 198:51–6. doi:10.1016/j.mseb.2015.04.004.

[61] Qian T, Zhou J, Xu N, Yang T, Shen X, Liu X, et al. On-chip supercapacitors with ultrahigh volumetric performance based on electrochemically co-deposited CuO/polypyrrole nanosheet arrays. *Nanotechnology* 2015; 26. doi:10.1088/0957-4484/26/42/425402.

[62] Zhou C, Zhang Y, Li Y, Liu J. Construction of high-capacitance 3D CoO@Polypyrrole nanowire array electrode for aqueous asymmetric supercapacitor. *Nano Lett* 2013; 13:2078–85. doi:10.1021/nl400378j.

[63] Wang F, Zhan X, Cheng Z, Wang Z, Wang Q, Xu K, et al. Tungsten oxide@polypyrrole core-shell nanowire arrays as novel negative electrodes for asymmetric supercapacitors. *Small* 2015; 11:749–55. doi:10.1002/smll. 201402340.

[64] Sun X, Li Q, Mao Y. Understanding the influence of polypyrrole coating over V2O5 nanofibers on electrochemical properties. *Electrochim Acta* 2015; 174. doi:10.1016/j.electacta.2015.06.026.

[65] Liu R, Duay J, Lane T, Bok Lee S. Synthesis and characterization of RuO2/poly(3,4- ethylene dioxythiophene) composite nanotubes for supercapacitors. *Phys Chem Chem Phys* 2010; 12:4309–16. doi:10.1039/b918589p.

[66] Long L, Yao Y, Yan M, Wang H, Zhang G, Kong M, et al. Ni3S2@polypyrrole composite supported on nickel foam with improved rate capability and cycling durability for asymmetric supercapacitor device applications. *J Mater Sci* 2017; 52:3642–56. doi:10.1007/s10853-016-0529-9.

[67] Huang K-J, Wang L, Liu Y-J, Wang H-B, Liu Y-M, Wang L-L. Synthesis of polyaniline/2-dimensional graphene analog MoS2 composites for high-performance supercapacitor. *Electrochim Acta* 2013; 109:587–94. doi:10.1016/j.electacta.2013.07.168.

[68] Ge Y, Jalili R, Wang C, Zheng T, Chao Y, Wallace GG. A robust free-standing MoS2/poly(3,4-ethylenedioxythiophene): poly(styrenesulfonate) film for supercapacitor applications. *Electrochim Acta* 2017; 235:348–55. doi:10.1016/j.electacta.2017.03.069.

[69] Ma G, Peng H, Mu J, Huang H, Zhou X, Lei Z. In situ intercalative polymerization of pyrrole in graphene analog of MoS2 as advanced electrode material in supercapacitor. *J Power Sources* 2013; 229:72–8. doi:10.1016/j.jpowsour.2012.11.088.

[70] Lei Z, Wang H, Mu J, Peng H, Ma G, Sun K. High-performance supercapacitor based on multi-structural CuS@polypyrrole composites prepared by in situ oxidative polymerizations. *J Mater Chem A* 2014; 2:3303–7. doi:10.1158/0008-5472.CAN-05-1818.

[71] Ren L, Zhang G, Yan Z, Kang L, Xu H, Shi F, et al. Three-dimensional tubular MoS2/PANI hybrid electrode for high rate performance supercapacitor. *ACS Appl Mater Interfaces* 2015; 7:28294–302. doi:10.1021/acsami.5b08474.

[72] Wang J, Wu Z, Hu K, Chen X, Yin H. High conductivity graphene-like MoS2/polyaniline nanocomposites and its application in supercapacitor. *J Alloys Compd* 2015; 619:38–43. doi:10.1016/j.jallcom.2014.09.008.

[73] Fu G, Ma L, Gan M, Zhang X, Jin M, Lei Y, et al. Fabrication of 3D Spongia-shaped polyaniline/MoS2 nanospheres

composite assisted by polyvinylpyrrolidone (PVP) for high-performance supercapacitors. *Synth Met* 2017; 224:36–45. doi:10.1016/j.synthmet.2016.12.022.

[74] Roa S, Punnoose D, Bae JH, Durga IK, Thulasi-Varma CV, Naresh B, et al. Preparation and electrochemical performances of NiS with PEDOT:PSS chrysanthemum petal like nanostructure for high performance supercapacitors. *Electrochim Acta* 2017; 254:269–79. doi:10.1016/j.electacta.2017.09.134.

[75] Zhu J, Sun W, Yang D, Zhang Y, Hoon HH, Zhang H, et al. Multifunctional architectures constructing of PANI nanonee-dle arrays on MoS2 thin nanosheets for high-energy supercapacitors. *Small* 2015; 11:4123–9. doi:10.1002/smll.201403744.

[76] Lee CC, Omar FS, Numan A, Duraisamy N, Ramesh K, Ramesh S. An enhanced performance of hybrid supercapacitor based on polyaniline-manganese phosphate binary composite. *J Solid State Electrochem* 2017; 21:3205–13. doi:10.1007/s10008-017-3624-1.

[77] Xie Y, Xia C, Du H, Wang W. Enhanced electrochemical performance of polyaniline/carbon/titanium nitride nano-wire array for flexible supercapacitor. *J Power Sources* 2015; 286:561–70. doi:10.1016/j.jpowsour. 2015.04.025.

[78] Qi K, Hou R, Zaman S, Qiu Y, Xia BY, Duan H. Construction of metal-organic framework/conductive polymer hybrid for all-solid-state fabric supercapacitor. *ACS Appl Mater Interfaces* 2018; 10:18021–8. doi:10.1021/acsami.8b05802.

[79] Zhang J, Shi L, Liu H, Deng Z, Huang L, Mai W, et al. Utilizing polyaniline to dominate the crystal phase of Ni (OH)$_2$ and its effect on the electrochemical property of polyaniline/Ni(OH)$_2$ composite. *J Alloys Compd* 2015; 651:126–34. doi:10.1016/j.jallcom.2015.08.090.

[80] Wang Z, Carlsson DO, Tammela P, Hua K, Zhang P, Nyholm L, et al. Surface modified nanocellulose fibers yield conducting polymer-based flexible supercapacitors with enhanced capacitances. *ACS Nano* 2015; 9:7563–71. doi:10.1021/acsnano.5b02846.

6 Polyaniline-Based Supercapacitor Applications

Nuzhat Jamil and Sadia Zafar Bajwa
Nanobiotechnology Group National Institute for Biotechnology and Genetic
Engineering (NIBGE)
Faisalabad, Pakistan

Muhammad Yasir
Department of Chemistry
University of Lahore
Lahore, Pakistan

Asma Rehman and Waheed S. Khan
Nanobiotechnology Group National Institute for Biotechnology and Genetic
Engineering (NIBGE)
Faisalabad, Pakistan

CONTENTS

6.1   INTRODUCTION

In the current scenario, constantly increasing demands for energy applications such as transportable electronic, electrically aided transportation, and energy storage due to depleting energy resources and search for green methods to store energy have stimulated the researchers to find alternatives of energy [1]. Transportable electronic industry has been transformed through the development of lithium-ion batteries (LIBs) for the last two decades because of having greater energy density and extended cycling stability. But, requirement of modern era is sustainable power and energy to develop new version of Li^+ ion batteries and alternatives of energy-storing device [2]. Therefore, lithium ion, lithium air, lithium sulfur, and supercapacitors, which are rechargeable, have gained much importance as energy storage alternatives in this current energy-depleting situation. The call for these energy storing devices requires plenty of materials for electrodes, with no harm to environment, and by possessing remarkable capability of energy storage [3].

6.2   POLYANILINE (PANI) AND ITS APPLICATION POTENTIAL

Conducting polymers (CPs) have gained much interest and focus by the researchers for their applications in different fields, owing to their high conductivity and excellent capacitive features. Among the most widely used CPs, PANI attracts great attention because of its highest specific capacitance due to multi-redox reactions, good electronic properties due to protonation, and its cost effective production. These features make PANI an attractive option for multifaceted applications as indicated in Figure 6.1.

6.3   SUPERCAPACITORS

Capacitors having longer cycling stability and highest power density are called supercapacitors. Electrodes with increased surface area and reduced space between charged layers make the capacitor able to store a lot of energy compared with other conventional capacitors. Supercapacitors are categorized as electrostatic double-layer supercapacitor (EDLC) and pseudocapacitor. [4] EDLC stores energy by creating two layers on the electrode and electrolyte interface because of ions adsorption and desorption electrostatically in conducting electrolyte [5]. Cost-effective porous materials of carbon are mostly considered as suitable electrodes for double-layer supercapacitors because of being mechanically and chemically stable and having increased surface area. During charging process, electrostatic double layer is formed because external load makes the electrons move from negative to

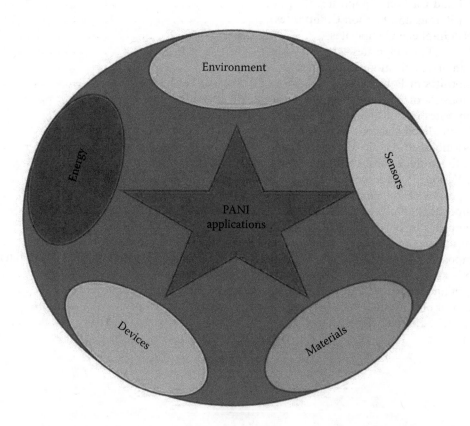

FIGURE 6.1   Application Potential of Polyaniline (PANI) in different areas.

positive electrode while moving cations and anions in electrolyte toward negative and positive electrode, respectively, as shown in Figure 6.2(a). This process is reversed in discharging. Therefore, in EDLC, no transfer of electrons across electrode/electrolyte interface and no exchange of ions between two electrodes occur and thus energy is saved in both capacitor layers [7].

On the other hand, pseudocapacitors work on the principle of storing energy through redox reactions occurring between electrode and electrolyte. Static double-layer capacitance creates pseudocapacitance, and absorption/intercalation of electrons together with quick faradic redox reaction in reversible direction on the surface of electrode causes electron charge transfer as described in Figure 6.2 (b). The adsorbed ions observe no chemical bonding and reaction with atoms of electrode but only transfer of charge takes place [8]. The pseudocapacitors possess higher range of capacitance than electrostatic double-layer capacitors having same surface area, as electrochemical process occurs in bulk and on the surface of electrode, which is solid in nature. On contrary, compared with electrostatic double-layer capacitor, pseudocapacitor exhibits low conductivity and cycling stability, which limits their applications on a wider scale. To deal with these bottlenecks, scaffolds made up of carbon materials are added into electrode materials and hence efficiency can be improved.

Chemical affinity of effective surface of electrodes with adsorbed ions is a crucial factor for performance of pseudocapacitor [9]. This redox potential of electrodes can be improved either by using transition metal oxide, such as RuO_2, IrO_2, V_2O_5, Fe_3O_4, Co_3O_4, MnO_2, NiO, MoS_2, and TiS_2, or CPs. Transition metal oxides have the potential for maximum capacitance due to repeated oxidizing cycles at a specified potential with minimum resistance. Oxide of Ruthenium metal (RuO_2) with aqueous solution of H_2SO_4 as electrolyte offers the outstanding model, as it exhibits charging/discharging capacity over a range of ~1.2 V per electrode. CPs function on the principle of switching between two doping phases while electrolyte ions are being intruded or extruded from polymer's core side for its capacitance enhancement. Entire exposure of polymer chains to charge/discharge phase enhances its capacitance, but causes structural degradation, thus reduces the shelf life [10]. However, carbon supports should be coupled with such CPs in order to cope with such structural damages, that is, hybrid capacitors. Batteries have far better energy density in contrast with supercapacitors such as EDLCs and pseudocapacitors. Scientists have great interest to formulate such capacitors that have high energy/power density and cycling life. In this regard, hybrid capacitors can be chosen as best candidates with relatively large capacitance and energy density as they exhibit far better electrochemical functions.

A number of modifications are being made for improved electrochemical efficiency and durability of rechargeable batteries and supercapacitors, but certain changes are still needed to be done [11]. Less electrochemical performance is mainly caused by small surface area and size of electrodes while loss of active ingredient of manufactured material and its structural instability pose reduced durability. Therefore, there is an increasing trend to explore more and more materials with electrochemical properties, nanocomposites with variable morphologies to overcome the performance deficiencies. In this regard, PANI has emerged as an attractive product and can be used as sole material or consortium with others to achieve good performance [12]. This is manufactured by using a strong oxidant for chemical polymerization of aniline. PANI can persist in leucoemeraldine (LE), emeraldine (EM), and pernigraniline (PE) states referred as entirely reduced, partially oxidized, and entirely oxidized, respectively. This reaction is performed in an acidic environment to formulate EM salt while its conduction capability greatly depends on doping and extent of oxidation. However, PANI is synthesized

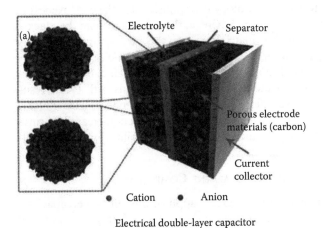
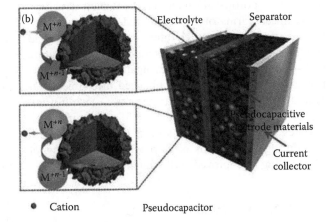

FIGURE 6.2 Schematic diagram of (a) the electrochemical double-layer capacitors and (b) the pseudocapacitors; adapted from ref [6], with permission from The Royal Society of Chemistry.

either based on template-dependent or -independent manner. Main features that are used to be focused while preparation of PANI include simple redox reactions, unique chemical mechanism, easy and convenient synthesis procedure, high electrical conductivity, improved mechanical properties, and enhanced stability of electrodes for Li$^+$ ion batteries and supercapacitors [13].

6.3.1 PANI IN SUPERCAPACITORS

Use of PANI is more profound for applications in supercapacitors because of its easy synthesis, high theoretical pseudocapacitance, and efficient electrical conductance. But certain limitations such as decrease in capacitance at increased charge/discharge rate and reduction in their durability because of long-term cycling pose major hindrance to achieve optimal output from such CPs. Reduced efficiency is caused by its morphology while repeated doping and de-doping phases and enhanced level of oxidation at elevated potential is responsible for their reduced durability. These limitations impose further hindrance at the polymer/electrolyte and electrode/current-collector interfaces [14].

Pure PANI cathodes with no additive material have elevated energy density. However, one-dimensional (1D) to three-dimensional (3D) interconnected nanostructure-based more advanced forms of PANI are being synthesized and practiced as supercapacitors, which have less structural instability and efficient electrochemical performance. This nanostructured PANI is most prominent feature in this regard, as it allows charge transfer through interface of electrode–electrolyte junction, also makes possible the occurrence of such reactions that are not possible in the presence of massive material. Nowadays, these 1D PANI supercapacitors are synthesized on the basis of template–dependent, that is, hollow nanofibers or template-independent nanotube techniques. Their tubular morphology and enhanced surface enable efficient transfer of charges between electrode–electrolyte interface, although a few PANI nanomaterials still represent loss in capacitance [15].

In contrast, porous 3D interconnected nanostructure-based advanced forms of PANI showed excellent capacitance, retention power at elevated current values, and beyond 10,000 cycles. Such 3D intertwined nanostructure network-based upgraded curtails facilitate the uncontrolled electric conductivity along with enhanced mechanical stability, and regulated porosity that permits faster diffusion of ions at higher rates. Moreover, the 3D structures with macropores reduce the stress associated with changes in volume during the periods of doping and de-doping of PANI, resulting in improved cyclic stability [16].

6.3.2 PANI AND CARBON COMPOSITES

Multiple redox reactions make PANI-based electrodes more effective conductors even in pure form with specific

capacitance of 600 F/g in liquid electrolytic state. It was claimed by Wang Kai and co-workers that its capacitance for energy storage can be more improved by using nanowires as parent material, which actually provides assistance in ion diffusion across electrolyte material through rapid doping and de-doping functions. As structural design plays significant contribution in overall performance of PANI, so any structural impurity causes instability in the reaction process [17].

Luckily, the increased suppleness makes it feasible for PANI to unite with other materials very effectively. In this regard, use of carbon in fabrication provides maximum benefit because of its durability, stability, and high surface area. Moreover, it also strengthens parent material during the doping and de-doping reaction [17,18].

6.3.3 PANI/POROUS CARBON COMPOSITES

Basic motive of enhancement of stability of PANI is also achieved through using porous parent material. Such materials offer enhanced chemical stability, high surface area, and convenient processability, which make PANI more feasible in its application. In addition, use of double layers make them more efficient capacitors by reducing specific capacitance.

Activated carbon (AC) nanomaterials have made fabrication of PANI/carbon composites very feasible, which results in high conductivity, enhanced stability, low cost, and high affordability. Such composites are mainly synthesized from different types of carbon-based substances by number of physiochemical conversions and activations according to their commercial application for synthesis of supercapacitors [18]. Such PANI composites of AC can be effortlessly generated either through chemical deposition or electrochemically designed polymerization. The whole reaction is carried out in an environment of carbon powder and aniline mixture and PANI's polymerization is brought about more effectively. *In situ* polymerization will take place with the addition of oxidant material. Morphology is responsible for the effectiveness of these composites where carbon particles can be packed inside PANI matrix or may be struck on the surface depending upon the size of carbon particles, which can be assisted through using various surfactants. Such composites of activated carbonaceous materials with PANI result in very specific capacitance in the range of 200–700 F/g due to conjugation of PANI that offers increased intrinsic pseudocapacitance and enhanced stabilization of ACs [19].

6.3.4 PANI/GRAPHENE COMPOSITES

Graphene offers great applications in supercapacitors because of high surface area (2630 m^2/g), good electronic properties, and excellent thermal stability. Such high surface area helps to increase dispersion of PANI, which improves its utilization in promoting specific capacitance.

Large graphene sheets provide large surface to hold each PANI component, which could also result in additional conductivity of the resultant composite. Specific capacitance of graphene in acidic, alkaline, neutral, organic, and even ionic medium is found to be about 100 F/g. However, conjugation of graphene with PANI promotes the capacitance up to 1046 F/g. [20] Graphene for such energy-storing devices is manufactured by Hummer's method after slight modifications, which is a cost effective and efficient method. PANI is also conjugated with graphene oxide (GO) through electrochemical condensation and *in situ* chemical polymerization. Different nanostructures of such composites are possible including nanofibers that offer the advantage of fast charge transfer, hence high values of specific capacitance can be achieved. The conjugation of GO and PANI, and growth of their composite depends upon the amount of triggering monomer, aniline. When concentration is <0.05 M, PANI favorably disperses over GO; however, at higher concentration of about >0.06 M, there is nucleation is initiated. Such concentration-dependent growth pattern helps to optimize the resulting composite products. The composites of pseudocapacitive PANI and GO-layered sheets present high value specific capacitance, 555 F/g. [21] GO/PANI composites obtained by electrochemical co-deposition method offer high specific capacitance (640 F/g), excellent electrochemical performance, and

stability for long cycling. Such proposed growth mechanism for composites is very helpful to design the required products, as shown in Figure 6.3. Connecting (noncovalent or covalent) mode between PANI and graphene is quite important factor for their composite. Covalent connecting is considered more favorable and strong as it may cause enhanced capacitance and cycle period compared with noncovalent connectivity. Recently, PANI and functional reduced GO (rGO) were covalently connected using nitrophenyl-rGO and amino phenyl-rGO; this was achieved via solvothermal reaction in heating furnace with flowing ammonia gas [21]. The substituent groups direct the shapes of nanostructures and conductivity and capacitance of resulting supercapacitors. Nanowire arrays of vertical PANI on nitrophenyl-substituted rGO depicted greater capacitance, enhanced thermal stability, and longer life cycles. Such nanocomposites are connected by van der Waals' forces. Increased rates of charge transfer and high values of specific capacitance were associated with conjugated systems of PANI and rGO. Now, some modern graphene foam (GF) and graphene papers have been designed to fabricate the PANI-based electrodes [23]. Yao and co-workers stacked PANI on graphite papers to obtain free-standing electrodes with supreme mechanical characteristics [24].

FIGURE 6.3 The schematic of (a) growth mechanism of PANI on the surface of GO and (b) nucleation of PANI in solution. Adapted with permission from ref [22]. Copyright (2010) American Chemical Society.

6.3.5 PANI/CNTs COMPOSITES

Beside other forms of carbon, carbon nanotubes (CNTs) have also attained much attention of researchers in the last decade because of having excellent mechanical, structural, and electronic characteristics. Side-selective interaction occurs in PANI with CNTs having single-walled structure. Functional groups attached to the surface of the single-walled CNTs form specific binding with active sites of PANI. CNTs may also observe polarization after contacting electrode of CNTs with electrolyte, thus reducing capacitance. The low capacitance is because of the low wettability of the electrode made up of CNT, leading to limited use of surface area for storage of charge on double layer of ions in the electrolyte [25]. Incorporation of oxides of transition metals (TMOs) and CPs such as PANI on the surface of CNT can reduce the polarization of CNT and increase the electrochemical properties. CNT's array that is linked with some collector for current, act as support and make PANI and CNT composite material for electrode with great porosity. PANI and CNT that may be single-walled and multi-walled form composites are developed by coating a PANI layer on CNT, thus forming core-shell-like structure. These composite electrodes exhibit remarkably excellent electrochemical properties because of pseudocapacitance of PANI and high conductance of CNT, which are also mechanically stable. This configuration also improves the diffusion of ions. PANI can also form nanofibers and nanowires surrounding the CNTs forming a caterpillar-like arrangement. Chemical polymerization usually needs already chemically treated CNTs to enhance their stable dispersion.

Polymerization of PANI on the surface of single-walled CNTs using grafted poly(4-vinylpyridine) as doping agent, poly(4-vinylpyridine) makes covalent bonds on the surface of single-walled CNTs in aqueous solution, thus grafted. The modification of CNTs can avoid the self-aggregation of CNTs by not harming the conductance and structures, which results in enhanced capacitance and longer cycling stability. Fabrication of graphene pyrrole/CNT-PANI hybrids in which GF is used as support has been reported recently. These graphene pyrrole aerogels have been prepared though hydrothermal process. In PANI–graphene–CNTs arrangement, graphene causes the increased utilization of PANI, CNTs serve as activating wires that link the nanosheets of graphene together leading to large conductivity and PANI takes part in increasing capacitance owing to its pseudocapacitance. PANI coated on graphene nanosheets surface and on the CNTs during *in situ* conditions of polymerization has been reported, which suggests that quantity of graphene and CNT can be increased for thin layer of PANI [26].

Carbonaceous nanofibers can be good alternative of CNTs because of having highest conductance, flexibility, being chemically and mechanically stable, and for interesting 3D arrangement. During polymerization of PANI electrochemically, they can act as collectors for current

and PANI nanoparticles are uniformly layered on the three-dimensionally organized carbon nanofibers. The resultant electrode is flexible and able to manage original shape with greater angle bending.

Electrochemically etched and commercially available carbon nanofiber cloths acting as substrate to polymerize the PANI have been reported. After etching of carbon nanofibers, they observe increased surface area and smooth diffusion of electrolyte ions. Therefore, PANI nanofibers uniformly layered on carbon nanofibers after etching electrochemically exhibit enhanced electrochemical properties as compared to without etching of carbon nanofibers. They also show good electrochemical stability. Development of carbon nanofibers with various organic precursors has been reported. Carbonized filter paper acting as substrate to develop PANI and carbon nanofiber composites and polyacrylonitrile, which is soluble in DMF acting as current collector of carbon nanofibers in one-step carbonizing method have been reported [27].

6.3.6 POLYANILINE ACTIVATION/CARBONIZATION

PANI is advantageous in a sense that it can act as a precursor for carbon for the development of carbon materials doped with nitrogen for double-layer capacitors. The porous materials of carbon are obtained with various morphologies such as nanorods, nanotubes, nanowires, and nanofibers, which remain same irrespective of being carbonizing and activating before. Carbonization of PANI and then activation with potassium hydroxide show remarkably increased surface area with small pores distribution. It shows better electrochemical properties with enhanced capacitance in KOH solution. Incorporation of graphene into PANI causes increase in capacitance retention after activating the carbon [28]. Therefore, carbon obtained from PANI carbon precursor after activation is a promising material in various electrochemical supercapacitor applications. The process of activation causes enhancement in the surface area of carbon material matrix due to unclogging of the closed pores by releasing the gas and forming new pores. Activation of PANI at raised temperature by adopting ammonia as activation gas has been reported. After the heat treatment, a number of functional groups of nitrogen present on the surface are maintained in pyrrolic and pyridinic states and surface area gets increased, thus used as good option in pseudocapacitors. NH_3 is a good activation gas but some others, including H_3PO_4, KOH, $ZnCl_2$, are also used commonly as activating additives for activating carbonaceous materials [29].

6.3.7 COMPOSITES OF POLYANILINE WITH VARIOUS CONDUCTIVE POLYMER BLENDS

PANI is much capable of developing co-polymer hybrids with various other CPs. They combine to show excellent electrochemical properties as compared to the separate components individually, especially in terms of cycling stability. Although they are less efficient than PANI and graphene composites and PANI with CNT composites, yet they attract great interest only because of easy synthesis, cost effectiveness, and greater yield. Various doping agents are used for the fabrication of PANI-based co-polymer hybrids such as paraphenylenediamine (PPD), citric acid, melamine, and camphor sulfonic acid. PPD possesses the ability to modify the electrochemical and structural properties of electrodes based on PANI. PANI nanofibers are long and have smooth structure in the presence of doping material PPD as compared to the PANI nanofibers without PPD. This copolymer also exhibit enhanced capacitance and rate capability as compared to material with only PANI [30].

PANI, when doped with melamine, exhibited increased cycling stability in comparison to PANI without doping. With increasing melamine content, cycling stability and specific capacitance can also be enhanced as compared to un-doped PANI. Although their electrochemical properties are not comparable to PANI with graphene and PANI with CNT composites but still they catch attention due to easy synthesizing methodologies and cost effectiveness [31].

6.3.8 Composites of Polyaniline with Transition Metal Oxides

Organic and inorganic hybrids that are based on CP PANI have attained much interest these days because of their combined cooperative effect. Apart from carbonaceous material, oxides of metals also have tendency to develop organic and inorganic composite materials of PANI for electrodes. RuO_2 possesses high conductance, which makes it an excellent material for supercapacitor but presence in low quantity and being highly priced limits its usage commercially. Currently, low-priced TMOs like iron, manganese, vanadium, nickel, tungsten, nickel, molybdenum, and titanium oxides are being used for electrodes in supercapacitors. But they all have low conductance and require coating layer of some conducting material. Therefore, PANI is a good option for this purpose. In this regard, mostly manganese oxide and manganese dioxide are used [32].

6.3.9 Composites of Polyaniline Core-shells with Metal Oxides

TMOs are very promising because of having good specific capacitance and being chemically stable but electronically poor due to slow charge transfer with lower rate capacity. PANI has good conductance and serves as a connection layer coating by giving protection to TMOs, thus improving conductance, capacitance, and stability. The PANI layer provides porous structure with no hindrance to ions of electrolyte to diffuse into the inner TMOs, which induce high capacitance. Fe_2O_3 nanowire arrays produced on carbon cloth in the presence of electrolyte having iron

salt and ammonium oxalate with good stability, quick ion transfer, and with greater surface area have been reported. Core-shell PANI with TiO_2 or PANI with TiN nanowire arrays in supercapacitors has also been reported. Both of these nanomaterials are very good for supercapacitors having high pseudocapacitance due to their greater surface area and electronic characteristics but are unstable because of reversible redox reactions occurring in aqueous medium [33]. The carbon materials and CPs can solve this problem by improving cycling stability. The combination of TiO_2 with PANI nanowires, which are encapsulated in anodized titania nanotube, with excellent cycling stability as compared to only PANI nanowire or PANI and inorganic composites has been reported. TiO_2 possesses larger surface area and acts as buffer against various changes during doping and de-doping process of PANI and make it electrically and mechanically stable. Core/shell nanowire alignments such as PANI/C/TiN-NWA having remarkable cycling stability, have also been reported.

Spinal ferrites are also very important mixed metal oxide materials because of being electrochemically stable and having various redox states. Pure ferrites are not good conductor but this problem can be solved by coating them with carbon and CPs. A number of ferrites such as nickel, manganese, and cobalt ferrite with carbon and PANI make hybrids that show good capacitive performance because of various redox states and being electrochemically stable. Spinal ferrites only with PANI do not show good conduction, cycling stability, and specific capacitance but blending of these two with carbon improves their properties because of synergistic effect. Therefore, electrodes made up of spinal ferrites hybrid materials are good option to improve performance.

Metal-organic frameworks (MOFs) are newly discovered materials with immense pores which can store energy and can be used in various applications of conversions because of having greater surface area, high porosity to facilitate diffusion of ions such as Co MOF (ZIF-67) and Zn MOFs. There is only problem of low conductivity, which can be solved with incorporation of CPs such as PANI [34].

Metal oxide and hydroxide synthesis requires alkaline medium, whereas polymerization of PANI needs acidic medium for good products. One step co-deposition strategy can still be applicable for some metal oxides and final composites cannot be soluble in electrolyte. An important example of electrically deposited PANI and MnO_2 has been reported recently. Oxides of tungsten and vanadium possess pseudocapacitive properties in the range of negative potential and are much attractive in research. Metal oxides are embedded in PANI matrix and exhibit enhanced stability and conductivity.

6.3.10 PANI–modified Cathode Materials

Lithium (Li)-abundant cathode materials are used to fabricate PANI modification electrodes for LIBs. $LiFePO_4$

(LFP) is a very good cathode material for electrode being thermally stable with good cycling stability but poor conductivity. Coating of CP such as PANI can enhance the performance. The type of inorganic acid doping agents decides the conductivity of the polymer. The cathode made up of C-LFP/PANI composite and doping with hydrochloric acid gives remarkable enhancement in capacitance as compared to the sulfuric acid doping agent. This composite material for cathode has two problems, which are low conduction and less diffusion of Li ions. PANI possesses good conductivity and polyethylene glycol is an excellent solvent for Li salts, hence PANI and polyethylene glycol co-polymer, after coating the cathode of the above-mentioned composite, can enhance conductivity and ionic diffusion [35].

Vanadium oxide is very important third-generation material for manufacturing of cathode. Layered lithium vanadium oxide, in LIBs, has gained much importance because of having good discharging capability with energy density and being chemically stable and low priced value. This coating of polymer decreases resistance for transfer of charge by linking the composite nanorods together. Monometallic materials for cathode have been already used but trimetallic materials can also be used by coating with PANI and improving the efficiency. CP like PANI can raise the conductivity, improve the stability, and serve as buffer for cathode material to soluble in electrolyte irrespective of their being mono, bi, or trimetallic nature.

Composite materials enriched with lithium possess greater reversible capacity and high operation voltage. But these Li-enriched materials for cathode have still some limitations such as low rate capability, strict capacity fade, and voltage destruction because activation of Li_2MnO_3 at high voltage causes rearrangement of structure of the surface during cycling. Recently, Li-enriched cathode material with a layer of spinal hybrid and PANI coating with remarkable performance has been reported. Coating of CP such as PANI forms depressed solid-electrolyte-interface (SEI) layer and improves coulombic efficiency and cycling stability by enhancing the conductivity and diffusion of ions [36].

6.3.11　PANI-MODIFIED ANODE MATERIALS

Conductive coatings of PANI with oxides and sulfides of transition metals have been widely applied for development of LIB anodes. TMOs and metal sulfides typically form nanocomposites, but the incorporation of carbon layer causes enhancement in conductivity of composites by increasing LIB transport kinetics and work as a buffer for various changes in lithium insertion and extraction process. This unique architecture of nanocomposite observes effective improvement in electrochemical characteristics by reversible and alloy reaction with SnO_2 in comparison to the composite without carbon.

Fe_2O_3 is very important material for electrode, which is cost effective and present in abundance on earth. Preparation of Fe_2O_3 with PANI makes unique hollow nanosphere arrangement with remarkably good capacitance, excellent cycling stability, and rate capability. MoS_2 possesses 2D layered structure but charge transfer properties are not so good, but incorporation of MoS_2 embedded in PANI matrix improves conductivity and stability. The MoS_2/PANI nanoflowers with greater specific surface area and enhanced conductivity are reported. Similarly, the SnS_2/PANI nanoplates having lamellar sandwiched nanostructures make conducting configuration in between nanoplates, thus shortening the ion transfer path and improve the expansion and contraction of material for electrode during mechanism of charging and discharging, and hence finally improve the electrochemical properties. PANI can also be utilized to form nitrogen-containing carbon, which is porous in nature, suitable for being used as a material for anode. It can be alternatively used in place of graphite but its charging and discharging capacity is quite high. Therefore, conducting PANI combines with metal chalcogenides and observe conductance, provide protection, and serve as connection layer of anode in LIBs [37].

6.4　REDOX-ACTIVE ELECTROLYTES FOR PANI SUPERCAPACITORS

The pseudocapacitance and stability of electrodes based on PANI can be increased through alteration of traditional electrolytes by using redox substances such as hydroquinone. These redox substances cause extra faradic reactions, which enhance the conductivity of electrolytes by decreasing resistance in charge transfer to improve pseudocapacitance and stability. A symmetric supercapacitor based on PANI with H_2SO_4 electrolyte altered with benzoquinone–hydroquinone (HQ–BQ) mediators with enhanced capacitance retention in comparison to PANI without redox agents has been reported [38].

6.5　EXAMPLES OF VARIOUS POLYANILINE-BASED SUPERCAPACITORS

6.5.1　COMPOSITES OF POLYANILINE DOPED WITH CoCL₂ AS MATERIALS FOR ELECTRODES

The development of these electrochemical capacitors requires dynamic chemical ingredients including inorganic nanofillers along with CPs. CPs exhibit outstanding electrochemical attributes and are reasonably priced. The synergistic conjugation of these CPs and nanofillers is highly crucial for hybrid composites used for various applications in rechargeable batteries, anticorrosion protection coatings, organic light-emitting diodes, electrolyte membrane, chemical/biological/gas sensors, value-added

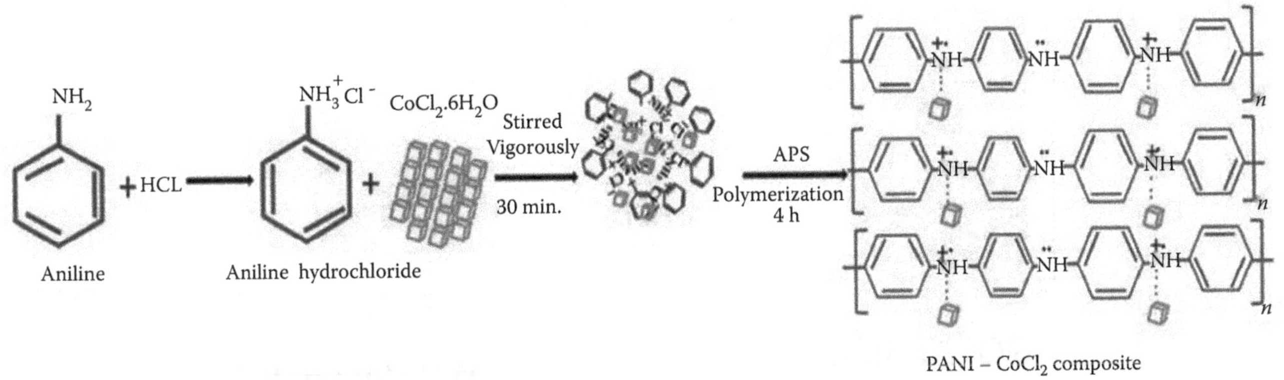

FIGURE 6.4 Schematic route for the synthesis procedure of PANI–CoCl$_2$ composites; adapted from ref. [39], with permission from Springer Nature.

catalyst, organic transistors, electromagnetic interference shielding, and organic solar cells. Out of these CPs, PANI is a potential candidate having remarkable electrochemical characteristics and tunable properties. It is present in various oxidation states, including EM salt, LE base, PE base, and EM base. Emeraldine salt is widely used in potential applications because of being highly electrically conductive with having cationic radicals in polymeric (PANI) chain. As PANI has net-like structure with high conductivity, it provides surface area for immobilization of metals and their oxides. TMOs can also be utilized as electrodes in supercapacitors. Their charging storing mechanisms work on pseudocapacitance. RuO$_2$ possesses high capacitance because of having valuable redox reactions but elemental Ru is costly, which is major limitation for commercial acceptance. But some other TMOs, including Co, Fe, Mn, Sn, and Ni are relatively inexpensive. Moreover, the addition of inorganic nanofiller such as cobalt chloride in the PANI polymeric chain improves the electronic properties. Simple schematics of composite development are shown in Figure 6.4. Cobalt chloride is mostly preferable in hexahydrate form and used in development of hybrid electrode materials. The cobalt chloride addition in polymeric matrix induces changes in electrochemical, electrical, thermal, and dielectric properties. Recently, various polymeric composites with electrochemical properties have been studied, which include MnO$_2$ nanorods–PANI, PANI–zinc acetate–graphene, and graphene–SnO$_2$-PANI. The cyclic voltammetric evaluation of PANI–CoCl$_2$ composites synthesized in laboratory by in situ polymerization process showed enhanced specific capacitance value with increasing PANI content percentage in composite as compared to pure PANI. Therefore, such a fascinating improvement in electrochemical properties was recognized because of synergistic contribution of inorganic nanofiller (CoCl$_2$) and polymer (PANI) [40, 41].

6.5.2 COMPOSITES OF POLYANILINE NANOFIBERS WITH GRAPHENE AS MATERIALS FOR ELECTRODES

In recent years, graphene having excellent electronic and mechanical characteristics has attained much importance in research. Graphene and modification of graphene with chemicals show high conductivity and mechanical characteristics with great surface area as compared to carbon CNTs. Some chemical processing of graphite can produce various graphene-based materials. Hence, utilization of graphene-derived materials in supercapacitors has attained great importance. Actually, the specific capacitance in the range from tens to 135 F/g has been evaluated for various graphene-derived materials. The development of graphene composite paper with PANI by *in situ* anodic electropolymerization of aniline on graphene paper reported specific capacitance up to 233 F/g. Simple schematics of graphene and PANI composite development are shown in Figure 6.5. Development of graphene/PANI composites having homogenous dispersion of graphene sheets in polymeric matrix has also been reported. The ratio of graphene and PANI can be adjusted to make composite with respective main component, such as graphene-doped PANIs or PANI-doped graphenes. PANI in the fibrous configuration shows large surface area and capacitance than the native PANI particles. So the composites with various compositions exhibit excellent electrochemical properties [41].

These composites possess high electrical conductivity, which is an important property for the material to be used as electrode material in supercapacitors. Hence, PANI-doped carbon materials or carbon-doped PANIs observes enhanced capacitance and cycling stability. Therefore, graphene-based electrode materials can be used in various supercapacitor applications by tuning and controlling the structures of component and ratios of composites.

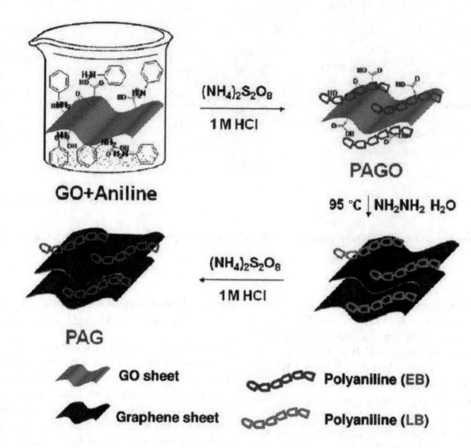

FIGURE 6.5 Illustration of the process for preparation of graphene–PANI composites; reprinted with permission from ref. [41], copyright (2010) American Chemical Society.

6.5.3 COMPOSITES OF POLYANILINE (PANI) WITH GRAPHENE OXIDE AS ELECTRODE MATERIALS

Previously, carbon materials, including CNTs, AC, and PANI, have been extensively applied as materials for electrodes in supercapacitors, but their usage is now restricted because of some limitations such as CNTs being costly, inadequate performance of AC, and poor PANI life cycle. Therefore, by combining carbon materials with PANI, their synergistic performance as electrode materials can be enhanced. Nanocomposites of graphene oxide-doped PANI with increased electrochemical capacitance have been reported. The composite ratio of PANI nanofibers and graphene oxide sheets, 100:1, is good for electrode material in capacitor applications. Nanocomposites show surprisingly enhanced conductivity and capacitance as compared to simple PANI. So, these products show great potential for applications in supercapacitors and other power sources [42].

6.5.4 HYBRID FILMS OF MANGANESE DIOXIDE AND POLYANILINE AS ELECTRODE MATERIALS

Manganese oxide is an excellent substance for electrode material in supercapacitors because of having low cost and being environment friendly as compared to other TMOs. Chemical and electrochemical polymerization techniques can be used to produce composites of PANI and manganese oxides. Previously these composites for electrode materials have been reported but with some limitation of low cell voltage because of using aqueous electrolyte. So, with increasing potential, stability of the electrode will be decreased. Therefore, there is a need to find some suitable electrolyte that can improve the stability and capacitance of $PANI/MnO_2$ composite as electrode material in supercapacitors. 1 M $LiClO_4$/AN electrolyte is considered as adequate electrolyte for supercapacitive performance. Therefore, nanocomposite thin film of PANI and manganese dioxide, which is co-deposited on AC substrate having high surface area, is suitable approach for enhancing electrical conductivity and stability of composite electrode. In addition, asymmetric capacitor was designed using $PANI/MnO_2$/AC as positive and pure AC as negative electrodes, and electrochemical properties were evaluated in $LiClO_4$/AN, an organic electrolyte. Therefore, nonaqueous organic electrolyte is appropriate for $PANI/MnO_2$ hybrid electrode material in supercapacitor applications [43].

6.5.5 Composites of Activated Carbon/Polyaniline with Tungsten Trioxide as Electrode Materials

Tungsten trioxide is gaining attraction in research as electrode material because of having low cost, excellent conductivity, enhanced theoretical capacity, and simple synthesis. WO_3/polypyrrole core-shell nanowires with outstanding supercapacitance have been reported in the near past. Recently, a new asymmetric supercapacitive device based on composites of AC derived from chestnut shell with PANI, as positive electrode and WO_3 nanowires as negative electrode in 1 M H_2SO_4 aqueous electrolyte with excellent electrochemical properties, such as wider operating voltage window, increased energy density, and remarkable cycling stability, have been reported. Therefore, there is a possibility to meet the rising demands of low cost and high energy in the next-generation storage batteries by using low-priced CP and metal oxides [44].

6.5.6 PANI- and MOF-based Flexible Solid-state Supercapacitors

MOFs are the new class of porous materials and have been used in various applications. The conductivity of the MOFs is poor and organic ligands are insulating in nature. Metal ions make coordination bond through *d*-orbital participation with organic ligands and cause limited delocalization of electrons across the framework. Flexible solid-state supercapacitors using MOFs based on PANI have been reported recently. Simple schematic of composite development is shown in Figure 6.6. Porous frame of MOFs with PANI as interlude lines makes an ideal architecture that avoids the insulation of MOFs. Porous structure of MOFs stores the charge, increases the charge path, so that the conductivity, capacity, and other electrochemical characteristics of PANI can be improved. UiO-66 is one of the important MOFs with remarkable aqueous, acidic, and thermal stability. The flexible solid-state supercapacitor that is based on PANI and UiO-66 MOF with excellent electrochemical properties has been reported recently [45].

The addition of PANI in the porous structure of UiO-66 excellently alters the charge transfer properties of PANI by enhancing the conjugation delocalization length of PANI to improve the capability of carrier transfer. Therefore, the accumulation of CP and MOFs is a favorable choice in supercapacitors and various electrochemical devices.

6.5.7 Polyaniline-based Nickel Electrodes for Electrochemical Supercapacitors

PANI is a remarkable material to be used in electrochemical capacitors because it has various oxidation states. PANI in fully reduced form is named as LE, which slightly oxidizes to form EM, and further oxidized at positive potential to form PE. Doping and de-doping of counter anions cause oxidation and reduction processes respectively. Facilitated storage of charge causes specific behavior, the pseudocapacitance (C_-). Charge separation can also take place at the interface of electrolyte and PANI, which produces double-layer capacitance (Cdl), and the total capacitance is $Ct = Cdl + C_-$. Both the factors depend on the surface area, hence, electrodes based on PANI hold increased porosity and large surface area and current collector should also be cost effective. Electrodeposition of PANI on nickel in the company of *p*-toluene sulfonic acid can solve this problem of surface area and cost, and can increase the capacitance [46].

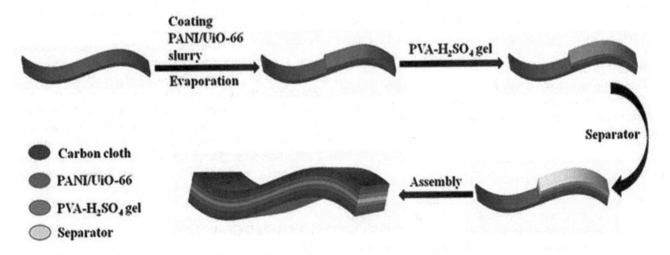

FIGURE 6.6 Schematic illustration for the preparation process of PANI/UiO-66 flexible solid-state supercapacitor. Reprinted from ref. [45], copyright (2018), with permission from Elsevier.

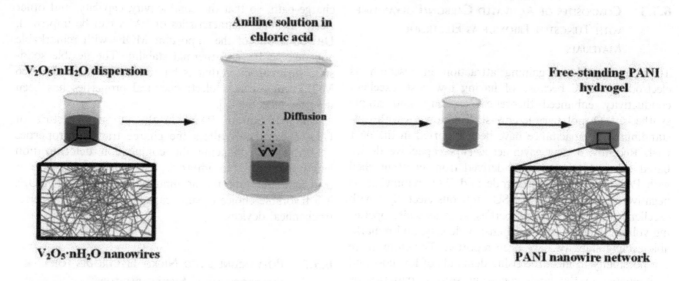

FIGURE 6.7 Schematic illustration of the fabrication route to a pure PANI hydrogel; adapted with permission from ref. [47], copyright (2018) American Chemical Society.

6.5.8 HYDROGEL OF ULTRATHIN PURE POLYANILINE NANOFIBERS IN SUPERCAPACITOR APPLICATION

The development of CPs in hydrogels is a rational approach to low-weight materials being soft and permeable to electrolyte with large surface area and biocompatibility. Therefore, to meet the demands of highly conductive and permeable material, CP hydrogels are suitable choice for the development of devices with improved performances. It is very important to find some reliable methods for the development of CP hydrogels possessing remarkable microstructure and electronic characteristics. Recently, an oxidant-templating methodology for the development of PANI-based hydrogels having interwoven PANI nanofibers with nanowires of $V_2O_5 \cdot nH_2O$ as oxidant to polymerize aniline has been reported, which gives up template to direct the arrangement of 3D network. Their conducting framework and ultrathin nanofiber make the PANI hydrogels as suitable material for electrode in supercapacitors. Simple schematics of preparation of PANI hyrogels are shown in Figure 6.7 [47].

This approach gives a way to develop hydrogels based on CP for various applications, including sensors, energy storing, and biomedicine, and also to be a method to fabricate CPs.

CONCLUSION

PANI is a promising material to be used in various applications, such as an excellent material for electrode, energy storing devices, and energy conversions, because of being environmentally stable, flexible, cost effective, and having unique electrical and redox properties with

easy-to-synthesize protocols. Therefore, PANI can itself be used as a remarkable good material for electrode in supercapacitors and charging batteries. PANI possesses great specific capacitance and cycling stability to be used as pseudocapacitors. It can also be applied as an excellent material for cathode in LIBs. After carbonization and subsequent activating processes, PANI produces porous carbon with enhanced surface area, appropriate pore size, and increased nitrogen content, thus, it is used as an outstanding carbonaceous material for storing energy and conversion applications along with as a support for electrocatalysts. The various applications of PANI largely depend on its novel structure of conjugated bonding and presence of large number of N-active sites that make it able to excellently link with other materials of electrodes such as carbon materials, metallic compounds, and other polymeric compounds, which result in the development of composites with enhanced outstanding performance in comparison to individual components due to combined effect.

Therefore, PANI-based supercapacitors, charging batteries, and other electrocatalysts, in which PANI serves as the porous support for conduction, protecting and connecting matrix on electrode surface material exhibit outstanding performance. A number of methodologies that have been used to develop composite materials with appropriate morphology, structure, and size resulted in wonderful advancements. In conclusion, PANI is a promising outstanding material for electrode development to be used in energy storing and conversion devices.

ACKNOWLEDGEMENTS

Authors acknowledge Higher Education Commission (HEC) of Pakistan for financial support under Research

Grant Nos. 6114 and 6115. Dr. Waheed S. Khan and Dr. Sadia Zafar Bajwa are also grateful to their parent department *National Institute for Biotechnology and Genetic Engineering (NIBGE)*, Faisalabad, Pakistan for providing excellent environment for such type of writing work.

REFERENCES

1. Simotwo, S.K. and Kalra, V., Polyaniline-based electrodes: Recent application in supercapacitors and next generation rechargeable batteries. *Current Opinion in Chemical Engineering*, 2016. **13**: pp. 150–160.

2. Wang, H., Lin, J., and Shen, Z.X., Polyaniline (PANi) based electrode materials for energy storage and conversion. *Journal of Science: Advanced Materials and Devices*, 2016. **1**(3): pp. 225–255.

3. Grixti, S., Mukherjee, S., and Singh, C.V., Two-dimensional boron as an impressive lithium-sulphur battery cathode material. *Energy Storage Materials*, 2018. **13**: pp. 80–87.

4. Choi, M.S., Park, S., Lee, H., and Park, H.S., Hierarchically nanoporous carbons derived from empty fruit bunches for high performance supercapacitors. *Carbon Letters*, 2018. **25**(1): pp. 103–112.

5. Li, S., Zhang, N.S., Zhou, H.H., Li, J.W., Gao, N., Huang, Z.Y., Jiang, L.L., and Kuang, Y.F., An all-in-one material with excellent electrical double-layer capacitance and pseudocapacitance performances for supercapacitor. *Applied Surface Science*, 2018. **453**: pp. 63–72.

6. Zhong, C., Deng, Y., Hu, W.B., Qiao, J.L., Zhang, L., and Zhang, J.J., A review of electrolyte materials and compositions for electrochemical supercapacitors. *Chemical Society Reviews*, 2015. **44**(21): pp. 7484–7539.

7. Zhang L., Hu X.S., Wang Z.P., Sun F.C., and Dorrell D.G., A review of supercapacitor modeling, estimation, and applications: A control/management perspective. *Renewable and Sustainable Energy Reviews*, 2018. **81**: pp. 1868–1878.

8. Kouchachvili, L., Yaïci, W., and Entchev, E., Hybrid battery/supercapacitor energy storage system for the electric vehicles. *Journal of Power Sources*, 2018. **374**: pp. 237–248.

9. Boota, M. and Gogotsi, Y., MXene—Conducting polymer asymmetric pseudocapacitors. *Advanced Energy Materials*, 2019. **9**(7): p. 1802917.

10. Wang, G., Oswald, S., Löffler, M., Müllen, K., and Feng, X., Beyond activated carbon: Graphite-cathode-derived Li-ion pseudocapacitors with high energy and high power densities. *Advanced Materials*, 2019. **31**(14): p. 1807712.

11. Sun, S., Zhai, T., Liang, C.L., Savilov, S.V., and Xia, H., Boosted crystalline/amorphous Fe2O3-δ core/shell heterostructure for flexible solid-state pseudocapacitors in large scale. *Nano Energy*, 2018. **45**: pp. 390–397.

12. Selvakumar, M., Multilayered electrode materials based on polyaniline/activated carbon composites for supercapacitor applications. *International Journal of Hydrogen Energy*, 2018. **43**(8): pp. 4067–4080.

13. Yu, J.H., Xie, F.F., Wu, Z.G., Huang, T., Wu, J.F., Yan, D.D., Huang, C.Q., and Li, L., Flexible metallic fabric supercapacitor based on graphene/polyaniline composites. *Lectrochimica Acta*, 2018. **259**: pp. 968–974.

14. Aydinli, A., Yuksel, R., and Unalan, H.E., Vertically aligned carbon nanotube–polyaniline nanocomposite supercapacitor electrodes. *International Journal of Hydrogen Energy*, 2018. **43**(40): pp. 18617–18625.

15. Huang, Z.Q., Li, L., Wangm, Y.F., Zhang, C., and Liu, T. X., Polyaniline/graphene nanocomposites towards high-performance supercapacitors: A review. *Composites Communications*, 2018. **8**: pp. 83–91.

16. Kandasamy, S.K. and Kandasamy, K., Recent advances in electrochemical performances of graphene composite (graphene-polyaniline/polypyrrole/activated carbon/carbon nanotube) electrode materials for supercapacitor: A review. *Journal of Inorganic and Organometallic Polymers and Materials*, 2018. **28**(3): pp. 559–584.

17. Zengin H., Zhou W., Jin J., Czerw R., Smith Jr. D.W., Echegoyen L., Carroll D.L., Foulger S.H., and Ballato J., Carbon nanotube doped polyaniline. *Advanced Materials*, 2002. **14**(20): pp. 1480–1483.

18. Chen, W.C. and Wen, T.C., Electrochemical and capacitive properties of polyaniline-implanted porous carbon electrode for supercapacitors. *Journal of Power Sources*, 2003. **117**(1–2): pp. 273–282.

19. Li, G.C., Li, G.R., Ye, S.H., and Gao, X.P., A polyaniline-coated sulfur/carbon composite with an enhanced high-rate capability as a cathode material for lithium/sulfur batteries. *Advanced Energy Materials*, 2012. **2**(10): pp. 1238–1245.

20. Li, S.Y., Gao, A.M., Yi, F.Y., Shu, D., Cheng, H.H., Zhou, X.P., He, C., Zeng, D.P., and Zhang, F., Preparation of carbon dots decorated graphene/polyaniline composites by supramolecular in-situ self-assembly for high-performance supercapacitors. *Electrochimica Acta*, 2019. **297**: pp. 1094–1103.

21. Moyseowicz, A. and Gryglewicz, G., Hydrothermal-assisted synthesis of a porous polyaniline/reduced graphene oxide composite as a high-performance electrode material for supercapacitors. *Composites Part B: Engineering*, 2019. **159**: pp. 4–12.

22. Xu, J.J., Wang, K., Zu, S.Z., Han, B.H., and Wei, Z.X., Hierarchical nanocomposites of polyaniline nanowire arrays on graphene oxide sheets with synergistic effect for energy storage. *ACS Nano*, 2010. **4**(9): pp. 5019–5026.

23. Li, J.P., Xiao, D.S., Ren, Y.Q., Liu, H.R., Chen, Z.X., and Xiao, J.M., Bridging of adjacent graphene/polyaniline layers with polyaniline nanofibers for supercapacitor electrode materials. *Electrochimica Acta*, 2019. **300**(20): pp. 193–201.

24. Lin Y.X., Zhang H.Y., Deng W.T., Zhang D.F., Li N., Qibai B., and He C.H., In-situ growth of high-performance all-solid-state electrode for flexible supercapacitors based on carbon woven fabric/polyaniline/graphene composite. *Journal of Power Sources*, 2018. **384**: pp. 278–286.

25. Liu, P.B., Yan, J., Gao, X.G., Huang, Y., and Zhang, Y.Q., Construction of layer-by-layer sandwiched graphene/polyaniline nanorods/carbon nanotubes heterostructures for high performance supercapacitors. *Electrochimica Acta*, 2018. **272**: pp. 77–87.

26. Agyemang, F.O., Tomboc, T.M., Kwofie, S., and Kim, H., Electrospun carbon nanofiber-carbon nanotubes coated polyaniline composites with improved electrochemical properties for supercapacitors. *Electrochimica Acta*, 2018. **259**: pp. 1110–1119.

27. Li, P.P., Ni, C.H., Shi, G., Zhang, D.Z., and Xu, Y.H., Fabricating composite supercapacitor electrodes of polyaniline and aniline-terminated silica by mechanical agitation and sonication. *Journal of Solid State Electrochemistry*, 2018. **22**(4): pp. 1249–1256.

28. Jabur, A.R., Effect of polyaniline on the electrical conductivity and activation energy of electrospun nylon films. *International Journal of Hydrogen Energy*, 2018. **43**(1): pp. 530–536.

29. Zornitta, R.L., Nogueira, F.G.E., and Ruotolo, L.A.M., Carbonization temperature as a key factor for ultrahigh performance activated carbon from polyaniline for capacitive deionization, in *Meeting Abstracts*. 2018, The Electrochemical Society.

30. Vighnesha, K., Sangeetha, D., and Selvakumar, M., Synthesis and characterization of activated carbon/conducting polymer composite electrode for supercapacitor applications. *Journal of Materials Science: Materials in Electronics*, 2018. **29**(2): pp. 914–921.

31. Ezzati, N., Asadi, E., Leblanc, R.M., Ezzati, M.H., and Sharma, S.K., Other applications of polyaniline-based blends, composites, and nanocomposites, in P.M. Visakh, C. Della Pina, and E. Falletta, editors, *Polyaniline Blends, Composites, and Nanocomposites*. 2018, Elsevier. pp. 279–303.

32. Rose, A., Prasad, K.G. Sakthivel, T., Gunasekaran, V., Maiyalagan, T., and Vijayakumar, T., Electrochemical analysis of graphene oxide/polyaniline/polyvinyl alcohol composite nanofibers for supercapacitor applications. *Applied Surface Science*, 2018. **449**: pp. 551–557.

33. Javed, M., Abbas, S.M., Siddiq, M., Han, D.H., and Niu, L., Mesoporous silica wrapped with graphene oxide-conducting PANI nanowires as a novel hybrid electrode for supercapacitor. *Journal of Physics and Chemistry of Solids*, 2018. **113**: pp. 220–228.

34. Reddy K.R., Hemavathi B., Balakrishna G.R., and Anjanapura R., Organic conjugated polymer-based functional nanohybrids: Synthesis methods, mechanisms and its applications in electrochemical energy storage supercapacitors and solar cells, in K. Pielichowski and T.M. Majka, editors, *Polymer Composites with Functionalized Nanoparticles*, 2019, Elsevier. pp. 357–379.

35. Shah, A.H.A., Khan, M.O., Bilal, S., Rahman, G., and Hoang, H.V., Electrochemical co-deposition and characterization of polyaniline and manganese oxide nanofibrous composites for energy storage properties. *Advances in Polymer Technology*, 2018. **37**(6): pp. 2230–2237.

36. Han, Y. and Dai, L., Conducting polymers for flexible supercapacitors. *Macromolecular Chemistry and Physics*, 2019. **220**: p. 1800355.

37. Chen, W.W., Liu, Z.L., Hou, J.X., Zhou, Y., Lou, X.G., and Li, Y.X., Enhancing performance of microbial fuel cells by using novel double-layer-capacitor-materials modified anodes. *International Journal of Hydrogen Energy*, 2018. **43**(3): pp. 1816–1823.

38. Luo, Y.X., Zhang, Q., Hong, W.J., Xiao, Z.Y., and Bai, H., A high-performance electrochemical supercapacitor based on a polyaniline/reduced graphene oxide electrode and a copper (ii) ion active electrolyte. *Physical Chemistry Chemical Physics*, 2018. **20**(1): pp. 131–136.

39. Majhi, M., Choudhary, R.B., Thakur, A.K., Omar, F.S., Duraisamy, N., Ramesh, K., and Ramesh, S., $CoCl_2$-doped polyaniline composites as electrode materials with enhanced electrochemical performance for supercapacitor application. *Polymer Bulletin*, 2018. **75**: pp. 1563–1578.

40. Kharangarh, P.R., Uapathy, S., Singh, G., Sharma, R. K., and Kumar, A., High-performance pseudocapacitor electrode materials: cobalt (II) chloride–GQDs electrodes. *Emerging Materials Research*, 2017. **6**(2): pp. 227–233.

41. Zhang, K., Zhang, L.L., Zhao, X.S., and Wu, J.S., Graphene/polyaniline nanofiber composites as supercapacitor electrodes. *Chemistry of Materials*, 2010. **22**(4): pp. 1392–1401.

42. Wang, H.L., Hao, Q.L., Yang, X.J., Lu, L., and Wang, X., Graphene oxide doped polyaniline for supercapacitors. *Electrochemistry Communications*, 2009. **11**(6): pp. 1158–1161.

43. Zou, W.Y., Wang, W., He, B.L., Sun, M.L., and Yin, Y.S., Supercapacitive properties of hybrid films of manganese dioxide and polyaniline based on active carbon in organic electrolyte. *Journal of Power Sources*, 2010. **195**(21): pp. 7489–7493.

44. Wang, H., Ma, G., Tong, Y., and Yang, Z., Biomass carbon/polyaniline composite and WO 3 nanowire-based asymmetric supercapacitor with superior performance. *Ionics*, 2018. **24**(10): pp. 3123–3131.

45. Shao, L., Wang, Q., Ma, Z., Ji, Z., Wang, X., Song, D., Liu, Y., and Wang, N., A high-capacitance flexible solid-state supercapacitor based on polyaniline and metal-organic framework (UiO-66) composites. *Journal of Power Sources*, 2018. **379**: pp. 350–361.

46. Girija, T. and Sangaranarayanan, M., Polyaniline-based nickel electrodes for electrochemical supercapacitors—Influence of Triton X-100. *Journal of Power Sources*, 2006. **159**(2): pp. 1519–1526.

47. Zhou, K., He, Y., Xu, Q., Zhang, Q., Zhou, A., Lu, Z., Yang, L.K., Jiang, Y., Ge, D.T., Liu, X.Y., and Bai, H., A hydrogel of ultrathin pure polyaniline nanofibers: Oxidant-templating preparation and supercapacitor application. *ACS Nano*, 2018. **12**(6): pp. 5888–5894.

7 Conductive Polymer-derived Materials for Supercapacitor

P.E. Lokhande and U.S. Chavan
Department of Mechanical Engineering
Sinhgad Institute of Technology Lonavala
Lonavala, India

CONTENTS

7.1 INTRODUCTION

Owing to the world's economical development and continuous increase in population, the consumption of fossil has increased tremendously. The demand for more energy is on rise. The depletion of fossil fuel is severely affecting the environment. Today, we not only need clean and renewable energy sources, but we also need environment friendly technologies for their conversion and storage (Yu et al. 2015; Yan et al. 2016; Ma et al. 2018). Energy-storage technologies such as batteries, supercapacitors, and fuel cells are important for sustaining renewable energy resources (Zhang, Li, and Pan 2012). Figure 7.1 shows the Ragone plot, which covers power against energy density change for the above-mentioned major energy storage technologies (Zhong et al. 2015). Among these electrochemical-based technologies, supercapacitors attracted more attention due to their higher power density, outstanding cycle life, high specific capacitance, safe functioning, and low maintenance cost (Lokhande and Panda 2015; Lokhande and Chavan 2019a). The supercapacitor, also called ultracapacitor, takes an important place in recent flexible and wearable electronic devices that require foldable, stretchable, and bendable power source. Supercapacitor is considered to be an ideal energy storage technology for flexible and wearable electronic devices (Han and Dai 2019).

The supercapacitor technology was patented first in 1957 by General Electric Motors, where activated charcoal was used as electrode. Second patent registered by the Standard Oil Company, Cleveland, Ohio (SOHIO) was based on energy storage device which stores energy in double-layer interface. Same technology was licensed by Nippon Electric Company and introduced the device for memory backup for computer. In 1970, SOHIO patented disc-shaped capacitor, which was prepared by using carbon paste socked in electrolyte. The first commercial capacitor under the name 'supercapacitor' was launched by NEC (Sharma and Bhatti 2010). Nowadays, many other companies are producing supercapacitors by using different materials as supercapacitor.

Supercapacitor is a device having higher specific capacitance and minimum internal resistance, which is composed of two electrodes separated by a separator and an electrolyte. The characteristic properties of supercapacitor such as energy density, power density, and specific capacitance depend on the electrode material and electrolyte used in cell (Iro, Subramani and Dash 2016). Basically, metal oxides, conducting polymers (CPs), and carbon-based materials are used as electrode materials for supercapacitor. Among them, CP has been widely researched due to its low cost and easy production. CP generally exhibits high conductivity and capacitance along with lower equivalent series resistance (Iro, Subramani and Dash 2016). Various conductive polymers and their composites used for supercapacitor electrode are discussed thoroughly in the present review.

7.2 TYPES OF SUPERCAPACITOR

Primarily, on the basis of principle mechanism for charge storage and ion movements, supercapacitor is

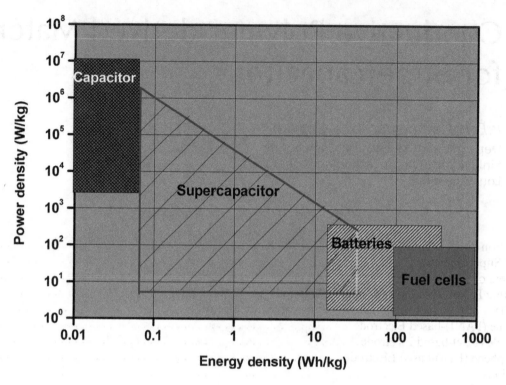

FIGURE 7.1 Ragone plot for various electrical energy-storage devices (specific power against specific energy).

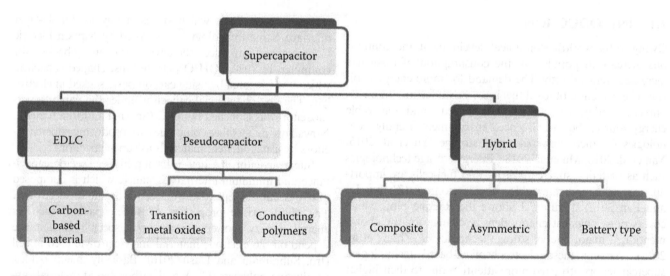

FIGURE 7.2 Classification of electrochemical supercapacitor.

classified in three categories: electrical double-layer capacitor (EDLC), pseudocapacitor, and hybrid capacitor (Figure 7.2). In EDLC, charges are stored electrostatically on opposite electrode forming a double layer. When voltage is applied, charges are accumulated on the electrode surface; potential difference attracts the opposite charge, which leads to ions from electrolyte diffusing over the separator and through the pores of the oppositely charged electrode. The double-layer charge is formed to avoid recombination (Iro, Subramani and Dash 2016) at the electrodes. The charge accumulation process in EDLC is pure physical. The movement of electrons from negative to positive electrode occurs during charging via external loop. During discharge process, electrons and ions move in reverse direction. Throughout the process, charges are not transferred across the interface, as faradaic current is absent (Shi et al. 2014). There is physical charge accumulation and no charge transfer, swelling, or volume change occurs in case of electrode material, hence number of cycles and stability of EDLC are higher than the pseudocapacitors (Simon and Gogotsi 2008). The carbon-based materials such as activated carbon (Li et al. 2010), carbon aerogels (Wang et al. 2011), carbon nanotube (CNT) (Pan, Li, and Feng 2010), and graphene (Zhang, Li and Pan 2012) are used as activated material, which gives high power density (up to 10 kW kg^{-1}) and excellent cyclic performance (over 100,000 cycles), but its low energy density limits its wide application.

In contrast to EDLC, pseudocapacitor involves the reversible redox reaction as energy-storage mechanism. When potential is applied, fast and reversible faradaic reaction takes place on and near the surface of electrodes, which generates charges and there is a transfer of charges across the double layer. The faradaic reaction is grouped in three categories: reversible adsorption (adsorption of hydrogen ion), redox reaction (transition metal), and electrochemical doping–dedoping (conductive polymer). The specific capacitance obtained for pseudocapacitance is higher than the EDLC as the reaction occurs on and near the surface (Simon and Gogotsi 2008). Pseudocapacitor suffers from lack of stability due to redox reaction and also low power density is observed (Wang, Zhang and Zhang 2012). The transition metal hydroxide/oxide (RuO_2 (Liu, Pell and Conway 1997), NiO (Ci et al. 2015; Kate, Khalate and Deokate 2018), $Ni(OH)_2$ (Lokhande and Chavan 2018b; Lokhande, Pawar and Chavan 2018; Lokhande and Chavan 2018a), ZnO (Saranya, Ramachandran and Wang 2016), CuO (Lokhande and Chavan 2019a), $Co(OH)_2$ (Gupta et al. 2007), Co_2O_3 (Tan et al. 2017), and MnO_2 (Qu et al. 2009), and CPs were used as active material in pseudocapacitors.

Hybrid capacitor combines the advantages and mitigates disadvantages of pseudocapacitor and EDLC. A pseudocapacitor offers high specific capacitance and energy density while an EDLC offers high power density, cyclability, and stability. With the correct electrode combination, it is possible to enhance the electrochemical performance of supercapacitor. The hybrid capacitor further can be classified into three groups on the basis of combination of electrodes as composite, asymmetric, and battery type (Kate, Khalate and Deokate 2018).

7.3 PARAMETERS OF SUPERCAPACITORS

To evaluate the quality performance of prepared electrode by using various advanced materials, several characteristic properties are defined for this material, such as specific capacitance, energy density, power density, cyclability, and rate capability (Zhang and Zhao 2009). There are three characterization techniques used for measuring specific capacitance, and from these other properties are measured using various equations derived below. The specific capacitance for the material can be evaluated via two techniques, namely cyclic voltammetry (CV) and galvanostatic charge–discharge (GCD) (Kate, Khalate and Deokate 2018). The specific capacitance is measured from CV curve and GCD profile by using following equations (Lokhande, Pawar and Chavan 2018; Lokhande and Chavan 2019b):

$$Cs = 1/(mv(v_f - v_i)) \int_{v_i}^{v_f} I(V)dV \qquad (7.1)$$

$$Cs = \frac{I\Delta t}{m\Delta V} \qquad (7.2)$$

where v is the scan rate, $(v_f - v_i)$ is the potential window and m, I, ΔV, and Δt refer to the weight of active electrode material, instantaneous current, voltage window, and discharge time, respectively.

The energy density and power density are the two important parameters for evaluating electrochemical performance of supercapacitors. Following equations are used for measuring the maximum energy density (E) (Wh kg^{-1}) and power density (P) (W kg^{-1}) (Zhang and Zhao 2009):

$$E = CV^2/2 \qquad (7.3)$$

$$P = V^2/4R \qquad (7.4)$$

where V is the cells operating voltage in V, C is the capacitance in F, and R is the equivalent series resistance in Ω. The cell voltage is dependent on the thermodynamic stability of electrolyte and electroactive materials of electrode. The equivalent series resistance (ESR) is composed of the following factors: (1) electroactive materials intrinsic resistance, (2) resistance generated in between electroactive material and substrate, (3) resistance offered by ions during diffusion in electrode material, and (4) separator and electrolytes ion resistance (Zhang and Zhao 2009). From Equations 7.3 and 7.4 it is clear that the electrochemical performance of supercapacitor can be enhanced by increasing specific capacitance and wide cell voltage while keeping minimum ESR. The major challenge that the scientists face today is to enhance the energy density of battery (Iro, Subramani and Dash 2016). In that context, parameters for increasing energy density are taken into consideration. The energy density can be increased effectively by enhancing the specific capacitance, which can be done by increasing the capacitance of both positive and negative electrolytes. By using advanced materials and with different approach, improved properties such as higher specific surface area, optimized pore size, and higher electric conductivity can be developed (Sun et al. 2015).

Second factor is the operating voltage, which is very crucial because energy density is directly proportional to the square of the operating voltage. To enhance the operating voltage, various strategies such as selecting electrolyte with wide operating voltage are being applied. Advanced

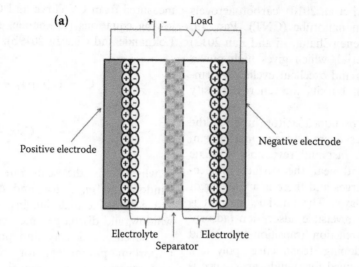

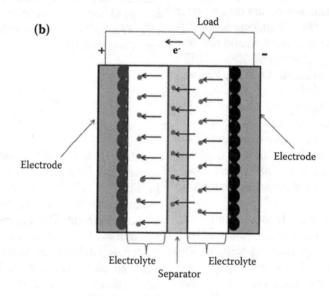

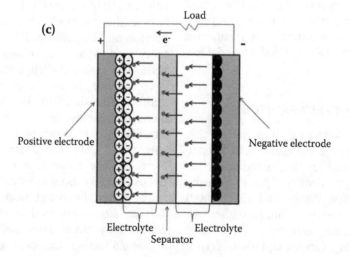

FIGURE 7.3 Schematic diagram of charge storage mechanism of various supercapacitors: (a) EDLC; (b) pseudocapacitor; and (c) hybrid capacitor.

electrolytes such as organic electrolyte (up to 2.5–3 V) and ionic liquid (up to 4 V) are used to achieve higher operating voltage (Lokhande et al. 2016). Another approach to enhance the operating voltage is by developing asymmetric supercapacitor. Asymmetric supercapacitor is prepared by using two different types of electrodes. It combines the advantages of both batteries and capacitors (Halper and Ellenbogen 2006; Raza et al. 2018). In the following section, in-depth progress of the electrode material and electrolyte material is discussed.

7.4 CONDUCTING POLYMERS (CPS) AS ELECTRODE MATERIALS

In 1977, conducting properties of CPs were discovered as a new class for the first time. Such materials exhibited excellent optical and electrical properties, and were generally found in inorganic systems. Several kinds of CPs have been researched, including an electrode material polyaniline (PANI), polypyrrole (PPy), and poly(3,4 ethylenedioxythiophene) (PEDOT) (Magu et al. 2019). CPs attracted attention as pseudocapacitor material because of their higher theoretical capacitance, good electrical conductivity, environmental stability, low cost, and easy mass production. At the same time, their poor cycling stability due to swelling/shrinkage of chain during charging–discharging resulting in inadequate drainage of electrolyte ions is the major obstacle to use CPs as electrode material. To overcome this drawback and to enhance such properties of CPs, many researches fabricated composites of CPs with transition metal oxide or carbon-based materials (Wang, Huang and Wei 2010; Zhou et al. 2010; Long et al. 2011; Das and Prusty 2012).

7.4.1 CLASS OF CONDUCTING POLYMER AS SUPERCAPACITOR ELECTRODE

On the basis of doping mechanism, CPs as active mass are classified into two groups: p-doping and n-doping. In p-doping, electrochemical doping takes place when through external circuit, electrons are separated from polymer backbone while anions from solution are incorporated into polymer in order to maintain charge neutrality as shown in Figure 7.4a. On the other hand, electrochemical n-doping takes place when through external circuit, electrons are transported on polymer backbone while through solution cations enter into the polymer to counter balance as shown in Figure 7.4b. A large number of p-doped and undoped CPs are available and doping process takes place at electrode potential occurred in aqueous solution. While limited number of n-doped CPs can doped, which virtually occurs at high electrode reducing potential and is possible at catholically stable and in nonaqueous system (Rudge et al. 1994). Because of these limitations, researchers focus their attention on p-doped CPs such as PPy and PANI, but overall voltage achieved from this is limited up to 1 V.

7.5 POLYANILINE (PANI)-BASED ELECTRODE

Among all the CPs used as electrode for supercapacitor, PANI is considered to be the most promising one due to its excellent characteristics such as high conductivity, low cost, easy synthesis, and high capacity to store energy. The electrochemical performance of PANI-based electrodes is affected by morphology of PANI nanostructure, hence efficient method for the synthesis of appropriate nanostructure of PANI has been researched on priority. General methods used for the synthesis of PANI are chemical or electrochemical polymerization. Also interfacial polymerization is a commonly used method, which is easy and is relatively less expensive. Dhawale et al. prepared one-step template-free nanostructured PANI at room temperature to be used as electrode material for supercapacitor application, which exhibited specific capacitance of 503 F g^{-1} (Dhawale, Vinu and Lokhande 2011). Sivaraman et al. fabricated PANI-based electrode for supercapacitor showed specific capacitance of 480 F g^{-1} (Sivaraman et al. 2010). The nanofibers of PANI prepared electrochemically by template-free method obtained pseudocapacitance of 1210 F g^{-1} in H$_2$SO$_4$ solution (Zhang et al. 2008). Gupta et al. demonstrated PANI nanowires deposited on stainless steel substrate used as supercapacitor electrode obtained specific capacitance of 775 F g^{-1} at scan rate of 10 mV s^{-1} (Gupta and Miura 2006). Zhou et al. utilized *in situ* polymerization approach to get pure PANI hydrogel, which showed high specific capacitance of 636 F g^{-1} and higher rate capability with good cyclability (Zhou et al. 2018a). Ran et al. prepared nanocomposite of gold nanoparticles and PANI for supercapacitor application. By increasing gold nanoparticles from 4.2% to 24.72%, specific capacitance of composite electrode increased from 334 to 392 F g^{-1} and after that it was decreased (Ran et al. 2018).

Even though higher specific capacitance was shown by PANI material, during repetitive cycles, swelling–shrinkage occurs repetitively leading to rapid degradation of electrodes performance. Hence to get over on this limitation, PANI material was reinforced with other carbon-based materials to form composites (Magu et al. 2019). Aydinli et al. electrodeposited PANI on vertically aligned CNTs, which were decorated on aluminum foils via CVD. The prepared nanocomposite electrode exhibited specific capacitance of 16.17 mF cm^{-2} at a current density of 0.25 mA cm^{-2} (Aydinli, Yuksel and Unalan 2018). Zhang et al. demonstrated the graphene and PANI nanofiber composites for supercapacitor application, and morphology study showed that PANI fibers were absorbed on graphene surface or were filled in between graphene sheets. The specific capacitance of 480 F g^{-1} was calculated for PANI/graphene composite (Zhang et al. 2010). To achieve good rate of capability and excellent cyclability, layer-by-layer sandwiched heterostructures of graphene/PANI nanorods/CNTs were synthesized. The sample exhibited intimate contact, synergetic effect, and specific capacitance of 638 F g^{-1} at 0.5 A g^{-1} was obtained with 82.2% rate capability

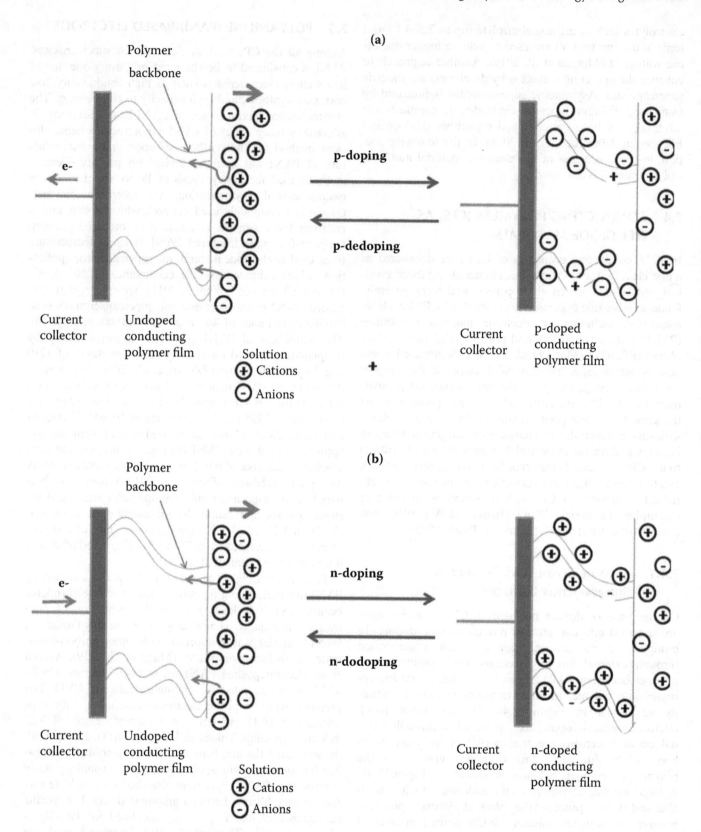

FIGURE 7.4 Schematic representation of charging–discharging process in conducting polymers (a) p-doped, (b) n-doped. (Reproduced with permission), (Rudge et al. 1994) copyright1994 Elsevier.

and 93% capacitance retention after 2000 cycles (Gao et al. 2018). Boddula et al. synthesized graphene-Mn_3O_4 core incorporated into PANI shell in order to enhance capacitance through sonochemical polymerization method. The specific capacitance of graphene-Mn_3O_4 was boosted from 44 to 660 F g^{-1} due to incorporation of PANI (Boddula, Bolagam and Srinivasan 2017). Wu et al. used self-assembly method for fabrication of PANI/rGO porous composite gel, where reduced graphene oxide sheet was covered by PANI thin coating, the composite delivered specific capacitance of 808 F g^{-1} at current density 53.33 A g^{-1} with excellent rate performance (Wu et al. 2018). In another case, PANI nanofibers confined in to graphene oxide architecture exhibited higher capacitance of 780 F g^{-1} at 0.5 A g^{-1} while symmetric supercapacitor achieved energy density of 30 Wh kg^{-1} at 216 W kg^{-1} power density (Zhou et al. 2018b).

To overcome structural stability, forming composite with transition metal oxide was also proved to be a good solution by researchers. Cao et al. demonstrated that PANI/MgFe LDH composite for supercapacitor achieved specific capacitance of 592.5 F g^{-1} at current density of 2 A g^{-1}, and most importantly 87% capacitance retention after 500 cycles. SEM images of sample composite before and after CV test are shown in Figure 7.5, which confirm that the particle size of the used LDH/PANI was smaller than that of LDH/PANI composite (Xiao et al. 2018). Yasoda et al. demonstrated that brush like PANI V3O7 decorated reduced graphene oxide composite as electrode material for supercapacitor application delivered specific capacitance of 579 F g^{-1} at current density 0.2 A g^{-1}, which also exhibited excellent capacitance retention (94% after 2500 cycles) as shown in Figure 7.6 (Yasoda et al. 2019).

7.6 POLYPYRROLE (PPY)-BASED ELECTRODE

PPy is one of the most important CP used as electrode material for electrochemical supercapacitor due to its excellent properties such as high conductivity, relatively high specific capacitance, easy synthesis, good cyclability, good thermal stability, and high energy density (Shown et al. 2015; Meng et al. 2017a). The electrochemical performance of PPy-based electrode is affected by electrode synthesis method and effective surface area of active mass (Zhi et al. 2016).

Fan et al. electrodeposited PPy on Ti foil via CV. The prepared highly porous structure showed pseudocapacitance of 480 F g^{-1} and excellent rate capability (Fan and Maier 2006). Zhao et al. demonstrated that PPy nanowires synthesized under mild condition, having width of 120 nm, and due to intertwined structure (Figure 7.7) formed many nanogaps and nanopores. The prepared sample was used for supercapacitor application, which exhibited pseudocapacitance of 420 F g^{-1} (Zhao et al. 2016). The poor cycling stability is the major hurdle for use of PPy as electrode material in supercapacitor. It occurs due to structural pulverization and counter ion drain effect. Song et al. demonstrated two solutions to enhance cyclic stability: one use of partial functionalized graphite film as substrate which stabilizes PPy and β-naphthalene sulfonate ions doping. The 97.5 % capacitance retention was achieved after 10,000 cycles for prepared PPy sample (Liu et al. 2015). Huang et al. fabricated stretchable supercapacitor by electrochemically depositing PPy on stretchable stainless steel meshes, which obtained specific capacitance of 170 F g^{-1} that enhanced up to 240 F g^{-1} for 20% strain. The prepared supercapacitor with 0% and 20% strains exhibited remarkable 98% and 87% capacitance retention, respectively, after 10,000 cycles (Huang et al. 2015).

The limited stability of CPs during cycling can also be improved by making composite with carbon-based material and transition metal oxides (Afzal et al. 2017). Chang et al. prepared PPy/GO composite film by electrooxidation, which delivered pseudocapacitance of 424 F g^{-1} at 1 A g^{-1} in 1 M H_2SO_4 electrolytic solution (Chang et al. 2012). Xu et al. synthesized graphene (GN)/activated carbon (AC)/PPy-based ternary composite by using vacuum filtration and anode constant current deposition method at different

FIGURE 7.5 SEM images of PANI/LDH composite film before and after CV test. (Reproduced with permission), (Xiao et al. 2018) copyright 2019 Elsevier.

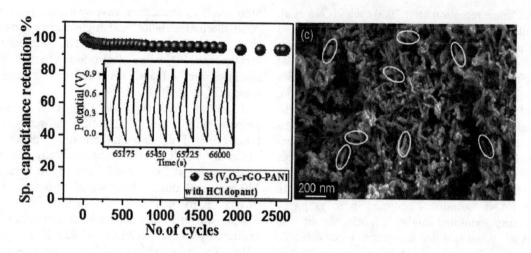

FIGURE 7.6 (a) Cyclic stability test of V_3O_7-rGO-PANi with HCl dopant. (b) SEM image of V_3O_7-rGO-PANi with HCl dopant and yellow oval rings shows the presence of some 1D structure. (Reproduced with permission), (Yasoda et al. 2019) copyright 2019 Elsevier.

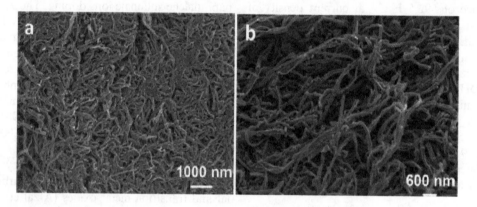

FIGURE 7.7 SEM images of polypyrrole nanowires. (Reproduced with permission), (Zhao et al. 2016) copyright 2016 Elsevier.

current deposition and time. Among these samples, GN/AC/PPy-200 s delivered specific capacitance of 906 mF cm^{-2} with a PPy loading mass of 2.75 g cm^{-2} in electrode and 64.4% capacitance retention after 5000 charge–discharge cycle (Xu et al. 2017). Wu et al. reported facile solution method to prepare nano-flower like PPy@NiCo(OH)$_2$ composite, which exhibited ultrahigh specific capacitance of 1469.25 F g^{-1} at 1 A g^{-1} and 95.2% capacitance was achieved after 10,000 cycles (Wu, Lian and Wang 2019). The PPy/PbOx composite on Pb intercalated graphite electrode also demonstrated good electrochemical performance such as specific capacitance of 377 F g^{-1} in 2 A g^{-1} in 100 mM H$_2$SO$_4$. The symmetric supercapacitor prepared by using these electrodes exhibited 0.29 kW kg^{-1} energy density and 22 Wh kg^{-1} power density along with 94% capacitance retention after 5000 cycles (Karaca et al. 2019). Shivakumara et al. reported MnO$_2$/PPy composite prepared using co-precipitation method having surface area of 232 m^2 g^{-1} and pore size of 6.6 nm. The prepared sample showed

specific capacitance of 258 and 232 F g^{-1} after 3000 cycles (Shivakumara and Munichandraiah 2019).

7.7 POLYTHIOPHENE (PTH)-BASED ELECTRODE

PTh and its derivatives based polymers are both p-doped and n-doped conductive polymers. Even though PTh has low conductivity, p-doped PTh is highly stable in air and humidified environment. The higher operating potential window is another advantage of PTh polymer when used as electrode in supercapacitor. Among all the thiophene derivatives, poly(3,4-ethylenedioxythiophene) is most popular electrode material (Park et al. 2004; Shown et al. 2015). In order to enhance the electrode performance, various synthesis techniques have been used till now (Meng et al. 2017b). Ni et al. used modified self-assembled micellar soft template method for the synthesis of flexible and free standing PEDOT nanowire film having high electric conductivity and exhibited excellent electrochemical performance (Ni et al. 2019). Patil et al. synthesized PTh thin film by SILAR method at room temperature for supercapacitor application, which delivered specific capacitance of 252 F g^{-1} in 0.1 M LiClO$_4$ electrolyte (Patil, Jagadale and Lokhande 2012). Higher stability during charging and discharging, higher reversibility, and low ohmic resistance are the major challenges for CP materials. In that context, PTh is combined with carbon-based materials, including CNT, graphene, and so on (Alvi et al. 2011; Fu et al. 2012; Alabadi et al. 2016). Rajesh et al. deposited PEDOT nanostructure on flexible 3D carbon fiber growth using *in situ* hydrothermal polymerization method. The symmetric supercapacitor using PEDOT/PEDOT showed high specific capacitance of 203 F g^{-1} at 5 mV s^{-1} scan rate along with the energy density of 4.4 Wh kg^{-1} and power density of 40.25 kW kg^{-1} in H$_2$SO$_4$ electrolyte (Rajesh et al. 2017). He et al. deposited PEDOT on CNT framework to form CNTs/PEDOT composite for compressible and flexible free standing electrode sponge for flexible supercapacitor. The highest mass specific capacitance of 147 F g^{-1} was obtained and 95% capacitance retention was achieved after 3000 cycles (He et al. 2018).

The electrical conductivity of metal oxides/hydroxide can be enhanced by combining them with CPs. Yin et al. prepared PEDOT/Ni-Mn-Co-O via solvothermal-co-precipitation method and also studied the effects of different doping of anions. The hybrid prepared by using this composite exhibited specific capacitance of 540 F g^{-1} at 1 A g^{-1} and 98.3% capacitance was achieved after 5000 cycles (Yin, Zhou, and Li 2019). The poor electrical conductivity of Bi$_2$O$_3$ material can be enhanced by combining it with conductive PEDOT. As-prepared CC/Bi$_2$O$_3$@PEDOT NAs showed vertical and interconnected network structure, which increases the contact area to connect with electrolyte, resulting in the enhanced electrochemical performance (Chen et al. 2019). Zhuzhelskii et al. synthesized PEDOT/WO$_3$ composite film delivered specific capacitance of 689 F g^{-1} in 0.5 M H$_2$SO$_4$ (Zhuzhelskii et al. 2019). Ambade et al. demonstrated controlled growth of PPy nanofiber in TiO$_2$ nanotube and such 1D PTh-TNT nanofibers showed great potential as supercapacitor electrode and exhibited ultrahigh specific capacitance of 1052 F g^{-1} and high coulombic efficiency (Ambade et al. 2017).

7.8 CONCLUSIONS

A comprehensive review of CPs for supercapacitor application has been examined. The CPs not only offer interesting characteristics such as doping–dedoping, excellent electrical conductivity, appropriate morphology, but also the changeable structures during charging–discharging process. Such structural changes reduce the cyclic stability and lead to reduced electrochemical performance. Such problems can be solved by synthesizing composite with carbon-based materials or transition metal oxide/hydroxides. In spite of such progress, there is still scope for the enhancement of energy density and power density of CP electrodes.

ACKNOWLEDGEMENT

We are sincerely thankful to Mr. Ram Dayal for his help during data collection.

REFERENCES

Afzal, Adeel, Faraj A. Abuilaiwi, Amir Habib, Muhammad Awais, Samaila B. Waje, and Muataz A. Atieh. 2017. "Polypyrrole/carbon Nanotube Supercapacitors Technological Advances and Challenges." *Journal of Power Sources* 352. Elsevier B.V.: 174–186. doi:10.1016/j.jpowsour.2017.03.128.

Alabadi, Akram, Shumaila Razzaque, Zehua Dong, Weixing Wang, and Bien Tan. 2016. "Graphene Oxide-Polythiophene Derivative Hybrid Nanosheet for Enhancing Performance of Supercapacitor." *Journal of Power Sources* 306. Elsevier B.V.: 241–247. doi:10.1016/j.jpowsour.2015.12.028.

Alvi, Farah, Manoj K. Ram, Punya A. Basnayaka, Elias Stefanakos, Yogi Goswami, and Ashok Kumar. 2011. "Graphene-Polyethylenedioxythiophene Conducting Polymer Nanocomposite Based Supercapacitor." *Electrochimica Acta* 56 (25). Elsevier Ltd: 9406–9412. doi:10.1016/j.electacta.2011.08.024.

Ambade, Rohan B., Swapnil B. Ambade, Nabeen K. Shrestha, Rahul R. Salunkhe, Wonjoo Lee, Sushil S. Bagde, Jung Ho Kim, Florian J. Stadler, Yusuke Yamauchi, and Soo Hyoung Lee. 2017. "Controlled Growth of Polythiophene Nanofibers in TiO$_2$ Nanotube Arrays for Supercapacitor Applications." *Journal of Materials Chemistry A* 5 (1): 172–180. doi:10.1039/c6ta08038c.

Aydinli, Alptekin, Recep Yuksel, and Husnu Emrah Unalan. 2018. "Vertically Aligned Carbon Nanotube – Polyaniline Nanocomposite Supercapacitor Electrodes." *International Journal of Hydrogen Energy* 43 (40). Elsevier Ltd: 18617–18625. doi:10.1016/j.ijhydene.2018.05.126.

Boddula, Rajender, Ravi Bolagam, and Palaniappan Srinivasan. 2017. "Incorporation of Graphene-Mn$_3$O$_4$ Core into Polyaniline Shell: Supercapacitor Electrode Material." *Ionics* 1–8. doi:10.1007/s11581-017-2300-x.

Chang, Hao Hsiang, Chih Kai Chang, Yu Chen Tsai, and Chien Shiun Liao. 2012. "Electrochemically Synthesized Graphene/Polypyrrole Composites and Their Use in Supercapacitor." *Carbon* 50 (6). Elsevier Ltd: 2331–2336. doi:10.1016/j.carbon.2012.01.056.

Chen, Jian, Yang Fu, Shizheng Zheng, Zhengyou Zhu, Zhiqiang Niu, Dachi Yang, and Lijun Zheng. 2019. "PEDOT Engineered Bi$_2$O$_3$ Nanosheet Arrays for Flexible Asymmetric Supercapacitors with Boosted Energy Density." *Journal of Materials Chemistry A* doi:10.1039/c9ta00854c.

Ci, Suqing, Zhenhai Wen, Yuanyuan Qian, Shun Mao, Shumao Cui, and Junhong Chen. 2015. "NiO-Microflower Formed by Nanowire-Weaving Nanosheets with Interconnected Ni-Network Decoration as Supercapacitor Electrode." *Scientific Reports* 5 (1). Nature Publishing Group: 11919. doi:10.1038/srep11919.

Das, Tapan K., and Smita Prusty. 2012. "Review on Conducting Polymers and Their Applications." *Polymer – Plastics Technology and Engineering* 51 (14): 1487–1500. doi:10.1080/03602559.2012.710697.

Dhawale, D. S., A. Vinu, and C. D. Lokhande. 2011. "Stable Nanostructured Polyaniline Electrode for Supercapacitor Application." *Electrochimica Acta* 56 (25). Elsevier Ltd: 9482–9487. doi:10.1016/j.electacta.2011.08.042.

Fan, Li Zhen, and Joachim Maier. 2006. "High-Performance Polypyrrole Electrode Materials for Redox Supercapacitors." *Electrochemistry Communications* 8 (6): 937–940. doi:10.1016/j.elecom.2006.03.035.

Fu, Chaopeng, Haihui Zhou, Rui Liu, Zhongyuan Huang, Jinhua Chen, and Yafei Kuang. 2012. "Supercapacitor Based on Electropolymerized Polythiophene and Multi-Walled Carbon Nanotubes Composites." *Materials Chemistry and Physics* 132 (2–3): 596–600. doi:10.1016/j.matchemphys.2011.11.074.

Gao, Xiaogang, Yiqing Zhang, Jing Yan, Panbo Liu, and Ying Huang. 2018. "Construction of Layer-by-Layer Sandwiched Graphene/Polyaniline Nanorods/Carbon Nanotubes Heterostructures for High Performance Supercapacitors." *Electrochimica Acta* 272. Elsevier Ltd: 77–87. doi:10.1016/j.electacta.2018.03.198.

Gupta, Vinay, Teruki Kusahara, Hiroshi Toyama, Shubhra Gupta, and Norio Miura. 2007. "Potentiostatically Deposited Nanostructured α-Co(OH)$_2$: A High Performance Electrode Material for Redox-Capacitors." *Electrochemistry Communications* 9 (9): 2315–2319. doi:10.1016/j.elecom.2007.06.041.

Gupta, Vinay, and Norio Miura. 2006. "High Performance Electrochemical Supercapacitor from Electrochemically Synthesized Nanostructured Polyaniline." *Materials Letters* 60 (12): 1466–1469. doi:10.1016/j.matlet.2005.11.047.

Halper, Marin S., and James C. Ellenbogen. 2006. "Supercapacitors: A Brief Overview." *Mitre Nanosystems* Group, 3. Retrieved 2015-02-16.

Han, Yongqin, and Liming Dai. 2019. "Conducting Polymers for Flexible Supercapacitors." *Macromolecular Chemistry and Physics* 1800355: 1800355. doi:10.1002/macp.201800355.

He, Xin, Wenyao Yang, Xiling Mao, Lu Xu, Yujiu Zhou, Yan Chen, Yuetao Zhao, Yajie Yang, and Jianhua Xu. 2018. "All-Solid State Symmetric Supercapacitors based on Compressible and Flexible Free-Standing 3D Carbon Nanotubes (CNTs)/poly(3,4-Ethylenedioxythiophene) (PEDOT) Sponge Electrodes." *Journal of Power Sources* 376: 138–146. doi:10.1016/j.jpowsour.2017.09.084.

Huang, Yan, Jiayou Tao, Wenjun Meng, Minshen Zhu, Yang Huang, Yuqiao Fu, Yihua Gao, and Chunyi Zhi. 2015. "Super-High Rate Stretchable Polypyrrole-based Supercapacitors with Excellent Cycling Stability." *Nano Energy* 11. Elsevier: 518–525. doi:10.1016/j.nanoen.2014.10.031.

Iro, Zaharaddeen S., C. Subramani, and S. S. Dash. 2016. "A Brief Review on Electrode Materials for Supercapacitor." *International Journal of Electrochemical Science* 11 (12): 10628–10643. doi:10.20964/2016.12.50.

Karaca, Erhan, Dinçer Gökcen, Nuran Özçiçek Pekmez, and Kadir Pekmez. 2019. "One-Step Electrosynthesis of polypyrrole/PbOx Composite in Acetonitrile as Supercapacitor Electrode Material." *Synthetic Metals* 247 (September 2018). Elsevier: 255–267. doi:10.1016/j.synthmet.2018.12.014.

Kate, Ranjit S., Suraj A. Khalate, and Ramesh J. Deokate. 2018. "Overview of Nanostructured Metal Oxides and Pure Nickel Oxide (NiO) Electrodes for Supercapacitors: A Review." *Journal of Alloys and Compounds* 734. Elsevier B. V.: 89–111. doi:10.1016/j.jallcom.2017.10.262.

Li, Xueliang, Changlong Han, Xiangying Chen, and Chengwu Shi. 2010. "Preparation and Performance of Straw Based Activated Carbon for Supercapacitor in Non-Aqueous Electrolytes." *Microporous and Mesoporous Materials* 131 (1–3). Elsevier Inc.: 303–309. doi:10.1016/j.micromeso.2010.01.007.

Liu, Tian-Yu, Dong-Yang Feng, Xin-Xin Xu, Yat Li, Xiao-Xia Liu, and Yu Song. 2015. "Pushing the Cycling Stability Limit of Polypyrrole for Supercapacitors." *Advanced Functional Materials* 25 (29): 4626–4632. doi:10.1002/adfm.201501709.

Liu, Tongchang, W.G. Pell, and B.E. Conway. 1997. "Self-Discharge and Potential Recovery Phenomena at Thermally and Electrochemically Prepared RuO$_2$ Supercapacitor Electrodes." *Electrochimica Acta* 42 (23–24): 3541–3552. doi:10.1016/S0013-4686(97)81190-5.

Lokhande, Prasad E., and Umesh S. Chavan. 2018a. "Conventional Chemical Precipitation Route to Anchoring Ni(OH)$_2$ for Improving Flame Retardancy of PVA." *Materials Today: Proceedings* 5 (8). Elsevier Ltd: 16352–16357. doi:10.1016/j.matpr.2018.05.131.

Lokhande, Prasad E., and Umesh S. Chavan. 2018b. "Nanoflower-like Ni(OH)$_2$ Synthesis with Chemical Bath Deposition Method for High Performance Electrochemical Applications." *Materials Letters* 218 (May): 225–228. doi:10.1016/j.matlet.2018.02.012.

Lokhande, Prasad E., and Umesh S. Chavan. 2019a. "Surfactant-Assisted Cabbage Rose-like CuO Deposition on Cu Foam by for Supercapacitor Applications." *Inorganic and Nano-Metal Chemistry* 0 (0). Taylor & Francis: 1–7. doi:10.1080/24701556.2019.1569685.

Lokhande, Prasad E., and Umesh S. Chavan. 2019b. "Materials Science for Energy Technologies Nanostructured Ni (OH) 2/rGO Composite Chemically Deposited on Ni Foam for High Performance of Supercapacitor Applications." *Materials Science for Energy Technologies* 2 (1). The Authors: 52–56. doi:10.1016/j.mset.2018.10.003.

Lokhande, Prasad E, and H.S. Panda. 2015. "Synthesis and Characterization of Ni.Co(OH)$_2$ Material for Supercapacitor Application." *IARJSET* 2 (8): 10–13. doi:10.17148/IARJSET.2015.2903.

Lokhande, Prasad E., Krishna Pawar, and Umesh S. Chavan. 2018. "Chemically Deposited Ultrathin α-Ni(OH)2 Nanosheet Using Surfactant on Ni Foam for High Performance Supercapacitor Application." *Materials Science for Energy Technologies* 1 (2). KeAi Communications Co., Ltd: 166–170. doi:10.1016/j.mset.2018.07.001.

Lokhande, Vaibhav C., Abhishek C. Lokhande, Chandrakant D. Lokhande, Jin Hyeok Kim, and Taeksoo Ji. 2016. "Supercapacitive Composite Metal Oxide Electrodes Formed with Carbon, Metal Oxides and Conducting Polymers." *Journal of Alloys and Compounds* 682. Elsevier Ltd: 381–403. doi:10.1016/j.jallcom.2016.04.242.

Long, Yun Ze, Meng Meng Li, Changzhi Gu, Meixiang Wan, Jean Luc Duvail, Zongwen Liu, and Zhiyong Fan. 2011. "Recent Advances in Synthesis, Physical Properties and Applications of Conducting Polymer Nanotubes and Nanofibers." *Progress in Polymer Science (Oxford)* 36 (10). Elsevier Ltd: 1415–1442. doi:10.1016/j.progpolymsci. 2011.04.001.

Ma, Jingyi, Xiaotian Guo, Yan Yan, Huaiguo Xue, and Huan Pang. 2018. "FeO X -Based Materials for Electrochemical Energy Storage." *Advanced Science* 5 (6): 1700986. doi:10.1002/advs.201700986.

Magu, Thomas Odey, Augustine U Agobi, Louis Hitler, and Peter Michael Dass. 2019. "A Review on Conducting Polymers-Based Composites for Energy Storage Application." *Journal of Chemical Reviews* 1 (1, pp.

1–77.): 19–34. www.jchemrev.com/article_77435.html%0Awww.jchemrev.com/pdf_77435_81cc25afb1dab7720b143850bf864697.html.

Meng, Qiufeng, Kefeng Cai, Yuanxun Chen, and Lidong Chen. 2017a. "Research Progress on Conducting Polymer based Supercapacitor Electrode Materials." *Nano Energy* 36 (February). Elsevier Ltd: 268–285: doi:10.1016/j.nanoen.2017.04.040.

Meng, Qiufeng, Kefeng Cai, Yuanxun Chen, and Lidong Chen. 2017b. "Research Progress on Conducting Polymer based Supercapacitor Electrode Materials." *Nano Energy* 36. Elsevier Ltd: 268–285. doi:10.1016/j.nanoen.2017.04.040.

Ni, Dan, Yuanxun Chen, Haijun Song, Congcong Liu, Xiaowei Yang, and Kefeng Cai. 2019. "Free-Standing and Highly Conductive PEDOT Nanowire Films for High-Performance All-Solid-State Supercapacitors." *Journal of Materials Chemistry A* 7 (3). Royal Society of Chemistry: 1323–1333. doi:10.1039/c8ta08814d.

Pan, Hui, Jianyi Li, and Yuan Ping Feng. 2010. "Carbon Nanotubes for Supercapacitor." *Nanoscale Research Letters* 5 (3): 654–668. doi:10.1007/s11671-009-9508-2.

Park, Yong Joon, Soon Ho Chang, Kwang Sun Ryu, Young-Sik Hong, Kwang Man Kim, Man Gu Kang, Young-Gi Lee, Nam-Gyu Park, and Xianlan Wu. 2004. "Poly(ethylenedioxythiophene) (PEDOT) as Polymer Electrode in Redox Supercapacitor." *Electrochimica Acta* 50 (2–3): 843–847. doi:10.1016/j.electacta.2004.02.055.

Patil, Bebi H., Ajay D. Jagadale, and Chandrakant D. Lokhande. 2012. "Synthesis of Polythiophene Thin Films by Simple Successive Ionic Layer Adsorption and Reaction (SILAR) Method for Supercapacitor Application." *Synthetic Metals* 162 (15–16). Elsevier B.V.: 1400–1405. doi:10.1016/j.synthmet.2012.05.023.

Qu, Qunting, Peng Zhang, Bin Wang, Yuhui Chen, Shu Tian, Yuping Wu, and Rudolf Holze. 2009. "Electrochemical Performance of MnO$_2$ Nanorods in Neutral Aqueous Electrolytes as a Cathode for Asymmetric Supercapacitors." *Journal of Physical Chemistry C* 113 (31): 14020–14027. doi:10.1021/jp8113094.

Rajesh, Murugesan, C. Justin Raj, Ramu Manikandan, Byung Chul Kim, Sang Yeup Park, and Kook Hyun Yu. 2017. "A High Performance PEDOT/PEDOT Symmetric Supercapacitor by Facile In-Situ Hydrothermal Polymerization of PEDOT Nanostructures on Flexible Carbon Fibre Cloth Electrodes." *Materials Today Energy* 6. Elsevier Ltd: 96–104. doi:10.1016/j.mtener.2017.09.003.

Ran, Fen, Yongtao Tan, Wenju Dong, Zhen Liu, Lingbin Kong, and Long Kang. 2018. "In Situ Polymerization and Reduction to Fabricate Gold Nanoparticle-Incorporated Polyaniline as Supercapacitor Electrode Materials." *Polymers for Advanced Technologies* 29 (6): 1697–1705. doi:10.1002/pat.4273.

Raza, Waseem, Faizan Ali, Nadeem Raza, Yiwei Luo, Ki Hyun Kim, Jianhua Yang, Sandeep Kumar, Andleeb Mehmood, and Eilhann E. Kwon. 2018. "Recent Advancements in Supercapacitor Technology." *Nano Energy* 52 (June). Elsevier Ltd: 441–473. doi:10.1016/j.nanoen.2018.08.013.

Rudge, Andy, John Davey, Ian Raistrick, Shimshon Gottesfeld, and John P. Ferraris. 1994. "Conducting Polymers as Active Materials in Electrochemical Capacitors." *J. Power Sources* 47 (1–2): 89–107. doi:10.1016/0378-7753(94)80053-7.

Saranya, Murugan, Rajendran Ramachandran, and Fei Wang. 2016. "Graphene-Zinc Oxide (G-ZnO) Nanocomposite for Electrochemical Supercapacitor Applications." *Journal of*

Science: Advanced Materials and Devices 1 (4). Elsevier Ltd: 454–460. doi:10.1016/j.jsamd.2016.10.001.

Sharma, Pawan, and T.S. Bhatti. 2010. "A Review on Electrochemical Double-Layer Capacitors." *Energy Conversion and Management* 51 (12). Elsevier Ltd: 2901–2912. doi:10.1016/j.enconman.2010.06.031.

Shi, Fan, Lu Li, Xiu-li Wang, Chang-dong Gu, and Jiang-ping Tu. 2014. "Metal Oxide/hydroxide-Based Materials for Supercapacitors." *RSC Advances* 4 (79): 41910–41921. doi:10.1039/C4RA06136E.

Shivakumara, S., and N. Munichandraiah. 2019. "In-Situ Preparation of Nanostructured α-MnO$_2$/polypyrrole Hybrid Composite Electrode Materials for High Performance Supercapacitor." *Journal of Alloys and Compounds* 787. Elsevier B.V.: 1044–1050. doi:10.1016/j.jallcom.2019.02.131.

Shown, Indrajit, Abhijit Ganguly, Li Chyong Chen, and Kuei Hsien Chen. 2015. "Conducting Polymer-Based Flexible Supercapacitor." *Energy Science and Engineering* 3 (1): 1–25. doi:10.1002/ese3.50.

Simon, Patrice, and Yury Gogotsi. 2008. "Materials for Electrochemical Capacitors." *Nature Materials* 7: 845–854.

Sivaraman, Padavattan, Ritesh Kumar Kushwaha, Kannakaje Shashidhara, Varsha R. Hande, Avinash P. Thakur, Asit Baran Samui, and Mahendra Khandpekar. 2010. "All Solid Supercapacitor Based on Polyaniline and Crosslinked Sulfonated Poly [ether Ether Ketone]." *Electrochimica Acta* 55 (7). Elsevier Ltd: 2451–2456. doi:10.1016/j.electacta.2009.12.009.

Sun, Wenping, Xianhong Rui, Mani Ulaganathan, Srinivasan Madhavi, and Qingyu Yan. 2015. "Few-Layered Ni(OH)$_2$ Nanosheets for High-Performance Supercapacitors." *Journal of Power Sources* 295. Elsevier B.V.: 323–328. doi:10.1016/j.jpowsour.2015.07.024.

Tan, Yongtao, Ying Liu, Lingbin Kong, Long Kang, and Fen Ran. 2017. "Supercapacitor Electrode of Nano-Co$_3$O$_4$ Decorated with Gold Nanoparticles via In-Situ Reduction Method." *Journal of Power Sources* 363. Elsevier B.V.: 1–8. doi:10.1016/j.jpowsour.2017.07.054.

Wang, Guoping, Lei Zhang, and Jiujun Zhang. 2012. "A Review of Electrode Materials for Electrochemical Supercapacitors." *Chemical Society Reviews* 41 (2): 797–828. doi:10.1039/C1CS15060J.

Wang, Kai, Jiyong Huang, and Zhixiang Wei. 2010. "Conducting Polyaniline Nanowire Arrays for High Performance Supercapacitors." *The Journal of Physical Chemistry C* 114 (Cv): 8062–8067. doi:10.1021/jp9113255.

Wang, Xingyan, Li Liu, Xianyou Wang, Li Bai, Hao Wu, Xiaoyan Zhang, Lanhua Yi, and Quanqi Chen. 2011. "Preparation and Performances of Carbon Aerogel Microspheres for the Application of Supercapacitor." *Journal of Solid State Electrochemistry* 15 (4): 643–648. doi:10.1007/s10008-010-1142-5.

Wu, Jifeng, Qin'E Zhang, Jingjing Wang, Xiaoping Huang, and Hua Bai. 2018. "A Self-Assembly Route to Porous Polyaniline/reduced Graphene Oxide Composite Materials with Molecular-Level Uniformity for High-Performance Supercapacitors." *Energy and Environmental Science* 11 (5): 1280–1286. doi:10.1039/c8ee00078f.

Wu, Xinming, Meng Lian, and Qiguan Wang. 2019. "A High-Performance Asymmetric Supercapacitors Based on Hydrogen Bonding Nanoflower-Like Polypyrrole and NiCo(OH)2 Electrode Materials." *Electrochimica Acta* 295. Elsevier Ltd: 655–661. doi:10.1016/j.electacta.2018.10.199.

Xiao, Gao-Fei, Jian Yuan, Xi Cao, Sheng Xu, Hong-Yan Zeng, and Jing Han. 2018. "Facile Fabrication of the Polyaniline/

layered Double Hydroxide Nanosheet Composite for Supercapacitors." *Applied Clay Science* 168 (November 2018). Elsevier: 175–183. doi:10.1016/j.clay.2018.11.011.

Xu, Lanshu, Mengying Jia, Yue Li, Shifeng Zhang, and Xiaojuan Jin. 2017. "Design and Synthesis of Graphene/Activated Carbon/Polypyrrole Flexible Supercapacitor Electrodes." *RSC Advances* 7 (50). Royal Society of Chemistry: 31342–31351. doi:10.1039/c7ra04566b.

Yasoda, K. Yamini, Alexey A. Mikhaylov, Alexander G. Medvedev, M. Sathish Kumar, Ovadia Lev, Petr V. Prikhodchenko, and Sudip K. Batabyal. 2019. "Brush Like Polyaniline on Vanadium Oxide Decorated Reduced Graphene Oxide: Efficient Electrode Materials for Supercapacitor." *Journal of Energy Storage* 22 (November 2018). Elsevier: 188–193. doi:S2352152X18307473.

Yan, Yan, Bing Li, Wei Guo, Huan Pang, and Huaiguo Xue. 2016. "Vanadium based Materials as Electrode Materials for High Performance Supercapacitors." *Journal of Power Sources* 329. Elsevier B.V.: 148–169. doi:10.1016/j.jpowsour.2016.08.039.

Yin, Chengjie, Hongming Zhou, and Jian Li. 2019. "Influence of Doped Anions on PEDOT/Ni-Mn-Co-O for Supercapacitor Electrode Material." *Applied Surface Science* 464: 220–228. doi:10.1016/j.apsusc.2018.09.028.

Yu, Zenan, Laurene Tetard, Lei Zhai, and Jayan Thomas. 2015. "Supercapacitor Electrode Materials: Nanostructures from 0 to 3 Dimensions." *Energy Environ. Sci.* 8 (3): 702–730. doi:10.1039/C4EE03229B.

Zhang, Haibin, Hanlu Li, Fengbao Zhang, Jixiao Wang, Zhi Wang, and Shichang Wang. 2008. "Polyaniline Nanofibers Prepared by a Facile Electrochemical Approach and Their Supercapacitor Performance." *Journal of Materials Research* 23 (9): 2326–2332. doi:10.1557/jmr.2008.0304.

Zhang, Kai, Zhang Li, X. S. Zhao, and Wu. Jishan. 2010. "Graphene/Polyaniline Nanofiber Composites as Supercapacitor Electrodes." *Chemistry of Materials* 22 (4): 1392–1401. doi:10.1021/cm902876u.

Zhang, Li Li, and X S Zhao. 2009. "Carbon-based Materials as Supercapacitor Electrodes." *Chemical Society Reviews* 38 (9): 2520. doi:10.1039/b813846j.

Zhang, Sanliang, Yueming Li, and Ning Pan. 2012. "Graphene based Supercapacitor Fabricated by Vacuum Filtration Deposition." *Journal of Power Sources* 206. Elsevier B.V.: 476–482. doi:10.1016/j.jpowsour.2012.01.124.

Zhao, Junhong, Jinping Wu, Bing Li, Weimin Du, Qingli Huang, Mingbo Zheng, Huaiguo Xue, and Huan Pang. 2016. "Facile Synthesis of Polypyrrole Nanowires for High-Performance Supercapacitor Electrode Materials." *Progress in Natural Science: Materials International* 26 (3). Elsevier: 237–242. doi:10.1016/j.pnsc.2016.05.015.

Zhi, Chunyi, Zengxia Pei, Yan Huang, Hongfei Li, Yang Huang, Zifeng Wang, Minshen Zhu, and Qi Xue. 2016. "Nanostructured Polypyrrole as a Flexible Electrode Material of Supercapacitor." *Nano Energy* 22. Elsevier: 422–438. doi:10.1016/j.nanoen.2016.02.047.

Zhong, Cheng, Yida Deng, Wenbin Hu, Jinli Qiao, Lei Zhang, and Jiujun Zhang. 2015. "A Review of Electrolyte Materials and Compositions for Electrochemical Supercapacitors." *Chemical Society Reviews* 44 (21): 7484–7539. doi:10.1039/c5cs00303b.

Zhou, Kun, Yuan He, Qingchi Xu, Qin'e Zhang, An'an Zhou, Zihao Lu, Li-Kun Yang, Yuan Jiang, Dongtao Ge, Xiang Yang Liu, and Hua Bai. 2018a. "A Hydrogel of Ultrathin Pure Polyaniline Nanofibers: Oxidant-Templating Preparation and Supercapacitor Application." *ACS Nano* 12 (6): 5888–5894. doi:10.1021/acsnano.8b02055.

Zhou, Qihang, Tong Wei, Jingming Yue, Lizhi Sheng, and Zhuangjun Fan. 2018b. "Polyaniline Nanofibers Confined into Graphene Oxide Architecture for High-Performance Supercapacitors." *Electrochimica Acta* 291. Elsevier Ltd: 234–241. doi:10.1016/j.electacta.2018.08.104.

Zhou, Yi, Zong Yi Qin, Li Li, Yu Zhang, Yu Ling Wei, Ling Feng Wang, and Mei Fang Zhu. 2010. "Polyaniline/Multi-Walled Carbon Nanotube Composites with Core-Shell Structures as Supercapacitor Electrode Materials." *Electrochimica Acta* 55 (12). Elsevier Ltd: 3904–3908. doi:10.1016/j.electacta.2010.02.022.

Zhuzhelskii, D. V., E. G. Tolstopjatova, S. N. Eliseeva, A. V. Ivanov, Shoulei Miao, and V. V. Kondratiev. 2019. "Electrochemical Properties of PEDOT/WO$_3$ Composite Films for High Performance Supercapacitor Application." *Electrochimica Acta* 299. Elsevier Ltd: 182–190. doi:10.1016/j.electacta.2019.01.007.

8 Conducting Polymer-Metal Based Binary Composites for Battery Applications

Aneela Sabir, Dania Munir, Samina Saleem and Faiqa Shakeel
Department of Polymer Engineering and Technology
University of the Punjab
Lahore, Pakistan

Muhammad Hamad Zeeshan
Institute of Chemical Engineering and Technology (ICET)
University of the Punjab
Lahore, Pakistan

Muhammad Shafiq
Department of Polymer Engineering and Technology
University of the Punjab
Lahore, Pakistan

Rafi Ullah Khan
Department of Polymer Engineering and Technology
University of the Punjab
Lahore, Pakistan
Institute of Chemical Engineering and Technology (ICET)
University of the Punjab
Lahore, Pakistan

Karl I Jacob
School of Materials Science and Engineering
Georgia Institute of Technology
Atlanta, GA, USA

CONTENTS

8.1   CONDUCTING POLYMER (CPS)

Conductive polymers, distinguished as the organic compounds are electrically conductive and were first discovered in 1977 (Skotheim and Reynolds 2007). These polymers possess excellent electrical and optical properties that were initially found only in organic polymers. These compounds can be semiconductors or may have metallic conductivity. Due to their insulating nature, these are generally thermosets (Le, Kim et al. 2017).

Over the past decades, CPs have gained considerable attention and have attracted many researchers and publishers. This has provided an elemental approach to chemistry, material science and physics of these materials which helps in the industrial growth of CPs products (Joo, Choi et al. 2006). Various CPs have been produced and extensively used due to their high conductive nature. CPs exhibit attractive metallic properties, including electronic, electrical, and optical, along with mechanical properties and processability of customary polymers (Abidian, Corey et al. 2010).

The fascinating characteristics of leading polymers are broadly investigated for the improvement of their practical application as nanomaterials and creation of cutting-edge innovative frameworks such as sensors, transducers energy components, natural sun-based cells, natural light-producing diodes, and field-impact transistors (Simotwo 2017). These polymers show a brilliant appealing uses of leading polymers in vehicle innovations and their utilization as straightforward anode materials. In contrast with other anode materials, leading polymers are profitable because of their minimal effort for energy storage, plastic substrate similarity, mechanical flexibility, and light weight (Jeong, Kim et al. 2019).

Although the polyacetylene doped with iodine was first announced in 1977 and it practically showed identical electrical conductivity to that of copper (Cu) and silver (Ag), but this leading polymer model was highly reactive. With the endeavors of numerous analysts in last three decades, diverse kinds of enhanced air-dependable CPs have been developed. Of all the polymers, polyaniline (PANI), poly (3,4-ethylene dioxythiophene) (PEDOT), and polypyrole have been recognized as the best three critical directing polymers (Macdiarmid, Chiang et al. 1987).

It is believed that the future persistence's and progress in the field of innovation such as microelectronics, sensors and biosensors, compound and biochemical designing that will be distressed about new information built nanostructured materials. Conducting and electroactive polymers (EAP) are generally novel materials, broadly achieved in the past decades. The study of advancement of CPs and their composites requires innovative work, coordinated to employing of novel materials in present and future (Meng, Cai et al. 2017).

Specifically, nanostructured electrochemistry-related polymer parts have pulled in a lot of intrigue. The uses of these CPs and their composites in the fields of electrochemical industries are mainly electric vitality frameworks, synthetic connections to-electric or the other way around vitality transformations, sensors and biosensors. In this chapter, the capability of leading polymer inadequate layer stored on nonconductive substrates was examined as straightforward anode materials for present applications. Specifically, the issues identified with the arrangements, straightforwardness, conductivity, and attachment of directing polymer slight films were attended to (Kim, McQuade et al. 2001).

8.2   CONDUCTING POLYMERS CONDUCTIVITY

Conducting polymers (CPs) are associated as energizing novel category of electronic materials since their revelation in 1977. To boost the properties, CPs are being utilized in muscles pretending, electronic gismo creation, daylight primarily based vitality modification, powered batteries, and sensors (Talikowska, Fu et al. 2019). This examination contains two primary items of examination. The primary polymer includes (center directional), e.g., polythiophene, polyparaphenylene vinylene, polycarbazole, polyaniline, and poly pyrole, and their functionality as, for example, star cells, light-emitting diodes (LEDs), supercapacitors, field impact semiconductor unit (FET), and biosensors (Patil, Heeger et al. 1988).

To gain conductivity, a polymer should have molecular d-orbitals overlapping and a large degree conjugation of π-bond. This sweeping conducting polymers π-bond associated system consists of uneven solitary and dual bonds among the polymer chain, as presented in Figure 8.1

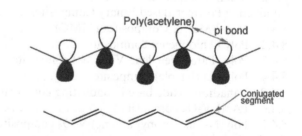

FIGURE 8.1   Conducting polymers π-conjugated system.

(Kivelson and Heeger 1988). In its neutral state, the exceedingly conjugated polymer is an insulating material.

During the polar formation (radical cation), π-bond electron of conjugated polymer backbone is removed, thus insulating polymer becomes conductive. In the conjugated polymer backbone, electrons in the d-orbitals become delocalized upon the removal of π-bond electron, resulting in polymer-free movement within the chain (Guerret-Legras, Audebert et al. 2019). The composition of the electric charge chemical compound backbone is in the middle of the amalgamation of a dopant ion that has the electrostatic impact on positive charges reconciliation (Cao, Smith et al. 1992).

Common conducting polymers such as polyaniline, polypyrrole, poly(3,4-ethylenedioxythiophe), polythiophene, and polyacetylene exhibit excellent conductivity of 30–200, 10–7500, 0.4–400, 10–1000, and 200–1000 Ωm, respectively (Kisliuk, Bocharova et al. 2019).

Upon doping, electrical conductivity of polyacetylene can largely be improved but is prone to atmospheric oxidation which limits its applications. Polyaniline and especially polypyrrole shows high electromagnetic (EMI) shielding properties. Their electrical conductivity can be tuned up by suitable doping from insulating to the metallic region. Poly(phenylene-vinylene) and Polythiophene are familiar for electroluminescent and photoluminescent properties (Diab, El-Ghamaz et al. 2017, Huang, Qian et al. 2018).

8.3 CONDUCTING POLYMER COMPOSITES

There are two essential types of CP composites with metals, as shown in Figure 8.2.

8.3.1 Metal Center Nanoparticles

It is verified within a main polymer shell. These composites are ordinarily masterminded on a colloid metal particle using the compound by the technique of electrochemical polymerization of an unstable, nanometer-sized main polymer layer. There are various methods distinguished in consideration of nanometer-sized coordinating polymer coat onto assorted substrates containing nanosized ones (Han, Wang et al. 2017).

8.3.2 Metal Nanoparticles

Metal nanoparticles are entrenched by a coordinating polymer cross section. These amalgams can be viably got through the metallic substance rebate of particles by their salt at the coordinating course of action interface. Numerous coordinating polymers, when exist in their abridged mode, have an adequately immense decreasing capability analogous to some metallic particles subsequently, but some metallic particles have acceptable great positive redox potential. A number of metallic particles that can be lessened at the layer of driving polymer include platinum (Pt), copper (Cu), silver (Ag), and gold (Au) surrounded by little particles inside penetrable coordinating polymer layer (Mansour, Poncin et al. 2019).

Owing to the immense metallic surface domain, these amalgams are now and again anticipated to be valuable in diverse electro catalytic structures for the electrochemical difference in solute species. This field was investigated before, however work does not try to demonstrate the properties of the metal particles entrenched in the polymer organize. In this regard, here we depict simply the nanodimension-related researches of the metal particles where these have definitely appeared (Kelly, Coronado et al. 2003).

8.4 CONDUCTING POLYMER-BASED BINARY COMPOSITES

Polyethylene/carbon black (PE/CB) or polycarbonate/carbon black (PC/CB) hot squeezed blends are one of the polymer-based composite materials that contain enough carbon atoms (15 wt%). This enough carbon verifies great conductivity and its conductivity was determined. These mass materials are effectively prepared in different shapes and are precisely very safe like metals which are lighter and less expensive substrates as presented in Figure 8.3 (Nasr Esfahani, Taheri Andani et al. 2016).

8.4.1 Metal Matrix Composites (MMC)

Metal network composites, at present, are manufacturing a good interest for analysis but do not seem to be

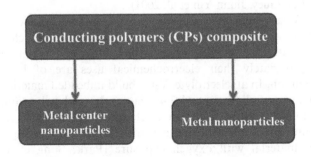

FIGURE 8.2 Conducting polymer composites.

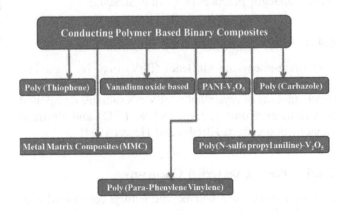

FIGURE 8.3 Conducting polymer based binary composites.

employed as MMC plastic partners. Top quality, crack strength, and firmness are offered by metal networks than those offered by their compound partners. These composites will stand up to hoisted temperature in damaging condition than compound composites (Behera, Dougherty et al. 2019). Most metals and compounds might be utilized as grids and need support materials that ought to be steady over a scope of temperature and should be nonresponsive. Furthermore, the dominant angle for the choice depends essentially on the framework material such as lightweight metals structure, the grid for temperature application thus, the fortifications reasons mentioned are pictured by high moduli (Clyne and Withers 1993).

Most metals and composites create good networks. But because of their intent and function, the choices for vasoconstrictor applications are found comparatively less. The lightweight metals are responsive; with their low thickness that demonstrate vantage. Titanium metal with atomic number 12 is the well-known framework metals, that are particularly useful for flying applications (Famodimu). On the off likelihood that metallic network materials provoke the top quality as they need high modulus fortifications. The standard to weight proportions of ongoing composites is beyond usual composite (Fudger, Klier et al. 2016).

The softening purpose, physical and mechanical properties of the composite at completely different temperatures decide the administration temperature of composites. Most metals, ceramics are produced, and mixtures are utilized for softening purpose. The choice of fortifications seems to be more and more hindered with increment within the liquefying temperature of framework materials (Rosso 2006).

8.4.2   Poly (Thiophene) Composite

Polythiophene has got considerable attention over the last 20 years because of its wide application. This reality has been made conceivable by its fascinating properties, for example, it's naturally available and safe to use. Various polyalkyl dependents of thiophene have been incorporated yet by synthetic and electrochemical methodologies, bringing about directing polymers with better solvency and higher capacitor practices (Fu, Shi et al. 2002).

8.4.3   Poly (Para-Phenylene Vinylene) Composite

Poly (para-phenylene vinylene) (PPV) may be a guiding compound of the inflexible bar polymer dynasty with elevated amounts of crystallinity. PPV is a valuable compound in various electronic gadgets such as LEDs and electrical phenomenon gadgets (Brédas and Heeger 1994).

8.4.4   Poly (Carbazole) Composite

Carbazole could be a heterocyclic natural compound with dynamic sides at three and six locations. The anodic chemical action of carbazole, significantly N-vinylcarbazole has been thought to be the chemical compound layer cathodes. The passion for polycarbazoles is likewise refreshed by their conceivable applications in electrochromic show gadgets and batteries (Nakabayashi and Mori 2012).

8.4.5   Vanadium Oxide-Based Conducting Composite

Vanadium oxide (V_2O_5) xerogel is fit for interpolate organic compound and metal particles or as encompassing composite substances with a wide scope of possible uses, particularly those which are related to electrochemistry (Sakamoto and Dunn 2002). Along with many species, pyrrole or aniline monomers can be interpolated into a xerogel of vanadium pentoxide (V_2O_5). This process of polymerization of these given monomers can happen either by the impairment of an existing V_2O_5 species inherent oxidation or by sub-nuclear oxygen. A couple of known processes were adopted to synthesize V_2O_5 xerogel (Xu, Sun et al. 2014).

8.4.6   PANI-V_2O_5 Composite

For instance, PANI that contains V_2O_5 xerogel could be synthesized by V_2O_5 softening in a solution of 10% made of H_2O_2, changing it to a wet gel by developing for a period of seven days. After that, wet gel is dried at raised temperature under the vacuum, then is imbedded to aniline in any organic mixture, and then calcining of this forerunner in an environment containing oxygen air at temperature of 460 °C for a couple of hours. Hence, a material for proper use as a cathode material for battery fueled Li-particles, could be obtained (Lira-Cantú and Gómez-Romero 1999).

Polyaniline imbedding into V_2O_5 resulted in basic innovation of the host system structure. It has been demonstrated with the process systems for changed frameworks. Because of its nanostructure, PANI-V_2O_5 composite exhibits characteristics such as charge trade between V^{5+} particles and those imbedded polyanilines along with intermolecular correspondence of the matrix of host. Studies by X-ray diffractions exposed the advancement of the V_2O_5 interlayer with interminable supply of polyaniline, by 0.48 nanometers (Chun-Guey, Jiunn-Yih et al. 2001).

A vast segment of the study managing nanocomposite V_2O_5 along with polyaniline or polypyrrole searched for battery-controlled applications and these materials appropriately their electrochemical uses are of basic attention. In an electrolyte, V_2O_5-build imbedded materials of poly(2,5-dimercapto-1,3,4-thiadiazole) and polyaniline demonstrate a basic electric discharge of 190 mA h g^{-1} at point of confinement and extended up to 220 mA h g^{-1} on material with oxygen exposure (Park, Song et al. 2010).

Charge accumulating capacity for V_2O_5 containing polyaniline composites from topical review are 1.86 to 2.25 mC/cm for the aggregate of the remote duties. This composite has been cultivated by composite layer by layer via electrostatic from poly aniline and V_2O_5 by a compactness of 2.5 nm per a bilayer. In development of the source polyaniline, its sulfonated backup, e.g., self-doped polyanilines have been used in composite synthesis (Liu, Yu et al. 2010).

8.4.7 POLY(N-SULFO PROPYL ANILINE)-V_2O_5 COMPOSITE

Those sulfonated dependents contain a sulfonate anion bounded to the backbone of that polymer, and thus over the range of electrochemical redox frames in this manner do not require an outside anion doping. The charge recompense was shown by using electrochemical quartz jewel microbalance (EQCM) during charging or discharging cycle for V_2O_5 composite with poly(N-sulfo propyl aniline) (Roy, Ray et al. 2018). This composite proceeding fantastically by positive lithium (Li^+) molecule transfer resulted in the camouflage of the swelling methodology of amalgam material cyclability, which is greater than that of V_2O_5. Vanadium pentoxide and versatile composite of poly(aniline-co-N-(sulfo phenyl)aniline) films were synthesized along very good firmness as compared to that of copolymer, and favored mechanical power than V_2O_5 xerogel (Huang, Virji et al. 2004).

8.5 CONDUCTING POLYMER COMPOSITE BATTERY APPLICATIONS

A general battery comprises a few electrochemical cells that are associated with each other to give voltage and ability to control the electronic gadgets. Every cell has a positive and a negative anode individually and they are isolated by an electrolyte arrangement containing separated salts, which empowers particle exchange between the two anodes (Naveen, Gurudatt et al. 2017). When the terminals are associated from outside gadgets, compound responses in the inner arrangements take place. By this capacity, the electrons in the arrangement are free and headed to make current. Some CP composite applications are given in Figure 8.4 (Zhao, Xia et al. 2012).

Among the most essential difficulties in the energy, investigate in the advancement of a vitality stockpiling gadget that can convey power for longer timeframes at higher power demand. An overseeing guideline, as described in Ragone plots of vitality thickness (ED) versus control thickness (PD), is that the deliverable vitality put away in a gadget diminishes with an expanding interest for power or current (Lepage, Savignac et al. 2019).

At high-power requests, the wastefulness of a gadget (loss of vitality) is because of the constraints including the mass exchange of particles or lazy response energy, and ionic or electric resistance inside materials or at interfaces between various stages of materials. Batteries and electric twofold layer capacitors (EDLCs) are the two types of a wide range of gadgets used to convey control. Batteries depend on faradic responses (e.g., electrochemical decrease and oxidation) and EDLCs depend on nonfaradaic responses (Macke, Schultz et al. 2012).

8.5.1 CONDUCTING POLYMER COMPOSITE FOR LITHIUM-ION (Li^+) BASED BATTERY

Lithium-particle batteries are the great encouraging, productive with basic great strength frameworks applied in the electrochemical vitality storing. The Li-particle battery innovation defines a reliable framework for power stockpiling showing essentially high vitality thickness and plan adaptability (Bhatt, Mehar et al. 2019). Generally used in "itinerant" electronic gadgets, such batteries are alike all the components of the CO_2 component discharges (as a promising force hotspot for cutting edge electric vehicles) as a potential cushion vitality stockpiling framework with discontinuous sustainable dealing with the power source assets (both on and off-the lattice) (Liu, Liu et al. 2015).

In this way, the generation and world utilization of Li-particle batteries depend on the expansion. Of late, the application of Li-particle batteries in a way such as electronic gadgets, and the enthusiasm for progressively viable and more secure batteries, has expanded (Liu, Huang et al. 2012).

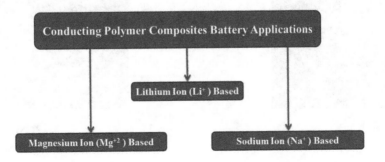

FIGURE 8.4 Conducting polymer composites battery applications.

In addition, there is an enthusiasm for polymer-built batteries that can be adaptable, delicate, and microelectronics. Such batteries have increased significant concern and the use of combustible natural solvents as electrolyte, lithium dendrites, and significant voltage fluctuations due to lack of basic soundness are amid the issues related to Li-particle batteries (Li, Ding et al. 2018). The primary hypothesis of Li-particle batteries is presented in Figure 8.5.

In Figure 8.5, negatively embedded anode materials of lithium with another positive embedded cathode material of lithium having positive redox potential in a lithium particle exchange cells. Through electrolyte process, anodic and cathodic materials are detached, which is an electronic cover yet in addition to a conductor of lithium particle. During charging process, lithium particles are discharged by the cathode material and when the cell is released, lithium particles are removed by the cathode materials (Aifantis, Hackney et al. 2010).

CPs are auspicious materials for natural inorganic half-breeding amalgams for Li-particle batteries because of their amazing characteristics. These include high electrical conductivity and Coulombic productivity, which cause them to be cycled. Conductive polymers display a few preferences, such as great processability, minimal effort, advantageous atomic adjustment, and light weight when connected as terminals. Low conductivity and poor dependability in cycling hindered their uses in Li-particle batteries. All these things are considered and the selection of nanostructured conductive polymers can incompletely conquest this issue, inferable from lithium-particles high surface and faster diffusion energy (Arruda, Kumar et al. 2012).

Inorganic CPs has got an expanding enthusiasm for Li-particle lattices batteries for repression. It is claimed that CPs synergistically mixed with inorganic, that increase anode life time, rates, voltage, mechanical and warm strength. Both applications of CP materials as anode and cathode exist but mostly they are applied as cathode in lithium-particle batteries (Meyer 1998). Various leading polymers show very different vitality and power densities in their synthetic structures and physical properties, e.g., PPy-built terminals have $10–50$ W h kg^{-1} vitality densities along with $5–25$ kW kg^{-1}, PANI-built anodes reveal $50–200$ W h kg^{-1} vitality density and $5–50$ kW kg^{-1} power densities, PTh-built cathodes reveal $20–100$ W h kg^{-1} vitality densities, and $5-50$ kW kg^{-1} power densities (Udayabhanu, Purty et al. 2019).

8.5.2 CONDUCTING POLYMER COMPOSITES FOR SODIUM-ION (Na⁺) BASED BATTERY

For green and manageable vitality, sodium (Na) particle batteries are examined to be the enhanced Li-particle innovation of the contradictory option, because of their material accessibility, minimal effort for storage, and ecological inevitability (Park, Son et al. 2015). Na-particle batteries currently in use are equipped with the most part of the nonintercalation inorganic material. Redox-dynamic polymers have been broadly investigated and observed to be good cathode dynamic materials for Na-particle batteries because of their auxiliary assorted variety and materials manageability (Kumar and Hashmi 2010). The working principle of sodium ion battery is presented in Figure 8.6.

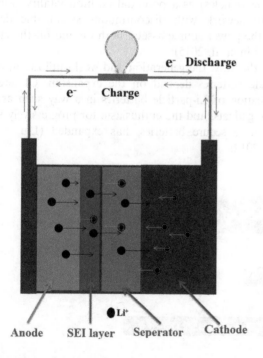

FIGURE 8.5 Li⁺-ion based battery principle.

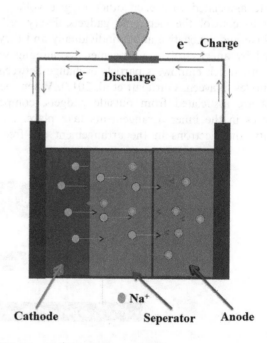

FIGURE 8.6 Na⁺-ion battery principle.

Basically, a Na-particle battery is used with a couple of natural cathodes and anodes, which are dependable on adequate potential contraction and can be combined with a battery response. All-natural Na-particle battery will be extremely easy-to-use for its scale and eco-cordiality behavior thus, has expansive scale electrical amassing applications (Wang, Liu et al. 2019). All-natural Na-particle batteries having p-doped poly triphenylamine as cathode material and n-doped redox-dynamic polyanthra-quinonyl sulfide without charging, as the progress materials are processed in electrochemical conventional batteries. Na-particle battery exhibited 1.8 V as an output voltage and 92 W h kg^{-1} vitality (Xiao, Zhang et al. 2019). Due to the auxiliary adaptability and redox dynamic polymers dependability, Na-particle battery was seen at a very high rate of 60% limit discharge information at a 16 C (3200 mA h g^{-1}) high rate and after 500 cycles at 8C rate along with great cycling security with 85% limit maintenance (Jia, Jin et al. 2019).

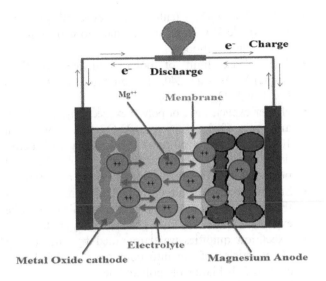

FIGURE 8.7 Mg^{++}- ion-based battery principle.

8.5.3 CONDUCTING POLYMER COMPOSITE FOR MG-ION (MG^{+2})-BASED BATTERY

In the ongoing improvements of progressive energy storage technologies, Li-ions batteries have been in focus globally. Until recently, these batteries were regarded as the best in terms of cost, long-term cycleability, and energy capacity per gram. However, many limitations with regard to high increase in cost of the materials required to prepare batteries and capacity hindrance by the mono-valent behavior of lithium-ion is now affecting their practical application (Pandey and Hashmi 2009).

From the theoretical standpoint, all the problems associated with lithium-ion batteries have been overcome by Mg-build batteries. Magnesium has double the theoretical capacity as compared to Li-ion cell, as it has a divalent ion. Furthermore, availability of magnesium (Mg) is much more than Li, thus lowering the manufacturing cost, handling of Mg is much more convenient than Li and Na, as it is less reactive in air (Chen, Niu et al. 2019). Mg has higher volumetric capacity of 3832 mA h cm^{-3}, as compared to those of Li and Na, which are 2062 and 1136 mA h cm^{-3}, respectively. Plating/stripping columbic efficiency (CE) of Mg was close to 100% and exhibited smooth, dendrite free Mg deposition as performed by some electrolyte. Since long, Li-metal battery development had been facing problems of low Coulombic efficiency and dendrite formation, as these discoveries are potentially transformative (Massé, Uchaker et al. 2015).

A remarkable breakthrough has been reported in the field of Mg rechargeable batteries, specifically with electrolyte fabrication and designing. Mg when comes in contact with conventional electrolyte (simple Mg salt mixture, Li-ion batteries aprotic solvent) and oxygen, forms a truly passivating film, resulting in Mg^{+2} migrations (Zeng, Cao et al.

2018). To enable the reversible Mg plating/stripping, no weakened passivation layer or solid electrolyte interphase (SEI) on Mg metal surface electrolyte associated with rechargeable Mg battery electrolyte has been applied, as shown in Figure 8.7. Mg(BH$_4$)$_2$ salt dissociation is increased by using nanocomposite polymer electrolyte, resulting in increased electrochemical performance.

8.6 CONDUCTING POLYMER-BASED COMPOSITES FOR ELECTRODE MATERIALS

- For Li-particles' battery, nanocomposite of polyaniline-V$_2$O$_5$ acts as the cathode material (Lira-Cantu and Gómez-Romero 1999).
- For auxiliary Li battery, cathode comprises of sulfonated polyaniline-V$_2$O$_5$ composite (Huguenin, Ticianelli et al. 2002).
- The initial release limit of polyaniline-V$_2$O$_5$ and polyaniline-poly(2,5-dimercapto-1,3,4-thiadiazole) is almost 190 mA h g^{-1} (versus Li/Li$^+$ range inside 4-2 V) (Guerra, Ciuffi et al. 2006).
- The cyclability of composite of poly(N-sulfopropyl) aniline-V$_2$O$_5$ is maximum as compared to V$_2$O$_5$ because of the lump procedure concealment.
- Poly(N-sulfopropyl) aniline-V$_2$O$_5$ amalgam has a greater explicit limit of 307 Ah kg^{-1}, and quicker decrease energy than V$_2$O$_5$ along with a specific charge of 200–302 Ah kg^{-1}, which relies on the sweep degree (Jeyakumari, Yelilarasi et al. 2010).
- V$_2$O$_5$ composites have a superior execution for polyaniline-V$_2$O$_5$ along with polypyrrole and polythiophene.
- The polypyrrole-V$_2$O$_5$ nanocomposite shows excellent execution after the postoxidative treatment (Huang, Virji et al. 2004).

- Polyaniline-V_2O_5 (enhanced revocability) and polypyrrole-V_2O_5 are among the more prominent, first-release composites.
- The release limit of poly(3,4-ethylenedioxythiophene)-V_2O_5 was 240, as compared to 140 mA h g^{-1} of V_2O_5.
- Higher explicit limit of poly (hexadecyl pyrrole)-V_2O_5 and poly (3-decyl pyrrole)-V_2O_5 was more than 100 Ah kg^{-1} after 50 revive cycles along with better dependability as compared with 50 Ah kg^{-1} of polypurrole-V_2O_5 (Huguenin, Girotto et al. 2002).
- Starting release limit of poly(2,5dimercapto-1,3,4-thiadizole)-V_2O_5 was 190 mA h g^{-1}, which could be expanded up to 220-mA h g^{-1} by the oxygen test.
- Execution qualities were upgraded by introducing nanoparticles of Ag into composite of 2,5-dimercapto-1,3,4-thiadiazole-polyaniline.
- Polypyrrole-maghemite (gamma-Fe_2O_3) composite has an expanded charge limit (Lira-Cantu and Gómez-Romero 1999).

REFERENCES

Abidian, Mohammad Reza, Joseph M. Corey, Daryl R. Kipke, and David C. Martin. "Conducting-Polymer Nanotubes Improve Electrical Properties, Mechanical Adhesion, Neural Attachment, and Neurite Outgrowth of Neural Electrodes." [In eng]. *Small (Weinheim an der Bergstrasse, Germany)* 6, no. 3 (2010/02 2010): 421–29.

Aifantis, Katerina E, Stephen A Hackney, and R. Vasant Kumar. *High Energy Density Lithium Batteries*. Wiley Online Library, 2010.

Arruda, Thomas M., Amit Kumar, Sergei V. Kalinin, and Stephen Jesse. "The Partially Reversible Formation of Li-Metal Particles on a Solid Li Electrolyte: Applications toward Nanobatteries." *Nanotechnology* 23, no. 32 (2012/07/23 2012): 325402.

Behera, Malaya, Troy Dougherty, and Sarat Singamneni. "Conventional and Additive Manufacturing with Metal Matrix Composites: A Perspective." *Procedia Manufacturing* 30 (01/01 2019): 159–66.

Bhatt, Pooja, Hemant Mehar, and Manish Sahajwani. "Electrical Motors for Electric Vehicle–a Comparative Study." *Available at SSRN 3364887* (2019).

Brédas, J. L., and A. J. Heeger. "Influence of Donor and Acceptor Substituents on the Electronic Characteristics of Poly(Paraphenylene Vinylene) and Poly(Paraphenylene)." *Chemical Physics Letters* 217, no. 5 (1994/01/28/ 1994): 507–12.

Cao, Yong, Paul Smith, and Alan J. Heeger. "Counter-Ion Induced Processibility of Conducting Polyaniline and of Conducting Polyblends of Polyaniline in Bulk Polymers." *Synthetic Metals* 48, no. 1 (1992/06/15/ 1992): 91–97.

Chen, Shuru, Chaojiang Niu, Hongkyung Lee, Qiuyan Li, Lu Yu, Wu Xu, Ji-Guang Zhang, *et al.* "Critical Parameters for Evaluating Coin Cells and Pouch Cells of Rechargeable Li-Metal Batteries." *Joule* 3, no. 4 (2019): 1094–105.

Chiang, C. K., C. R. Fincher, Y. W. Park, A. J. Heeger, H. Shirakawa, E. J. Louis, S. C. Gau, and Alan G. MacDiarmid. "Electrical Conductivity in Doped Polyacetylene." *Physical Review Letters* 40, no. 22 (05/29 1978): 1472–72.

Chun-Guey, Wu, Hwang Jiunn-Yih, and Hsu Shui-Sheng. "Synthesis and Characterization of Processible Conducting Polyaniline/V2o5 Nanocomposites." *Journal of Materials Chemistry* 11, no. 8 (2001): 2061–66.

Clyne, T. W., and P. J. Withers. *An Introduction to Metal Matrix Composites*. Cambridge Solid State Science Series. Cambridge: Cambridge University Press, 1993. doi: 10.1017/CBO9780511623080.

Diab, M. A., N. A. El-Ghamaz, F. Sh Mohamed, and E. M. El-Bayoumy. "Conducting Polymers Viii: Optical and Electrical Conductivity of Poly(Bis-M-Phenylenediaminosulphoxide)." *Polymer Testing* 63 (2017): 440–47.

Famodimu, Omotoyosi Helen. "Additive Manufacturing of Aluminium-Metal Matrix Composite Developed through Mechanical Alloying."

Fu, Mingxiao, Gaoquan Shi, Fengen Chen, and Xiaoyin Hong. "Doping Level Change of Polythiophene Film During Its Electrochemical Growth Process." *Physical Chemistry Chemical Physics - PHYS CHEM CHEM PHYS* 4 (05/29 2002): 2685–90.

Fudger, Sean, Eric Klier, Prashant Karandikar, Brandon McWilliams, and Chaoying Ni. "Mechanical Properties of Steel Encapsulated Metal Matrix Composites." In *Advanced Composites for Aerospace, Marine, and Land Applications Ii*, edited by T. Sano and T. S. Srivatsan, 121–36. Cham: Springer International Publishing, 2016.

Guerra, Elidia, Kaia Ciuffi, and Herenilton Oliveira. "V2o5 Xerogel–Poly(Ethylene Oxide) Hybrid Material: Synthesis, Characterization, and Electrochemical Properties." *Journal of Solid State Chemistry* 179 (12/01 2006): 3814–23.

Guerret-Legras, Laetitia, Pierre Audebert, Jean-Frédéric Audibert, Claude Niebel, Thibaut Jarrosson, Françoise Serein-Spirau, and Pierre Lère-Porte. "New Tbt Based Conducting Polymers Functionalized with Redox-Active Tetrazines." [In English]. *Journal of Electroanalytical Chemistry* 840 (2019–05 2019): 60–66.

Han, Jie, Minggui Wang, Yimin Hu, Chuanqiang Zhou, and Rong Guo. "Conducting Polymer-Noble Metal Nanoparticle Hybrids: Synthesis Mechanism Application." *Progress in Polymer Science* 70 (04/01 2017).

Huang, Congliang, Xin Qian, and Ronggui Yang. "Thermal Conductivity of Polymers and Polymer Nanocomposites." *Materials Science and Engineering: R: Reports* 132 (2018): 1–22.

Huang, Jiaxing, Shabnam Virji, Bruce H Weiller, and Richard B Kaner. "Nanostructured Polyaniline Sensors." *Chemistry – A European Journal* 10, no. 6 (2004): 1314–19.

Huguenin, Fritz, E. M. Girotto, Roberto Torresi, and Dan Buttry. "Transport Properties of V2o5/Polypyrrole Nanocomposite Prepared by a Sol-Gel Alkoxide Route." *Journal of Electroanalytical Chemistry* 536 (11/01 2002): 37.

Huguenin, Fritz, Edson A Ticianelli, and Roberto M Torresi. "Xanes Study of Polyaniline–V2o5 and Sulfonated Polyaniline–V2o5 Nanocomposites." *Electrochimica acta* 47, no. 19 (2002): 3179–86.

Jeong, Su-Hun, Hobeom Kim, Min-Ho Park, Yeongjun Lee, Nannan Li, Hong-Kyu Seo, Tae-Hee Han, *et al.* "Ideal Conducting Polymer Anode for Perovskite Light-Emitting Diodes by Molecular Interaction Decoupling." *Nano Energy* 60 (2019/06/01 2019): 324–31.

Jeyakumari, J., A. Yelilarasi, Balakrishnan Sundaresan, V. Dhanalakshmi, and R. Anbarasan. "Chemical Synthesis of Poly(Aniline-Co-O/M-Toluidine)/V2o5 Nano Composites and Their Characterizations." *Synthetic Metals* 160 (12/01 2010): 2605–12.

Jia, Miao, Yuhong Jin, Peizhu Zhao, Chenchen Zhao, Mengqiu Jia, Li Wang, and Xiangming He. "Hollow Nicose2 Microspheres@N-Doped Carbon as High-Performance Pseudocapacitive Anode Materials for Sodium Ion Batteries." *Electrochimica Acta* 310 (04/01 2019).

Joo, Won-Jae, Tae-Lim Choi, Jaeho Lee, Sang Kyun Lee, Myung-Sup Jung, Nakjoong Kim, and Jong Min Kim. "Metal Filament Growth in Electrically Conductive Polymers for Nonvolatile Memory Application." *The Journal of Physical Chemistry B* 110, no. 47 (2006/11/01 2006): 23812–16.

Kelly, K. Lance, Eduardo Coronado, Lin Lin Zhao, and George C. Schatz. "The Optical Properties of Metal Nanoparticles: The Influence of Size, Shape, and Dielectric Environment." *The Journal of Physical Chemistry B* 107, no. 3 (2003/01/01 2003): 668–77.

Kim, J., D. T. McQuade, A. Rose, Z. Zhu, and T. M. Swager. "Directing Energy Transfer within Conjugated Polymer Thin Films." [In eng]. *Journal of the American Chemical Society* 123, no. 46 (11/21 2001): 11488–89.

Kisliuk, A., V. Bocharova, I. Popov, C. Gainaru, and A. P. Sokolov. "Fundamental Parameters Governing Ion Conductivity in Polymer Electrolytes." *Electrochimica Acta* 299, no. Angew. Chem., Int. Ed. 55 2016 (2019): 191–96.

Kivelson, S., and A. J. Heeger. "Intrinsic Conductivity of Conducting Polymers." *Synthetic Metals* 22, no. 4 (1988/02/01 1988): 371–84.

Kumar, Deepak, and S. A. Hashmi. "Ionic Liquid Based Sodium Ion Conducting Gel Polymer Electrolytes." *Solid State Ionics* 181, no. 8–10 (2010): 416–23.

Le, Thanh-Hai, Yukyung Kim, and Hyeonseok Yoon. "Electrical and Electrochemical Properties of Conducting Polymers." *Polymers* 9, no. 4 (2017): 150.

Lepage, D., L. Savignac, M. Saulnier, S. Gervais, and S. B. Schougaard. "Modification of Aluminum Current Collectors with a Conductive Polymer for Application in Lithium Batteries." *Electrochemistry Communications* 102 (2019/05/01/ 2019): 1–4.

Li, Yang, Fei Ding, Zhibin Xu, Lin Sang, Libin Ren, Wang Ni, and Xingjiang Liu. "Ambient Temperature Solid-State Li-Battery Based on High-Salt-Concentrated Solid Polymeric Electrolyte." *Journal of Power Sources* 397 (09/01 2018): 95–101.

Lira-Cantú, M., and P. Gómez-Romero. "Synthesis and Characterization of Intercalate Phases in the Organic–Inorganic Polyaniline/V2o5 System." *Journal of Solid State Chemistry* 147, no. 2 (1999/11/01 1999): 601–08.

Lira-Cantu, Monica, and Pedro Gómez-Romero. "The Hybrid Polyaniline/V2o5 Xerogel and Its Performance as Cathode in Rechargeable Lithium Batteries." *Journal of New Materials for Electrochemical Systems* 2 (04/01 1999): 141–44.

———. "The Organic-Inorganic Polyaniline/V2o5 System - Application as a High-Capacity Hybrid Cathode for Rechargeable Lithium Batteries." *Journal of the Electrochemical Society* 146 (06/01 1999): 2029–33.

Liu, Chenguang, Zhenning Yu, David Neff, Aruna Zhamu, and Bor Z. Jang. "Graphene-Based Supercapacitor with an Ultrahigh Energy Density." *Nano Letters* 10, no. 12 (2010/12/08 2010): 4863–68.

Liu, Wei, Nian Liu, Jie Sun, Po-Chun Hsu, Yuzhang Li, Hyun-Wook Lee, and Yi Cui. "Ionic Conductivity Enhancement of Polymer Electrolytes with Ceramic Nanowire Fillers." *Nano Letters* 15, no. 4 (2015/04/08 2015): 2740–45.

Liu, Xian-Ming, Zhen dong Huang, Sei woon Oh, Biao Zhang, Peng-Cheng Ma, Matthew M. F. Yuen, and Jang-Kyo Kim. "Carbon Nanotube (Cnt)-Based Composites as

Electrode Material for Rechargeable Li-Ion Batteries: A Review." *Composites Science and Technology* 72, no. 2 (2012/01/18 2012): 121–44.

Macdiarmid, A. G., J. C. Chiang, A. F. Richter, and A. J. Epstein. "Polyaniline: A New Concept in Conducting Polymers." *Synthetic Metals* 18, no. 1 (1987/02/01 1987): 285–90.

Macke, Anthony, Benjamin Schultz, and P. K. Rohatgi. "Metal Matrix Composites Offer the Automotive Industry an Opportunity to Reduce Vehicle Weight, Improve Performance." *Advanced Materials and Processes* 170 (03/01 2012): 19–23.

Mansour, Agapy, Fabienne Poncin, and D. Debarnot. "Distribution of Metal Nanoparticles in a Plasma Polymer Matrix According to the Structure of the Polymer and the Nature of the Metal." *Thin Solid Films* (04/01 2019).

Massé, Robert C, Evan Uchaker, and Guozhong Cao. "Beyond Li-Ion: Electrode Materials for Sodium-and Magnesium-Ion Batteries." *Science China Materials* 58, no. 9 (2015): 715–66.

Meng, Qiufeng, Kefeng Cai, Yuanxun Chen, and Lidong Chen. "Research Progress on Conducting Polymer Based Supercapacitor Electrode Materials." *Nano Energy* 36 (2017): 268–85.

Meyer, Wolfgang H. "Polymer Electrolytes for Lithium-Ion Batteries." *Advanced Materials* 10, no. 6 (1998): 439–48.

Nakabayashi, Kazuhiro, and Hideharu Mori. "Novel Complex Polymers with Carbazole Functionality by Controlled Radical Polymerization." *International Journal of Polymer Science* 2012 (02/28 2012).

Nasr Esfahani, Sajedeh, Mohsen Taheri Andani, Narges Shayesteh Moghaddam, Reza Mirzaeifar, and Mohammad Elahinia. "Independent Tuning of Stiffness and Toughness of Additively Manufactured Titanium-Polymer Composites: Simulation, Fabrication, and Experimental Studies." *Journal of Materials Processing Technology* 238 (06/01 2016).

Naveen, Malenahalli Halappa, Nanjanagudu Ganesh Gurudatt, and Yoon-Bo Shim. "Applications of Conducting Polymer Composites to Electrochemical Sensors: A Review" *Applied Materials Today* 9 (2017): 419–33.

Pandey, G. P., and S. Hashmi. "Experimental Investigations of an Ionic-Liquid-Based, Magnesium Ion Conducting, Polymer Gel Electrolyte." *Journal of Power Sources - J POWER SOURCES* 187 (02/01 2009): 627–34.

Park, A. Reum, Dae-Yong Son, Jung Sub Kim, Jun Young Lee, Nam-Gyu Park, Juhyun Park, Joong Kee Lee, and Pil J. Yoo. "Si/Ti2o3/Reduced Graphene Oxide Nanocomposite Anodes for Lithium-Ion Batteries with Highly Enhanced Cyclic Stability." *ACS Applied Materials & Interfaces* 7, no. 33 (2015/08/26 2015): 18483–90.

Park, Kyung-Il, Hahn-Mok Song, Youna Kim, Sun-il Mho, Won Cho, and In-Hyeong Yeo. "Electrochemical Preparation and Characterization of V 2o 5/Polyaniline Composite Film Cathodes for Li Battery." *Electrochimica Acta - ELECTROCHIM ACTA* 55 (11/01 2010): 8023–29.

Patil, A. O., A. J. Heeger, and Fred Wudl. "Optical Properties of Conducting Polymers." *Chemical Reviews* 88, no. 1 (1988/01/01 1988): 183–200.

Rosso, M. "Ceramic and Metal Matrix Composites: Routes and Properties." *Journal of Materials Processing Technology* 175 (06/01 2006): 364–75.

Roy, Atanu, Apurba Ray, Priyabrata Sadhukhan, Samik Saha, and Sachindranath Das. "Morphological Behaviour, Electronic Bond Formation and Electrochemical Performance Study of V2o5-Polyaniline Composite and Its Application in Asymmetric Supercapacitor." *Materials Research Bulletin* 107 (2018/11/01/ 2018): 379–90.

Sakamoto, Jeffrey S, and Bruce Dunn. "Vanadium Oxide-Carbon Nanotube Composite Electrodes for Use in Secondary Lithium Batteries." *Journal of the Electrochemical Society* 149, no. 1 (2002): A26–A30.

Simotwo, Silas K. *Polyaniline Based Nanofibers for Energy Storage and Conversion Devices.* Drexel University, 2017.

Skotheim, Terje A, and John Reynolds. *Handbook of Conducting Polymers, 2 Volume Set.* CRC Press, 2007.

Talikowska, M., X. Fu, and G. Lisak. "Application of Conducting Polymers to Wound Care and Skin Tissue Engineering: A Review." [In eng]. *Biosens Bioelectron* 135 (6/15 2019): 50–63.

Udayabhanu, G., Bela Purty, Ram Choudhary, and Amrita Biswas. "Temperature Dependent Supercapacitive Performance of Nh3 Modified Tio2 Decorated Ppy Nanohybrids in Various Electrolytes." *Synthetic Metals* 249 (02/15 2019): 1–13.

Wang, Ying, Xiaoxu Liu, Chen Yang, Na Li, Kai Yan, Tianyi Ji, Hongyang Chi, *et al.* "Low Cost Fabrication of Three-Dimensional Hierarchical Porous Graphene Anode Material

for Sodium Ion Batteries Application." *Surface and Coatings Technology* 360 (2019/02/25/ 2019): 110–15.

Xiao, Bin, Wen-hai Zhang, Peng-Bo Wang, Lin-Bo Tang, Chang-Sheng An, Zhen-Jiang He, Hui Tong, Jun-chao Zheng, and Bo Wang. "V2(Po4)O Encapsulated into Crumpled Nitrogen-Doped Graphene as a High-Performance Anode Material for Sodium-Ion Batteries." *Electrochimica Acta* 306 (03/01 2019).

Xu, Jie, Huajun Sun, Zhaolong Li, Shan Lu, Xiaoyan Zhang, Shanshan Jiang, Quanyao Zhu, and Galina Zakharova. "Synthesis and Electrochemical Properties of Graphene/V2o5 Xerogels Nanocomposites as Supercapacitor Electrodes." *Solid State Ionics* 262 (09/01 2014): 234–37.

Zeng, Jing, Zulai Cao, Yang Yang, Yunhui Wang, Yueying Peng, Yiyong Zhang, Jing Wang, and Jinbao Zhao. "A Long Cycle-Life Na-Mg Hybrid Battery with a Chlorine-Free Electrolyte Based on Mg(Tfsi)2." *Electrochimica Acta* 284 (2018/09/10 2018): 1–9.

Zhao, Xiuyun, Dingguo Xia, and Kun Zheng. "Fe3o4/Fe/Carbon Composite and Its Application as Anode Material for Lithium-Ion Batteries." *ACS Applied Materials & Interfaces* 4, no. 3 (2012/03/28 2012): 1350–56.

9 Novel Conducting Polymer-Based Battery Application

Aneela Sabir
Department of Polymer Engineering and Technology
University of the Punjab
Lahore, Pakistan

Faiza Altaf
Department of Environmental Sciences (Chemistry)
Fatima Jinnah women university
Rawalpindi, Pakistan

Zahid Bashir, Muhammad Hassan Tariq, Muhammad Khalil and Muhammad Shafiq
Department of Polymer Engineering and Technology
University of the Punjab
Lahore, Pakistan

Muhammad Hamad Zeeshan
Institute of Chemical Engineering and Technology (ICET)
University of the Punjab
Lahore, Pakistan

Rafi Ullah Khan
Department of Polymer Engineering and Technology
University of the Punjab
Lahore, Pakistan

Institute of Chemical Engineering and Technology (ICET)
University of the Punjab
Lahore, Pakistan

Karl I Jacob
School of Materials Science and Engineering
Georgia Institute of Technology
Atlanta, GA, USA

CONTENTS

9.1   CONDUCTING POLYMERS (CPs)

In conductive polymers, power is conducted by delocalization of electrons. Such polymer conductors may either be a semiconductor or have metallic conductivity, and these materials are natural insulating materials. Their mechanical properties are not analogous with the financially accessible polymers but they offer high electrical conductivities (Zheng et al. 2019). Cutting-edge scattering method and naturally blend strategies calibrate the electrical properties of the conducting polymers. Conducting polymers incorporated as nanomaterial's are of maneuvering because of their applications essentially changes due to the properties of their mass partner (Shklovsky et al. 2018). Some novel conducting polymers are presented in Figure 9.1.

9.1.1   POLY(ACETYLENE)

Poly(acetylene) was created by Shirakawa during 1970s. Polyacetylene was coincidentally created by Shirakawa during the 70's. It is the most suitable polymer used for conduction. It is a naturally occurring polymer along with the $(C_2H_2)n$ rehashing unit. Conducting poly(acetylene) development profoundly prompted the fast journey in study due to the revelation of novel conducting polymers. Because of these polymers high electrical conductivity has driven thoughtful enthusiasm for the utilization of natural mixes in microelectronics (Wang, Sun et al. 2018).

9.1.2   POLY(THIOPHENE)

Of all the available conducting polymers, Polythiophene (PTh) is the one that fabricates the earth and electrical supercapacitor as heat stable materials, PLEDs, electrochromic,

photo resistive, non-straight optics, sensors, batteries. Moreover, it also fabricate the supercapacitor with the properties of electromagnet insulating materials, antistatic coatings, sun-oriented cells, transistors, memory gadgets and envisaging materials. Another Poly (thiophene) subsidiary was produced in 1980 by two researchers in the labs at the Bayer AG in Germany (Venkatesan and Cindrella 2018, Wang, Baek et al. 2018).

9.1.3   POLY(ANILINE)

Polyaniline (PANI) is a semi-flexible conducting polymer. It is the most seasoned polymer, found in the nineteenth century by Henry Lethe who observed the oxidation of aniline as a result of electrochemical in acidic media (Yelilarasi et al. 2019). PANI is of much significance because of its novel properties, for example, high electrical conductivity, light weight, high toughness, and adaptability. It is also called aniline black (Li et al. 2014).

9.1.4   POLY(PYRROLE)

Polypyrrole (PPy) is prepared by the polymerization of pyrrole and is a naturally existing polymer. It is the part of conducting polymer family and was first found in 1968. PPy is used with various conducting polymers mostly due to its easy availability, unrivaled redox properties, balanced out oxidized structure, ability to give transcending conductivity, water solvency, affordability, and profitable optical and electrical properties (Mao and Zhang 2016).

These polymers are known to be the part of pyrrole but these secretive polymers have not been portrayed in immense detail that fundamentally hold α,α' carbons. In 1963, Weiss and colleagues clarified the tetraiodopyrrole pyrolysis to manufacture shockingly conductive materials. In 1979, an enhanced electrochemical method was utilized to incorporate unsupported films with sufficiently great mechanical properties to contemplate this framework as a conducting polymer (Mizera et al. 2019).

9.1.5   POLY(PARAPHENYLENE) AND POLY(PHENYLENE)

Polyparaphenylene (PPP) and polyphenylene are the forerunners of conducting polymer family composed of rehashing p-phenylene units and utilizing an oxidant or a dopant to change its conducting structure. To achieve the conductivity tantamount to PA, it was doped in year 1980 (Geetha and Trivedi 2005, Gil Herrera et al. 2008).

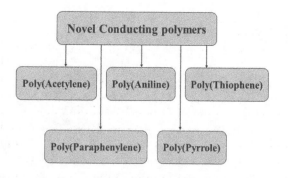

FIGURE 9.1   Novel conducting polymers.

Conducting properties of nanocetylenic hydrocarbon polymer were gained by doping it with either an electron acceptor or an electron contributor. Electrochemical oxidation delivers very high crystalline PPP films from benzene/H_2SO_4 (96%) arrangements. Some conductive polymers such as polyaniline, polyparavinylene, polythiophene, polyacetylene, and polyparaphenylene, with conductivity of 200, 10^4, 10^3, 10^4, and 10^3 S cm^{-1}, respectively, exhibit excellent conductive properties (Sahiner and Demirci 2016).

9.2 BATTERY APPLICATIONS OF CONDUCTING POLYMERS

A variety of applications in metal-based batteries have been found so far. Some important applications are given in Figure 9.2.

9.2.1 LITHIUM SULFIDE BATTERIES

Lithium–sulfur batteries are one of the striking types of powered batteries for their better explicit vitality. Li–S batteries are generally found to be light (about the thickness of water) due to the low nuclear load of lithium (Li) and reasonable load of sulfur (S) (Nair et al. 2016). In August 2008, they were utilized on the commercially and most noteworthy sun powered controlled plane trip. Li–S batteries may achieve Li-particle cells considering their higher vitality thickness and diminished expense from the utilization of sulfur (S) (Zeng et al. 2019). At present, the better lithium–sulfur batteries offer explicit powers of 500 Wh kg^{-1}, which are much higher than 150–250 Wh kg^{-1} of most Li-particle batteries. Li–S batteries with up to 1500 charges have been illustrated along with the release cycles. Started in mid-2014, none were industrially accessible. The key issue of Li–S batteries is the low sulfur cathode electrical conductivity, thus requiring an additional mass for a conducting operator (Zhou et al. 2019).

9.2.2 BINDER FOR LITHIUM SULFIDE BATTERY CATHODE

In lithium-sulfide batteries, binder is the essential part of sulfur cathode. It incorporates the dynamic materials (conductive and sulfur enclosed stuff carbon) composed of homogenous amalgam. Within the interim, it influences this compound which carries flair ability, for instance, tin foil, through the solid cement compel among the binder and therefore act as the gift gatherer (Zhang et al. 2017). The associated properties should run by an authorized folio. Within the initial place, the skillful scattering of sulfur and carbon is predicted to produce identical S/C compound. Second, proper attachment amid sulfur and carbon accomplish continuous conductive negatron pathways. Third, cowl should be insoluble to stay up during the formation of cathode within the solution (Qi et al. 2019).

Nevertheless, it might bulge within the solution arrangement so metallic element charge will move to the electro-dynamic material amid charge/release procedure. The chemical compound cowl should be electrochemically and unnaturally stable amid the phone's running procedure between 1-3 V vs Li/Li$^+$ within the potential window (Akhtar et al. 2018).

Conducting polymers for sulfur cathode have been investigated as binders to enhance the conductivity of anode. Poly(ethylenedioxythiophene) (PEDOT) conjugate was initially conferred as associate electrically conducting binder for lithium-sulfide cells. This binder cover up consolidating PAA enlargement properties such as electronic conductivity and compound body process capability by Li_2Sn doped poly styrene sulfonate (PEDOT-PSS) (Zheng et al. 2018).

This doped PEDOT-PSS strengthens the lithium-sulfide cell execution, structured a versatile, conductive, electro-dynamic nanocomposite binder created out of polyurethane (PU) and polypyrrole (PPy) for an adjustable sulfur cathode on carbon sensed, which diminished initiation potential and increased terminal execution (Radha et al. 2017).

A semiconducting binder sulfur corrosive-doped PANI (polyaniline) was effectively created in m-cresol, presented in Figure 9.4, and sulfur particles due to the electricity and Van der Waal's collaborations, prompting the incorporated terminal with low matter of binder (~2 wt%). The heteroatoms and electrically semiconducting network might facilitate electrostatically sorbs the contrarily charged metal sulfides and polysulfides at intervals with charging-releasing methodology (Li et al. 2018).

With the exceptional trademark, the cathode with PANI binder incontestable associate in nursing underlying the limit of 872 mAh g^{-1} and omitted 439 mAh g^{-1} once in 50 cycles, which was twofold as compared to that with PVDF cowl. Due to high physical phenomenon of cathode created by PANI, the responsibility and better limit over past folios at completely different unleash flows were acquired (Li et al. 2019).

9.2.3 SULFUR ENCAPSULATION FOR ELECTRODE MATERIALS

To bind the polysulfide dispersion, the polymers have been broadly connected as covering layer which limits

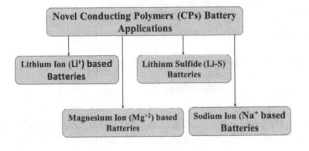

FIGURE 9.2 Conducting polymers battery applications.

the dynamic material inside the cathode for better limit maintenance. The polymers ought to be synthetically and electrochemically stable amid Li–S battery activity (Sovizi and Fahimi 2018).

Furthermore, important volumetric differences that was gained by high mechanical properties in the Li-S redox items. Besides, it ought to be anything but difficult to get ready and manufacture such a covering material. Moreover, the volume and mass stacking of the polymer coatings ought to be insignificant. These prerequisites have to be mulled over to guarantee that the complete electrodynamic sulfur stacking and vitality thickness are expanded (Ren et al. 2019).

9.2.4 Sulfur Encapsulation Through Conductive Polymers

To enhance conduction of the conducting sulfur cathode while embodying semi conductive macromolecules is commonly preferred by specialists. The sulfur polypyrrole was founded by the artificial polymerization technique with (dopant) Na (p-toluenesulfonate), 4-styrenesulphonic Na salt (surfactant), and $FeCl_3$ (oxidant). The polypyrrole network diminishes the molecule to molecule contact resistance, thus upgraded the composite's conduction(Cai et al. 2018).

In view of the conduction and S-restricting capability, PPy was focused as an S-coating material. In Figure 9.3(a), 1polymer sulfur composite comprising polypyrrole incorporated by poly(2-acryamido-2-methyl-1-1propane-sulfonic acid) (PAAMPSA) boosted the particle lepton conduction. In nanoparticles complex, the PAAMPSA maintain a powerful nonetheless permeable terminal structure by mixed ion electron conductor (MIEC) as shown in Figure 9.3(b), that essentially increased the chemical science execution through modification and chemical science resistivity presented in Figure 9.3(c) (Liu et al. 2018).

Then again, to boost the polysulfide catching capability as presented in Figure 9.4(a), the $PPy-MnO_2$ nanotubes were incorporated during pyrrole unaltered polymerization by mistreatment of MnO_2 nanowires due to each oxidization instigator and format (Bahloul, Nessark et al. 2013). Contrasted and pure PPy-typified sulfur, the $S/PPy-MnO_2$ amalgams incontestable better improved chemical science execution, with columbic productivity, cyclic security and rate capability as shown in Figure 9.4(b). The solid collaboration among polysulfides, MnO_2, and the versatile and conducting compound with rounded structure was accountable for the improved properties (Wang et al. 2015).

In the interim, apart from PPy, the conducting polymers such as PANI, (PEDOT) can be booted through heat filtration or compound covering and used as the host materials for sulfur. Visible of various varieties of unpretentious structure arrange of the PANI nanostructure, Chow et al. created the PANI yolk shell structure covered with sulfur for Lithium-Sulfide batteries. Pilar cyst et al. planned a sulfur cathode with brilliant

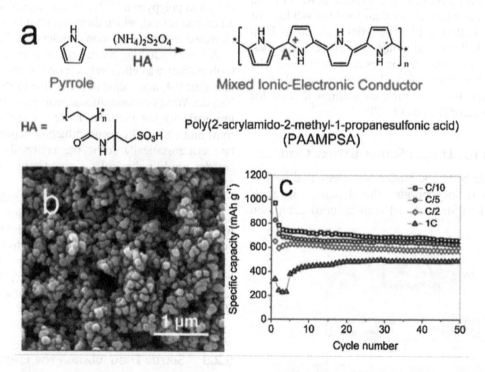

FIGURE 9.3 (a) PAAMPSA mixture doped with PPy intermingled MIEC schematic delineation; (b) Integrated MIEC SEM images; (c) S-MIEC composite cycleability at complete different C-rate. Leaky surface topology, polypyrrole nanoparticles covered onto the sulfur powder skin, which has ability to ingest polysulfide.

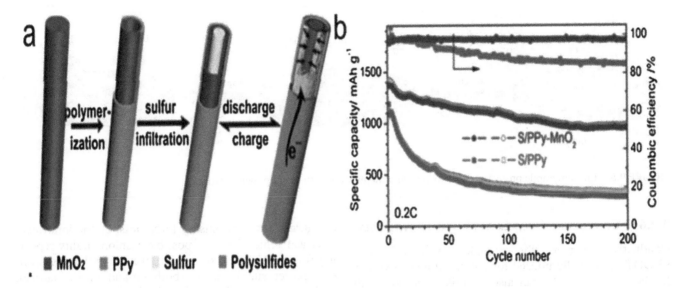

FIGURE 9.4 (a) PPy-MnO$_2$ composite nanotubes. (b) S/PPy and S/PPy-MnO$_2$ improved properties.

chemical science execution enthusiastic about empty PANI circles that might smother the impact and support the degree development adequately (Suma et al. 2019).

Cui and associates preferred three outstanding semi-conductive polymers, PPy, PANI, and PEDOT, singly covered by a simple, flexible, and functional polymerization process over empty monodisperse nanosphere sulfur-(S). The competency of those polymers diminished in the order PEDOT>PPy>PANI. Afterward, a pair of alternative structures enthusiastic about PEDOT was accounted (Javed et al. 2018).

A layer of PEDOT was accustomed to handle the sulfur-depleted electrical conduction and went concerning to stay the polysulfide disintegration as a defensive layer. The PEDOT functionalized with MnO$_2$ nanosheets gave more Associate in nursing and passing dynamic contact territory to upgrade the humidity of the terminal components by electrolytes and booted interconnected with the compound chains which boost the conduction and strengthen the composite (Agnihotri et al. 2017).

9.2.5 CONDUCTING POLYMER ANODES FOR LITHIUM SULFIDE BATTERY

Lithium metal has been viewed the very encouraging anode component for the up and coming back age of high vitality reposting framework notably Li-S batteries. Be that because it could, the Li metal as anode is heretofore endeavor some difficulties.

- Li may respond to border an SEI layer that initiates permanent capability damage, and therefore the Li deposition forward and backward on the SEI layer could prompt anode pounding.

- The development of Lithium dendrites starting after a rough Li deposition will cause battery disenchantment and eudemonia malfunctioning.
- Li soluble in polysulfide's spread during the apparatus might respond through the lithium anode to develop the inexplicable lagging sulfides e.g. Li$_2$S and Li$_2$S$_2$ onto the Li anode surface that hinder the lithium movement (Zhao et al. 2018). To solve these malfunctions, polymers assume a rare job in modification of Lithium anode apparatus. The defensive layer on the Li associate degree founded by an actinic radiation improved chemical change technique, as shown in Figure 9.5.

A treatable, amalgamated arrangement comprising the chemical compound glycol dimethacrylate, fluid solution, and photo-initiator (methyl benzoylformate) was utilized to create a cover coating on the Li anode with actinic radiation light that improved the cycle execution of Li–S cell (Li et al. 2016).

In addition, conductive polymers have a critical job in the lithium-sulfide battery anodes. Wen and colleagues utilized the PEDOT-co-PEG conductive as the lithium metal anode covering coat. The lithium-sulfide battery with an ensured lithium anode keeps up 875.6 mA h g^{-1} of maximum release limit after 200 cycles at 0.2 C with a great 73.45% maintenance limit (Zhao et al. 2018), whereas, the lithium-sulfide battery with an unblemished lithium anode demonstrates a release limit of 399.8 mA h g^{-1} with a weak 32.8% maintenance limit at 0.2 C. The interaction zone concerning the PEDOT-co-PEG and the Li$_2$S/Li$_2$S$_2$ layer are both plainly obvious generally. After delayed cycles, the interaction zone even stays intact, showing excellent mechanical properties and solid glue over the surface of Li smothering the lithium dendrite development (Huang et al. 2019).

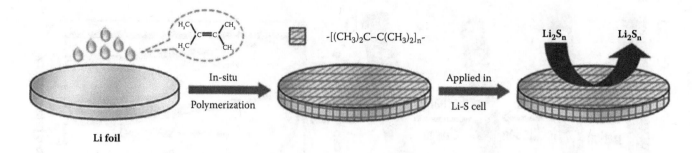

FIGURE 9.5 Lithium anode protective layer formation scheme via in situ olefin polymerization.

9.2.6 CONDUCTING POLYMER AS MATERIALS INTERLAYER

Conductive polymers, for example, PANI, PPy, and PEDOT are broadly utilized for sulfur cathode covering. In the meantime, CPs can also be utilized as materials interlayer. PPy nanotube film (PNTF) was found to be a useful interlayer for lithium-sulfide batteries and was inserted between sulfur cathode and separator (Fan et al. 2019). The (T-PPy) tubular polypyrrole was blended with an enhanced self-debased layout technique. Through drying and vacuum filtration, nanotube films of PPy were obtained. PNTF smothers the movement and disintegration of polysulfide (Psf) in the electrolyte due to the crude surface of PPy nanotubes and the hydrogen bond (Fu et al. 2018).

A pseudo upper current gatherer such as PNTF conducting interlayer decreased the resistance of the exceptionally lagging sulfur cathodes. After 200 cycles, the release limit of lithium-sulfide PNTF batteries was found to be 890 mA h kg^{-1} at 0.2 C, which is lot more than 287 mA h g^{-1} of cell without the interlayer after 200 cycles (Beek et al. 2005). All the while, the normal Coulombic effectiveness of lithium-sulfide batteries with PNTF was observed as high as 90.89%, which was maximum without PNTF cell. Electrodepositing of the sulfur terminal covering a PANI nanowires coat at the best of Lithium-Sulfide batteries surface was adjusted by tooth presented. The battery trial outcome showed that the concealing interlayer of the PANI assumed a job which hindered to control the polysulfides disintegration and gave distance just as sulfur class conductive media (Deng et al. 2019).

9.3 LI$^+$-ION-BASED BATTERY APPLICATIONS OF CONDUCTING POLYMERS

Li-particle batteries are among the foremost promising, productive, and basic high vitality thickness frameworks used in chemical science vitality reposting. The Li-particle battery innovation outlines a reliable framework for power reposting showing basically high vitality thickness and arranges ability typically used in "itinerant" electronic gadgets (Bhatt and Lee 2019). Such batteries as being an important part to moderate carbon dioxide discharges (as a promising force hotspot for innovative electrical vehicles) with a possible cushion vitality reposting framework to accommodate the discontinuous property power supply assets (both on-and off-the lattice). Today, the generation and utilization of Li-particle batteries are continuously increasing. As of late, the employment of Li-particle batteries in electronic gadgets, has become enthusiasm for increasingly viable and sounder batteries (Sawas et al. 2019).

Batteries that have outstanding mechanical properties, maximum effectiveness, and are nanosized are needed for hand-held devices to proceed with the short cumulative process power, bigger screens, and additional slender with easier plans of such gadgets. Moreover, polymer-built batteries have a cumulative enthusiasm to be handled with versatile, and delicate. In addition, there has been an important increment in considerations about the problems connected with such batteries. The utilization of combustible natural diluents as solution, advancement of atomic number 3 dendrites, and immense volume amendment because of poor basic soundness are the problems encountered by lithium-particle batteries (Chen et al. 2018).

The primary working principle of lithium-particle batteries is shown in Figure 9.6 along a negative anode material by additional Li imbedded cathode material with a progressively positive oxidation-reduction potential which delivered a Lithium-particle exchange unit. Anode associated cathode are detached through the solution that is an electronic cowl nonetheless a Lithium-particle conductor. On charging, lithium particles are discharged from the cathode and are placed on the anode. Once the anode is free, lithium particles are removed from the cathode and embedded into the anode (Mindemark et al. 2018).

Conducting polymers are useful resources for natural inorganic 0.5 breed amalgams for lithium-particle batteries because of their wonderful attributes, including great Coulombic productivity and electrical physical phenomenon, which causes them to be recycled (Dong et al. 2018). Conductive polymers show some preferences, such as processability, accessibility, advantageous atomic adjustment, and light weight once connected as terminals. Limitations

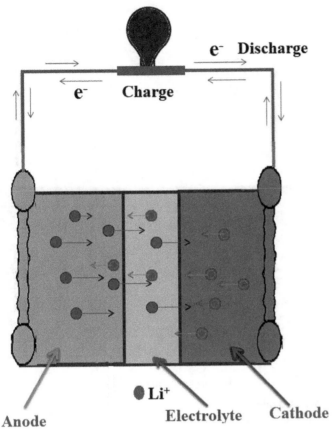

FIGURE 9.6 Working principal of lithium-ion battery.

batteries. In any case, the limit and power thickness remains typically low, and the soundness of natural materials remains a troublesome issue (Peng et al. 2014).

9.4 NA⁺-ION-BASED BATTERY APPLICATIONS OF CONDUCTING POLYMERS

For eco-friendly and manageable vitality, sodium particle batteries are deliberated as an excellent option in contrast to present lithium-particle innovation, because of their material accessibility, minimal effort, and environment friendliness (Jamesh 2019). In ongoing research on Na-particle battery, most of the part is prepared to create the inorganic Sodium (Na) embedded materials. The basic working principle of Na⁺-ion battery is presented in Figure 9.7. Polymer of redox-dynamic has been broadly inquired and is observed with a decent decision. Such polymeric material is used for cathode dynamic for sodium particle batteries because of their auxiliary assorted variety and materials manageability (Guerfi et al. 2016).

Owing to their flexibility, natural polymers can hold bigger Na-particles revocability absent three-dimensional (3D) issues, subsequently accomplishing quick energy for

such as poor reliability and low physical phenomenon in diminished state hinder the applications of conductive polymers in lithium-particle batteries. However, the use of nanostructured conductive polymers will help to overcome this issue, inferable from their large surface and quick diffusion energy of lithium particles (Chen et al. 2019).

Likewise, cycle inorganic composites assortments have gotten secondary cumulative enthusiasm as brilliant frameworks for the repression of Lithium-particle batteries. It is often claiming cycle synergistically with inorganic mixes, basically enhanced the anode existence, degree, voltage, mechanical and heat strength. Cycle is used as each anodic and cathode material nonetheless is for the foremost half utilized as cathodes in Li-particle batteries. Distinct CPs show different power densities, artificial structure, physical properties and vitality, e.g., PPy-built terminals have around 10–50 W h kg⁻¹ vitality densities and 5–25 kW kg⁻¹ power densities (Khati et al. 2018).

PANI-built anodes, PTh-built cathodes exhibit vitality densities of 50–200 and 200 Wh kg⁻¹, respectively. The power densities of both materials are found in between 5 and 50 kW kg⁻¹. HClO₄-doped PANI nanotubes emulsified by Cheng and collaborators were used as positive anode materials, which displayed desirable execution over the business PANI powders in lithium-particle

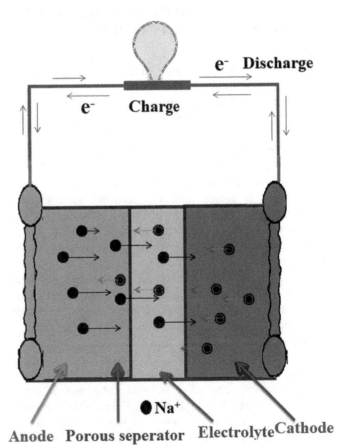

FIGURE 9.7 Schematic illustration of sodium-ion battery.

Na inclusion and extraction responses. On a fundamental level, an Na-particle battery can be planned to employ a couple of natural materials as anode and cathode, which have adequate potential contrast thus, to complete a battery response, these materials are properly aligned. Thus, all-natural sodium particle battery would be enormously appealing for its expansive electric stockpiling scale uses due to its ease and eco-cordiality (Eftekhari and Kim 2018).

The cathode used in an all-natural Na-particle battery is p-doped poly triphenylamine, while an n-type redox dynamic poly(anthraquinonyl sulfide) is used as an anode material, exclusively litigating the progress metals as were utilized in conventional electrochemical batteries. Thus, Na-particle battery demonstrated a 1.8-V yield voltage and achieved a 92-W h kg^{-1} critical explicit vitality (Li et al. 2003). Because of the auxiliary redox dynamic polymers dependability and adoptability, such batteries were observed with a brilliant degree and ability of 60% limit discharge information at an exceptionally high degree of 16 C (3200 mAh g^{-1}), which demonstrate a great cycling security with 85% limit maintenance after 500 cycles at 8 C rate (Vo et al. 2019).

9.5 MG^{+2}-ION-BASED BATTERY APPLICATIONS OF CONDUCTING POLYMERS

Lithium-particle batteries have received considerable attention worldwide as part of the ongoing improvement of progressive energy storing technologies. Until recently, these batteries proposed the foremost effective energy consolidation capability per gram along with long cycle stability and cost (Dong et al. 2019). Moreover, lithium-ion batteries have numerous drawbacks including the high cost of material, and their capability is typically restricted due to the monovalent nature of lithium ion. The metallic batteries have overcome the obstacles of lithium-ion battery, which are revealed by many researchers (Bobba et al. 2019). Mg^{+2}-ion battery working principle is shown in Figure 9.8.

Metals have double the theoretical capability than lithium-ion cell because of being bivalent particles. To boost the capability, metal is very easily accessible, thus decreasing the assembly worth. Mg-metal is found to be less reactive in air than any other element such as Li; thus, Mg is convenient to handle (Zhang et al. 2018). Lithium and Na elements have the power capacity of 1136 and 2062 mA h cm^{-3}, respectively. Mg has higher meter capability, reported to be 3832 mA h cm^{-3}, which is higher than Na and lithium elements. Fiber configuration and low Coulombic efficiency are the big hindrance with Li-metal battery evolution; thus, these inventions are probably transformative (Manjuladevi et al. 2017).

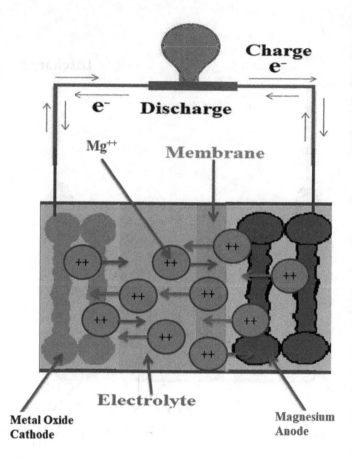

FIGURE 9.8 Magnesium-ion battery principle.

REFERENCES

Agnihotri, Nidhi, Pintu Sen, Amitabha De, and Manabendra Mukherjee. "Hierarchically Designed Pedot Encapsulated Graphene-Mno2 Nanocomposite as Supercapacitors." *Materials Research Bulletin* 88 (2017): 218–25.

Akhtar, Naseem, Hongyuan Shao, Fei Ai, Yuepeng Guan, Qifan Peng, Hao Zhang, Weikun Wang, *et al.* "Gelatin-Polyethylenimine Composite as a Functional Binder for Highly Stable Lithium-Sulfur Batteries." *Electrochimica Acta* 282 (2018): 758–66.

Bahloul, Ahmed, Belkacem Nessark, E. Briot, Henri Groult, Alain Mauger, K. Zaghib, and Christian Julien. "Polypyrrole-Covered Mno2 as Electrode Material for Hybrid Supercapacitor." *Journal of Power Sources* 240 (10/01 2013): 267–72.

Beek, Waldo J. E., Martijn M. Wienk, Martijn Kemerink, Xiaoniu Yang, and René A. J. Janssen. "Hybrid Zinc Oxide Conjugated Polymer Bulk Heterojunction Solar Cells." *The Journal of Physical Chemistry B* 109, no. 19 (2005/05/01 2005): 9505–16.

Bhatt, Mahesh Datt, and Jin Yong Lee. "High Capacity Conversion Anodes in Li-Ion Batteries: A Review." *International Journal of Hydrogen Energy* 44, no. 21 (2019): 10852–905.

Bobba, Silvia, Fabrice Mathieux, and Gian Andrea Blengini. "How Will Second-Use of Batteries Affect Stocks and Flows in the Eu? A Model for Traction Li-Ion Batteries." *Resources, Conservation and Recycling* 145 (2019/06/01/ 2019): 279–91.

Cai, Junjie, Zengyao Zhang, Shaoran Yang, Yonggang Min, Guangcheng Yang, and Kaili Zhang. "Self-Conversion Templated Fabrication of Sulfur Encapsulated inside the N-Doped Hollow Carbon Sphere and 3d Graphene Frameworks for High-Performance Lithium–Sulfur Batteries." *Electrochimica Acta* 295 (11/01 2018).

Chen, Tzu-Ling, Rui Sun, Carl Willis, Brian F. Morgan, Frederick L. Beyer, and Yossef A. Elabd. "Lithium Ion Conducting Polymerized Ionic Liquid Pentablock Terpolymers as Solid-State Electrolytes." *Polymer* 161 (2019/01/14 2019): 128–38.

Chen, Yazhou, Yunsheng Tian, Zhong Li, Nan Zhang, Danli Zeng, Guodong Xu, Yunfeng Zhang, *et al.* "An Ab Alternating Diblock Single Ion Conducting Polymer Electrolyte Membrane for All-Solid-State Lithium Metal Secondary Batteries." *Journal of Membrane Science* 566 (2018/11/15 2018): 181–89.

Deng, Chao, Zhuowen Wang, Shengping Wang, and Jingxian Yu. "Inhibition of Polysulfide Diffusion in Lithium–Sulfur Batteries: Mechanism and Improvement Strategies." *Journal of Materials Chemistry A* 7, no. 20 (2019): 12381–413.

Dong, Hui, Yanliang Liang, Oscar Tutusaus, Rana Mohtadi, Ye Zhang, Fang Hao, and Yan Yao. "Directing Mg-Storage Chemistry in Organic Polymers toward High-Energy Mg Batteries." *Joule* 3, no. 3 (2019): 782–93.

Dong, Jiaming, Yunfeng Zhang, Jiaying Wang, Zehui Yang, Yubao Sun, Danli Zeng, Zhihong Liu, and Hansong Cheng. "Highly Porous Single Ion Conducting Polymer Electrolyte for Advanced Lithium-Ion Batteries via Facile Water-Induced Phase Separation Process." *Journal of Membrane Science* 568 (2018/12/15 2018): 22–29.

Eftekhari, A., and D. W. Kim. "Sodium-Ion Batteries: New Opportunities beyond Energy Storage by Lithium." *Journal of Power Sources* 395 (08/15 2018): 336–48.

Fan, Linlin, Matthew Li, Xifei Li, Wei Xiao, Zhongwei Chen, and Jun Lu. "Interlayer Material Selection for Lithium-Sulfur Batteries." *Joule* (2019).

Fu, Xuewei, Yu Wang, Louis Scudiero, and Wei-Hong Zhong. "A Polymeric Nanocomposite Interlayer as Ion-Transport-Regulator for Trapping Polysulfides and Stabilizing Lithium Metal." *Energy Storage Materials* 15, ChemElectroChem 4 2017 (2018): 447–57.

Geetha, S., and D. dinesh Trivedi. "Electrochemical Synthesis and Characterization of Conducting Polyparaphenylene Using Room-Temperature Melt as the Electrolyte." *Synthetic Metals* 148 (01/31 2005): 187–94.

Gil Herrera, Luz, Eval Baca, O. Morán, C. Quinayas, and Gilberto Bolaños. "Influence of Polyparaphenylene on the Magnetotransport of Manganite/Polymer Composites." *Physica B: Condensed Matter* 403 (05/01 2008): 1813–18.

Guerfi, A., J. Trottier, C. Gagnon, F. Barray, and K. Zaghib. "High Rechargeable Sodium Metal-Conducting Polymer Batteries." *Journal of Power Sources* 335 (2016): 131–37.

Huang, Sheng, Ruiteng Guan, Shuanjin Wang, Min Xiao, Dongmei Han, Luyi Sun, and Yuezhong Meng. "Polymers for High Performance Li-S Batteries: Material Selection and Structure Design." *Progress in Polymer Science* 89 (2019): 19–60.

Jamesh, Mohammed-Ibrahim. "Recent Advances on Flexible Electrodes for Na-Ion Batteries and Li–S Batteries." *Journal of Energy Chemistry* 32 (2019): 15–44.

Javed, Mohsin, Syed Mustansar Abbas, Mohammad Siddiq, Dongxue Han, and Li Niu. "Mesoporous Silica Wrapped with Graphene Oxide-Conducting Pani Nanowires as a Novel Hybrid Electrode for Supercapacitor." *Journal of Physics and Chemistry of Solids* 113 (2018): 220–28.

Khati, Komal, Ila Joshi, and Mohammad Ghulam Haider Zaidi. "Electro-Capacitive Performance of Haemoglobin/ Polypyrrole Composites for High Power Density Electrode." *Journal of Analytical Science and Technology* 9, no. 1 (2018/10/27 2018): 24.

Li, Fang, Mohammad Rejaul Kaiser, Jianmin Ma, Zaiping Guo, Huakun Liu, and Jiazhao Wang. "Free-Standing Sulfur-Polypyrrole Cathode in Conjunction with Polypyrrole-Coated Separator for Flexible Li-S Batteries." *Energy Storage Materials* 13 (2018/07/01 2018): 312–22.

Li, Jun, Yifei Yuan, Huile Jin, Huihang Lu, Aili Liu, Dewu Yin, Jichang Wang, Jun Lu, and Shun Wang. "One-Step Nonlinear Electrochemical Synthesis of Texsy@Pani Nanorod Materials for Li-Texsy Battery." *Energy Storage Materials* 16 (2019/01/01 2019): 31–36.

Li, Nian-Wu, Ya-Xia Yin, Chun-Peng Yang, and Yu-Guo Guo. "An Artificial Solid Electrolyte Interphase Layer for Stable Lithium Metal Anodes." *Advanced Materials* 28, no. 9 (2016): 1853–58.

Li, Qingfeng, Ronghuan He, Jens Oluf Jensen, and Niels J. Bjerrum. "Approaches and Recent Development of Polymer Electrolyte Membranes for Fuel Cells Operating above 100 °C." *Chemistry of Materials* 15, no. 26 (2003/ 12/01 2003): 4896–915.

Li, Xiao-Qiang, Wan-Wan Liu, Shui-Ping Liu, Meng-Juan Li, Yong-Gui Li, and Ming-Qiao Ge. "In Situ Polymerization of Aniline in Electrospun Microfibers." *Chinese Chemical Letters* 25, no. 1 (2014): 83–86.

Liu, Ning, Lu Wang, Yan Zhao, Taizhe Tan, and Yongguang Zhang. "Hierarchically Porous Tio2 Matrix Encapsulated Sulfur and Polysulfides for High Performance Lithium/ Sulfur Batteries." *Journal of Alloys and Compounds* 769 (2018): 678–85.

Manjuladevi, R., M. Thamilselvan, S. Selvasekarapandian, R. Mangalam, M. Premalatha, and S. Monisha. "Mg-Ion Conducting Blend Polymer Electrolyte Based on Poly (Vinyl Alcohol)-Poly (Acrylonitrile) with Magnesium Perchlorate." *Solid State Ionics* 308 (2017): 90–100.

Mao, Jifu, and Ze Zhang. "Conductive Poly(Pyrrole-Co-(1-(2-Carboxyethyl)Pyrrole)) Core-Shell Particles: Synthesis, Characterization, and Optimization." *Polymer* 105 (2016/ 11/22 2016): 113–23.

Mindemark, Jonas, Matthew J. Lacey, Tim Bowden, and Daniel Brandell. "Beyond Peo—Alternative Host Materials for Li+-Conducting Solid Polymer Electrolytes." *Progress in Polymer Science $V* 81 (2018): 114–43.

Mizera, Adam, Sławomir J. Grabowski, Paweł Ławniczak, Monika Wysocka-Żołopa, Alina T. Dubis, and Andrzej Łapiński. "A Study of the Optical, Electrical and Structural Properties of Poly(Pyrrole-3,4-Dicarboxylic Acid)." *Polymer* 164 (2019/02/15 2019): 142–53.

Nair, Jijeesh R., Federico Bella, Natarajan Angulakshmi, Arul Manuel Stephan, and Claudio Gerbaldi. "Nanocellulose-Laden Composite Polymer Electrolytes for High Performing Lithium–Sulphur Batteries." *Energy Storage Materials* 3 (2016): 69–76.

Peng, Hui, Guofu Ma, Jingjing Mu, Kanjun Sun, and Ziqiang Lei. "Low-Cost and High Energy Density Asymmetric Supercapacitors Based on Polyaniline Nanotubes and Moo3 Nanobelts." *Journal of Materials Chemistry A* 2, no. 27 (2014): 10384–88.

Qi, Qi, Xiaohui Lv, Wei Lv, and Quan-Hong Yang. "Multifunctional Binder Designs for Lithium-Sulfur Batteries." *Journal of Energy Chemistry* (2019).

Radha, Mukkable, Praveen Meduri, Melepurath Deepa, Sonnada Shivaprasad, and P. Ghosal. "Sulfur Enriched Carbon Nanotubols with a Poly(3,4-Ethylenedioxypyrrole) Coating as Cathodes for Long-Lasting Li-S Batteries." *Journal of Power Sources* 342 (02/01 2017): 202–13.

Ren, Juan, Yibei Zhou, Huali Wu, Fengyu Xie, Chenggang Xu, and Dunmin Lin. "Sulfur-Encapsulated in Heteroatom-Doped Hierarchical Porous Carbon Derived from Goat Hair for High Performance Lithium–Sulfur Batteries." *Journal of Energy Chemistry* 30 (2019/03/01 2019): 121–31.

Sahiner, Nurettin, and Sahin Demirci. "Conducting Semi-Interpenetrating Polymeric Composites via the Preparation of Poly(Aniline), Poly(Thiophene), and Poly(Pyrrole) Polymers within Superporous Poly(Acrylic Acid) Cryogels." *Reactive and Functional Polymers* 105 (05/01 2016).

Sawas, Abdulrazzag, Ganguli Babu, Naresh Kumar Thangavel, and Leela Mohana Reddy Arava. "Electrocatalysis Driven High Energy Density Li-Ion Polysulfide Battery." *Electrochimica Acta* 307 (2019): 253–59.

Shklovsky, Jenny, Amir Reuveny, Yelena Sverdlov, Slava Krylov, and Yosi Shacham-Diamand. "Towards Fully Polymeric Electroactive Micro Actuators with Conductive Polymer Electrodes." *Microelectronic Engineering* 199 (2018/11/05 2018): 58–62.

Sovizi, Mohammad Reza, and Zohre Fahimi. "Honeycomb Polyaniline-Dodecyl Benzene Sulfonic Acid (Hpani-Dbsa)/Sulfur as a New Cathode for High Performance Li–S Batteries." *Journal of the Taiwan Institute of Chemical Engineers* 86 (2018/05/01 2018): 270–80.

Suma, B. P., Prashanth Shivappa Adarakatti, Suresh Kumar Kempahanumakkagari, and Pandurangappa Malingappa. "A New Polyoxometalate/Rgo/Pani Composite Modified Electrode for Electrochemical Sensing of Nitrite and Its Application to Food and Environmental Samples." *Materials Chemistry and Physics* 229, no. Am. J. Clin. Nutr. 90 2009 (2019): 269–78.

Venkatesan, R., and L. Cindrella. "Methyl Substituted, Azine Bridged Poly Thiophenes and Their Structure Related Surface Characteristics." *Synthetic Metals* 246 (2018/12/01 2018): 150–63.

Vo, Duy Thanh, Hoang Nguyen, Thien Trung Nguyen, Thi Tuyet Hanh Nguyen, Tran Van Man, Shigeto Okada, and My Loan Phung Le. "Sodium Ion Conducting Gel Polymer Electrolyte Using Poly(Vinylidene Fluoride Hexafluoropropylene)." *Materials Science and Engineering B* 241, no. Nature 451 2008 (2019): 27–35.

Wang, Jian-Gan, Feiyu Kang, and Bingqing Wei. "Engineering of Mno2-Based Nanocomposites for High-Performance Supercapacitors." *Progress in Materials Science* 74 (2015): 51–124.

Wang, Min, Paul Baek, Lenny Voorhaar, Eddie Chan, Andrew Nelson, David Barker, and Jadranka Travas-Sejdic. "Long Side-Chain Grafting Imparts Intrinsic Adhesiveness to Poly(Thiophene Phenylene) Conjugated Polymer." *European Polymer Journal* 109 (10/01 2018a).

Wang, Xiao, Jing Zhi Sun, and Ben Zhong Tang. "Poly(Disubstituted Acetylene)S: Advances in Polymer Preparation and Materials Application." *Progress in Polymer Science* 79 (2018b): 98–120.

Yelilarasi, A., R. Anbarasan, and K. M. Manikandan. "Electrical Conductivity Studies on the Nanocomposites of Poly (Aniline) with Various Initiator and Oxide Nanoparticles." *Vacuum* 163 (2019): 172–75.

Zeng, Fang-Lei, Ning Li, Yan-Qiu Shen, Xin-Yu Zhou, Zhao-Qing Jin, Ning-Yi Yuan, Jian-Ning Ding, et al. "Improve the Electrodeposition of Sulfur and Lithium Sulfide in Lithium-Sulfur Batteries with a Comb-Like Ion-Conductive Organo-Polysulfide Polymer Binder." *Energy Storage Materials* 18 (2019/03/01 2019): 190–98.

Zhang, Liang, Min Ling, Jun Feng, Gao Liu, and Jinghua Guo. "Effective Electrostatic Confinement of Polysulfides in Lithium/Sulfur Batteries by a Functional Binder." *Nano Energy* 40, no. Nature 451 2008 (2017): 559–65.

Zhang, Yufei, Hongbo Geng, Weifeng Wei, Jianming Ma, Libao Chen, and Cheng Li. "Challenges and Recent Progress in the Design of Advanced Electrode Materials for Rechargeable Mg Batteries." *Energy Storage Materials* (12/01 2018).

Zhao, Huijuan, Nanping Deng, Jing Yan, Weimin Kang, Jingge Ju, Yanli Ruan, Xiaoqing Wang, et al. "A Review on Anode for Lithium-Sulfur Batteries: Progress and Prospects." *Chemical Engineering Journal* $V 347 (2018a): 343–65.

Zhao, Jinbao, Yiyong Zhang, Yunhui Wang, He Li, and Yueying Peng. "The Application of Nanostructured Transition Metal Sulfides as Anodes for Lithium Ion Batteries." *Journal of Energy Chemistry* 27, no. 6 (2018b): 1536–54.

Zheng, Min, Yu Wang, Jacqueline Reeve, Hamid Souzandeh, and Wei-Hong Zhong. "A Polymer-Alloy Binder for Structures-Properties Control of Battery Electrodes." *Energy Storage Materials* 14 (2018): 149–58.

Zheng, Tianyue, Ting Zhang, Mauricio Solis de la Fuente, and Gao Liu. "Aqueous Emulsion of Conductive Polymer Binders for Si Anode Materials in Lithium Ion Batteries." *European Polymer Journal* 114 (2019/05/01 2019): 265–70.

Zhou, Jinqiu, Haoqing Ji, Jie Liu, Tao Qian, and Chenglin Yan. "A New High Ionic Conductive Gel Polymer Electrolyte Enables Highly Stable Quasi-Solid-State Lithium Sulfur Battery." *Energy Storage Materials* (02/01 2019).

10 Conducting Polymer–Carbon-Based Binary Composites for Battery Applications

Christelle Pau Ping Wong, Joon Ching Juan and Chin Wei Lai
Nanotechnology & Catalysis Research Centre (NANOCAT), Institute for Advanced Studies
University of Malaya (UM)
Kuala Lumpur, Malaysia

CONTENTS

ABBREVIATIONS

AC	Activated carbon
CB	Carbon black
CF	Carbon fiber
CNTs	Carbon nanotubes
EDLC	Electrical double-layer capacitance
GO	Graphene oxide
HRPSoC	High rate partial state of charge
LIBs	Lithium-ion batteries
PANI	Polyaniline
PDPA	Polydiphenylamine
PEDOT	Poly(3,4-ethylenedioxythiophene)
PG	Porous graphene
PI	Polyimide
PMT	Poly(3-methyl thiophene)
PPP	Polyparaphenylene
PPy	Polypyrrole
PTh	Polythiophene
PTTh	Polyterthiophene
S	Sulfur
SWNT	Single-walled carbon nanotube
VC	Vinylene carbonate
WMCNTs	Multiwalled carbon nanotubes

10.1   INTRODUCTION

Battery is a promising energy storage device since 1800s until now. It directly converts chemical energy to electrical energy, which is the power source for electric vehicles, portable electronics, communication devices, sensors, and other power tools. Batteries can be grouped into two types: (i) primary batteries, used once and disposed after they have reached their service life; and (ii) secondary batteries, so-called rechargeable battery that can charge/discharge many times (Yoshino, 2012). A battery is composed of electrodes, separator, and electrolytes. Among them, electrodes play a vital role in a battery as their properties mainly influence the electrochemical performance of the battery.

Electrode materials commonly used in battery fields are transition metal oxides and conducting polymers. Transition metal oxides have high valence state but they suffer from low conductivity. In contrast, conducting polymers have been extensively investigated because they are electrochemically active and conductive, which allow penetration of electrolyte into the polymer. Conducting polymers including PANI, PI, PEDOT, PPy, PPP, PA, PTh, and their derivatives have been used as an electrode material for batteries (Hossain and Hoque, 2018).

However, volumetric expansion and aggregated morphologies are the major drawbacks of conducting polymers, leading to the poor cycling life in battery applications. To eradicate this problem, conducting polymers have been composited with some carbon-based materials including activated carbon (AC), graphene, carbon aerogel, and carbon nanotubes (CNTs). Carbon materials possess relatively large porosity, superior electrical conductivity, and large specific surface area, which serve as efficient supporting substrates to the formation of conducting polymers.

The aim of this chapter is to review conducting polymer–carbon-based binary composite for battery applications since past decades. Here, we also briefly discuss the types of batteries and electrode materials that are widely used in a battery.

10.2 BATTERIES

10.2.1 TYPES OF BATTERIES

Batteries and supercapacitors are energy storage devices that store energy through electrochemical reactions. Among them, nowadays batteries are irreplaceable owing to their highest energy density (120–200 Wh kg^{-1}) when compared with supercapacitors (Thounthong et al., 2009). Batteries can be grouped into two types, namely, primary and secondary batteries. Primary batteries are used once and are disposed off because the electrode materials undergo irreversible change during the discharge process such as alkaline battery. In contrast, secondary batteries can charge/discharge many a times; they are lead acid batteries, nickel metal hydride batteries, metal-ion batteries, and metal–air batteries (Yoshino, 2012).

10.2.1.1 Lead Acid Batteries

Of the four listed types of batteries, lead acid batteries are the most matured energy storage devices because of their excellent recycling efficiency, low initial cost, and good safety performance. Generally, the redox reaction occurs on the negative plate, shown as follows:

$$Pb - 2e \leftrightarrow Pb^{2+} \tag{10.1}$$

$$Pb^{2+} + HSO_4^- \leftrightarrow PbSO_4 + H^+ \tag{10.2}$$

Nonetheless, it suffers from poor cycle life because of sulfation that happens at the negative plate under HRPSoC cycle conditions (Moseley et al., 2007). The $PbSO_4$ crystals cannot be fully reduced to Pb, resulting in the accretion of nonreversible $PbSO_4$ crystals on the negative electrode and depletion of Pb in the battery.

10.2.1.2 Metal-Ion Batteries

Metal-ion batteries, such as lithium-ion batteries (LIBs), potassium-ion batteries, and sodium-ion batteries, possess the merit of no memory effect, excellent energy storage density, long cyclability, and low self-discharge (Sengodu and Deshmukh, 2015). LIBs are commercialized and are widely used in cellular phones, laptop, electric vehicles, and other portable electronic devices. The redox reaction for commercialized $LiCoO_2$/graphite battery is as follows:

Positive electrode:

$$LiCoO_2 \leftrightarrow Li_{1-x}CoO_2 + xLi + xe^- \tag{10.3}$$

Negative electrode:

$$6C + xLi^+ + xe^- \leftrightarrow C_6Li_x \tag{10.4}$$

The charging and discharging process of LIBs is based on the repeated intercalation/deintercalation of Li ions into/from the positive and negative electrode (Zhang et al., 2018b). However, the rapid growing of energy demand has urged researchers to work out on better LIBs in terms of power density, cost, safety, and cycle life.

10.2.1.3 Metal–Air Batteries

Metals including Zn, Fe, Mg, Li, and Al have been investigated as an electrode material for air batteries. Among them, Zn–air batteries showed a promising attention because of their superior theoretical energy density, high working voltage, and cost effectiveness. During the discharge process, oxygen from atmosphere is circulated into the cathode and reduced to OH$^-$. The hydroxide ions diffuse into the zinc anode (i.e., Zn in metal, paste, and powder form) and react with it to form zincate ions. Then, the electrons are transported to the cathode (Xu et al., 2015). The detailed redox reaction of Zn–air battery in alkaline solutions is shown as follows (Caramia and Bozzini, 2014):

Anode:

$$Zn + 4OH^- \rightarrow Zn(OH)_4^{2-} + 2e^- \tag{10.5}$$

$$Zn(OH)_4^{2-} \rightarrow ZnO + H_2O + 2OH^- \tag{10.6}$$

$$Zn + 2H_2O \rightarrow Zn(OH)_2 + H_2 \uparrow \tag{10.7}$$

Cathode:

$$O_2 + 2H_2O + 4e^- \rightarrow 4OH^- \tag{10.8}$$

However, the large overpotential, low energy efficiency, and poor cycle life are far from providing satisfactory results and thus limit the practical applications of these batteries (Zhu et al., 2015).

To date, short cycle life, less safety, and low rate capability (i.e., power density, energy efficiency) are the major demerits of rechargeable batteries. Therefore, design and engineering of the electrode materials are in great demand for battery applications.

10.2.2 Electrode Materials

Electrode material is the most vital part for determining the electrochemical performance of batteries. The most common electrode materials are carbonaceous, transition metal oxides, and conducting polymers. Transition metal oxides such as Co_3O_4, ZnO, MnO_2, NiO, Fe_2O_3, WO_3, TiO_2, and others are alternative potential electrode materials because of their high theoretical capacities,

high valence states, low cost, good safety performance, and high chemical stability. Nonetheless, they show poor electrical conductivity and large volume change upon repeated charge/discharge, leading to rapid capacity fading, low coulombic efficiency, and low rate capability (Liu et al., 2015). Consequently, these limited the utilization of transition metal oxides in the field of battery.

Since the discovery of polyacetylene (PA) by MacDiarmid et al. (Shirakawa et al., 1977), conducting polymers in doped state have received enormous attention owing to their comparable conductivity with metals. Conducting polymers exhibit high electrical conductivity, which is attributed to the interchange of single and double bonds along the backbone chains (Magu et al., 2019). Table 10.1 lists the conductivity of conducting polymers

TABLE 10.1

The electrical conductivity of different CPs in the doped state and their structure (Magu et al., 2019)

Polymer	Structure	Electrical Conductivity (S cm^{-1})	Doping
PANI		30–200	n, p
PPy		40–750	p
PEDOT		300–500	n, p
PTh		300–400	p
PA		200–1000	n, p
PPP		500	n, p

ranging from 1 to 1000 S cm^{-1}. The known conducting polymers used as electrode materials of battery are PEDOT, PANI, PPy, PI, PTh, and their derivatives. Moreover, conducting polymers also provide advantages of high flexibility, tunable conductivity, high thermal and chemical stability, light weight, and easy processability and some of them can cycle hundreds to thousands of times with only little degradation (Mike and Lutkenhaus, 2013).

Furthermore, the carbonaceous material is a unique electrode material as it stores charge based on electrical double-layer capacitance (EDLC) method. The family of EDLC materials includes graphene, AC, graphene oxide (GO), carbon aerogel, and CNTs. They are three types of CNTs, namely, single-walled, double-walled, and multiwalled nanotubes (Meer et al., 2016). In literature, these carbons exhibit excellent specific surface area, superior electrical conductivity, and large volume of pores (Wang et al., 2016). Because of this, they tend to be a supporting substrate. However, it shows low energy density owing to the EDLC storage method.

10.3 CONDUCTING POLYMER–CARBON-BASED BINARY COMPOSITE IN BATTERY APPLICATIONS

A lot of attention has been directed toward conducting polymers because of their unique properties since the last decade. Unfortunately, volumetric shrinkage and strong agglomeration are the major challenges that need to be addressed to improve the cycling and diffusion issues of CPs in battery applications. To alleviate these problems, the incorporation of CPs with carbonaceous to form a composite is quite common, which takes the advantage of each component.

10.3.1 POLYANILINE PANI–CARBON-BASED COMPOSITE

Among the conducting polymers, polyaniline PANI was first commercialized in lithium button cells in 1987 owing to its high conductivity due to protonation (Silva et al., 2012), corrosion resistance, and excellent electrochromic performance. However, the aggregated morphology of PANI has restricted its applications in the field of battery. Therefore, Zhang et al., (2018c) modified PANI by reduced graphene oxide (rGO) by adsorption double oxidant route, which formed a layered structure. The prepared rPANI/rGO composite delivered high specific capacity of 175 mAh g^{-1} at 0.2 C of current rate and degraded only 9.8% of capacity after 100 cycles.

Moreover, Sivakkumar et al., (2007) modified PANI by multiwalled carbon nanotube (WMCNT) by the in situ chemical polymerization route. At current rate of 0.2 C, the PANI/CNT composite used as a cathode in metal–polymer cells showed a specific capacity of 139 mAh g^{-1}. Ionic liquid was used as the electrolyte. The good performance is because of the high conductivity of CNT

and its porous structure allows the rapid diffusion of ions into the inner layer of the active polymer material.

Liu et al., (2017) also prepared PANI/WMCNTs by the in situ chemical oxidative polymerization. At 0.2 C, the reduced PANI/WMCNTs composite displayed excellent capacity of 181.8 mAh g^{-1}. As for pristine PANI, the discharge capacity is much lower than that of the composite, which showed only 74.8 mAh g^{-1}. It can confirm that the introduction of WMCNTs into the composite increased the electrical conductivity and crystallinity of PANI. The reduced PANI/WMCNTs composite retained 76.75% of the initial capacity after 100 cycles. In future, PANI/carbon-based composite can utilized as a cathode material in lithium–polymer cell because of its excellent electrochemical performance.

10.3.2 POLYPYRROLE (PPY)–CARBON-BASED COMPOSITE

Other than PANI, PPy is also one of the promising electrode materials owing to its conductive properties, stability in air, high theoretical capacity, and easy synthesis. However, the chemical oxidative polymerization of PPy has a major disadvantage: it cannot closely attach on the current contact. This issue can be solved by adding GO or carbon fiber (CF) as a supporting substrate or binder. Su et al., (2017) successfully synthesized PPy/GO and PPy/CF composite cathodes of LIB by in situ electrochemical polymerization. This demonstrated that PPy/CF and PPy/GO exhibited higher specific capacity of 144.3 and 176.1 mAh g^{-1} than that by pristine PPy (41 mAh g^{-1}) at a current density of 20 mA g^{-1}. The introduction of GO into PPy could affect the morphology of the sample, from aggregated morphology (PPy) to multilayer structure (PPy/GO). This multilayer structure helps to increase the specific capacity as it increases the contact area for electrons. The PPy/CF composite shows a fibrillary shape, where surrounding of PPy is attached to the surface of CF.

In addition, PPy–GO composite also showed an improved electrochemical performance in lead–acid batteries. Yang et al. (2016) reported that by adding PPy/GO composite as additives to the negative plate of lead–acid battery can increase their specific surface area and total porosity; thus (i) hinders the hydrogen evolution of negative plate; (ii) restrains the growth rate of PbSO$_4$ crystal; and (iii) prolongs the cycle life of the battery.

In lithium–sulfur batteries, there are several challenges such as the insulator of sulfur, the formation of Li$_2$S on the cathode, and the volumetric shrinkage of sulfur cathode owing to several times of charge/discharge. To solve these problems, PPy and graphene were used. Zhang et al., (2018d) fabricated a novel 3D hierarchical S/PPy/PG composite by template-assisted technique. The S/PPy/PG composite showed the specific capacity of 477 mAh g^{-1} at 2 C as the cathode material for lithium–sulfur batteries. The unique properties of graphene can increase the electrical conductivity and can prevent the volumetric changes of the composite electrode. The composite electrode also showed a capacity retention of 78.6% after 200 cycles. The excellent cyclic stability is attributed to the presence of PPy, which functions as a binder to enhance the interactions between sulfur and PPy/graphene composite.

10.3.3 Poly(3,4-ethylenedioxythiophene) (PEDOT)–Carbon-Based Composite

Poly(3,4-ethylenedioxythiophene) (PEDOT) is a conducting polymer that shows high catalytic activities in Li–air batteries because of its excellent electrical conductivity and high stability under oxygen reduction and evolution conditions. Nonetheless, the electrical conductivity of PEDOT is not high enough to achieve a large capacity of Li–air battery. Hence, Yoon and Park, (2017) proposed to incorporate of graphene, a superior conductive material with PEDOT. The PEDOT microflower/graphene composite presented superior capacity of 1500 mAh g^{-1}, while pure PEDOT exhibited a discharge capacity of 430 mAh g^{-1} at a current density of 400 mA g^{-1}. This is due to the enhanced electrical conductivity. The composite electrode also showed an excellent cyclic stability owing to the PEDT microflower that is able to suppress the formation of undesired products.

PEDOT is highly electrical conductive in the oxidized form, but it is insoluble. This issue can be solved by adding a charge balance dopant such as polystyrene sulfonic acid (PSS) to form PEDOT:PSS. Yoon et al., (2016) also proposed PEDOT:PSS as an electrode material for Li–air battery. PEDOT:PSS was coated on graphene by ultrasonic treatment. The prepared composite electrode yielded a lower capacity (almost 8000 mAh g^{-1}) than that of pristine graphene (almost 9000 mAh g^{-1}). This is due to the fact that the surface area of graphene is larger than the composite. Unfortunately, PEDOT:PSS/graphene composite showed a longer cyclic performance than that by graphene. This can be ascribed to PEDOT:PSS, which acts as a protecting layer to reduce the side reactions. PEDOT:PSS also can work as a conductive binder to minimize the used of inactive material (i.e., PVDF) during preparation of electrode.

10.3.4 Others Conducting Polymer–Carbon-Based Composite

Polythiophene (PTh) is seldom used as an active material in LIBs due to its low doping level, which means low charge storage capability. Terthiophene oligomer is expected to store more charge owing to more α,α′ linkages of the thiophene unit. A PTTh/WMCNT composite was prepared by the in situ chemical polymerization and was assembled with the ionic liquid electrolyte (Sivakkumar et al., 2009). The charge/discharge test showed that the capacity of PTTh/WMCNT composite in lithium cells was 50 mAh g^{-1} at C/5 rate. The capacity retention rate was 78% after 100 cycles. However, the capacity value was still lower than that of PANI and PPy.

In addition, PMT/WMCNT composite was prepared by the in situ chemical polymerization (Kim et al., 2008). A small quantity of vinylene carbonate (VC) was added to the ionic liquid electrolyte to study the effects of VC. The discharge capacities of the composite electrode in VC-added cell and VC-free cell were 80 and 12 mAh g^{-1}, respectively, at the 1 C rate. The addition of VC helps to form a solid electrolyte interphase (SEI) layer on the lithium anode surface. The SEI functions as a protecting layer to prohibit the decomposition of imidazolium cations.

PDPA/SWNT composites were successfully prepared by electrochemical polymerization of diphenylamine (Baibarac et al., 2011). The SWNT film was used as the current collector. As the cathode material in LIBs, the composite electrode manifested an excellent discharge capacity of 245 mAh g^{-1} at a current density of 10 mA g^{-1}, which is seven-fold higher than that of pristine PDPA (35 mAh g^{-1}). This is most probably because of the synergistic effects between PDPA and SWNT.

Zhang et al., (2018a) fabricated the composite of PI/CB by the in situ polymerization approach and assembled as a cathode in LIBs. At a current density of 0.1 A g^{-1}, the PI/CB composite demonstrated a superior capacity of 182 mAh g^{-1}. The incorporation of CB with PI increases the conductivity of PI and suppresses the crystallization of polymer chains, leading to a full utilization of the active side of PI.

Furthermore, PI is also coated on the SWNT film b in situ polymerization method to form a PI/SWNT film and is used as a cathode in LIBs. The PI/SWNT film showed a specific capacity of 226 mAh g^{-1} at a current rate of 0.1 C and retained 85% of the capacity retention after 200 cycles. In contrast, pristine PI resulted in a specific capacity of 170 mAh g^{-1} at 0.1 C of current rate. This is because the SWNT increased the conductivity of PI from 2.5×10^{-9} to 0.6 S cm^{-1} (PI/SWNT). In the reduction process, Li^{+} is attached to the PI to form a radical anion followed by dianion. In the oxidation process, Li^{+} is removed from the PI (Wu et al., 2014).

10.4 CONCLUSIONS

In conclusion, conducting polymer–carbon-based binary composite has shown promising electrochemical performance in lead acid batteries, metal-ion batteries and metal–air batteries. This is because of the synergistic effects between conducting polymers and carbonaceous materials. However, there is still room for improvement in terms of power density, rate capability, cyclability, safety and cost. The future perspective of battery applications are as follows: (i) searching a new electrode material; (ii) designing a suitable structure and morphology to enhance the transportation of ions at the electrode and electrolyte interface; (iii) adopting environmental-friendly approach to fabricate batteries; and (iv) better understanding toward the storage mechanisms of materials and its kinetic transport.

ACKNOWLEDGEMENTS

This research work was financially supported by the Impact-Oriented Interdisciplinary Research Grant (No. IIRG018A-2019) and Global Collaborative Programme – SATU Joint Research Scheme (No.ST012-2019).

REFERENCES

Baibarac, M., Baltog, I., Lefrant, S. & Gomez-Romero, P. 2011. Polydiphenylamine/carbon nanotube composites for applications in rechargeable lithium batteries. *Materials Science and Engineering: B*, 176: 110–120.

Caramia, V. & Bozzini, B. 2014. Materials science aspects of zinc–air batteries: a review. *Materials for Renewable and Sustainable Energy*, 3: 28.

Hossain, S. K. S. & Hoque, M. E. 2018. Polymer nanocomposite materials in energy storage: properties and applications. *Polymer-based Nanocomposites for Energy and Environmental Applications* (pp. 239–282). Woodhead Publishing.

Kim, D. W., Sivakkumar, S., Macfarlane, D. R., Forsyth, M. & Sun, Y. K. 2008. Cycling performance of lithium metal polymer cells assembled with ionic liquid and poly (3-methyl thiophene)/carbon nanotube composite cathode. *Journal of Power Sources*, 180: 591–596.

Liu, P., Han, J. J., Jiang, L. F., Li, Z. Y. & Cheng, J. N. 2017. Polyaniline/multi-walled carbon nanotubes composite with core-shell structures as a cathode material for rechargeable lithium-polymer cells. *Applied Surface Science*, 400: 446–452.

Liu, X., Chen, T., Chu, H., Niu, L., Sun, Z., Pan, L. & Sun, C. Q. 2015. Fe_2O_3-reduced graphene oxide composites synthesized via microwave-assisted method for sodium ion batteries. *Electrochimica Acta*, 166: 12–16.

Magu, T. O., Agobi, A. U., Hitler, L. & Dass, P. M. 2019. A review on conducting polymers-based composites for energy storage application. *Journal of Chemical Reviews*, 1: 19–34.

Meer, S., Kausar, A. & Iqbal, T. 2016. Trends in conducting polymer and hybrids of conducting polymer/carbon nanotube: a review. *Polymer-Plastics Technology and Engineering*, 55: 1416–1440.

Mike, J. F. & Lutkenhaus, J. L. 2013. Recent advances in conjugated polymer energy storage. *Journal of Polymer Science Part B: Polymer Physics*, 51: 468–480.

Moseley, P., Bonnet, B., Cooper, A. & Kellaway, M. 2007. Lead–acid battery chemistry adapted for hybrid electric vehicle duty. *Journal of Power Sources*, 174: 49–53.

Sengodu, P. & Deshmukh, A. D. 2015. Conducting polymers and their inorganic composites for advanced Li-ion batteries: a review. *RSC Advances*, 5: 42109–42130.

Shirakawa, H., Louis, E. J., Macdiarmid, A. G., Chiang, C. K. & Heeger, A. J. 1977. Synthesis of electrically conducting organic polymers: halogen derivatives of polyacetylene, (CH) x. *Journal of the Chemical Society, Chemical Communications*, 16:578–580.

Silva, C. H., Galiote, N. A., Huguenin, F., Teixeira-Neto, É., Constantino, V. R. & Temperini, M. L. 2012. Spectroscopic, morphological and electrochromic characterization of layer-by-layer hybrid films of polyaniline and hexaniobate nanoscrolls. *Journal of Materials Chemistry*, 22: 14052–14060.

Sivakkumar, S., Howlett, P. C., Winther-Jensen, B., Forsyth, M. & Macfarlane, D. R. 2009. Polyterthiophene/CNT composite as a cathode material for lithium batteries employing an ionic liquid electrolyte. *Electrochimica Acta*, 54: 6844–6849.

Sivakkumar, S., Macfarlane, D. R., Forsyth, M. & Kim, D. W. 2007. Ionic liquid-based rechargeable lithium metal-polymer cells assembled with polyaniline/carbon nanotube composite cathode. *Journal of the Electrochemical Society*, 154: A834–A838.

Su, C., He, H., Xu, L., Zhou, N., Zhu, X., Wang, G. & Zhang, C. 2017. Preparation of the nanocarbon/polypyrrole composite electrode by in situ electrochemical polymerization and its electrochemical performances as the cathode of the lithium ion battery. *Journal of Nanoscience and Nanotechnology*, 17: 1963–1969.

Thounthong, P., Rael, S. & Davat, B. 2009. Energy management of fuel cell/battery/supercapacitor hybrid power source for vehicle applications. *Journal of Power Sources*, 193: 376–385.

Wang, H., Lin, J. & Shen, Z. X. 2016. Polyaniline (PANi) based electrode materials for energy storage and conversion. *Journal of Science: Advanced Materials and Devices*, 1: 225–255.

Wu, H., Shevlin, S. A., Meng, Q., Guo, W., Meng, Y., Lu, K., Wei, Z. & Guo, Z. 2014. Flexible and binder-free organic cathode for high-performance lithium-ion batteries. *Advanced Materials*, 26: 3338–3343.

Xu, M., Ivey, D., Xie, Z. & Qu, W. 2015. Rechargeable Zn-air batteries: progress in electrolyte development and cell configuration advancement. *Journal of Power Sources*, 283: 358–371.

Yang, H., Qiu, Y. & Guo, X. 2016. Effects of PPy, GO and PPy/GO composites on the negative plate and on the high-rate partial-state-of-charge performance of lead-acid batteries. *Electrochimica Acta*, 215: 346–356.

Yoon, D. H., Yoon, S. H., Ryu, K. S. & Park, Y. J. 2016. PEDOT: PSS as multi-functional composite material for enhanced Li-air-battery air electrodes. *Scientific Reports*, 6: 19962.

Yoon, S. & Park, Y. 2017. Air electrode based on poly(3,4-ethylenedioxythiophene) microflower/graphene composite for superior $Li-O_2$ batteries with excellent cycle performance. *RSC Advances*, 7: 56752–56759.

Yoshino, A. 2012. The birth of the lithium-ion battery. *Angewandte Chemie International Edition*, 51: 5798–5800.

Zhang, G., Xu, Z., Liu, P., Su, Y., Huang, T., Liu, R., Xi, X. & Wu, D. 2018a. A facile in-situ polymerization strategy towards polyimide/carbon black composites as high performance lithium ion battery cathodes. *Electrochimica Acta*, 260: 598–605.

Zhang, M., Song, X., Ou, X. & Tang, Y. 2018b. Rechargeable batteries based on anion intercalation graphite cathodes. *Energy Storage Materials*, 16: 65–84.

Zhang, R. T., Han, J. J., Liu, P., Bao, C. Y. & Cheng, J. N. 2018c. Synthesis of PANI/rGO composite as a cathode material for rechargeable lithium-polymer cells. *Ionics*, 24: 3367–3373.

Zhang, Y., Bakenov, Z., Tan, T. & Huang, J. 2018d. Three-dimensional hierarchical porous structure of PPy/porous-graphene to encapsulate polysulfides for lithium/sulfur batteries. *Nanomaterials*, 8: 606.

Zhu, Y. G., Jia, C., Yang, J., Pan, F., Huang, Q. & Wang, Q. 2015. Dual redox catalysts for oxygen reduction and evolution reactions: towards a redox flow $Li-O_2$ battery. *Chemical Communications*, 51: 9451–9454.

11 Polyethylenedioxythiophene-Based Battery Applications

G. Nasymov, J. Swanson, A. M. Numan-Al-Mobin, C. Jahnke, and A. Smirnova
Department of Materials and Metallurgical Engineering
South Dakota School of Mines and Technology
Rapid City, South Dakota

CONTENTS

11.1 CHEMISTRY OF PEDOT

High electric conductivity of polyethylenedioxythiophene (PEDOT) and its derivatives has been extensively studied due to the dioxane ring attached to the thiophene backbone, which decreases the bandgap and sterically aids in retaining the conjugation. Through alteration of synthetic and post synthetic methodology, more desirable chemical and physical properties are achievable, such as alternate morphologies, increased ionic and electrical conductivity, transparency, fewer defects, and high tensile strength (McCullough 1998).

11.1.1 PEDOT SYNTHESIS AND MORPHOLOGY

11.1.1.1 Synthetic Techniques to Achieve Desired Morphologies

The classical method of PEDOT polymer synthesis is attributed to Dr. Freidrich Jonas (Heywang and Jonas 1992) who described a chemical and an electrochemical oxidation methods for PEDOT polymerization. Chemical oxidizing agents, such as iron (III) chloride, were shown to induce polymerization by a free radical mechanism. However, in electrochemical oxidation approach, 0.01–100% PEDOT monomer is polymerized under applied electric current (0.0001–100 mA/cm^2) in the presence of electrochemically stable solvents (e.g., water, aromatic and aliphatic hydrocarbons, etc.) mixed with free acids or conducting salts (e.g., alkyl sulfonic acids, tetrafluoroborate salts, etc.). Low concentration of polythiophene monomers leads to thin, high surface area porous films, while high concentration results in relatively dense films with conductivities reaching 200 S/cm (Jonas and Krafft 1994).

To improve PEDOT mechanical, electrochemical, and physical properties, verified in energy storage devices, for example, lithium ion and lithium–metal batteries, various copolymers have been used. It was validated that the addition of pyrrole improves mechanical properties without sacrificing electric conductivity (Jonas and Krafft 1994). Furthermore, copolymerization of PEDOT with polystyrene sulfonate (PSS; Figure 11.1) increases the polymer solubility in water. As a result, polymer dispersions require simple heat treatment to release water and create colorless, stable, transparent, and highly conductive (<1 S/cm) PEDOT:PSS films (Jonas, Krafft, and Muys 1995).

One of the most efficient approaches in enhancing the PEDOT:PSS electric conductivity is based on the addition of ethylene glycol during copolymerization that allows for higher-order packing of PEDOT nanocrystals and PSS chains and increases both the carrier ion mobility (from 0.045 to 1.7 cm^2/V s) and electronic conductivity (from

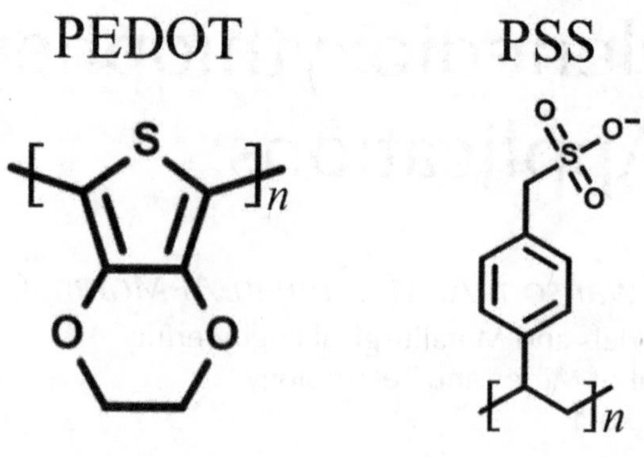

PEDOT   PSS

FIGURE 11.1   Structure of PEDOT and PSS. Adapted with permission from Takano et al. (2012).

1.2 to 830 S/cm) (Wei et al. 2013). A similar treatment with ethylene glycol, either during or after copolymerization, decreases the size of PEDOT crystals (~4.5 nm) coated with PSS and creates micelle-like structures (Takano et al. 2012). Furthermore, numerous other studies have demonstrated that the addition of alcohol during polymerization can significantly increase the PEDOT:PSS conductivity (Döbbelin et al. 2007). As an example, PEDOT:PSS films with a 50 nm micelle diameter and 5 nm thickness were produced (Figure 11.2). Higher conductivities were achieved by decreasing the micelle diameter to 5 nm and increasing the molecular orientation and polymer order of the corresponding polymer films (Nardes et al. 2007, Lang et al. 2009).

Another way to increase PEDOT:PSS electric conductivity is to use ionic liquids that induce a beneficial three-dimensional morphology to the PEDOT:PSS network, which is similar to the effects produced by alcohols (Döbbelin et al. 2007). Specifically, the addition of 0.7–1.5 wt% ionic liquid (e.g., alkyl imidazole-tetrafluoroborate) to PEDOT:PSS dispersions results in significant electric conductivity increase from 14 to 136 S/cm. In the electropolymerization approach (Poverenov et al. 2010), changes in polymer film morphology leading to the corresponding changes in physical and electrochemical properties were observed in the presence of propylene carbonate with tetrabutyl ammonium tetrafluoroborate ionic liquid.

In general, PEDOT and PEDOT:PSS morphology, for example, films, nanofibers, or nanotubes, plays a critical role in many applications, including catalysis an electrochemical devices. One of the methods to produce micron-long PEDOT-based nanofibers is to use V_2O_5 as a seed, which induces polymerization and can be easily removed afterward by dissolution in acidic media. It is envisioned that more precise PEDOT film morphology can optimize ionic and electronic transport – leading to highly tunable conductivities (Zhang, Macdiarmid, and Manohar 2005). Another polymer morphology of interest, specifically nanotubes, can be produced by adding dioctyl sulfosuccinate sodium salt to the reaction media prior to PEDOT polymerization. This synthetic approach results in PEDOT nanotubes with conductivity of ~10^{-2} S/cm (Zhang et al. 2006). An alternate route for one-dimensional (1-D) PEDOT polymer nanostructures synthesis involves growing them within the pores of metal oxide templates, for example, Al_2O_3 (Figure 11.3). In this case, the polymerization process is terminated by dissolving the membrane and yields 5 μm long PEDOT tubes (Han and Foulger 2005).

Spin-casted PEDOT has also been shown to produce high-conductivity thin films (>900 S/cm) (Zhang, Macdiarmid, and Manohar 2005) with pancake-like structure. However, an extra-thick PSS layer forming on the top of the PEDOT films causes significant electric conductivity fluctuations between the top layer and the polymer bulk (Kemerink et al. 2004).

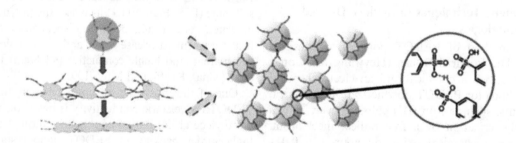

FIGURE 11.2   Micelles of PEDOT:PSS copolymer. PEDOT is the yellow core while the surrounding blue is PSS.

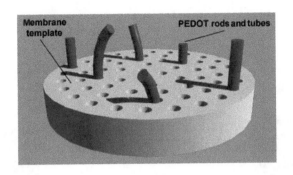

FIGURE 11.3 A "hard" Al_2O_3 template used to synthesize 1-D PEDOT structures (Han and Foulger 2005). Chemical communications by Royal Society of Chemistry (Great Britain). Reproduced with permission of Royal Society of Chemistry in the format Book via Copyright Clearance Center.

11.1.2 PEDOT-Based Nanocomposites

In recent years, doped PEDOT and PEDOT:PSS polymers demonstrated significant progress, such as broad alterations in chemical and physical properties. Most commonly, the polymers are functionalized to increase the conductivity while retaining the beneficial properties of the PSS copolymer. Functionalized PEDOT and PEDOT:PSS polymers have found use in catalysis, nanochemistry, and energy storage (Zhang et al. 2006). Specifically, PEDOT functionalized by silver nanoparticles resulted in even dispersion of the metal nanoparticles (18–22 nm in diameter) on the nanotube surface (Zhang et al. 2006). Besides silver, 50 nm gold nanoparticles were incorporated into the PEDOT polymer structure (Figure 11.4), which was achieved through simultaneous nanoparticle synthesis and PEDOT polymerization or by mixing the gold nanoparticle dispersion with PEDOT. As a result, incorporation of gold nanoparticles into PEDOT significantly alters the band gap of the polymer, increases the optical absorption, and improves the PEDOT conductivity (Selvaganesh et al. 2007, Salsamendi et al. 2008).

Similar to noble metal nanoparticles, like silver or gold, metal oxides have been also incorporated into the PEDOT or PEDOT:PSS polymers. Nanoparticles of metal oxides, for example, iron oxide, have been incorporated within PEDOT nanotubes to improve chemical and physical properties of the PEDOT nanocomposites (Zhang et al. 2006). Besides metals and metal oxides, PEDOT–silicon nanocomposites were produced by in situ thermal polymerization of 2,5-dibromo-3,4-ethylenedioxythiophene (DBEDOT) with silicon nanoparticles (SiNPs) in application for lithium-ion battery anodes (Figure 11.5). This approach led to high specific capacity of the corresponding lithium-ion battery anodes, reaching 2300 mA h/g (McGraw et al. 2016, Smirnova, Mcgraw, and Kolla 2017).

11.2 PEDOT-BASED POLYMERS IN LITHIUM–SULFUR BATTERIES

Lithium–sulfur (Li–S) rechargeable batteries have notable theoretical specific capacity of 1675 mA h/g and high energy density of 2600 W h/kg due to the two-electron reaction of Li_2S formation. Low cost of sulfur makes these batteries economically feasible; however, they have limited applications due to low electrical conductivity of sulfur cathodes. Furthermore, the shuttle mechanism due to dissolution of lithium polysulfides from sulfur-based

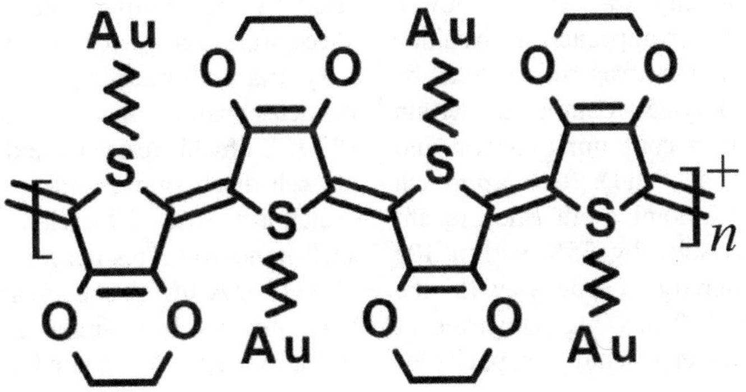

FIGURE 11.4 Scheme of gold nanoparticle binding with PEDOT. Adapted from Selvaganesh et al. (2007).

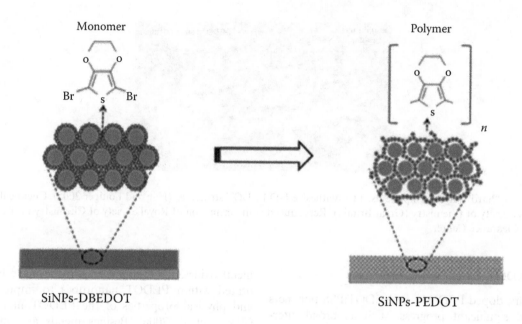

FIGURE 11.5 One-step in situ thermal polymerization of SiNPs-DBEDOT on Cu substrate resulting in electrically conducting SiNPs–PEDOT polymer nanocomposite. SiNPs are represented by spheres.

cathodes causes low charge rates of the electrochemical cells. The lithium-soluble polysulfides, such as Li_2S_3, Li_2S_4, Li_2S_6, and Li_2S_8 formed as a result of the discharge process, diffuse to the anode and accumulate in the form of lithium insoluble salts (e.g., Li_2S and Li_2S_2) leading to poor Li–S battery cyclability (Lee and Choi 2015). Improvement of the Li–S battery performance can be achieved by application of conducting polymers, especially PEDOT and PEDOT-based conjugated polymers that minimize dissolution of polysulfides and facilitate lithium-ion transport (Li et al. 2014). Specifically, sulfur–graphene cathode nanocomposites coated with PEDOT: PSS demonstrate porous structure resulting in high surface area and strong interaction with sulfur particles (Figure 11.6). As a result, the composite cathode with 60.1% sulfur demonstrates initial discharge capacity of 1198 mA h/g. However, after 200 cycles at 0.1 C the specific capacity dropped to 845 mA h/g, whereas at higher C-rate (2 C) the specific capacity was slightly lower, reaching 718 mA h/g. In a similar approach, polysulfide diffusion from the carbon matrix was minimized by deposition of PEDOT:PSS polymer composite within mesoporous CMK-3 architectures containing carbon and sulfur (Figure 11.7) (Yang et al. 2011). This approach resulted in a 4% increase in Coulombic efficiency reaching 97% reduced decay of capacity by 25% within 100 cycles, and <10% higher discharge capacity with the starting discharge values of ~1140 mA h/g. However, at C/5 within 150 cycle rate, a lower discharge capacity of <600 mA h/g was observed.

Besides sulfur–graphene cathode nanocomposites, functionalization of sulfur–metal oxide nanocomposites with conducting polymers could be considered as an alternative strategy to improve the stability and the electrochemical performance of Li–S battery cathodes. For instance, MnO_2 nanosheets functionalized by PEDOT for application in Li–S battery cathodes provide active contact area, improve electrolyte wettability, and interlink the polymer chains. As a result, an enhanced conductivity and electrochemical stability can be achieved. As a result, at 0.2 C and after 200 cycles, the capacity values of 827 mA h/g and <50% battery performance improvement compared to sulfur–PEDOT nanocomposites without MnO_2 (551 mA h/g) have been observed. In particular, at 0.5 C and after 200 cycles the reported discharge capacity values reach 545 mA h/g (Yan et al. 2016).

The application of core/shell nanostructures (Chen et al. 2013) with 10–20 nm sulfur nanoparticles synthesized by precipitation and coated by PEDOT is another successful strategy for enhancement of electrical conductivity and sulfur utilization in Li–S batteries. Small sulfur particles ensure efficient encapsulation of sulfur within the PEDOT shield, restrict the diffusion of polysulfides, alleviate self-discharging, and enhance the electrochemical cell cyclability. After 50 cycles, the resulting S/PEDOT core/shell nanostructures demonstrate initial discharge capacity of 1117 mA h/g that decreased to 930 mA h/g in the following cycles. In a similar approach (Lee and Choi 2015), sulfur particles are coated with commercial PEDOT–PSS polymer blend by means of a wet mixing in aqueous

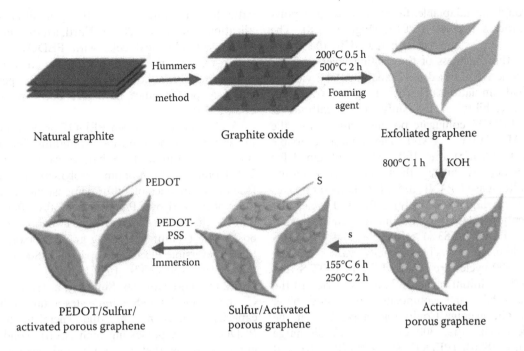

FIGURE 11.6 Schematic diagram of the formation of the PEDOT/S/activated porous graphene composite. Adapted from Li et al. (2014).

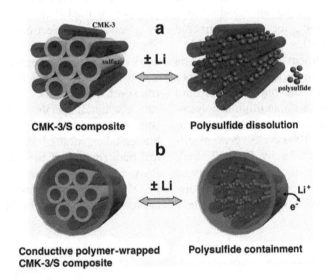

FIGURE 11.7 Scheme of PEDOT:PSS-coated CMK3/sulfur composite for improving the cathode performance. (a) In bareCMK-3/S particles (gray: CMK-3, yellow: sulfur), polysulfides (green color) still diffuse out of the carbon matrix during lithiation/delithiation. (b) With conductive polymer coating layer (blue color), polysulfides could be confined within the carbon matrix. Lithium ions and electrons can move through this polymer layer.

media. In comparison to unmodified sulfur cathodes (292 mA h/g), this surface functionalization results in lower resistance, significantly higher capacity of 566 mA h/g at 0.2 C discharge after 50 cycles, and gravimetric energy density reaching 1113 W h/kg. These experimental efforts confirm that PEDOT–PSS coatings restrict the polysulfide

diffusion to the lithium metal anode, suppress mechanical strain, and provide active sites and surfaces for detention and reuse of the polysulfides.

Another application of conductive polymers in lithium–sulfur batteries is focused on sulfur nanospheres coated with polypyrrole (PPY), polyaniline (PANI), or

PEDOT through an adaptable, facile, and scalable polymerization process (Li et al. 2013) (Figure 11.8). The sulfur cathodes are compared with regard to the chemical bonding, the thickness of the coating layer, and the polymer conductivity based on experimental evidence and theoretical simulations. The best long-term rate capability and cyclability of the sulfur-based cathode is obtained for PEDOT conjugate polymers and follow the trend: PEDOT > PPY > PANI. The highest reversible capacity (1071 mA h/g) after 100 cycles is obtained for PEDOT-S, which is higher than the corresponding values for sulfur-based PANI and PPY nanocomposites (876 and 923 mA h/g, respectively). After 100 cycles, the capacity retention relative to the initial cycle decreased as following: PEDOT-S (83%) > PPY-S (70%) > PANI-S (65%). The highest reversible capacity values of 860 mA h/g after 300 cycles are reported for PEDOT-S that retains the 67% of initial capacity. This is superior to the PANI-S and PPY-S nanocomposites (61% and 52%, respectively). The average Coulombic efficiencies of the electrochemical cells after 300 cycles at C/5 discharge rate are equal to 99.6% (PEDOT-S), 99.2% (PPY-S), and 98.5% (PANI-S), confirming that PEDOT-based nanocomposites have the highest positive impact on the overall cell performance.

To minimize the lithium polysulfides dissolution and control large volume changes during charge–discharge cycles, PEDOT coatings are used for multichambered carbon nanocube-sulfur composites (Chen et al. 2015). In the proposed dual confinement strategy, sulfur is impregnated into the interconnected multichamber carbon nanocubes that serve as the physical enclosure and multilayer containers for the soluble lithium polysulfides. The micro/mesoporous carbon architectures retain their morphology within at least 1000 cycles, prevent diffusion of soluble lithium polysulfides by PEDOT, and, thus, deliver at 1 C discharge rate high initial capacity of 1086 mA h/g. The polysulfides temporal and spatial distribution in charge/discharge cycles within Li–S battery

cathodes was evaluated directly by in situ/*operando* visualization (Sun et al. 2015). Furthermore, in the case of Li–S electrochemical cells with PEDOT-coated hollow sulfur nanoparticles and Nafion-coated separator, minimized dissolution of polysulfides and superior electrochemical performance have also been emphasized (Sun et al. 2015).

A comparison between PEDOT or polyaniline (PANI) in ternary nanocomposites containing graphene oxide (GO) and sulfur as Li–S battery cathodes (Wang et al. 2015) prepared by monomer polymerization on GO surface in the presence of sulfur demonstrates obvious advantages of PEDOT over PANI. Specifically, after 200 cycles at 0.5 C, the sulfur-based PEDOT nanocomposites with 66.2 wt% of sulfur-containing graphene oxide show much higher reversible discharge capacity of 800.2 mA h/g than that of PANI/GO/S (599.1 mA h/g) or PANI/S (407.2 mA h/g) cathodes. Similarly, at the highest rate of 4 C, the PEDOT/GO/S composites retained high specific capacity of 632.4 mA h/g due synergistic effect between PEDOT matrix possessing high electric conductivity and graphene oxide that physically and chemically encloses sulfur and polysulfides inside the cathode architecture (Wang et al. 2015).

A different approach focused on a synergistic effect of the dissimilar functional groups belonging to different polymers is used (Pan et al. 2016) to improve specific capacity and Li–S electrochemical cell cyclability. In this approach, polyacrylic acid (PAA) and PEDOT–PSS polymers are mixed to counteract polysulfide dissolution and assist electron and ion transfer and promote lithium-ion transport. As a result, after 80 cycles at 0.5 C discharge rate, the initial specific capacity of 1121 and 830 mA h/g was demonstrated for the cells with PEDOT/PAA:PSS binder taken in a 2:3 ratio. In this case, the electrochemical performance of the sulfur cathode with the PEDOT/PAA composite binder was higher than either PAA or PEDOT in a single-component binder. A more rigorous approach based on four binders with different chemical

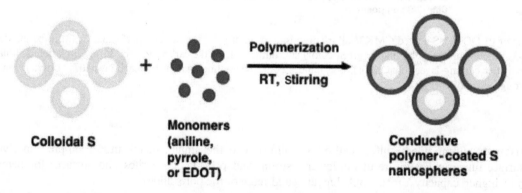

FIGURE 11.8 Schematic illustration of the fabrication process of conductive polymer-coated hollow sulfur nanospheres. RT, room temperature.

and electrical properties (Ai et al. 2015) is focused on the interfacial properties, phase transformations, and the corresponding reaction mechanisms depending on the polymer group functionality. It is emphasized that conductive binders promote sulfur species precipitation. Specifically, it is stated that more sulfur solid species precipitate in the presence of binders with carbonyl groups, for example, poly(9,9-dioctylfluorene-*co*-fluorenone-*co*-methylbenzoic ester) and poly(vinylpyrrolidone) (PVP). These outcomes imply that the carbonyl functional groups attract and bind well the particles within the structure. As a result, more sulfur species in solid state are produced minimizing the effect of the shuttle mechanism and, consequently, enhancing the cell performance (Ai et al. 2015).

Besides graphite or polymer matrices, synthetic pigments, such as Prussian blue derivatives (PBAs), have been explored as derivatives for sulfur accommodation in Li–S batteries (Su et al. 2017). Because of the mobile two-electron reduction–oxidation-active centers in PBA-based structures, sulfur particles stored in the PBA interstitial sites help to insert and extract lithium ions and transport electrons. One of such derivatives with large open framework, $Na_2Fe[Fe(CN)_6]$, stores sulfur and acts as a polysulfide diffusion inhibitor based on the Lewis acid–base bonding effect. The produced $S/Na_2Fe[Fe(CN)_6]$/PEDOT composites demonstrate rapid kinetics and electrochemical stability. Their electrochemical characteristics are attributed to the transport of lithium ions and electrons, enhancing the use of sulfur. Likewise, the metal centers function as the Lewis acidic sites with high attraction to the negative polysulfides, diminishing their diffusion and decreasing the influence of shuttle mechanism. Functionalization of $Na_2Fe[Fe(CN)_6]$ nanocrystals containing 82 wt% of sulfur with PEDOT polymer (Su et al. 2017) resulted in high specific capacity (1291 mA h/g) at 0.1 C discharge. The performance of the $Na_2Fe[Fe(CN)_6]$–sulfur-based nanocomposites with and without PEDOT has been compared. It is demonstrated that in a 100 cycle experiment at 0.1 C, the

PEDOT-based electrodes demonstrated over 30% capacity improvement in comparison to PEDOT-free nanocomposites (763 mA h/g) with degradation rates of 0.15%. In another example, sulfur nanoparticles encapsulated within carbon clusters and coated with PEDOT as a binder demonstrated steady-state cyclability, improved volumetric energy density, and outstanding rate capability (Figure 11.9) (Li et al. 2015). The cells with sulfur–carbon–PEDOT clusters treated with toluene show 707 mA h/g specific capacity in 300 cycles at 0.5 C discharge rate and capacity retention of 70%, indicating that PEDOT functionalization coating and free volume in-between carbon clusters improve cyclability behavior by reducing the active material loss in surface layers in the beginning of lithiation/delithiation. The PEDOT-C/S electrodes demonstrate the 99.5% Coulombic efficiency within 300 cycles and exhibit very little shuttle effect. This achievement is the result of several factors, such as small sulfur particle for high active material utilization, hollow structure of carbon particles that provide good contact between particles, and conductive carbon support functionality as a path for rapid electron exchange. Besides, controlled empty volumes in carbon clusters allow accommodation of sulfur volume change and uniform Li_2S accumulation and distribution within the sulfur nanocomposite. Furthermore, the continuous and connected channels within the structure allow physical confinement of polysulfides, while carbon coating provides limited electrolyte accessibility. Finally, efficient packing of small sulfur particles within carbon clusters in contact with PEDOT promotes high volumetric energy density.

The active sulfur particle size effect on the cell performance in PEDOT-based cathodes is compared to those containing a polyvinylidene difluoride binder (Wang et al. 2014). In addition to the binder and sulfur particle size variation, two different electrolytes have been evaluated, comprising lithium salt of bis(trifluoromethanesulfonyl) imide that was investigated in two different solvents,

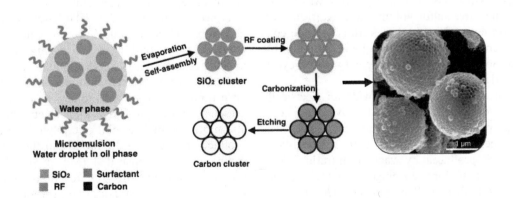

FIGURE 11.9 Schematic diagram of the fabrication process of carbon clusters by a bottom-up microemulsion approach. Adapted from Li et al. (2015).

specifically polyethylene glycol dimethyl ether (PEGDME) and 1,3-dioxolane-dimethoxy ethane. The highest capacity retention (68%) and specific capacity exceeding 570 mA h/g in 100 cycles performance were demonstrated from micrometer-sized sulfur with PEDOT in PEGDME electrolyte (Wang et al. 2014).

Besides broad application of PEDOT:PSS for cathode enhancement, the dual shielding effect of PEDOT:PSS conjugated polymers in a separator (Abbas et al. 2016) was used to minimize the shuttle phenomenon and confined lithium polysulfides within the cathode. In this case, the negatively charged sulfonate groups (SO_3^-) in PSS chain act as an electrostatic protection for soluble lithium polysulfide species due to electrostatic repulsion, while PEDOT serves chemical interactions with insoluble Li_2S and Li_2S_2 species. Furthermore, the PEDOT:PSS polymer film changes the hydrophobic surface of the separator membrane into hydrophilic, which improves the overall electrochemical performance. The observed decay in capacity (0.0364% per cycle) for these cathodes in 1000 cycles at 0.25 C was significantly lower than that for the nonmodified separator (Abbas et al. 2016). Another known application (Niu et al. 2016) is in thin polysulfide blocking layers (MPBLs), which wrap the cathode of a Li–S battery and provide fast ion diffusion. These multifunctional layers allow high battery performance in impeding polysulfides due to a synergistic combination of physical and chemical properties originating from carbon and conductive polymer, respectively. MPBLs also function as the current collector to reutilize the confined polysulfides and thus increase sulfur utilization. Coated with MPBL, the carbon–sulfur cathodes exhibit 0.042% capacity decrease per cycle at 1 C rate after 1000 cycles and 615 mA h/g capacity at 3 C discharge (Niu et al. 2016).

Mixing PEDOT–PSS polymer with sulfur/carbon black produced core/shell structures that were coated onto a porous carbon paper to produce Li–S battery cathodes (Gong et al. 2015). In this case, PEDOT/S/carbon black composites promote electron transport and minimize polysulfide diffusion, while the porous carbon paper retains the electrolyte with dissolved polysulfides within the cathode and alleviates sulfur volumetric expansion. Following this strategy, a Li–S battery with a cathode containing 3 mg/cm^2 of sulfur yielded 99.4% Coulombic efficiency and 868 mA h/g specific capacity in 100 cycles (Gong et al. 2015).

Another successful attempt to provide an electrically conductive network and close contact between the components in graphene/PEDOT/PSS nanocomposites was achieved in free-standing cathodes with a multi-layer flexible structure produced by vacuum filtration. This attempt resulted in uniform coverage of sulfur particles by graphene and PEDOT:PSS polymer (Xiao et al. 2017). These cathodes show high reversible volumetric capacity of 1432 A h/L C and 1038 A h/L (at 0.1 and 1 C, respectively), cyclability with a negligible decay

rate of 0.04% per cycle in a 500 cycle experiment at 1 C, and 701 mA h/g capacity at 4 C discharge rate (Xiao et al. 2017).

In comparison to PEDOT-functionalized cathodes, a bottom-up method achieved in one step at ambient conditions is presented to manufacture well-dispersed hollow sulfur nanospheres coated with PVP, which allows to control cathode architectures within a broad nanoscale/macroscale range. Interestingly, in this case, the cell performance characteristics for PVD- based cathodes were better than for those with PEDOT. Specifically, the discharge capacity at the highest C/2 rate level reached the value of 1179 mA h/g capacity and about 73% capacity retention after 300–500 cycles (Li et al. 2013).

In addition to numerous publications focused on PEDOT and PEDOT:PSS polymers and polymer composites, an application of polythiophene (PTh) performing a role of an absorbent and a conducting additive for Li–S battery applications has been reported (Wu et al. 2011). In order to enhance the electrochemical properties and cyclability of the Li–S electrochemical cells, PTh was evenly deposited on the surface of the sulfur nanoparticles to form a sulfur core/PTh shell architectures (Wu et al. 2011). These structures were produced by in situ polymerization in which chloroform was used as a solvent and PTh as a monomer in presence of iron (III) chloride for oxidation at 0°C. An optimized composition comprising of sulfur (71.9%) and PTh (18.1%) is reported to result in the initial high discharge capacity of 1119.3 mA h/g, but significantly lower capacity (830.2 mA h/g) in the following 80 cycles. The cycle life of the manufactured electrochemical cells, utilization of sulfur, and the rate performance of the S core/PTh shell structures demonstrate improved characteristics compared to the conventional sulfur electrode. It was achieved by optimization of the pore size and shell thickness, providing efficient lithium-ion diffusion channels (Wu et al. 2011). Another example of a PTh application in Li–S battery cathodes is related to the in situ oxidative polymerization of PTh monomers with chloroform as the solvent and FeCl$_3$ for oxidation (Wu et al. 2010). The chemical reaction between sulfur and polythiophene has not been detected. The PTh-based cathode showed acceptable electrochemical performance with the discharge capacity of 1168 mA h/g sulfur in the first cycle and the remaining 819.8 mA h/g in 50 cycles.

Besides Li–S battery cathodes, conductive polymers have been used to create protective anode layers. In the presence of the protective film, electrochemically stable and more conductive solid electrolyte interface layers are produced between the Li anode and the ether-based electrolyte. It is demonstrated (Ma et al. 2014) that such layers can inhibit corrosion reactions between the lithium anode and polysulfides and prevent Li dendrites formation. With 2.5–3 mg/cm^2 of sulfur in the presence of commercial electrolyte, the corresponding value of 815 mA h/g after 300 cycles at C/2 with 91.3% efficiency was observed. Another application of PEDOT in Li-ion battery anodes involves S/carbon/Si nanocomposites. In this approach, a porous Si/S-doped carbon composite was prepared by a magnesio-thermic reaction of meso-porous SiO_2, coated with a sulfur-containing PEDOT, and subsequently carbonized. The as-prepared Si composite was coated with S-doped carbon containing 2.6 wt% sulfur, which resulted in S/carbon/Si nanocomposites with 58.8 m^2/g surface area. As an anode material in lithium ion cells, the Si/S-doped carbon composites showed better electrochemical performance, long-term cyclability, and 450 mA h/g specific capacity (Yue et al. 2013).

11.3 LITHIUM–AIR BATTERY BASED ON PEDOT OR PEDOT:PSS

In recent years, lithium–air (Li–O_2) batteries have been broadly investigated due to high expectations with regard to the high-energy/high-power density electric power generation and storage. These batteries have considerably high gravimetric energy density (3500 W h/kg), at least by one order of magnitude higher than the state-of-the-art Li-ion batteries (Amanchukwu et al. 2016). In contrast to this promising behavior, there are numerous challenges that should be resolved before the lithium–air and lithium–oxygen batteries can be used commercially. In this regard, the progress in Li–O_2 battery based on PEDOT and/or PEDOT:PSS polymers should be emphasized.

Several studies on nanoporous gold (Peng et al. 2012), carbides (Thotiyl et al. 2013), and inorganic metals (Kundu et al. 2015) have been reported for the operation of Li–O_2 batteries. For effective operation of Li–O_2 batteries, the cathodes require high surface area and high catalytic activity during the oxygen reduction and oxygen evolution reactions (OERs). At the same time, high electronic conductivity and low-cost cathodes are also desired for efficient and cost-effective battery manufacturing. However, the cost of precious metals limits Li–O_2 batteries applications. To minimize these disadvantages, a number of electrically conducing polymers, such PEDOT (Yao et al. 2012), poly(pyrrole) (Kuwabata, Masui, and Yoneyama 1999), and poly(aniline), have shown a great promise as binders or electrode materials due to their desired electrical properties.

11.3.1 PEDOT-BASED NANOCOMPOSITES FOR Li–O_2 BATTERIES

Investigation of PEDOT and PPEDOT-PSS polymers in electrodes exposed to oxygen, rather than air, have been performed due to the higher catalytic activity in 100% oxygen environment and for understanding the fundamental catalytic and transport mechanisms (Geng et al. 2016). Pure PEDOT-based electrode performance was compared to PEDOT-PSS-containing nanocomposite for better understanding of the electrocatalytic Li–O_2 reactions. Based on this strategy, an isolated from of micro-structured PEDOT electrode containing no carbon or binders in a Li–O_2 battery can be used to demonstrate the formation of lithium peroxide (Shui et al. 2013) as well as the performance of Li–O_2 cycling capability. In PEDOT, the thiophene ring (Figure 11.10) degrades due to sulfone that forms during this reaction and contains

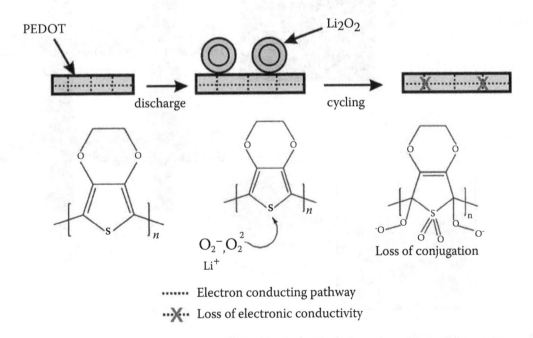

······ Electron conducting pathway

··✗·· Loss of electronic conductivity

FIGURE 11.10 PEDOT structure showing loss of electronic conductivity and degradation of thiophene ring in PEDOT. Reprinted (adapted) with permission from Amanchukwu et al. (2016). Copyright © 2016 American Chemical Society.

a sulfonyl functional group attached to two carbon atoms (Amanchukwu et al. 2016). The formation of sulfone eventually results in the loss of the PEDOT conjugated behavior (Marciniak et al. 2004) and diminishes the electronic conductivity resulting in poor electrochemical cell cyclability.

The free-standing or isolated PEDOT films can be manufactured by an evaporative vapor-phase polymerization, which causes PEDOT films to form and yield polymer fibers with nano- and microscale architectures (**Error! Hyperlink reference not valid.**). Besides, the PEDOT films possessing specific nanostructure and high electronic conductivity can be produced using in situ polymerization (Hohnholz et al. 2001) and oxidative chemical vapor deposition (Lock, Im, and Gleason 2006).

In state-of-the-art lithium-ion batteries with PEDOT-based electrodes, low performance was observed due to the poor lithium ions insertion/deinsertion in the PEDOT matrix. In contrast, lithium intercalation is absent for PEDOT used in Li–O_2 batteries; however, an electronically conducting PEDOT surface helps in diffusion and adsorption of O_2 as well as the following oxygen reduction reactions (ORRs) (Singh, Crispin, and Zozoulenko 2017). Moreover, the PEDOT electrode in Li–O_2 batteries supports Li–O_2 discharge-and-charge cycles leading to the formation and oxidation of Li_2O_2, respectively (Lu et al. 2011). However, after a few discharge-and-charge cycles,

the thiophene ring in the PEDOT polymer starts to oxidize that leads to release of sulfone (Figure 11.10) along with the fragmentation of the PEDOT polymer chain (Amanchukwu et al. 2016). The loss of PEDOT conjugation results in the electronic conductivity drop and poor cycling behavior confirming that PEDOT application in Li–O_2 battery cathodes is not feasible.

11.3.2 PEDOT:PSS-Based Li–O_2 Battery Cathodes

Carbon surface modification is a new approach to minimize carbon from oxidation in air without substantial decrease in battery capacity (Kim and Park 2014, Yoon and Park 2014, Lee and Park 2015). As a conducting polymer blend, PEDOT:PSS offers stable and high electronic conductivity. It can be used to form nanocomposites (Friedel et al. 2009, Li, Liu, and Gao 2010) enabling a simple cathode manufacturing process (Figure 11.11). Furthermore, Li–air cells with PEDOT:PSS reveals redox activity (Nasybulin et al. 2013), indicating its use as the oxygen reduction/OER media. At the same time, it performs as a protective material to suppress electrochemical parasitic reactions or as an electrically conducting binder.

It has been reported that the mixture of PEDOT–PSS and graphene at a ratio of 1:1 by weight has better performance when compared to that of pristine graphene as an air electrode (Yoon et al. 2016). Highly concentrated

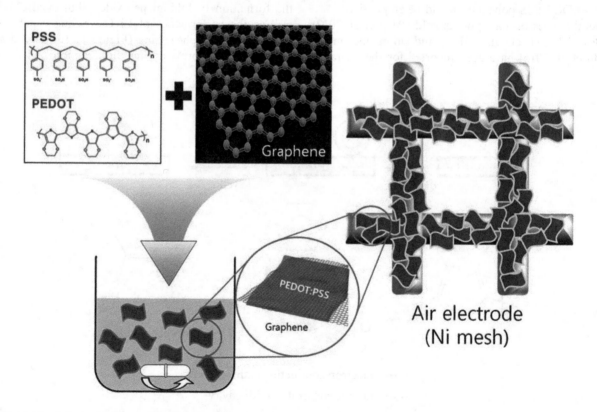

FIGURE 11.11 Schematic diagram showing Graphene/PEDOT:PSS composite.

PEDOT–PSS blends in graphene-based composites have a multidimensional functionality including their operation as redox reaction matrix, protective electrically conductive coating, and a conducting binder. It is should be noted that the protective layers of PEDOT:PSS on carbon surfaces significantly improve the cyclability of the PEDOT cathode (Kim and Park 2014). In addition, the protective PEDOT:PSS layer prevents and eliminates the parasitic reactions, such as the development of Li_2CO_3 because of the reaction of Li_2O_2 and carbon along with electrolyte decomposition promoted by the carbon surface (McCloskey et al. 2012, Ottakam Thotiyl et al. 2012). However, the PEDOT electrode shows similar C-rate capability as that of the nonmodified PEDOT graphene-based cathode. It is experimentally confirmed that PEDOT–PSS polymer blends have conductivity (10^1–10^2 S/cm) that is significantly lower than the conductivity of graphene (10^2–10^4 S/m) and defines the lower rate capability of the PEDOT–PSS cathodes. It is noteworthy to mention that the surface resistivity. It is noteworthy to mention that the surface resistivity of the PEDOT -PSS-based cathode (5 Ω/sq) is much smaller than that of the carbon electrode without PEDOT (145 Ω/sq). (Yoon et al. 2016).

The most significant limitation that prevents the practical deployment of $Li-O_2$ batteries is the large charge/discharge voltage hysteresis (Mo, Ong, and Ceder 2011). The theoretical standard reduction potential for $2Li^+ + O_2 + 2e \leftrightarrow Li_2O_2$ reaction (2.96 V) versus lithium metal (Lu et al. 2010) is considered as a limiting factor in lithium–air batteries. This is due to the fact that in $Li-O_2$ batteries the discharge process takes place at 2.7 V; however the battery charges at over 4.0 V. The ORR takes place at the potential near the thermodynamic value, whereas the OER necessitates substantial overpotential resulting in a slow OER kinetics (Lu et al. 2010, Harding et al. 2012).

11.4 LITHIUM AND ALKALI ION POLYTHIOPHENE BATTERIES

11.4.1 CATHODES

Polythiophene-based polymers are used as a cathode material in lithium-ion and lithium–metal batteries (Etacheri et al. 2011, Yang et al. 2011, Wu et al. 2016). In general, the polythiophene-based cathodes show a discharge voltage of 3.7 V versus Li^+/Li and good cyclability at high current densities reaching of 900 mA h/g. However, the discharge capacity in long-term cyclability tests 500 cycles was low (50.6 mA h/g) indicating that certain electrochemical transformations can take place within the PEDOT-based nanocomposites during battery operation (Liu et al. 2012). On the contrary, the polythiophene-based oligomers consisting of tetrahydrothiophene and aromatic thiophene groups demonstrated high specific discharge capacity of 400 mA h/g. The observed electrochemical performance was explained by the differences in charge/discharge mechanisms that were

different from the conventional insertion–deinsertion process (Rozier et al. 2018) and could be the result of the electron–electrode reactions due to the conjugated structure of the PEDOT polymer (Tang et al. 2008).

11.4.1.1 Cathode Binders and Composites

Application polythiophenes and polyimides with electron conductivity provide physical contact and superior conduction pathways for reversible redox reactions in composite cathodes. The polyimide polymer blend containing 30 wt% PTh demonstrates the best lithium-ion kinetics combined with high electronic conductivity. The effect of polyimide–polythiophene blending approach results in 216.8 mA h/g specific capacity at C/10 discharge rate, which, however, decreases to 89.6 mA h/g at 20 C, but demonstrates high capacity retention (94 %) after 1000 cycles (Lyu et al. 2017).

Another example of PTh in 15 wt% LiV_3O_8 composite cathode for lithium-ion battery demonstrates that a uniform 3–5 nm thick polythiophene coatings on the surface of LiV_3O_8 particles can be produced. The discharge capacity of LiV_3O_8–polythiophene nanocomposites was relatively high (200.3 mA h/g with virtually no capacity loss after 50 cycles) (Guo et al. 2014). Among many traditional cathode materials for lithium-ion batteries, the most well-known $LiNi_{1/3}Mn_{1/3}Co_{1/3}O_2$ (Cho et al. 2019) has been also used for cathode functionalization by PEDOT. The PEDOT-modified NMC-111 showed both enhanced cycle and rate capabilities as compared with the pure $LiNi_{1/3}Mn_{1/3}Co_{1/3}O_2$ cathode; specifically, the discharge capacity was shown to improve by 60% up to 73.9 mA h/g for the NMC-111/2 wt% PEDOT after the heat- treatment at 300 °C. The decreased battery polarization due to the PEDOT layer deposition was explained by the enhanced electron transfer at the electrode-electrolyte interface (Liu et al. 2013). A dual role of PEDOT:PSS as a binder and conducting polymer is demonstrated for carbon-free $LiFePO_4$ electrodes (Das et al. 2015). As PEDOT:PSS content within the cathode goes up, an increase in the rate capability and a corresponding decrease in the overvoltage are observed. These cathodes with 8% PEDOT-PSS demonstrate improved cyclability and similar capacity in comparison to the commercially available electrode materials.

Besides Li-ion batteries, polythiophene has been also tested in cathode of the Na-ion electrochemical cells based on $NaFePO_4$. The composite cathode manufactured by using $NaFePO_4$ particles without PTh demonstrates a discharge capacity, which is 31% lower than for the polythiophene-layered $NaFePO_4$ electrodes (142 mA h/g) tested within 100 cycles. At relatively high current densities of 300 mA h/g, these cathodes exhibit reversible behavior at 42 mA h/g, compared to negligibly small capacity values without polythiophene (Ali et al. 2016). The in situ XAS spectroscopy explained the observed behavior of the $NaFePO_4$ cathode by the valence change of the iron atoms and reversible volume deviations of the unit cell and the iron–oxygen octahedra during charge/discharge.

Thiophene-based derivatives have been examined as a conductive polymer phases and binders for high-voltage lithium cobaltite (LCO) cathodes. Specifically, bithiophene and terthiophene were used as protective layers to prevent electrolyte decomposition at high voltages reaching 4.4.V and improve electrochemical cell cyclability (Xia, Xia, and Liu 2015). For these thiophene-based LCO cathodes, the capacity retention in 100 cycles at discharge reached 50%. Interestingly, the LCO cathodes with deposited films of thiophene-based polymer in the presence of the electrolyte containing 0.1 wt% terthiophene displayed much over 30% higher capacity retention at 0.25 C, which could be considered as potentially valuable observation for creating new electrolyte compositions for high-voltage energy storage devices (Xia, Xia, and Liu 2015).

Another example of a composite cathode containing PTh derivative is a mesoporous polythiophene–MnO_2 cathode with interconnected mesoporous structure as confirmed by type IV hysteresis of the nitrogen adsorption/desorption isotherms, surface area of 226 m^2/g, and pore diameter of 5.2 nm. Because of large surface area and quick lithium-ion diffusion, the corresponding cathodes demonstrated close-to-theoretical discharge capacity (305 mA h/g) at a current density of 10 mA h/g and good cyclability at large current densities (Thapa et al. 2014).

11.4.2 Anodes

Besides their application in lithium-ion battery anodes, PEDOT- and PTh-based conjugated polymers have been used for ultra-long-life anodes (Xu et al. 2015). However, some publications report small redox activity and low rate capability, which could be related to different synthesis approaches and nonoptimized testing conditions. Redox-active crosslinked PTh-based porous anodes with enhanced surface area of 696 m^2/g demonstrated significantly enhanced electrochemical performance (Zhang et al. 2018). The corresponding electrochemical cells show over 1200 mA h/g discharge capacity, high rate capability, good cyclability, and specific capacity exceeding 650 mA h/g in 1000 cycles (Xu et al. 2015).

11.4.2.1 Anode Binders and Composites

Improvement of the lithium storage ability within metal oxide nanostructured anodes can be achieved by deposition of PTh thin film (5 nm) on their surface. The MnO_2 particles coated by PTh reveal improved capacity in high-voltage discharging/charging cycles (Xiao et al. 2010). In silicon-based anodes with silicon core-PTh shell porous architectures produced by oxidative polymerization, PTh shell flexibility allow to minimize detrimental effects of silicon volume expansion/contraction during lithiation/

delithiation. As a result, the lifetime and the rate capability/current density was improved reaching 1130.5 mA h/g at 1 A/g current density in 500 cycles, which can be explained by the synergistic combination of properties specific for flexible polythiophene film and the silicon nanospheres (Zheng et al. 2016). Further conductivity enhancement of PEDOT–PSS films was achieved by using aqueous zwitterion solution (Xia, Zhang, and Ouyang 2010) causing over 2.5 orders of magnitude conductivity improvement. Largely defined by the temperature and zwitterion concentration, this result is explained by significantly lower energy barrier for the electric charge transfer along the PEDOT chains and polymer chemical instability in presence of zwitterions. A novel polythiophene/carbon composite for organic Li-ion batteries, where n-doped poly(3,4-dihexylthiophene) is in situ polymerized on carbon nanofibers has been investigated (Zhu et al. 2013). The cells demonstrate relatively high capacity (300 mA h/g/200 A h/L) and over 95% capacity retention after a hundred cycles. Likewise, n-type redox-active poly(3,4-dihexylthiophene) was synthesized by oxidative coupling polymerization and exhibited large reversible electric capacity of 300 mA h/g through Li-ion insertion/extraction reactions with an application as a large capacity anodic material for organic lithium-ion electrochemical cells (Zhu et al. 2013).

11.4.3 All-Polythiophene and Metal-Free Batteries

The biggest challenge of the green energy storage is in creating the next generation of safe, reliable, and inexpensive metal-free battery technology. To function properly, both electrodes in these batteries can be made of alternating multilayer polymer films, for example, PEDOT and poly(N-methylpyrrole) (Aradilla et al. 2014). In all-polythiophene batteries, PEDOT is blended with polyethyleneimine to prevent the PTh in its low oxidation state, while the high oxidation state of PEDOT (**Error! Hyperlink reference not valid.**) serves as a cathode. During discharge, polyethyleneimine is consumed at the anode, while oxygen reacts with the PEDOT on the cathode (Figure 11.12). In the absence of electric load, the difference in PEDOT oxidation states on both sides of the electrochemical cell results in the electrochemical potential difference of about 0.5 V (Xuan et al. 2012).

All-polymer flexible battery can be manufactured using a PEDOT oxidative polymerization approach in vapor phase by redox polymers, such as substituted acrylamides. Cyclic voltammograms for the obtained composite electrodes exhibited redox behavior with two peaks at 0.62 and 0.10 V versus Ag/AgCl, Cl^- reference electrode (Suga, Winther-Jensen, and Nishide 2011). Galvanostatic charge/discharge profiles exhibited a plateau at 0.7 V, which corresponds to the gap of the redox potentials for both n- and p-type radical polymers. The charge capacity of the composite electrodes was in the range of 55–120 mA h/g, even for the higher content (70 wt%) of the polythiophene-based

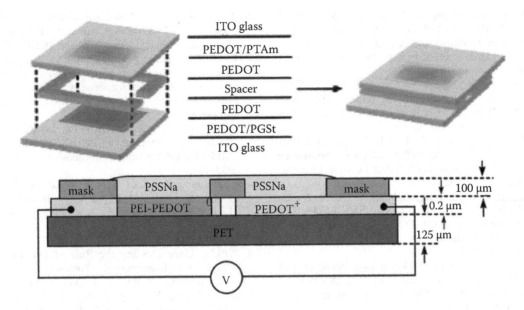

FIGURE 11.12 Diagram of all-polythiophene battery with PEDOT:PSS on a polyethylene interface with an overlying layer of polysodiumstyrenesulfonate (PSSNa).

derivative. Another example of flexible solid-state Li-ion battery supported on fabric matrix and modified with PEDOT allows fast-charging ability with specific capacity of 68 mA h/g. This new flexible battery design offers new solutions for the future generation of eco-friendly energy storage applicable to wearable as well as woven electronics (Wei, Cotton, and Ryhänen 2012).

In conclusion, it is necessary to emphasize that despite numerous achievements in the area of sustainable energy storage, more efficient electronically and ionically conducting materials for rechargeable batteries should be considered. One of the possible directions could be focused on vast group of functionalized and nonfunctionalized conducting polymers, their blends, and the corresponding nanocomposites, which unique properties are yet to be discovered.

REFERENCES

Abbas, S. A., M. A. Ibrahem, L. H. Hu, C. N. Lin, J. Fang, K. M. Boopathi, P. C. Wang, L. J. Li, and C. W. Chu. 2016. Bifunctional Separator as a Polysulfide Mediator for Highly Stable Li–S Batteries. *J. Mater. Chem. A.* 4 (24): 9661–9669.

Ai, G., Y. Dai, Y. Ye, W. Mao, Z. Wang, H. Zhao, Y. Chen, J. Zhu, Y. Fu, and V. Battaglia. 2015. Investigation of Surface Effects through the Application of the Functional Binders in Lithium Sulfur Batteries. *Nano Energy.* 16: 28–37.

Ali, G., J.-H. Lee, D. Susanto, S.-W. Choi, B. W. Cho, K.-W. Nam, and K. Y. Chung. 2016. Polythiophene-Wrapped Olivine Nafepo₄ as a Cathode for Na-ion Batteries. *ACS Appl. Mater. Interfaces.* 8 (24): 15422–15429.

Amanchukwu, C. V., M. Gauthier, T. P. Batcho, C. Symister, Y. Shao-Horn, J. M. D'Arcy, and P. T. Hammond. 2016. Evaluation and Stability of PEDOT Polymer Electrodes for Li–O₂ Batteries. *J. Phys. Chem. Lett.* 7 (19): 3770–3775.

Aradilla, D., F. Estrany, F. Casellas, J. I. Iribarren, and C. Alemán. 2014. All-Polythiophene Rechargeable Batteries. *Org. Electron.* 15 (1): 40–46.

Chen, H., W. Dong, J. Ge, C. Wang, X. Wu, W. Lu, and L. Chen. 2013. Ultrafine Sulfur Nanoparticles in Conducting Polymer Shell as Cathode Materials for High Performance Lithium/Sulfur Batteries. *Sci. Rep.* 3: 1910.

Chen, S., B. Sun, X. Xie, A. K. Mondal, X. Huang, and G. Wang. 2015. Multi-Chambered Micro/Mesoporous Carbon Nanocubes as New Polysulfides Reservoirs for Lithium–Sulfur Batteries with Long Cycle Life. *Nano Energy.* 16: 268–280.

Cho, S. J., C. C. Chung, S. Podowitz-Thomas, and J. L Jones. 2019. Understanding the Lithium Deficient Li$_x$Ni$_y$Mn$_z$Co$_{1-y-z}$O₂ (x<1) Cathode Materials Structure. *Mater. Chem. Phys.* 228: 32–36.

D'Arcy, J. M., M. F. El-Kady, P. P. Khine, L. Zhang, S. H. Lee, N. R. Davis, D. S. Liu, M. T. Yeung, S. Y. Kim, C. L Turner, and A. T. Lech. 2014. Vapor-Phase Polymerization of Nanofibrillar Poly (3, 4-Ethylenedioxythiophene) for Supercapacitors. *ACS Nano.* 8 (2): 1500–1510.

Das, P. R., L. Komsiyska, O. Osters, and G. Wittstock. 2015. PEDOT:PSS as a Functional Binder for Cathodes in Lithium-ion Batteries. *J. Electrochem. Soc.* 162 (4): A674–A678.

Döbbelin, M., R. Marcilla, M. Salsamendi, C. Pozo-Gonzalo, P. M. Carrasco, J. A. Pomposo, and D. Mecerreyes. 2007. Influence of Ionic Liquids on the Electrical Conductivity and Morphology of PEDOT:PSS Films. *Chem. Mater.* 19 (9): 2147–2149.

Etacheri, V., R. Marom, R. Elazari, G. Salitra, and D. Aurbach. 2011. Challenges in the Development of Advanced Li-ion Batteries: A Review. *Energy Environ. Sci.* 4 (9): 3243–3262.

Friedel, B., P. E. Keivanidis, T. J. Brenner, A. Abrusci, C. R. Mcneill, R. H. Friend, and N. C. Greenham. 2009. Effects of Layer Thickness and Annealing of PEDOT:PSS Layers in Organic Photodetectors. *Macromol.* 42 (17): 6741–6747.

Geng, D., N. Ding, T. A. Hor, S. W. Chien, Z. Liu, D. Wuu, X. Sun, and Y. Zong. 2016. From Lithium-Oxygen to Lithium-Air Batteries: Challenges and Opportunities. *Adv. Energy Mater.* 6 (9): 1502164.

Gong, Z., Q. Wu, F. Wang, X. Li, X. Fan, H. Yang, and Z. Luo. 2015. PEDOT-PSS Coated Sulfur/Carbon Composite on Porous Carbon Papers for High Sulfur Loading Lithium–Sulfur Batteries. *RSC Adv.* 5 (117): 96862–96869.

Guo, H., L. Liu, H. Shu, X. Yang, Z. Yang, M. Zhou, J. Tan, Z. Yan, H. Hu, and X. Wang. 2014. Synthesis and Electrochemical Performance of LiV_3O_8/Polythiophene Composite as Cathode Materials for Lithium Ion Batteries. *J. Power Sources.* 247: 117–126.

Han, M., and S. Foulger. 2005. 1-Dimensional Structures of Poly(3,4-Ethylenedioxythiophene) (PEDOT): A Chemical Route to Tubes, Rods, Thimbles, and Belts. *Chem. Commun.* 24: 3092–3094.

Harding, J. R., Y.-C. Lu, Y. Tsukada, and Y. Shao-Horn. 2012. Evidence of Catalyzed Oxidation of Li_2O_2 for Rechargeable Li–Air Battery Applications. *Phys. Chem. Chem. Phys.* 14 (30): 10540–10546.

Heywang, G., and F. Jonas. 1992. Poly (Alkylenedioxythiophene)s—New, Very Stable Conducting Polymers. *Adv. Mater.* 4 (2): 116–118.

Hohnholz, D., A. G. Macdiarmid, D. M. Sarno, and W. E. Jones, Jr. 2001. Uniform Thin Films of Poly-3, 4-Ethylenedioxythiophene (PEDOT) Prepared by In-Situ Deposition. *Chem. Commun.* 23: 2444–2445.

Jonas, F., and W. Krafft. 1994. Polythiophene Dispersions, their Production and their Use. Google Patents.

Jonas, F., W. Krafft, and B. Muys. 1995. Poly(3, 4-Ethylenedioxythiophene): Conductive Coatings, Technical Applications and Properties. *Macromol. Symp.* 100 (1): 169–173.

Kemerink, M., S. Timpanaro, M. M. De Kok, E. A. Meulenkamp, and F. J. Touwslager. 2004. Three-Dimensional Inhomogeneities in PEDOT:PSS Films. *J. Phys. Chem. B.* 108 (49): 18820–18825.

Kim, D. S., and Y. J. Park. 2014. A Simple Method for Surface Modification of Carbon by Polydopamine Coating for Enhanced Li–Air Batteries. *Electrochim. Acta.* 132: 297–306.

Kundu, D., R. Black, E. J. Berg, and L. F. Nazar. 2015. A Highly Active Nanostructured Metallic Oxide Cathode for Aprotic $Li–O_2$ Batteries. *Energy Environ. Sci.* 8 (4): 1292–1298.

Kuwabata, S., S. Masui, and H. J. Yoneyama. 1999. Charge–Discharge Properties of Composites of $LiMn_2O_4$ and Polypyrrole as Positive Electrode Materials for 4 V Class of Rechargeable Li Batteries. *Electrochim. Acta.* 44 (25): 4593–4600.

Lang, U., E. Müller, N. Naujoks, and J. Dual. 2009. Microscopical Investigations of PEDOT:PSS Thin Films. *Adv. Funct. Mater.* 19 (8): 1215–1220.

Łapkowski, M., and A. Proń. 2000. Electrochemical Oxidation of Poly (3, 4-Ethylenedioxythiophene)—"In Situ" Conductivity and Spectroscopic Investigations. *Synth. Met.* 110 (1): 79–83.

Lee, C. K., and Y. J. Park. 2015. Polyimide-Wrapped Carbon Nanotube Electrodes for Long Cycle Li–Air Batteries. *Chem. Commun.* 51 (7): 1210–1213.

Lee, J., and W. Choi. 2015. Surface Modification of Sulfur Cathodes with PEDOT:PSS Conducting Polymer in Lithium-Sulfur Batteries. *J. Electrochem. Soc.* 162 (6): A935–A939.

Li, H., M. Sun, T. Zhang, Y. Fang, and G. Wang. 2014. Improving the Performance of PEDOT-PSS Coated Sulfur@ Activated Porous Graphene Composite Cathodes for Lithium–Sulfur Batteries. *J. Mater. Chem. A.* 2 (43): 18345–18352.

Li, J., J. C. Liu, and C. J. Gao. 2010. On the Mechanism of Conductivity Enhancement in PEDOT/PSS Film Doped with Multi-Walled Carbon Nanotubes. *J. Polym. Res.* 17 (5): 713–718.

Li, W., Z. Liang, Z. Lu, H. Yao, Z. W. Seh, K. Yan, G. Zheng, and Y. Cui. 2015. A Sulfur Cathode with Pomegranate-Like Cluster Structure. *Adv. Energy Mater.* 5 (16): 1500211.

Li, W., Q. Zhang, G. Zheng, Z. W. Seh, H. Yao, and Y. Cui. 2013a. Understanding the Role of Different Conductive Polymers in Improving the Nanostructured Sulfur Cathode Performance. *Nano Lett.* 13 (11): 5534–5540.

Li, W., G. Zheng, Y. Yang, Z. W. Seh, N. Liu, and Y. Cui. 2013b. High-Performance Hollow Sulfur Nanostructured Battery Cathode through a Scalable, Room Temperature, One-Step, Bottom-Up Approach. *Proc. Natl. Acad. Sci. U S A.* 110 (18): 7148–7153.

Liu, L., F. Tian, X. Wang, Z. Yang, M. Zhou, and X. Wang. 2012. Porous Polythiophene as a Cathode Material for Lithium Batteries with High Capacity and Good Cycling Stability. *React. Funct. Polym.* 72 (1): 45–49.

Liu, X., H. Li, M. Ishida, and H. Zhou. 2013. PEDOT Modified $LiNi_{1/3}Co_{1/3}Mn_{1/3}O_2$ with Enhanced Electrochemical Performance for Lithium Ion Batteries. *J. Power Sources.* 243: 374–380.

Lock, J. P., S. G. Im, and K. K. Gleason. 2006. Oxidative Chemical Vapor Deposition of Electrically Conducting Poly(3, 4-Ethylenedioxythiophene) Films. *Macromolecules.* 39 (16): 5326–5329.

Lu, Y. C., H. A. Gasteiger, M. C. Parent, V. Chiloyan, and Y. Shao-Horn. 2010. The Influence of Catalysts on Discharge and Charge Voltages of Rechargeable Li–Oxygen Batteries. *Electrochem. Solid-State Lett.* 13 (6): A69–A72.

Lu, Y. C., D. G. Kwabi, K. P. Yao, J. R. Harding, J. Zhou, L. Zuin, and Y. Shao-Horn. 2011. The Discharge Rate Capability of Rechargeable $Li–O_2$ Batteries. *Energy Environ. Sci.* 4 (8): 2999–3007.

Lyu, H., J. Liu, S. Mahurin, S. Dai, Z. Guo, and X. G. Sun. 2017. Polythiophene Coated Aromatic Polyimide Enabled Ultrafast and Sustainable Lithium Ion Batteries. *J. Mater. Chem. A.* 5 (46): 24083–24090.

Ma, G., Z. Wen, Q. Wang, C. Shen, J. Jin, and X. Wu. 2014. Enhanced Cycle Performance of a Li–S Battery Based on a Protected Lithium Anode. *J. Mater. Chem. A.* 2 (45): 19355–19359.

Marciniak, S., X. Crispin, K. Uvdal, M. Trzcinski, J. Birgerson, L. Groenendaal, F. Louwet, and W. R. Salaneck. 2004. Light Induced Damage in Poly (3, 4-Ethylenedioxythiophene) and Its Derivatives Studied by Photoelectron Spectroscopy. *Synth. Met.* 141 (1–2): 67–73.

McCloskey, B. D., A. Speidel, R. Scheffler, D. Miller, V. Viswanathan, J. Hummelshøj, J. Nørskov, and A. Luntz. 2012. Twin Problems of Interfacial Carbonate Formation in Nonaqueous $Li–O_2$ Batteries. *J. Phys. Chem. Lett.* 3 (8): 997–1001.

McCullough, R. D. 1998. The Chemistry of Conducting Polythiophenes. *Adv. Mater.* 10 (2): 93–116.

McGraw, M., P. Kolla, B. Yao, R. Cook, Q. Quiao, J. Wu, and A. Smirnova. 2016. One-Step Solid-State In-Situ Thermal Polymerization of Silicon-PEDOT Nanocomposites for the Application in Lithium-Ion Battery Anodes. *Polym.* 99: 488–495.

Mo, Y., S. P. Ong, and G. Ceder. 2011. First-Principles Study of the Oxygen Evolution Reaction of Lithium Peroxide in the Lithium-Air Battery. *Phys. Rev. B.* 84 (20): 205446.

Nardes, A. M., M. Kemerink, R. A. J. Janssen, J. A. M. Bastiaansen, N. M. M. Kiggen, B. M. W. Langeveld, A. J. J. M. Van Breemen, and M. M. De Kok. 2007. Microscopic Understanding of the Anisotropic Conductivity of PEDOT:PSS Thin Films. *Adv. Mater.* 19 (9): 1196–1200.

Nasybulin, E., W. Xu, M. H. Engelhard, X. S. Li, M. Gu, D. Hu, and J.-G. Zhang. 2013. Electrocatalytic Properties of Poly (3, 4-Ethylenedioxythiophene)(PEDOT) in Li–O_2 Battery. *Electrochem. Commun.* 29: 63–66.

Niu, S., W. Lv, G. Zhou, H. Shi, X. Qin, C. Zheng, T. Zhou, C. Luo, Y. Deng, and B. Li. 2016. Electrostatic-Spraying an Ultrathin, Multifunctional and Compact Coating onto a Cathode for a Long-Life and High-Rate Lithium-Sulfur Battery. *Nano Energy.* 30: 138–145.

Ottakam Thotiyl, M. M., S. A. Freunberger, Z. Peng, and P. G. Bruce. 2012. The Carbon Electrode in Nonaqueous Li–O_2 Cells. *J. Am. Chem. Soc.* 135 (1): 494–500.

Pan, J., G. Xu, B. Ding, Z. Chang, A. Wang, H. Dou, and X. Zhang. 2016. PAA/PEDOT:PSS as a Multifunctional, Water-Soluble Binder to Improve the Capacity and Stability of Lithium–Sulfur Batteries. *RSC Adv.* 6 (47): 40650–40655.

Peng, Z., S. A. Freunberger, Y. Chen, and P. G. Bruce. 2012. A Reversible and Higher-Rate Li–O_2 Battery. *Science.* 337 (6094): 563–566.

Poverenov, E., M. Li, A. Bitler, and M. Bendikov. 2010. Major Effect of Electropolymerization Solvent on Morphology and Electrochromic Properties of PEDOT Films. *Chem. Mater.* 22 (13): 4019–4025.

Rozier, P., E. Iwama, N. Nishio, K. Baba, K. Matsumura, K. Kisu, J. Miyamoto, W. Naoi, Y. Orikasa, and P. Simon. 2018. Cation-Disordered Li_3VO_4: Reversible Li Insertion/Deinsertion Mechanism for Quasi Li-Rich Layered Li_{1+x} [$V_{1/2}Li_{1/2}$]O_2 (x= 0–1). *Chem. Mater.* 30 (15): 4926–4934.

Salsamendi, M., R. Marcilla, M. Döbbelin, D. Mecerreyes, C. Pozo-Gonzalo, J. A. Pomposo, and R. Pacios. 2008. Simultaneous Synthesis of Gold Nanoparticles and Conducting Poly(3,4-Ethylenedioxythiophene) towards Optoelectronic Nanocomposites. *Phys. Status Solidi A.* 205 (6): 1451–1454.

Selvaganesh, S. V., J. Mathiyarasu, K. L. N. Phani, and V. Yegnaraman. 2007. Chemical Synthesis of PEDOT–Au Nanocomposite. *Nanoscale Res. Lett.* 2 (11): 546.

Shui, J. L., J. S. Okasinski, P. Kenesei, H. A. Dobbs, D. Zhao, J. D. Almer, and D. J. Liu. 2013. Reversibility of Anodic Lithium in Rechargeable Lithium–Oxygen Batteries. *Nat. Commun.* 4: 2255.

Singh, S. K., X. Crispin, and I. V. Zozoulenko. 2017. Oxygen Reduction Reaction in Conducting Polymer PEDOT: Density Functional Theory Study. *J. Phys. Chem.* 121 (22): 12270–12277.

Smirnova, A., M. Mcgraw, and P. Kolla. 2017. Self-Organized and Electrically Conducting Pedot Polymer Matrix for Applications in Sensors and Energy Generation and Storage. Google Patents.

Su, D., M. Cortie, H. Fan, and G. Wang. 2017. Prussian Blue Nanocubes with an Open Framework Structure Coated with PEDOT as High-Capacity Cathodes for Lithium–Sulfur Batteries. *Adv. Mater.* 29 (48): 1700587.

Suga, T., B. Winther-Jensen, and H. Nishide. 2011. Redox-Active Radical Polymer/Vapor-Phase Polymerized Polythiophene Composite Electrodes for a Totally Organic Rechargeable Battery. *Polym. Prepr.* 52: 1087–1088.

Sun, Y., Z. W. Seh, W. Li, H. Yao, G. Zheng, and Y. Cui. 2015. In-Operando Optical Imaging of Temporal and Spatial Distribution of Polysulfides in Lithium-Sulfur Batteries. *Nano Energy.* 11: 579–586.

Takano, T., H. Masunaga, A. Fujiwara, H. Okuzaki, and T. Sasaki. 2012. PEDOT Nanocrystal in Highly Conductive PEDOT:PSS Polymer Films. *Macromol.* 45 (9): 3859–3865.

Tang, J., L. Kong, J. Zhang, L. Zhan, H. Zhan, Y. Zhou, and C. Zhan. 2008. Solvent-Free, Oxidatively Prepared Polythiophene: High Specific Capacity as a Cathode Active Material for Lithium Batteries. *React. Funct. Polym.* 68 (9): 1408–1413.

Thapa, A. K., B. Pandit, R. Thapa, T. Luitel, H. S. Paudel, G. Sumanasekera, M. K. Sunkara, N. Gunawardhana, T. Ishihara, and M. Yoshio. 2014. Synthesis of Mesoporous Birnessite-MnO_2 Composite as a Cathode Electrode for Lithium Battery. *Electrochim. Acta.* 116: 188–193.

Thotiyl, M. M. O., S. A. Freunberger, Z. Peng, Y. Chen, Z. Liu, and P. G. Bruce. 2013. A Stable Cathode for the Aprotic Li–O_2 Battery. *Nat. Mater.* 12 (11): 1050.

Wang, X., Z. Zhang, X. Yan, Y. Qu, Y. Lai, and J. Li. 2015. Interface Polymerization Synthesis of Conductive Polymer/Graphite Oxide@ Sulfur Composites for High-Rate Lithium-Sulfur Batteries. *Electrochim. Acta.* 155: 54–60.

Wang, Z., Y. Chen, V. Battaglia, and G. Liu. 2014. Improving the Performance of Lithium–Sulfur Batteries Using Conductive Polymer and Micrometric Sulfur Powder. *J. Mater. Res.* 29 (9): 1027–1033.

Wei, D., D. Cotton, and T. Ryhänen. 2012. All-Solid-State Textile Batteries Made from Nano-Emulsion Conducting Polymer Inks for Wearable Electronics. *Nanomaterials.* 2 (3): 268–274.

Wei, Q., M. Mukaida, Y. Naitoh, and T. Ishida. 2013. Morphological Change and Mobility Enhancement in PEDOT:PSS by Adding Co-solvents. *Adv. Mater.* 25 (20): 2831–2836.

Wu, F., J. Chen, R. Chen, S. Wu, L. Li, S. Chen, and T. Zhao. 2011. Sulfur/Polythiophene with a Core/Shell Structure: Synthesis and Electrochemical Properties of the Cathode for Rechargeable Lithium Batteries. *J. Phys. Chem. C.* 115 (13): 6057–6063.

Wu, F., J. Liu, L. Li, X. Zhang, R. Luo, Y. Ye, and R. Chen. 2016. Surface Modification of Li-Rich Cathode Materials for Lithium-Ion Batteries with a PEDOT:PSS Conducting Polymer. *ACS Appl. Mater. Interfaces.* 8 (35): 23095–23104.

Wu, F., S. Wu, R. Chen, J. Chen, and S. Chen. 2010. Sulfur–Polythiophene Composite Cathode Materials for Rechargeable Lithium Batteries. *Electrochem. Solid-State Lett.* 13 (4): A29–A31.

Xia, L., Y. Xia, and Z. Liu. 2015. Thiophene Derivatives as Novel Functional Additives for High-Voltage $LiCoO_2$ Operations in Lithium Ion Batteries. *Electrochim. Acta.* 151: 429–436.

Xia, Y., H. Zhang, and J. Ouyang. 2010. Highly Conductive PEDOT:PSS Films Prepared through a Treatment with Zwitterions and their Application in Polymer Photovoltaic Cells. *J. Mater. Chem.* 20 (43): 9740–9747.

Xiao, P., F. Bu, G. Yang, Y. Zhang, and Y. Xu. 2017. Integration of Graphene, Nano Sulfur, and Conducting Polymer into

Compact, Flexible Lithium–Sulfur Battery Cathodes with Ultrahigh Volumetric Capacity and Superior Cycling Stability for Foldable Devices. *Adv. Mater.* 29 (40): 1703324.

Xiao, W., J. S. Chen, Q. Lu, and X. W. Lou. 2010. Porous Spheres Assembled from Polythiophene (Pth)-coated Ultrathin MnO_2 Nanosheets with Enhanced Lithium Storage Capabilities. *J. Phys. Chem. C.* 114 (27): 12048–12051.

Xu, G. L., Y. Li, T. Ma, Y. Ren, H. H. Wang, L. Wang, J. Wen, D. Miller, K. Amine, and Z. Chen. 2015. PEDOT-PSS Coated ZnO/C Hierarchical Porous Nanorods as Ultralong-Life Anode Material for Lithium Ion Batteries. *Nano Energy.* 18: 253–264.

Xuan, Y., M. Sandberg, M. Berggren, and X. Crispin. 2012. An All-Polymer-Air PEDOT Battery. *Org. Electron.* 13 (4): 632–637.

Yan, M., Y. Zhang, Y. Li, Y. Huo, Y. Yu, C. Wang, J. Jin, L. Chen, T. Hasan, and B. Wang. 2016. Manganese Dioxide Nanosheet Functionalized Sulfur@ PEDOT Core–Shell Nanospheres for Advanced Lithium–Sulfur Batteries. *J. Mater. Chem. A.* 4 (24): 9403–9412.

Yang, Y., G. Yu, J. J. Cha, H. Wu, M. Vosgueritchian, Y. Yao, Z. Bao, and Y. Cui. 2011. Improving the Performance of Lithium–Sulfur Batteries by Conductive Polymer Coating. *ACS Nano.* 5 (11): 9187–9193.

Yao, Y., N. Liu, M. T. Mcdowell, M. Pasta, and Y. Cui. 2012. Improving the Cycling Stability of Silicon Nanowire Anodes with Conducting Polymer Coatings. *Energy Environ. Sci.* 5 (7): 7927–7930.

Yoon, D. H., S. H. Yoon, K. S. Ryu, and Y. J. Park. 2016. PEDOT:PSS as Multi-Functional Composite Material for Enhanced Li-Air-Battery Air Electrodes. *Sci. Rep.* 6: 19962.

Yoon, T. H., and Y. J. Park. 2014. New Strategy toward Enhanced Air Electrode for Li–Air Batteries: Apply a Polydopamine Coating and Dissolved Catalyst. *RSC Adv.* 4 (34): 17434–17442.

Yue, L., H. Zhong, D. Tang, and L. Zhang. 2013. Porous Si Coated with S-Doped Carbon as Anode Material for Lithium Ion Batteries. *J. Solid State Electrochem.* 17 (4): 961–968.

Zhang, C., Y. He, P. Mu, X. Wang, Q. He, Y. Chen, J. Zeng, F. Wang, Y. Xu, and J. X. Jiang. 2018. Toward High Performance Thiophene-Containing Conjugated Microporous Polymer Anodes for Lithium-Ion Batteries through Structure Design. *Adv. Funct. Mater.* 28 (4): 1705432.

Zhang, X., J. S. Lee, G. S. Lee, D. K. Cha, M. J. Kim, D. J. Yang, and S. K. Manohar. 2006. Chemical Synthesis of PEDOT Nanotubes. *Macromol.* 39 (2): 470–472.

Zhang, X., A. G. Macdiarmid, and S. K. Manohar. 2005. Chemical Synthesis of PEDOT Nanofibers. *Chem. Commun.* 42: 5328–5330.

Zheng, H., S. Fang, Z. Tong, H. Dou, and X. Zhang. 2016. Porous Silicon@ Polythiophene Core–Shell Nanospheres for Lithium-Ion Batteries. *Part. Part. Syst. Char.* 33 (2): 75–81.

Zhu, L., W. Shi, R. Zhao, Y. Cao, X. Ai, A. Lei, and H. Yang. 2013. N-Dopable Polythiophenes as High Capacity Anode Materials for All-Organic Li-ion Batteries. *J. Electroanal. Chem.* 688: 118–122.

12 Polythiophene-Based Supercapacitor Applications

Shankara S. Kalanur
Department of Materials Science and Engineering
Ajou University
Suwon, Republic of Korea

Veerendra Kumar A. Kalalbandi
Department of Chemistry, Jain College of Engineering
Belagavi, Karnataka, India

Hyungtak Seo
Department of Materials Science and Engineering
Ajou University
Suwon, Republic of Korea

Department of Energy Systems Research
Ajou University
Suwon, Republic of Korea

CONTENTS

12.1   INTRODUCTION

The future of modern society demands portable, flexible, and cutting-edge electronic systems for sustainable energy production and storage. In this context, extensive research and development are underway to invent efficient, sustainable, and stable energy storage systems possessing considerable energy and power density (Tarascon and Armand 2001; Nishide and Oyaizu 2008; Rogers, Someya, and Huang 2010; Lipomi and Bao 2011). Currently, lithium-ion batteries have shown dominance in the energy storage industry by powering numerous kinds of consumer electrons and electric vehicles. However, lithium-ion batteries suffer from poor life cycle and more importantly are prone to fire hazards. In view of this, supercapacitors are favored as alternate energy storage systems capable of holding ~10 times more power density, have longer charge–discharge cycles, and are mainly safe from fire hazards, making them an ideal choice in

terms of commercial utilization (Conway 1999; Snook, Kao, and Best 2011; Kumar et al. 2018). Hence, the electrochemical supercapacitors (ESs) are considered as a promising energy storage device for the future (Miller and Simon 2008; Jennings et al. 2009). In an electrochemical capacitor, the charge storage mechanism involves ionic and electronic transport during the charging and discharging process. Hence, the ESs exhibit high power and low energy densities due to the fast charge and discharge mechanism. Practically, the supercapacitors can be combined with a rechargeable battery or fuel cells to meet peak energy demand that maintains the load leveling and also increases the lifetime of the battery/fuel cells (Bélanger 2009).

The efficient working of a supercapacitor mainly depends on the type of materials used. Till now, various materials have been explored and optimized to achieve high power density, rapid charge and discharge capability, and stability of the supercapacitors. Among the employed materials, conducting polymers are evolving as an important class of components for supercapacitor devices. Moreover, research and development on conducting π-conjugated organic systems have initiated and shown significant advancement in organic electronics, especially in supercapacitors. The unique property of conducting polymers is a result of its double and single bonds arranged alternatively in the structural backbone, thereby allowing the effective delocalization of π-electrons. The delocalization makes the π-electrons freely mobile over the entire length of the molecule and hence the polymers are intrinsically conducting. Importantly, the presence of π-conjugation in the polymer backbone results in low-energy unoccupied and high-energy occupied molecular orbitals. Such a molecular orbital system in polymers could allow effective oxidation and reduction of the surface by the diffusion of ions when the supercapacitor works (Shirakawa et al. 1977; Seidel 1987). Furthermore, the conducting polymers exhibit an unique property of transiting between its semiconducting and conducting states when modified. This allows controlled tuning of its mechanical and electrical properties by chemical alteration, doping, and composite formation. Importantly, the ability to engineer the properties of such conducting polymers could make their effective integration in a number of advanced optical devices and electronic systems. Hence, employing the conducting polymers as an active material in supercapacitors could be an effective strategy for fabricating future energy storage devices.

Till now, various conducting polymers have been explored and widely utilized in supercapacitor systems including polythiophenes (PThs). The PTh supercapacitors have mainly fascinated significant attention due to its superior electrical properties and high chemical stability. PThs are the conducting polymer made up of polymerized thiophene, a five-membered sulfur heterocycle. Notably, the unique and distinguished features of PTh and its derivatives make them an exciting moiety in the field of ESs. The presence of PThs in supercapacitor systems allows the storage of charges via the electrical double layer formation at the electrode/electrolyte interface (Pringle et al. 2005; Frackowiak et al. 2006; Österholm et al. 2013). Considerable research has been done by using PTh in the ES system, and exciting milestones were achieved. This chapter focuses on the utilization of PTh and its few derivatives (all derivatives are not included in this chapter as some of the derivatives are discussed in other chapters) as active electrode materials for supercapacitor applications.

12.2 PROPERTIES OF POLYTHIOPHENE (PTh)

PTh is a five-membered heteroaromatic ring that is highly stable from both an electrochemical and environmental standpoint. The presence and prolonged π-conjugated system in PThs give promising characteristics that include high conductivity, optical, electrochemical, thermoelectrical,

and electrochromic properties. In particular, the chain length of conjugation in the PTh polymer backbone affects its optical properties. That is, PTh shows a significant red shift in their absorption properties with the increase in the conjunction chain length. Importantly, an array of functional groups that are required to tune its chemical, biological (Barisci et al. 2001), optical (Nilsson et al. 2004; Ribeiro et al. 2004), and electrical properties (Andreani et al. 1998; Chan, On, and Ng 1998; Emge and Bäuerle 1999; Skabara et al. 2001; Nilsson et al. 2004; Ribeiro et al. 2004) can be covalently put together into PThs using readily accessible chemistries (Nilsson et al. 2004; Ribeiro et al. 2004), thus allowing an efficient tuning of its properties. Mainly, the tunable properties of PThs could allow its effective utilization in electrical systems including ESs.

The structure of conducting polymers shows a dominant role in determining their physical properties. For example, a simple poly(3-alkylpolythiophene) system shows different conducting property based on its regioregular and regioirregular configuration. The regioregular poly(3-hexylthiophene) shows significantly high mobility of 0.2 cm^2 V^{-1} s^{-1}, whereas the regioirregular poly (3-hexylthiophene) exhibits poor mobility of $\sim 10^{-5}$ cm^2 V^{-1} s^{-1} (McCullough et al. 1993; McCulloch et al. 2006). The experimental findings revealed that the introduction of dopants in PThs creates polarons/bipolarons that significantly influence its electronic properties (Chiu et al. 2005). Furthermore, the experimental and DFT calculations (Zade and Bendikov 2007; Colella et al. 2015) have demonstrated that the electronic properties of PThs can be easily tuned by various arrangements including adsorption or intercalation of layers or by the formation of multilayers or stacking of layers in bulk (Kaloni, Schreckenbach, and Freund 2015). Such a stacking in the PThs allows an efficient transport of charge and energy for long ranges within the polymer backbone and hence finds its effective utilization in various nanoscale electronic devices (Niles et al. 2012; Gao et al. 2013). Furthermore, an investigation by Oh et al. showed the applicability of PTh microparticles as a redox couple by conducting a charge and discharge experiment by employing the nonflow system. Importantly, PTh shows electrochemical redox activity between the potential range of -2.0 (in n-doping mode) and $+0.5$ V (in p-doping mode) that allows its utilization in all-organic redox flow battery exhibiting 2.5 V of cell potential (Oh et al. 2014). The redox reaction of PTh also displays excellent electrochromic property (electrochromism is a phenomenon shown by some materials exhibiting a reversible change in color stimulated by redox reactions/applied potential) possessing the parameters such as transformation time of 10–20 ms with a significant lifetime and several minutes of color maintenance (Haitao et al. 1986). Moreover, it is also evident from the literature reports that the substituted PThs display better electrochromic effect than with that of naked PTh films.

12.3 SYNTHESIS OF POLYTHIOPHENE

The first ever synthesis of PThs was conducted accidentally by Meyer in 1883 (Meyer 1883), more than a century ago. Later, the distinct properties of PThs prompted researchers to report new routes for the synthesis of both the naked and substituted PThs. Notably, the chemical polymerization was routinely used for the synthesis of unsubstituted PThs (Lin and Dudek 1980; Yamamoto, Sanechika, and Yamamoto 1980). Thus, the synthesized 2,5-coupled PTh was found to be environmentally, thermally stable and was highly conductive with poor solubility due to intra- and interchain aggregations. To obtain soluble and conducting PThs, alkyl was substituted at 3-position of PThs, that is, poly(3-alkyl thiophene)s were prepared by various methods including, metal catalyzed, metal-catalyzed-C–C coupling, cross-coupling catalysis, oxidative polymerization, and so on (Yamamoto, Hayashi, and Yamamoto 1978; Kobayashi et al. 1984; McCullough and Lowe 1992). Figure 12.1 shows the general synthetic route to obtain PThs.

Apart from chemical synthesis, PThs can also be synthesized using electrochemical methods. A solution containing thiophene and a suitable electrolyte in an electrochemical cell can produce a conductive PTh film on the anode surface. The initial reaction steps involved during the electropolymerization of PTh in an electrochemical cell is shown in Figure 12.2.

One of the advantages of employing electrochemical polymerization route is that the PTh polymer could be directly deposited on the substrate, thus avoiding the time-consuming steps such as isolation and purification. However, the electrochemical polymerization route could yield polymers having different proportions of regioregularity due to the formation of α-β linkages at unwanted positions. It is worth to note that the regioregularity in PThs is essential for the photovoltaic and other electronic applications as it influences the conductivity, liquid crystallinity, mobility, and chirality. Various methods have been reported, demonstrating the synthesis of regioregular PThs (Barbarella, Bongini, and Zambianchi 1994). However, the unsymmetrical molecular property of poly (3-alkylthiophene)s leads to the three district positionings by the coupling at 2- and 5-positions during the polymerization (Osaka and McCullough 2008). The three possible orientations of the thiophene ring couplings are shown in Figure 12.3, that is, the 2-5′ coupling forming head-to-tail configuration, the 2-2′ coupling leading to head-to-head (HH) system, and the 5-5′ coupling resulting in tail-to-tail (TT) orientation. Therefore, the properties of synthesized

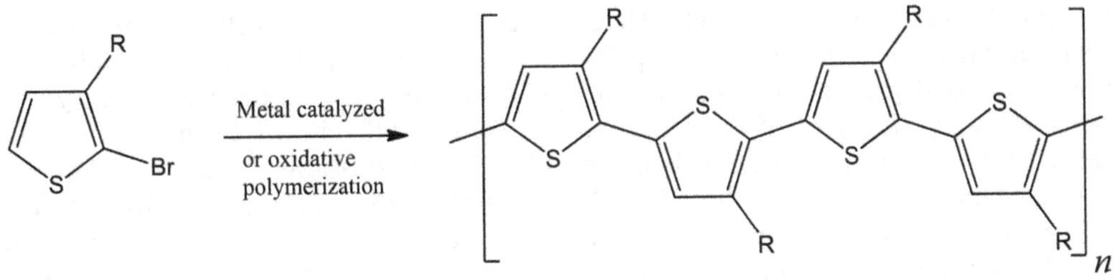

FIGURE 12.1 The general synthetic route for the synthesis of poly(3-alkylthiophene).

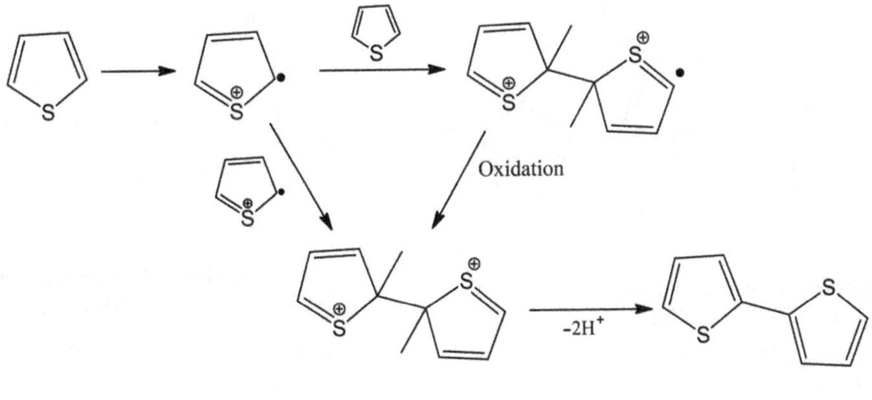

FIGURE 12.2 The mechanistic pathway for the initial steps involved in the electropolymerization of PTh on the electrode surface.

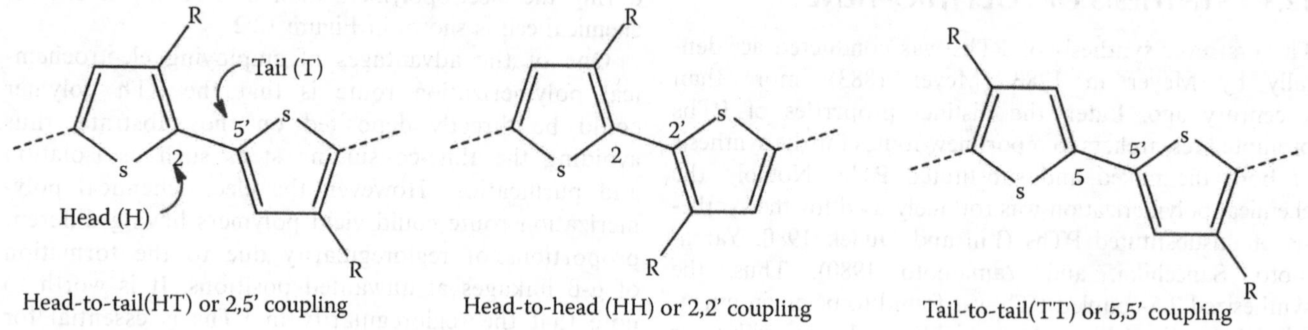

FIGURE 12.3 Regioisomeric couplings of 3-alkylthiophenes, yielding head-to-tail coupling, head-to-head coupling, and tail-to-tail coupling.

PThs significantly vary based on the presence of possible linkages involved in the PThs.

Generally, the chemical and electrochemical methods employed in the synthesis of PThs produce random couplings in PThs and poly(3-alkylthiophene)s, directing to merely 50–80% of HT couplings. The presence of multiple HH and TT coupling in PThs leads to a significant decrease in regioregularity that causes a sterically curled system in the PTh polymer structure. Such a curled structure leads to the loss of π-conjugation and thus shows decreased conductivity. The typical regioregularity consisting of planar backbone and regioirregular nonplanar backbone in the PThs is depicted in Figure 12.4a and b, respectively (Osaka and McCullough 2008). Here, the rise in steric twisting within the PTh moiety in the polymer results in a widened band gap and also significantly worsens the various necessary properties of PThs. Hence, exploring a suitable synthetic route to obtain a desirable regioselective PThs is highly essential. Given this, in 1992, McCullough et al. invented a new regioselective synthetic method that includes lithiation of bromo- and alkyl-substituted thiophene. This was achieved by using lithium diisopropylamide involving the transfer of ligands from one metal to another with magnesium bromide diethyl

etherate, which yields intermediate substituted with 3-alkylPTh having the bromo and bromomagnesio at 2- and 5-positions, respectively. Thus, the generated intermediate was polymerized with [1,3-bis(diphenylphosphino)propane]dichloronickel(II) yielding poly(3-alkylthiophene)s having a significant percentage of HT–HT coupling (McCullough and Lowe 1992). Subsequently, the second synthetic approach to HT regioselective poly(3-alkylTh)s was demonstrated by Chen and Rieke (1992), producing anticipated HT-poly(3-alkylthiophene)s. The aforementioned approach could be further modified as the GRIM (Grignard metathesis) method to improve the percentage of HT–HT couplings (Loewe, Khersonsky, and McCullough 1999). Conclusively, the discovery of the suitable synthesis routes of regioregular poly(3-alkylthiophenes) produced a diverse category of PThs having numerous different functional groups having superior electrical possessions. Notably, improved properties of the synthesized regioregular poly(3-alkylthiophenes) were achieved by holding the polymer backbone in the single plane structures of π-conjugation (McCullough et al. 1993) that yields superior conductivity due to the presence of effective and efficient charge pathways along the single plane.

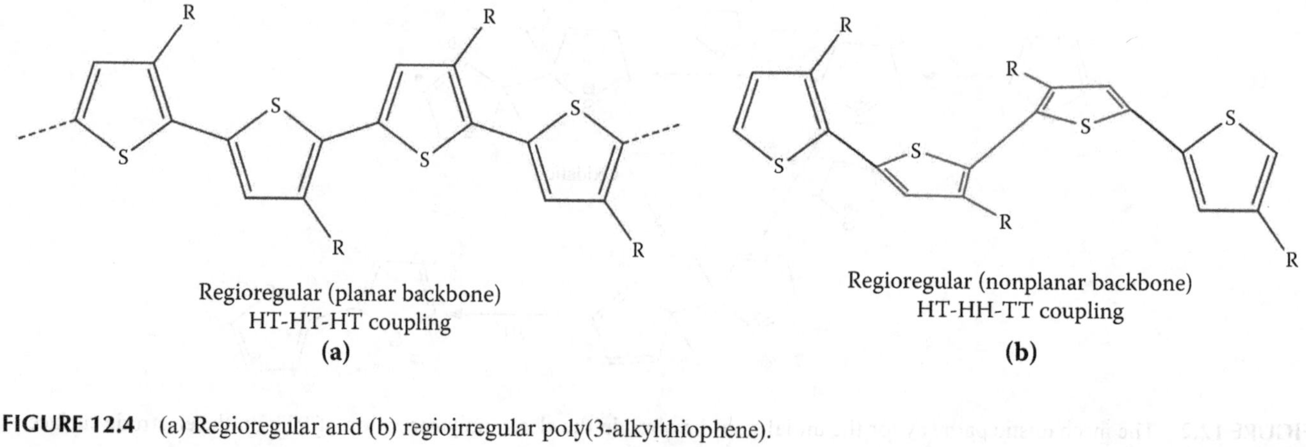

FIGURE 12.4 (a) Regioregular and (b) regioirregular poly(3-alkylthiophene).

Since the characteristic properties of PThs find extensive applications in many fields, suitable synthesis routes to obtain a variety of substituted, water-soluble, nanocomposites and regioregular PThs were explored. In 1987, Patil et al. (1987) synthesized the sodium salt of poly(3-(4-butanesulfonate) TP) (Figure 12.5a) and poly (3-(2-ethanesulfonate) TP) polymers (Figure 12.5b) that are soluble in water as well as conducting and also demonstrated a pathway of doping, which was designated as self-doping.

Furthermore, the poly(alkoxythiophene) derivatives having long-range π-conjugation system can be obtained by forming the polyanion and polycation system such as poly-3-(3′-thienyloxy)propyltriethylammonium (Figure 12.5c) and poly-3-(3′-thienyloxy)propanesulfonate (Figure 12.5d) (Lukkari et al. 2001). Followed by this, Ho et al. described cationic poly (3-alkoxy-4-methylthiophene) (Figure 12.5e) having superior electrical and photocatalytic properties (Ho

et al. 2002). Martinez et al. demonstrated the polymerization of 3-phenylthiophene, 2-phenyl-thiophene, 2,5-diphenylthiophene, 2-(2-thienyl) pyridine, and 3,6-bis(2-thienyl)pyridazine in the electrochemical and chemical systems. Here, the electrochemical polymerization was achieved using galvanostatic conditions (platinum electrodes), whereas the chemical polymerization was carried out by oxidation with iron trichloride (Martínez et al. 1995). Zhai et al. demonstrated the scheme of obtaining regioregular PThs derivatives using the Grignard metathesis route. In particular, poly(3-alkylthiophenes) having 3-bromo (Figure 12.6a), 6-carboxylic acid (Figure 12.6b), 6-amine (Figure 12.6c), and 6-thiol (Figure 12.6d) functionalities (Zhai et al. 2003) have been synthesized with a significant yield. Correspondingly, there are a number of synthetic methods for PThs in literature now. Based on the functionality of the monomer, the synthetic methods can be altered from the basic methodologies that

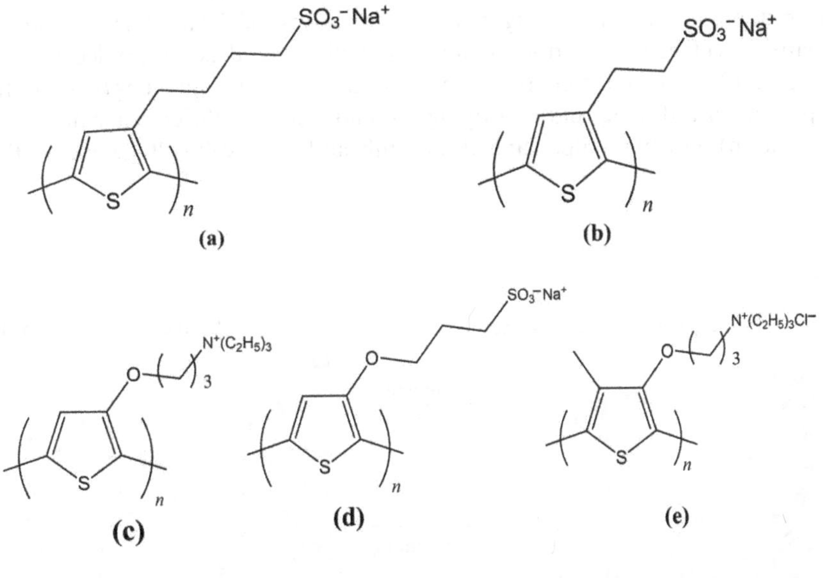

FIGURE 12.5 Chemical structures of water-soluble polythiophenes: Sodium salt of poly(3-(2-ethanesulfonate) thiophene) (a) and poly(3-(4-butanesulfonate) thiophene) (b); poly-3-(3′-thienyloxy)propyltriethylammonium (c); poly-3-(3′-thienyloxy)propanesulfonate (d) (Lukkari et al. 2001); and poly(3-alkoxy-4-methylthiophene) (e).

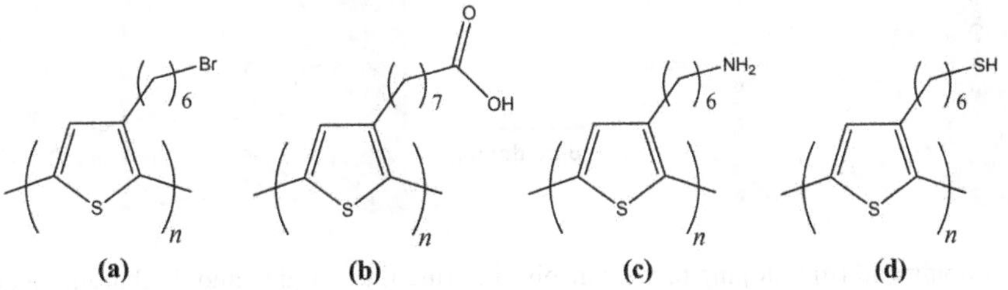

FIGURE 12.6 Chemical structures of poly(3-bromohexylthiophenes) (a), poly(3-alkylthiophenes) bearing 6-carboxylic acid (b), 6-amine (c), or 6-thiol (d) functionalities.

utilize either oxidation or metal catalysis to yield desired PThs for various applications.

12.4 CHARGE STORAGE IN POLYTHIOPHENE ELECTROCHEMICAL CAPACITORS

In polymer-based ESs such as PThs, the storage of charge takes place in the bulk of the material. The driving force for charge storage mechanism in polymer supercapacitors is the redox reaction taking place in the presence of applied potential in a suitable electrolyte solution. The charge storage mechanism in a PTh supercapacitor takes place in the bulk of the electrode by the diffusion of ions. Such a diffusion of ions leads to a decrease in the rate of charging and discharging, and hence the polymer-based supercapacitors show a reduced self-discharge reaction. Most of the active polymers employed in supercapacitors could be operated as p-doped or n-doped system separately. However, the PThs-based polymers could be employed as both n- and p-doped redox processes individually or in the same system. This means that the PThs could be utilized as a positive electrode via p-doping and p-de-doping mechanisms and as a negative electrode by n-doping and n-de-doping courses. Figure 12.7 shows the charging and discharging processes present in a doping and de-doping system in (as both p and n) polymer supercapacitors

(Rudge et al. 1994). The minus sign (without the circle) in Figure 12.7 indicates the reduced form of the polymer, whereas the plus sign shows the oxidized polymer, and the plus signs inside the circle represent the cations, whereas the minus signs inside the circle represent the anions of the electrolyte. Such an ability of PThs operating as both positive and negative electrodes makes it unique and favorable for supercapacitor applications. When used as a positive electrode, that is, during p-doping, the partial oxidation of the PThs backbone occurs via the removal of electrons from the undoped PTh. Similarly, using PThs as a negative electrode causes n-doping via the partial reduction of the PTh backbone. The positive charges formed after the oxidation of polymer will be repaid by incorporating the anions present in the solution.

Hence, during p-de-doping, the electrons were ejected back to the polymer with the removal of diffused anions, thus generating back the undoped polymer. Likewise, during the n-doping, the cation insertion takes place by the ionic solution, whereas the de-doping consequently removes the cations electrolyte from the polymer film. The derivatives of PThs exhibit both n- and p-doping and therefore could be assembled in a single capacitor system as both positive and negative electrodes and importantly could yield an efficient capacitor capable of generating significantly high cell voltage of 3 V (Bélanger 2009).

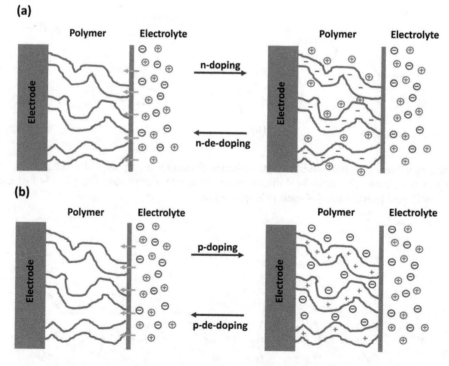

FIGURE 12.7 The (a) n-doping and (b) p-doping process involved during the charging and discharging process at conducting polymer. The minus sign (without the circle) indicates the reduced form of polymer, whereas the plus sign indicates the oxidized polymer. The plus signs inside the circle represent the cations, whereas the minus signs inside the circle represent the anions of the electrolyte.

Compared to p-doped system of the PTh, the n-doped polymer shows low conductivity and low stability to oxygen and water (Levi et al. 2000; Laforgue et al. 2003; Skompska et al. 2005). The lower stability of n-doped PThs causes easy oxidation back to the original state of PThs, which in turn results in a significant self-discharge of PTh. Hence, the device containing n-doped PThs exhibits low cycle life. Note that these n-doped PThs can be obtained by applying significantly low potentials during electropolymerization. Therefore, the derivatives of n-doped PThs were majorly explored, which show improved stability when compared to bare PThs. These derivatives of n-doped PThs were obtained at a comparatively lesser negative potential and exhibit a lower band gap (Lefrant et al. 1999; Villers et al. 2003; Ryu et al. 2004). In particular, the stability of PThs was significantly enhanced by substituting ethyl, phenyl, and alkoxy groups thiophene ring at 3-position. Importantly, by such substitutions, the oxygen and water stability could be enhanced (Arbizzani et al. 1995). Furthermore, the simultaneous addition of electron-withdrawing group in the derivatives of PThs could further improve the stability. In view of this, various derivatives of PThs containing ethylenedioxy, fluorophenyl, difluorophenyl, and thienyl-vinylene substituents were widely studied (Villers et al. 2003). Note that, in this chapter, we have focused only on the supercapacitance of unsubstituted PThs.

12.5 POLYTHIOPHENE ELECTRODE FABRICATION

The electrical, chemical, and morphological properties of PThs depend on the utilized synthetic method, and hence the electrochemical capacitance behavior was expected to change based on the synthetic routes. Note that the supercapacitor performances of electrodes containing PThs were also dependent on the size of the counter-ion present in the electrolyte, that is, an electrolyte or a gel electrolyte. For the fabrication of ESs, the PThs need to be assembled in an electrode system. For example, the PTh electrode can be constructed by first chemically synthesizing PThs and then assembling in the electrode systems (Senthilkumar, Thenamirtham, and Kalai Selvan 2011; Balakrishnan, Kumar, and Angaiah 2014; Rafiqi and Majid 2015) or by directly obtaining the electrodes by methods, namely, successive ionic layer adsorption and reaction deposition (SILAR) (Patil, Jagadale, and Lokhande 2012), electropolymerization (Zhang et al. 2014; Ambade et al. 2016), and chemical bath deposition methods (CBD) (B. H. Patil, Patil, and Lokhande 2014), which are direct and single-step fabrication routes. Here, the fabrication process can be eased by using direct or single-step method but it is challenging/difficult to obtain regioregular PThs. In contrast, the selective chemical synthesis yields regioregular PThs but requires multistep and time-consuming procedures for electrode fabrication.

During the synthesis of PThs via dilute polymerization method, the $FeCl_3$ was routinely used as an oxidant. During the polymerization, the presence of surfactants in addition to the oxidant was found to influence the properties of PThs including morphology. Some of the reported surfactants utilized in the synthesis are N-cetyl-N,N,N-trimethylammonium bromide (CTAB) (Senthilkumar, Thenamirtham, and Kalai Selvan 2011), sodium dodecyl sulphate (SDS) (Balakrishnan, Kumar, and Angaiah 2014), and Triton-100 (Senthilkumar, Thenamirtham, and Kalai Selvan 2011). After the synthesis and purification, the PThs were usually combined with conducting additive and a binder material to obtain a slurry for supercapacitor fabrication. Generally, various carbon materials were employed as conductive additives and poly(vinylidenefluoride) (PVDF) dissolved in N-methylpyrrolidone as a binder material. Subsequently, the slurry containing a mixture of PThs, conductive additives, and binding material will be assembled/placed/deposited onto various substrates of a fixed area to produce an electrode for supercapacitors. However, the use of such binders and the formation of the slurry could be avoided when the electrode is fabricated by a single-step process of forming PThs directly on the substrate using electropolymerization, SILAR, and CBD methods. The fabricated electrodes could be employed as either negative or positive electrodes based on the requirements. Once the electrode is fabricated, the next task is to measure its supercapacitance properties, which can be done with either a two-electrode or a three-electrode system in a suitable electrolyte.

12.6 ELECTROCHEMICAL SUPERCAPACITANCE OF POLYTHIOPHENE

The performance of the supercapacitance of PThs can be enhanced by compositing with metal oxides and carbon-based materials. However, in this section, we highlight the critical synthesis routes and the performance of the supercapacitance of a few bare PThs electrodes reported in the literature. As discussed previously, the chemical synthesis of PThs generally proceeds in the presence of an oxidant derived from a metal. Interestingly, during the chemical synthesis, the introduction of surfactants alters the chemical, optical, and structural properties of PThs. For example, the use of SDS surfactant results in PThs having fibrous morphology with a small globular structure (Balakrishnan, Kumar, and Angaiah 2014). The synthesized PThs could be annealed at 1400°C (2 h) to yield a carbonaceous PThs. To fabricate the supercapacitor, the synthesized nanofibrous PThs/carbonaceous PThs were dispersed in pearl carbon and PVDF in N-methylpyrrolidone resulting in a slurry. The slurry was then deposited on a coin cell to obtain supercapacitor systems. Here, the PVDF dissolved in N-methylpyrrolidone was used as a binder for the fabrication of the supercapacitor. In the final assembly of the asymmetric supercapacitor cell, the PTh nanofibers were utilized as a cathode electrode. In contrast, the carbonaceous PTh nanofibers were employed as an anode with polypropylene as a separator. This electrode system showed a superior capacitance of 252 F g^{-1}, with a working voltage of 1.2 V. The galvanostatic charge–discharge study showed the 54.6 W h kg^{-1} of specific energy and a higher power density of 1.7 kW kg^{-1}. Notably, the supercapacitor showed superior stability up to 2000 cycles, retaining its 93% specific capacitance.

During the chemical synthesis of PThs, H$_2$O$_2$ can also be utilized as a co-oxidant in addition to the FeCl$_3$ oxidant as demonstrated by Senthilkumar et al. (2011). Additionally, the introduction of various surfactants such as CTAB and Triton X-100 results in irregular plate-like morphology, whereas the presence of SDS resulted in uniformly distributed spherical microparticles. After the PThs synthesis, the slurry for electrode fabrication was formed by mixing with a conductive additive (carbon black) and a PVDF binder. The electrode was constructed by placing a suitable amount of slurry on to the pretreated graphite strip. The PThs synthesized using CTAB and SDS exhibited a red shift in the absorbance values ascribed to the n-π* electronic transition in the polymer, confirming the longer conjugation length of the prepared PThs. Among the used surfactants, Triton X-100 showed an improved capacitance of 117 F g^{-1}. However, this result contradicts with the report by Balakrishnan et al. (2014), which suggested that the PThs synthesized in the presence of SDS show comparatively better capacitance values.

The synthesis of PThs by electropolymerization can be achieved on various substrates/surfaces by using numerous electrolytic solutions containing thiophene monomer. During electropolymerization, the choice of a suitable electrolyte is very crucial as it strongly influences the properties and assembly of PThs on the substrates. Using an aqueous solution generally produces nonconducting or passive PTh films on the substrate, whereas using organic solvents could be toxic. Hence, Zhang et al. (2014) demonstrated the electropolymerization of small spheroid-shaped PThs on a carbon paper using an oil-in-ionic liquid microemulsion solution. The employed oil-in-ionic liquid microemulsion solution was noted to contain an oil phase of n-hexane dispersed in thiophene as the ionic liquid phase. Notably, the Triton X-100 (nonionic surfactant) and an ionic liquid BMIM-PF$_6$ were used. Figure 12.8 shows the schematic representation of the electropolymerization of PThs in the oil-in-ionic liquid microemulsion electrolyte. Here, the proposed electrolyte contains a micelle of oil droplet phase in which the thiophene monomer dissolves evenly. With the coverage of Triton X-100, the oil micelle will be distributed in the ionic liquid phase. When the current is applied, the electromigration of micelle takes place toward the working electrode. At the electrode surface, the thiophene monomer gets oxidized, thereby initiating the growth on the substrate. The opted conditions allowed the polymerization thiophene ring at the 2- and 5-positions to yield PTh film of α–α linkages. The electrode characteristics were explored in a three-electrode electrochemical cell employing the PTh deposited on

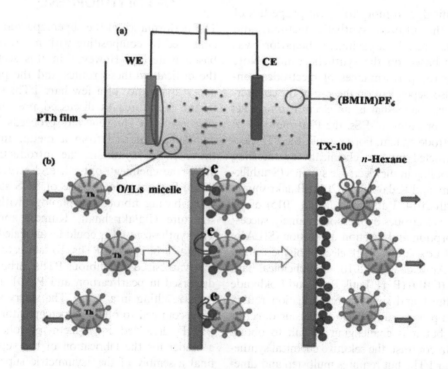

FIGURE 12.8 (a) The electrode cell configuration of polymerization and (b) the scheme of electropolymerization of thiophene-containing micelle at the substrate mediated by an oil-in-ionic liquid microemulsion electrolyte. Reprinted with permission from Elsevier (Zhang et al. 2014).

the carbon paper as the working electrode. In the operating electrochemical system, the graphite rod was used as a counter electrode and an Ag/AgCl as a reference electrode. The PTh coated on carbon paper electrode showed a 103 F g^{-1} of specific capacitance, with significant stability of 500 cycles. Note that the specific capacitance values of electropolymerized PThs were significantly lower when compared to those of the previous results obtained by using chemical synthesis routes. This could be due to the production of low degree of regioregular PThs.

The electropolymerization of PThs can also be achieved on Ti nanowires as demonstrated by Ambade et al. (2016) in the presence of tetrabutylammonium tetrafluoroborate in an acetonitrile solvent. Here, the thickness of deposited PThs on the Ti wire was controlled by scan cycles. A two-electrode system was used to measure capacitance characteristics, where PThs coated on the Ti wire were utilized as both a positive and a negative electrode. The polymer gel obtained by mixing H$_2$SO$_4$ and polyvinyl alcohol was used as both an electrolyte and a separator for electrodes. The two electrodes were immersed in the polymer gel to ensure complete coverage. Finally, the two gel-coated electrodes were intertwined to produce a wire-shaped all-solid-state symmetric supercapacitor that provided a specific capacitance of 1357.31 mF g^{-1} and energy density and power density of 23.11 µW h cm^{-2} and 90.44 µW cm^{-2}, respectively. Notably, the supercapacitor showed excellent mechanical flexibility (bending at 360°) without deteriorating its electrochemical properties. Furthermore, the electropolymerization of thiophene on a pencil graphite electrode could be achieved by using a solution containing monomer in acetonitrile, lithiumperchlorate, and perchloric acid. The PThs deposited on a pencil graphite supercapacitor showed a specific capacitance of 1503.23 mF g^{-1} and a charge/discharge stability over 1000 cycles.

Patil et al. (2014) proposed a facile fabrication method of PTh thin films on glass and stainless steel substrates using the CBD technique that utilizes ammonium peroxodisulfate as an oxidant. The deposition of PThs on the substrate was carried out for 48 h. The electrochemical capacitance properties were evaluated in a three-electrode system in a 0.1 M LiClO$_4$ electrolyte, showing a specific capacitance of 300 F g^{-1}. The thin films of PTh on a glass and a stainless-steel substrate can also be synthesized using SILAR method (Patil, Jagadale, and Lokhande 2012) with FeCl$_3$ as an oxidizing agent. In the SILAR method, the substrates were dipped in a solution containing monomer and oxidizing agent sequentially for specific time intervals for the direct polymerization of thiophene on the substrate. The synthesized PThs showed p-type behavior with lower resistivity and a wide band gap of 2.90 eV, and a specific capacitance of 252 F g^{-1} was recorded.

12.7 COMPOSITES OF A POLYTHIOPHENE WITH METAL OXIDES

The metal oxides can also be used as active electrode materials in supercapacitors owing to their essential properties such as high-rate performance, high specific capacitance, and stability. However, the choice of a suitable metal oxide material has to be based on parameters such as material cost, toxicity, abundance, durability, and chemical and electrical properties. Importantly, the process of charging and discharging takes place by the surface adsorption and desorption of the charged electrolyte ion on the oxide layers, and the reversible intercalation into a crystal lattice plays an important role. However, most of the abundant and cost-effective metal oxides do not possess good electrical conductivity, leading to a slow charge transfer that hinders the high-power capabilities of supercapacitors. To overcome these limitations, the metal oxides were generally decorated or mixed with superior conducting materials such as conducting polymer or carbon-based materials. Such a composite of a polymer and metal oxides shows synergic effects due to the advantages of individual elements by compensating the shortcoming of each other. Based on this strategy, the PThs can also be composited with suitable metal oxide nanostructures to yield superior supercapacitance properties. In this section, we highlight a few attempts made in the literature by combining the PThs with metal oxides for supercapacitors applications.

Given this, Lu and Zhou (2011) synthesized submicron-sphere/nanosheet hierarchical microstructure composite containing MnO$_2$ and PTh for supercapacitor applications. Figure 12.9a and b shows the SEM and TEM images of submicron-sphere/nanosheet hierarchical microstructure composite of MnO$_2$/PTh. Specifically, the nanosheet morphology is essential as it provides a larger surface area along with the shorter solid-state transport length that significantly enhances the kinetics and activity charges. The electrode for supercapacitance applications was constructed by mixing MnO$_2$/PTh with the polyvinylidene difluoride binder and acetylene black to yield a slurry that was coated on to the specific area of the nickel mesh. Finally, the supercapacitance characteristics were determined in the three-electrode system in 1 M Na$_2$SO$_4$ electrolyte. The MnO$_2$/PTh composite exhibited the highest specific capacitance of 273 F g^{-1}. Superior stability was noticed until 1000 cycles that showed a stability of 97.3% (Figure 12.9c). Furthermore, using the MnO$_2$/PTh composite, Yu et al. (2016) managed to produce a superior specific capacitance of 619 F g^{-1}, with a stability cycle of 10^5.

Similarly, PThs was composited with TiO$_2$ using in situ chemical oxidative polymerization (Li et al. 2018). Specifically, Li et al. (2018) demonstrated the M4EOT synthesis and the formation of a composite with TiO$_2$. The route of M4EOT synthesis is shown in Figure 12.10a, whereas Figure 12.10b shows the formation of

FIGURE 12.9 (a) SEM and (b) TEM images of submicron-sphere/nanosheet hierarchical microstructure MnO₂/PTh composite. (c) The stability of MnO₂/PTh composite showing a stability of 1000 cycles. Inset in (c) shows galvanostatic charge/discharge cycles measured at 2 A g⁻¹. Reprinted with permission from Elsevier (Lu and Zhou 2011).

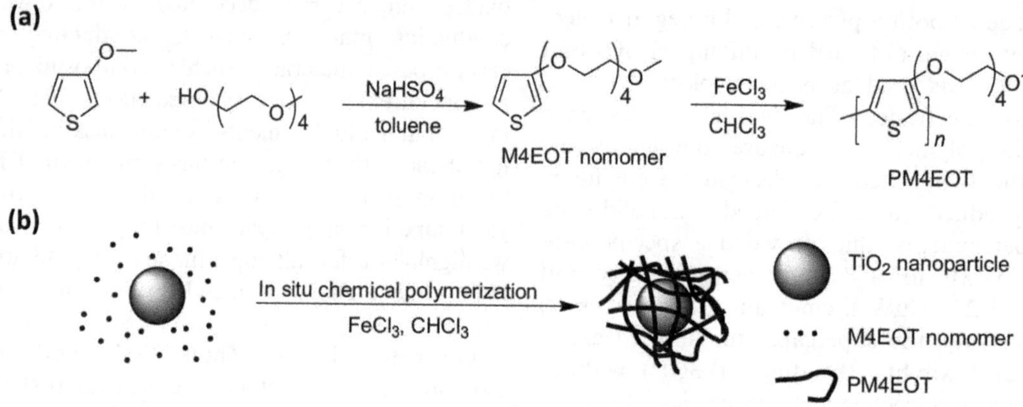

FIGURE 12.10 The synthesis scheme of (a) M4EOT and (b) M4EOT/TiO₂ composite. Reprinted by permission from Springer (Li et al. 2018).

the M4EOT/TiO₂ composite. Briefly, this was achieved by oxidative polymerization, that is, by the addition of an oxidant $FeCl_3$ into the solution containing TiO_2 nanoparticle and M4EOT monomer. Here the synthesis was performed at 0°C for one day, and the ratio of M4EOT to TiO_2 was optimized to obtain better performing capacitors. The proposed synthesis approach produced a core-shell morphology containing TiO_2 nanoparticles in the core, with a 2–10 nm of the PM4EOT shell structure. The supercapacitor was fabricated by combining the M4EOT/TiO₂ with acetylene black and polytetrafluoroethylene (PTFE) in ethanol to yield a slurry that was coated on to the specific area of the Ni foam. The composite electrode demonstrated a specific capacitance of 111 F g⁻¹ at a current density of 0.5 A g⁻¹. Furthermore, other polymers can also be substituted into the PTh–TiO₂ composite, such as polyaniline as demonstrated by Thakur et al. (2018). In this case, the oxidative polymerization procedure was used to form a mixture of polymer composite, followed by mixing it with metal oxide nanoparticles that yielded the final material. The addition of polyaniline into PTh/TiO₂ enhanced the

specific capacitance value to 265 F g⁻¹, which was double the value obtained for only PTh/TiO₂, as discussed earlier. The aforementioned results indicate that compositing the PTh with metal oxides is an important strategy to enhance the specific capacitance values. However, the limited literature suggests that various other metal oxides could be considered and tested for much higher and stable capacitance performance. Importantly, this gives an opportunity to explore, research, and optimize a wider range of metal oxide materials for superior capacitance applications.

12.8 COMPOSITES OF A POLYTHIOPHENE WITH CARBON-BASED MATERIALS

Despite having numerous advantageous, the supercapacitors containing PThs molecules solely do not show notably high specific capacitance. Because in the course of charging and discharging process of PThs, the insertion and removal of ions take place, causing a change in the volume of polymers by swelling and shrinking. Moreover, the redox location present in the backbone of the PTh polymer structure does not exhibit any significant stability due to the frequent redox processes. Eventually, in some electrolytes, the routine diffusion of ions may cause mechanical degradation of polymers that in turn affects the stability and working cycles of PTh supercapacitors. Overall, the low stability, poor reversibility, and high ohmic polarization of PThs will hamper the performance of the supercapacitance. Such shortcomings of PTh supercapacitors can be overcome by forming a composite with appropriate materials. Notably, the carbon materials are considered most suitable because of their excellent electronic properties, stability, increased area of activity, and porosity. The higher specific area possessed by carbon materials allows significantly higher charge accumulation, therefore producing effective electrical double-layer capacitance between the carbon material and the electrolyte. In this section, we highlighted the recent attempts reported to improve the capacitance of PThs by compositing with different carbon-based materials.

Among various carbon materials, graphene was commonly used as an active supercapacitor component owing to its two-dimensional honeycomb lattice structure exhibiting superior physical, chemical, and mechanical properties. The incorporation of graphene into conducting polymers shows significant advantages including stability, high surface-to-

volume ratio, efficient and rapid charge–discharge phenomenon, and wider electrochemical operating range. Moreover, incorporating such active and stable materials could lead to portable, flexible, and stretchable devices. Alvi et al. (2011) demonstrated the fabrication of graphene thiophane composite electrode for efficient supercapacitor applicants. The polymerization of PTh was done using an $FeCl_3$ oxidant in the solution containing graphene. The electrode was fabricated by mixing with nafion and was coated on the graphite electrode. Figure 12.11a and b shows charging and discharging mechanism, respectively, of a supercapacitor consisting of graphene thiophene composite in an acidic electrolyte. During charging, the diffusion of anions and cations takes place, thus separating and producing electrochemical double-layer capacitance through the separator. After applying the load, the system acquires its original configuration by discharging. The PTh graphene composite electrode exhibited a specific capacitance of 176 F g^{-1}.

The carbon-containing substrates can also be used as effective conductive platforms for depositing conducting polymers (Memon et al. 2016). The initial process involves the fabrication of graphene-based substrates using the process called ice-crystal-induced phase

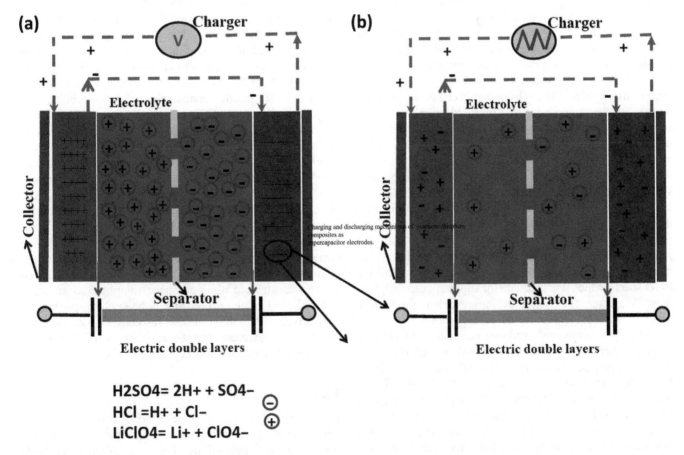

FIGURE 12.11 (a) Charging and (b) discharging mechanisms of graphene–thiophene composites as supercapacitor electrodes. Republished with permission of ECS transactions by Electrochemical Society (Alvi et al. 2011).

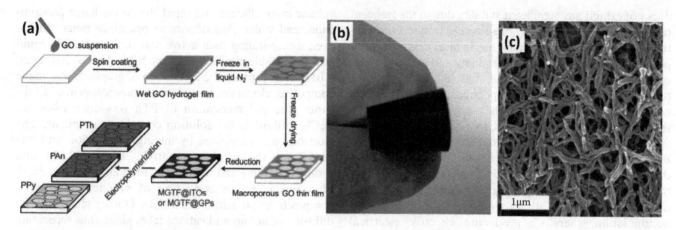

FIGURE 12.12 (a) A schematic presentation of the fabrication process of PThs (represented as red thin film with the name PTh)-coated macroporous-structured graphene oxide thin films. (b) A photographic image showing the flexibility of macroporous structured graphene oxide thin films on a graphene paper. (c) The SEM image showing the interconnected macroporous configuration consisting of exfoliated graphene oxide. Reprinted with permission from Memon et al. (2016). Copyright (2016) American Chemical Society.

separation. Briefly, the graphene paper substrates required for the study were first fabricated by filtration of reduced graphene oxide using PTFE membranes. Then the graphene oxide hydrogel was spin coated onto the substrate and was frozen in liquid N_2. Subsequently, the graphene-coated frozen substrates were freeze-dried using the lyophilize machine to produce macroporous-structured graphene oxide thin films. Note that, in addition to the graphene paper substrate, the indium tin oxide substrate can also be used. Lastly, the reduction of graphene oxide was carried out by thermal treatment at 200°C (400°C for indium tin oxide substrates) under argon atmosphere. The electropolymerization of thiophene on the macroporous structured graphene oxide thin films was carried out using an electrolyte containing monomer and tetrabutylammonium hexafluorophosphate. Figure 12.12a shows a schematic presentation of the fabrication process of PThs-coated macroporous-structured graphene oxide thin films. Figure 12.12b and c shows the flexibility and interconnected macroporous configuration consisting of exfoliated graphene oxide on graphene paper, respectively. Such a graphene-based substrate system includes an interconnected arrangement of the macroporous graphene oxide and sufficiently exfoliated RGO sheets, which are adequate for supercapacitor applications. Moreover, the system exhibits excellent mechanical flexibility feasible for portable device applications.

Similarly, multiwalled carbon nanotube (MWCNT)-modified carbon paper could be used as substrates for the electropolymerization of PThs (Zhang et al. 2014). Before the preparation of an electrode, the MWCNTs are functionalized by acid treatment. Then the functionalized MWCNTs were coated onto the carbon paper to form a flexible substrate for PTh deposition. The PTh was deposited via electropolymerization onto the MWCNT-modified carbon paper substrate using an oil-in-ionic liquid

microemulsion electrolyte as discussed previously in Section 12.6. The electropolymerization was carried out in the electrolyte containing n-hexane as an oil phase in the presence of a thiophene monomer. In this, Triton X-100 and (BMIM)PF_6 were taken as a nonionic surfactant and an ionic liquid, respectively. The mechanism of polymerization involves the diffusion of microemulsion toward a carbon paper electrode, initiating nucleation for the polymerization under the applied current. The outcome of the characterization revealed the formation of PTh thin layer having 2–3 nm thickness encapsulated onto the MWCNT wall. The MWCNT/PTh on a carbon paper electrode showed 216 F g^{-1} of specific capacitance and 80% stability after 500 cycles. The functionalization of thiophene and composite formation with graphene can be simultaneously achieved by the in situ polymerization method (Alabadi et al. 2016). The in situ polymerization was performed using poly[(Thiophene-2,5-diyl)-co-(benzylidene)], thiophene monomer, and GO using Maghnite-H$^+$ as a catalyst (Alabadi et al. 2016). The synthesized composite appeared to consist of high-thickness stacked sheets forming a large area-connected graphene, including pores and wrinkles on edges. Such a morphology could be beneficial for charge storage applications including capacitors. The supercapacitor was constructed using graphene–thiophene as an active material, and black carbon and polytetrafluoroethylene as conducting and binding agents, respectively. The electrode exhibited an excellent specific capacitance of 296 F g^{-1} with a stable charges/discharge cycles of 4000.

12.9 CONCLUSION

In the last few years, some crucial contribution and advancement have been made in the field of PThs for supercapacitor applications. Till now, different methods of synthesis/deposition of PThs have been reported.

Specifically, metal-catalyzed oxidation polymerization, that is, chemical polymerization route and electropolymerization are widely employed methods for the synthesis or deposition of PTh. The comparatively poor capacitance properties exhibited by unfunctionalized PTh (obtained via naked PTh) triggered more focus on acquiring functionalized PTh, which shows significantly superior capacitance values and stability. In fact, this allowed the exploration of a variety of new and important class of PTh derivatives achieving promising results and milestones. Notably, the electrical properties of PThs considerably vary based on the polymerization mechanism, which results in either regioisomeric couplings of 3-alkylthiophenes, yielding different coupling configuration. Importantly, for effective supercapacitor properties, regioregular PThs are recommended because of their superior electrical properties when compared to that of regioirregular systems. The charging and discharging properties of PTh are advantageous as they can be used both as a positive electrode and a negative electrode in the same system or with other electrodes in composite systems. The recent trends in published articles indicate that, when compared to other conducting polymers, the research on PTh is going on in a slower pace. Therefore, the limitations such as stability, mechanical strength, and electrical properties of PTh need to be addressed. However, the unique features of PThs reveal that they could still be considered as the material of choice for efficient supercapacitor applications. Compositing PThs with various metal oxides and carbon-based materials were found to significantly enhance the supercapacitance properties including stability. Even though compositing PThs improve capacitance properties, still a significant research contribution is essential in this direction. For example, wide variety of new and unexplored materials could be composited with PThs as effective electrodes systems for ESs applications. This opens more opportunities to study the supercapacitance activity of PThs with numerous compositing materials. Importantly, for the effective and efficient working of PThs-based supercapacitors, an achievable goal will be production of high cell voltage up to 3 V that in turn can increase the energy and power density with respect to low-voltage capacitors.

REFERENCES

Alabadi, Akram, Shumaila Razzaque, Zehua Dong, Weixing Wang, and Bien Tan. 2016. "Graphene Oxide-Polythiophene Derivative Hybrid Nanosheet for Enhancing Performance of Supercapacitor." *Journal of Power Sources* 306: 241–247. https://doi.org/10.1016/j.jpowsour.2015.12.028.

Alvi, Farah, Manoj Kumar Ram, Punya Basnayaka, Elias Stefanakos, Yogi Goswami, Andrew Hoff, and Ashok Kumar. 2011. "Electrochemical Supercapacitors Based on Graphene-Conducting Polythiophenes Nanocomposite." *ECS Transactions* 35 (34): 167–174. https://doi.org/10.1149/1.3654215.

Ambade, Rohan B., Swapnil B. Ambade, Rahul R. Salunkhe, Victor Malgras, Sung-Ho Jin, Yusuke Yamauchi, and Soo-Hyoung Lee. 2016. "Flexible-Wire Shaped All-Solid-State Supercapacitors Based on Facile Electropolymerization of Polythiophene with Ultra-High Energy Density." *Journal of Materials Chemistry A* 4 (19): 7406–7415. https://doi.org/10.1039/C6TA00683C.

Andreani, Franco, Luigi Angiolini, Daniele Caretta, and Elisabetta Salatelli. 1998. "Synthesis and Polymerization of 3,3″-Di[(S)-(+)-2-Methylbutyl]-2,2′:5′,2″-Terthiophene: A New Monomer Precursor to Chiral Regioregular Poly (Thiophene)." *Journal of Materials Chemistry* 8 (5): 1109–1111. https://doi.org/10.1039/A801593G.

Arbizzani, C., M. Catellani, M. Mastragostino, and C. Mingazzini. 1995. "N- and P-Doped Polydithieno[3,4-B:3′,4′-D] Thiophene: A Narrow Band Gap Polymer for Redox Supercapacitors." *Electrochimica Acta* 40 (12): 1871–1876. https://doi.org/10.1016/0013-4686(95)00096-W.

Balakrishnan, K., Manish Kumar, and Subramania Angaiah. 2014. "Synthesis of Polythiophene and Its Carbonaceous Nanofibers as Electrode Materials for Asymmetric Supercapacitors." *Advanced Materials Research* 938: 151–157. https://doi.org/10.4028/www.scientific.net/AMR.938.151.

Barbarella, Giovanna, Alessandro Bongini, and Massimo Zambianchi. 1994. "Regiochemistry and Conformation of Poly (3-Hexylthiophene) via the Synthesis and the Spectroscopic Characterization of the Model Configurational Triads." *Macromolecules* 27 (11): 3039–3045. https://doi.org/10.1021/ma00089a022.

Barisci, Joseph N., Rita Stella, Geoffrey M. Spinks, and Gordon G. Wallace. 2001. "Study of the Surface Potential and Photovoltage of Conducting Polymers Using Electric Force Microscopy." *Synthetic Metals* 124 (2): 407–414. https://doi.org/10.1016/S0379-6779(01)00387-3.

Bélanger, Daniel. 2009. "Polythiophenes as Active Electrode Materials for Electrochemical Capacitors." In *Handbook of Thiophene-Based Materials*, 577–594. John Wiley & Sons, Ltd. https://doi.org/10.1002/9780470745533.ch15.

Chan, Hardy Sze On, and Siu Choon Ng. 1998. "Synthesis, Characterization and Applications of Thiophene-Based Functional Polymers." *Progress in Polymer Science* 23 (7): 1167–1231. https://doi.org/10.1016/S0079-6700(97)00032-4.

Chen, Tian An, and Reuben D. Rieke. 1992. "The First Regioregular Head-to-Tail Poly(3-Hexylthiophene-2,5-Diyl) and a Regiorandom Isopolymer: Nickel versus Palladium Catalysis of 2 (5)-Bromo-5(2)-(Bromozincio)-3-HexylthiophenePolymerization." *Journal of the American Chemical Society* 114 (25): 10087–10088. https://doi.org/10.1021/ja00051a066.

Chiu, William W., Jadranka Travaš-Sejdić, Ralph P. Cooney, and Graham A. Bowmaker. 2005. "Spectroscopic and Conductivity Studies of Doping in Chemically Synthesized Poly(3,4-Ethylenedioxythiophene)." *Synthetic Metals* 155 (1): 80–88. https://doi.org/10.1016/j.synthmet.2005.06.012.

Colella, Nicholas S., Lei Zhang, Thomas McCarthy-Ward, Stefan C. B. Mannsfeld, H. Henning Winter, Martin Heeney, James J. Watkins, and Alejandro L. Briseno. 2015. "Controlled Integration of Oligo- and Polythiophenes at the Molecular Scale." *Physical Chemistry Chemical Physics* 17 (40): 26525–26529. https://doi.org/10.1039/C4CP02944E.

Conway, Brian E. 1999. *Electrochemical Supercapacitors: Scientific Fundamentals and Technological Applications*. Springer US. www.springer.com/kr/book/9780306457364.

Emge, A., and P. Bäuerle. 1999. "Molecular Recognition Properties of Nucleobase-Functionalized Polythiophenes."

Synthetic Metals 102 (1): 1370–1373. https://doi.org/10.1016/S0379-6779(98)00274-4.

Frackowiak, Elzbieta, Volodymyd Khomenko, Krzysztof Jurewicz, Katarzyna Lota, and Francois Béguin. 2006. "Supercapacitors Based on Conducting Polymers/Nanotubes Composites." *Journal of Power Sources* 153 (2): 413–418. https://doi.org/10.1016/j.jpowsour.2005.05.030.

Gao, Jian, John D. Roehling, Yongle Li, Hua Guo, Adam J. Moulé, and John K. Grey. 2013. "The Effect of 2,3,5,6-Tetrafluoro-7,7,8,8-Tetracyanoquinodimethane Charge Transfer Dopants on the Conformation and Aggregation of Poly(3-Hexylthiophene)." *Journal of Materials Chemistry C* 1 (36): 5638–5646. https://doi.org/10.1039/C3TC31047G.

Haitao, Huang, Chen Yanzhen, Tian Zhaowu, Huang Haitao, Chen Yanzhen, and Tian Zhaowu. 1986. "Electrochromic Properties of Polythiophene." *Acta Physico-Chimica Sinica* 2 (05): 417–423. https://doi.org/10.3866/PKU.WHXB19860505.

Ho, Hoang-Anh, Maurice Boissinot, Michel G. Bergeron, Geneviève Corbeil, Kim Doré, Denis Boudreau, and Mario Leclerc. 2002. "Colorimetric and Fluorometric Detection of Nucleic Acids Using Cationic Polythiophene Derivatives." *Angewandte Chemie International Edition* 41 (9): 1548–1551. https://doi.org/10.1002/1521-3773(20020503)41:9<1548::AID-ANIE1548>3.0.CO;2-I.

Jennings, Aaron A., Hise Sara, Kiedrowski Bryant, and Krouse Caleb. 2009. "Urban Battery Litter." *Journal of Environmental Engineering* 135 (1): 46–57. https://doi.org/10.1061/(ASCE)0733-9372(2009)135:1(46).

John, R. R., Barry, C. T., T. Skotheim. 2019. *Handbook of Conducting Polymers*, Fourth edition. CRC Press/Taylor & Francis.

Kaloni, Thaneshwor P., Georg Schreckenbach, and Michael S. Freund. 2015. "Structural and Electronic Properties of Pristine and Doped Polythiophene: Periodic versus Molecular Calculations." *The Journal of Physical Chemistry C* 119 (8): 3979–3989. https://doi.org/10.1021/jp511396n.

Kobayashi, M., J. Chen, T. -C. Chung, F. Moraes, A. J. Heeger, and F. Wudl. 1984. "Synthesis and Properties of Chemically Coupled Poly(Thiophene)." *Synthetic Metals* 9 (1): 77–86. https://doi.org/10.1016/0379-6779(84)90044-4.

Kumar, Kowsik Sambath, Nitin Choudhary, Yeonwoong Jung, and Jayan Thomas. 2018. "Recent Advances in Two-Dimensional Nanomaterials for Supercapacitor Electrode Applications." *ACS Energy Letters* 3 (2): 482–495. https://doi.org/10.1021/acsenergylett.7b01169.

Laforgue, A., P. Simon, J. F. Fauvarque, M. Mastragostino, F. Soavi, J. F. Sarrau, P. Lailler, M. Conte, E. Rossi, and S. Saguatti. 2003. "Activated Carbon/Conducting Polymer Hybrid Supercapacitors." *Journal of The Electrochemical Society* 150 (5): A645–51. https://doi.org/10.1149/1.1566411.

Lefrant, S., I. Baltog, M. Lamy de la Chapelle, M. Baibarac, G. Louarn, C. Journet, and P. Bernier. 1999. "Structural Properties of Some Conducting Polymers and Carbon Nanotubes Investigated by SERS Spectroscopy." *Synthetic Metals* 100 (1): 13–27. https://doi.org/10.1016/S0379-6779(98)00175-1.

Levi, M. D., Y. Gofer, D. Aurbach, M. Lapkowski, E. Vieil, and J. Serose. 2000. "Simultaneous Voltammetric and In Situ Conductivity Studies of N-Doping of Polythiophene Films with Tetraalkylammonium, Alkali, and Alkaline–Earth Cations." *Journal of The Electrochemical Society* 147 (3): 1096–1104. https://doi.org/10.1149/1.1393319.

Li, Yueqin, Minya Zhou, Yun Li, Qiang Gong, Yiting Wang, and Zongbiao Xia. 2018. "Structural, Morphological and Electrochemical Properties of Long Alkoxy-Functionalized Polythiophene and TiO2 Nanocomposites." *Applied Physics A* 124 (12): 855. https://doi.org/10.1007/s00339-018-2277-y.

Lin, John W.-P., and Lesley P. Dudek. 1980. "Synthesis and Properties of Poly(2,5-Thienylene)." *Journal of Polymer Science: Polymer Chemistry Edition* 18 (9): 2869–2873. https://doi.org/10.1002/pol.1980.170180910.

Lipomi, Darren J., and Zhenan Bao. 2011. "Stretchable, Elastic Materials and Devices for Solar Energy Conversion." *Energy & Environmental Science* 4 (9): 3314–3328. https://doi.org/10.1039/C1EE01881G.

Loewe, Robert S., Sonya M. Khersonsky, and Richard D. McCullough. 1999. "A Simple Method to Prepare Head-to-Tail Coupled, Regioregular Poly(3-Alkylthiophenes) Using Grignard Metathesis." *Advanced Materials* 11 (3): 250–253. https://doi.org/10.1002/(SICI)1521-4095(199903)11:3<250::AID-ADMA250>3.0.CO;2-J.

Lu, Qing, and Yikai Zhou. 2011. "Synthesis of Mesoporous Polythiophene/MnO2 Nanocomposite and Its Enhanced Pseudocapacitive Properties." *Journal of Power Sources* 196 (8): 4088–4094. https://doi.org/10.1016/j.jpowsour.2010.12.059.

Lukkari, Jukka, Mikko Salomäki, Antti Viinikanoja, Timo Ääritalo, Janika Paukkunen, Natalia Kocharova, and Jouko Kankare. 2001. "Polyelectrolyte Multilayers Prepared from Water-Soluble Poly(Alkoxythiophene) Derivatives." *Journal of the American Chemical Society* 123 (25): 6083–6091. https://doi.org/10.1021/ja0043486.

Martínez, F., J. Retuert, G. Neculqueo, and H. Naarmann. 1995. "Chemical and Electrochemical Polymerization of Thiophene Derivatives." *International Journal of Polymeric Materials and Polymeric Biomaterials* 28 (1–4): 51–59. https://doi.org/10.1080/00914039508012087.

McCulloch, Iain, Martin Heeney, Clare Bailey, Kristijonas Genevicius, Iain MacDonald, Maxim Shkunov, David Sparrowe, Tierney, Steve Tierney, Robert Wagner, Weimin Zhang, Michael L. Chabinyc, R. Joseph Kline, Michael D. McGehee, and Michael F. Toney 2006. "Liquid-Crystalline Semiconducting Polymers with High Charge-Carrier Mobility." *Nature Materials* 5 (4): 328–333. https://doi.org/10.1038/nmat1612.

McCullough, Richard D., and Renae D. Lowe. 1992. "Enhanced Electrical Conductivity in Regioselectively Synthesized Poly(3-Alkylthiophenes)." *Journal of the Chemical Society, Chemical Communications* 1: 70–72. https://doi.org/10.1039/C39920000070.

McCullough, Richard D., Renae D. Lowe, Manikandan Jayaraman, and Deborah L. Anderson. 1993. "Design, Synthesis, and Control of Conducting Polymer Architectures: Structurally Homogeneous Poly(3-Alkylthiophenes)." *The Journal of Organic Chemistry* 58 (4): 904–912. https://doi.org/10.1021/jo00056a024.

McCullough, Richard D., Stephanie Tristram-Nagle, Shawn P. Williams, Renae D. Lowe, and Manikandan Jayaraman. 1993. "Self-Orienting Head-to-Tail Poly(3-Alkylthiophenes): New Insights on Structure-Property Relationships in Conducting Polymers." *Journal of the American Chemical Society* 115 (11): 4910–4911. https://doi.org/10.1021/ja00064a070.

Memon, Mushtaque A., Wei Bai, Jinhua Sun, Muhammad Imran, Shah Nawaz Phulpoto, Shouke Yan, Yong Huang, and Jianxin Geng. 2016. "Conjunction of Conducting

Polymer Nanostructures with Macroporous Structured Graphene Thin Films for High-Performance Flexible Supercapacitors." *ACS Applied Materials & Interfaces* 8 (18): 11711–11719. https://doi.org/10.1021/acsami.6b01879.

Meyer, Victor. 1883. "Ueber Den Begleiter Des Benzols Im Steinkohlentheer." *Berichte Der Deutschen Chemischen Gesellschaft* 16 (1): 1465–1478. https://doi.org/10.1002/cber.188301601324.

Miller, John R., and Patrice Simon. 2008. "Electrochemical Capacitors for Energy Management." *Science* 321 (5889): 651–652. https://doi.org/10.1126/science.1158736.

Niles, Edwards T., John D. Roehling, Hajime Yamagata, Adam J. Wise, Frank C. Spano, Adam J. Moulé, and John K. Grey. 2012. "J-Aggregate Behavior in Poly-3-Hexylthiophene Nanofibers." *The Journal of Physical Chemistry Letters* 3 (2): 259–263. https://doi.org/10.1021/jz201509h.

Nilsson, K. Peter R., Johan D. M. Olsson, Peter Konradsson, and Olle Inganäs. 2004. "Enantiomeric Substituents Determine the Chirality of Luminescent Conjugated Polythiophenes." *Macromolecules* 37 (17): 6316–6321. https://doi.org/10.1021/ma048859e.

Nishide, Hiroyuki, and Kenichi Oyaizu. 2008. "Toward Flexible Batteries." *Science* 319 (5864): 737–738. https://doi.org/10.1126/science.1151831.

Oh, S. H., C.-W. Lee, D. H. Chun, J.-D. Jeon, J. Shim, K. H. Shin, and J. H. Yang. 2014. "A Metal-Free and All-Organic Redox Flow Battery with Polythiophene as the Electroactive Species." *Journal of Materials Chemistry A* 2 (47): 19994–19998. https://doi.org/10.1039/C4TA04730C.

Osaka, Itaru, and Richard D. McCullough. 2008. "Advances in Molecular Design and Synthesis of Regioregular Polythiophenes." *Accounts of Chemical Research* 41 (9): 1202–1214. https://doi.org/10.1021/ar800130s.

Österholm, Anna M., D. Eric Shen, Aubrey L. Dyer, and John R. Reynolds. 2013. "Optimization of PEDOT Films in Ionic Liquid Supercapacitors: Demonstration as a Power Source for Polymer Electrochromic Devices." *ACS Applied Materials & Interfaces* 5 (24): 13432–13440. https://doi.org/10.1021/am4043454.

Patil, A. O., Y. Ikenoue, N. Basescu, N. Colaneri, J. Chen, F. Wudl, and A. J. Heeger. 1987. "Self-Doped Conducting Polymers." *Synthetic Metals* 20 (2): 151–159. https://doi.org/10.1016/0379-6779(87)90554-6.

Patil, B. H., A. D. Jagadale, and C. D. Lokhande. 2012. "Synthesis of Polythiophene Thin Films by Simple Successive Ionic Layer Adsorption and Reaction (SILAR) Method for Supercapacitor Application." *Synthetic Metals* 162 (15): 1400–1405. https://doi.org/10.1016/j.synthmet.2012.05.023.

Patil, B. H., S. J. Patil, and C. D. Lokhande. 2014. "Electrochemical Characterization of Chemically Synthesized Polythiophene Thin Films: Performance of Asymmetric Supercapacitor Device." *Electroanalysis* 26 (9): 2023–2032. https://doi.org/10.1002/elan.201400284.

Pringle, Jennifer Mary, Maria Forsyth, Douglas Robert MacFarlane, Klaudia Wagner, Simon B. Hall, and David L. Officer. 2005. "The Influence of the Monomer and the Ionic Liquid on the Electrochemical Preparation of Polythiophene." *Polymer* 46 (7): 2047–2058. https://doi.org/10.1016/j.polymer.2005.01.034.

Rafiqi, Ferooze Ahmad, and Kowsar Majid. 2015. "Role of Gadolinium(III) Complex in Improving Thermal Stability of Polythiophene Composite." *Chemical Papers* 69 (10): 1331–1340. https://doi.org/10.1515/chempap-2015-0149.

Ribeiro, Adriana S., Wilson A. Gazotti, Pedro F. Dos Santos Filho, and Marco-A De Paoli. 2004. "New Functionalized 3-(Alkyl)Thiophene Derivatives and Spectroelectrochemical Characterization of Its Polymers." *Synthetic Metals* 145 (1): 43–49. https://doi.org/10.1016/j.synthmet.2004.04.005.

Rogers, John A., Takao Someya, and Yonggang Huang. 2010. "Materials and Mechanics for Stretchable Electronics." *Science* 327 (5973): 1603–1607. https://doi.org/10.1126/science.1182383.

Rudge, Andy, John Davey, Ian Raistrick, Shimshon Gottesfeld, and John P. Ferraris. 1994. "Conducting Polymers as Active Materials in Electrochemical Capacitors." *Journal of Power Sources* 47 (1): 89–107. https://doi.org/10.1016/0378-7753(94)80053-7.

Ryu, Kwang Sun, Young-Gi Lee, Young-Sik Hong, Yong Joon Park, Xianlan Wu, Kwang Man Kim, Man Gu Kang, Nam-Gyu Park, and Soon Ho Chang. 2004. "Poly(Ethylenedioxythiophene) (PEDOT) as Polymer Electrode in Redox Supercapacitor." *Electrochimica Acta* Polymer Batteries and Fuel Cells: Selection of Papers from First International Conference. 50 (2): 843–847. https://doi.org/10.1016/j.electacta.2004.02.055.

Seidel, Ch. 1987. "Handbook of Conducting Polymers. Volumes 1 and 2. Hg. von Terje A. Skotheim. ISBN 0-8247-7395-0 Und 0-8247-7454-X. New York/Basel: Marcel Dekker Inc. 1986. XVIII + XVII, 1417 S., Gcb. $ 150.00." *Acta Polymerica* 38 (1): 101. https://doi.org/10.1002/actp.1987.010380126.

Senthilkumar, B., P. Thenamirtham, and R. Kalai Selvan. 2011. "Structural and Electrochemical Properties of Polythiophene." *Applied Surface Science* 257 (21): 9063–9067. https://doi.org/10.1016/j.apsusc.2011.05.100.

Shirakawa, Hideki, Edwin J. Louis, Alan G. MacDiarmid, Chwan K. Chiang, and Alan J. Heeger. 1977. "Synthesis of Electrically Conducting Organic Polymers: Halogen Derivatives of Polyacetylene, (CH)X." *Journal of the Chemical Society, Chemical Communications* (16): 578–580. https://doi.org/10.1039/C39770000578.

Skabara, Peter J., Igor M. Serebryakov, Igor F. Perepichka, N. Serdar Sariciftci, Helmut Neugebauer, and Antonio Cravino. 2001. "Toward Controlled Donor−Acceptor Interactions in Noncomposite Polymeric Materials: Synthesis and Characterization of a Novel Polythiophene Incorporating π-Conjugated 1,3-Dithiole-2-Ylidenefluorene Units as Strong D−A Components." *Macromolecules* 34 (7): 2232–2241. https://doi.org/10.1021/ma0015931.

Skompska, Magdalena, Józef Mieczkowski, Rudolf Holze, and Jürgen Heinze. 2005. "In Situ Conductance Studies of P- and n-Doping of Poly(3,4-Dialkoxythiophenes)." *Journal of Electroanalytical Chemistry* 577 (1): 9–17. https://doi.org/10.1016/j.jelechem.2004.11.008.

Snook, Graeme A., Pon Kao, and Adam S. Best. 2011. "Conducting-Polymer-Based Supercapacitor Devices and Electrodes." *Journal of Power Sources* 196 (1): 1–12. https://doi.org/10.1016/j.jpowsour.2010.06.084.

Tarascon, J.-M., and M. Armand. 2001. "Issues and Challenges Facing Rechargeable Lithium Batteries." *Nature* 414 (November): 359–367. https://doi.org/10.1038/35104644.

Thakur, Anukul K., Ram Bilash Choudhary, Mandira Majumder, and Malati Majhi. 2018. "Fairly Improved Pseudocapacitance of PTP/PANI/TiO2 Nanohybrid Composite Electrode Material for Supercapacitor Applications." *Ionics* 24 (1): 257–268. https://doi.org/10.1007/s11581-017-2183-x.

Villers, Dominique, Donald Jobin, Chantal Soucy, Daniel Cossement, Richard Chahine, Livain Breau, and Daniel Bélanger. 2003. "The Influence of the Range of Electroactivity and Capacitance of Conducting Polymers on the Performance of Carbon Conducting Polymer Hybrid Supercapacitor." *Journal of The Electrochemical Society* 150 (6): A747–A752. https://doi.org/10.1149/1.1571530.

Yamamoto, Takakazu, Yasuhiro Hayashi, and Akio Yamamoto. 1978. "A Novel Type of Polycondensation Utilizing Transition Metal-Catalyzed C–C Coupling. I. Preparation of Thermostable Polyphenylene Type Polymers." *Bulletin of the Chemical Society of Japan* 51 (7): 2091–2097. https://doi.org/10.1246/bcsj.51.2091.

Yamamoto, Takakazu, Kenichi Sanechika, and Akio Yamamoto. 1980. "Preparation of Thermostable and Electric-Conducting Poly(2,5-Thienylene)." *Journal of Polymer Science: Polymer Letters Edition* 18 (1): 9–12. https://doi.org/10.1002/pol.1980.130180103.

Yu, Mei Hui, Hui Min Meng, and Ying Xue. 2016. "Nano-Mesh Structured Mn-Based Oxide/Conducting Polymer Composite Electrode for Supercapacitor." *Materials Science Forum* 859: 104–108. https://doi.org/10.4028/www.scientific.net/MSF.859.104.

Zade, Sanjio S., and Michael Bendikov. 2007. "Theoretical Study of Long Oligothiophene Polycations as a Model for Doped Polythiophene." *The Journal of Physical Chemistry C* 111 (28): 10662–10672. https://doi.org/10.1021/jp071277p.

Zhai, Lei, Richard L. Pilston, Karen L. Zaiger, Kristoffer K. Stokes, and Richard D. McCullough. 2003. "A Simple Method to Generate Side-Chain Derivatives of Regioregular Polythiophene via the GRIM Metathesis and Post-Polymerization Functionalization." *Macromolecules* 36 (1): 61–64. https://doi.org/10.1021/ma0255884.

Zhang, Haiqiang, Liwen Hu, Jiguo Tu, and Shuqiang Jiao. 2014. "Electrochemically Assembling of Polythiophene Film in Ionic Liquids (ILs) Microemulsions and Its Application in an Electrochemical Capacitor." *Electrochimica Acta* 120 (February): 122–127. https://doi.org/10.1016/j.electacta.2013.12.091.

Zhang, Haiqiang, Zongqian Hu, Mao Li, Liwen Hu, and Shuqiang Jiao. 2014. "A High-Performance Supercapacitor Based on a Polythiophene/Multiwalled Carbon Nanotube Composite by Electropolymerization in an Ionic Liquid Microemulsion." *Journal of Materials Chemistry A* 2 (40): 17024–17030. https://doi.org/10.1039/C4TA03369H.

13 Conducting Fibers for Energy Applications

Narendra Reddy
Center for Incubation, Innovation, Research and Consultancy
Jyothy Institute of Technology
Bengaluru, India

Roshan Paul
Department of Textile Science and Technology
University of Beira Interior
Covilha, Portugal

Asimananda Khandual
Department of Textile Engineering
College of Engineering & Technology (BPUT)
Bhubaneswar, India

CONTENTS

13.1 POLYANILINE (PANI)-BASED CONDUCTIVE FIBERS

Polyaniline (PANI) is one of the most popular conducting polymers. Since PANI is soluble in common solvents, attempts have been made to develop blends of PANI and other polymers. PANI and nylon 6 were separately dissolved in formic acid or sulfuric acid. A desired proportion of the two polymers was mixed together and extruded as fibers using a wet spinning apparatus (Mirmohseni, Salari, and Nabavi, 2006) into a coagulation bath containing 7.5% $LiSO_4$. Electrical resistance of the blend fibers containing different levels of PANI is listed in Table 13.1. Highest resistance of 2.26 MΩ/cm was obtained using 5% PANI. The electrical resistance of the fibers was dependent on the moisture content. Cyclic voltammetry studies suggested that the blend fibers were electroactive with 25% PANI, providing high electroactive stability and improved tensile properties (Mirmohseni, Salari, and Nabavi, 2006).

Poly(ethylene terephthalate) (PET) fabrics were made electrically conductive by immersing them in a solution of PANI dodecyl benzene sulfonic acid. A multistep sol–gel process of coating was adopted using a custom-built equipment (Kim, Koncar, and Dufour, 2006). In addition to electrical conductivity, the mechanical properties of the treated fabrics were also found to improve. Electrical resistance of the fabrics ranged between 600 and 1200 ohms depending on the treatment conditions. In addition to developing conductive fibers using conventional conductive polymers/metals, researchers have also shown that natural fibers such as cotton can be made conductive using various approaches.

Electrospun-conductive fiber blends were made by doping PANI with p-toluene sulfonic acid and poly (L-lactic acid). A desired amount of PANI and p-toluene was dissolved using HFIP for 2 h. After dissolution of the two polymers, PLA was added in different ratios and the blends were converted into fibers using electrospinning. The average diameter of the fibers obtained ranged from as low as 131 to 860 nm depending on the electrospinning conditions. Resistivity of the fibers decreased almost linearly with an increase in content of PANI. The fibers were suggested

TABLE 13.1

Changes in the electrical resistance (MΩ/cm) of PANI/ nylon blend fibers containing various levels of PANI and treated for different times (Mirmohseni, Salari, and Nabavi, 2006). Open access

Time (h)	PANI (%)				
	5	11.5	14.2	20	25
2	0.665	0.412	0.353	0.053	0.015
4	1.09	0.712	0.493	0.076	0.048
14	1.32	1.07	0.715	0.105	0.064
24	2.26	1.34	0.825	0.124	0.111
36	–	2.11	0.865	0.188	0.127
48	–	2.41	0.915	0.247	0.213
72	–	–	1.48	0.310	0.255
180	–	–	–	0.570	0.372

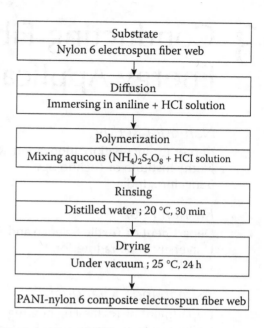

FIGURE 13.1 Flowchart of the steps used to prepare the conductive nylon electrospun fibers

to be suitable for biomedical, sensors, and other applications (Picciani, Medeiros, Pan, Orts, Mattoso, and Soares, 2009). An in situ polymerization approach was used to develop PANI/nylon electrospun fibers with high electrical conduction and good mechanical properties (Hong, Oh, and Kang, 2005). Electrospun nylon fibers were immersed in aqueous hydrochloride solution of aniline at 40°C, followed by treatment with ammonium peroxydisulfate in 0.35 M HCl. Polymerization of the electrospun fibers was done at 5°C for 1 h. A schematic representation of the process used to develop the conducting nanofibers is shown in Figure 13.1. Fibers obtained had diameters between 56 and 240 nm. There was an increase in volume conductivity, whereas the surface conductivity of the nanofibers decreased with an increase in diffusion time. The electrospun fibers had considerably higher conductivity (surface conductivity of 0.2 S/cm and volume conductivity of 1.3 S/cm) when compared to the nylon plain weave processed under similar conditions (Hong, Oh, and Kang, 2005).

Instead of blending with other polymers, pure PANI microfibers were obtained using a modified electrospinning set-up (Yu, Shi, Deng, Wang, and Chen, 2008). Prior to electrospinning, aniline was dissolved in 1 M HCl/H_2SO_4 and was treated with ammonium persulfate in HCl/H_2SO_4 to obtain HCl/H_2SO_4-doped PANI powder. The obtained powder was dissolved using hot sulfuric acid for 5 h and was later electrospun in the coagulation bath made of various concentrations of sulfuric acid. The surface morphology (Figure 13.2) of the fibers obtained was dependent on the concentration of sulfuric acid in the coagulation bath and also on the electrospinning conditions (Yu, Shi, Deng, Wang, and Chen, 2008). A high conductivity of 52.9 S/cm was obtained under optimized electrospinning conditions.

Highly porous and conductive PANI–carbon fibers were developed using poly(vinyl alcohol) as a binder. In this study, PANI and carbon black were dispersed in chloroform and later added into PVA dissolved in water, and this mixture was electrospun into fibers using a voltage of 15 kV and a flow rate of 0.5 mL/h (Sujith, Asha, Anjali, Sivakumar, Subramanian, Nair, and Balakrishnan, 2012). After electrospinning, the fibrous mats were heated at 230°C to sublimate the PVA and porous fibers were obtained. The diameter of the heated fibers was 170 nm when compared to 250 nm before heating. Heating provided higher porosity (79%) but with larger pore size (1.9 μm). The fibers had behavior between that of an insulator and a conductor but turned into good conductors after removing PVA. Such porous and conducting fibers were suggested to be suitable as capacitive electrodes, strain sensors, and so on (Sujith, Asha, Anjali, Sivakumar, Subramanian, Nair, and Balakrishnan, 2012). In a complementary study, ultrafine fibers of PANI and poly(ethylene oxide) (PEO) with diameters less than 30 nm were manufactured using electrospinning (Zhou, Freitag, Hone, Staii, Johnson, Pinto, and MacDiarmid, 2003). PANI was first combined with camphorsulfonic acid and was later blended with PEO. This blend was characterized for mechanical properties and applicability. Single fibers developed had diameters of 20–70 nm and conductivity between 10^{-3} and 10^{-2} S/cm, with the conductivity decreasing with an increase in fiber diameters. In yet another study, camphorsulfonic acid (HSCA)-doped PANI and PEO were developed using electrospinning. The concentration of doped PANI was between 0.5% and 11% when compared to the concentration of PEO of 2–4% in the solution. Fibers obtained had diameters between 950 nm to 2.1 μm, with no formation of beads. The electrical conductivity of the electrospun fibers increased with increasing percentage of PANI-HSCA (Figure 13.3)

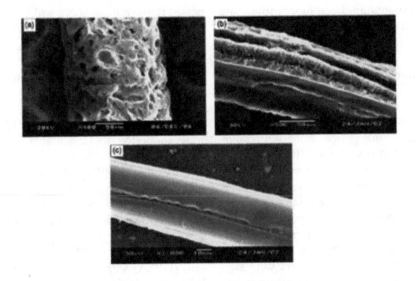

FIGURE 13.2

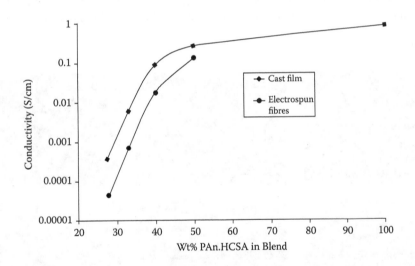

FIGURE 13.3

but was lower when compared to cast films prepared using the same blend solution. The lower conductivity of the fibers when compared to that of films was suggested to be due to the highly porous structure formed during electrospinning (Norris, Shaker, Ko, and Mac-Diarmid, 2000). Using the same approach, single fibers with considerably high conductivity of 3.294 S/cm were obtained when compared to 0.19 for the fibrous mats in a study conducted by Khan, 2006. A PANI: HSCA ratio of 70:30 was used and fibers obtained had a diameter of 700 nm. The absence of voids and poor connectivity between the individual fibers were considered to be the reason for the lower conductivity of the

mats when compared to those of the individual fibers. PANI was converted to carbon with high conductivity using the electrospinning approach. PANI was dissolved in DMF and was electrospun into fibers that were pyrolyzed in a vacuum furnace at 800°C for 30 min (Wang, Serrano, and Santiago-Aviles, 2002). The PAN graphite nanofibers had a substantially high conductivity of 490 S/m.

13.2 DEVELOPING CONDUCTIVE FIBERS USING POLYPYRROLE (PPy)

In one of the earliest studies, cotton, silk, and wool were treated with pyrrole/thiophene in 1,2-dichloroethane containing tetrabutylammonium hexachloroantimonate as the supporting electrolyte (Bhadani, Gupta, and Gupta, 1993). Fibers treated with pyrrole had electrical resistance of 20×10^{-3}, 5×10^{-3}, and 25×10^{-3} ohm-cm for cotton, silk, and wool, respectively, when compared to 7, 35, and 65 ohm-cm for thiophene-treated fibers, indicating that the chemical treatment could make the fibers conductive. In a similar approach, cotton and silk yarns were coated with pyrrole using both vapor and liquid phases. A conductivity of 6.4×10^{-4} and 3.2×10^{-4} S/cm was obtained for cotton and silk even after several washings (Hosseini, 2005). To further improve the conductivity of the fibers, modifications were done using aniline with HBF_4 as an electrolyte (Bhadani, Sen Gupta, Sahu, and Kumari, 1996). Weight gain of the fibers increased almost linearly with the increase in electrolysis time but decreased surface resistivity for all the fibers. Both fibers were suggested to be useful for electromagnetic and antistatic applications. In another study by Granato et al., pyrrole molecules were directly polymerized on the surface of the polyamide 6 (nylon) nanofibers to impart conductivity. Initially, nylon and anhydrous ferric chloride were combined in a solution and made into electrospun fibers. Vapors of pyrrole were passed through these fibers, which led to uniform deposition of pyrrole particles on the surface of the fibers through polymerization. Amount of pyrrole deposited and consequently the diameter of the fibers increased with an increase in polymerization time (Figure 13.4) (Granato, Bianco, Bertarelli, and Zerbi, 2009). I–V curves showed that the loss of conductivity was less than 5% even after testing for 1280 cycles. The fibers were suggested to be suitable for applications in sensors and other areas.

Polypyyrole (PPY) is another polymer that has very high electrical conductivity. However, PPY is insoluble in most solvents and hence it is difficult to be made into fibers or other structures. In a polymerization process, ammonium persulfate was used as an oxidant to obtain electrospun PPY, where dodecylbenzene sulfonic acid (DBSA) was used as a doping agent. The modified PPY was soluble in various solvents and was dissolved in chloroform for electrospinning using a voltage of 30–45 kV. Solubility of PPY and the conductivity were dependent on the conditions during polymerization (Table 13.2) (Kang, Lee, Joo, and Lee, 2005). Unlike the powder, PPy nanofibers had electrical conductivity of 0.5 S/cm, making them

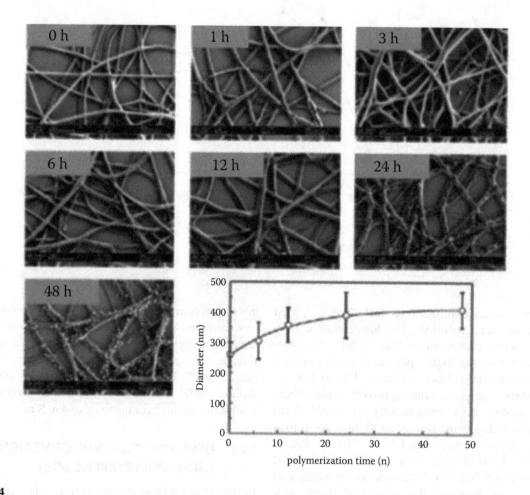

FIGURE 13.4

suitable as an electrode material for rechargeable battery or supercapacitor. Electrospun pure PPy fibers and those containing PEO as a carrier were prepared by Chronakis et al. High molecular PPy (62,300 g/mol) was made water soluble by doping with di (2-ethylhexyl) sulfosuccinate sodium salt (NaDEHS). PEO and PPy blends and pure PPy solution were electrospun at a voltage of 30 kV. After electrospinning, the PEO in the fibers was removed by treating it with ethanol as a solvent for 20 min at a temperature of 60°C and a pressure of 2000 psi. The diameters of the nanofibers obtained ranged between 125 and 300 nm depending on the amount of PPy or PEO. The electrical conductivity of the doped PPY was between 3.5×10^{-4} and 1.1×10^{-4} S/cm depending on the concentration of the polymer. Comparatively, the conductivity of the pure PPy fibers was considerably high at 2.7×10^{-2} S/cm due to higher PPy content and also molecular orientation during electrospinning (Chronakis, Grapenson, and Jakob, 2006). Using PEO as a substrate, PPy nanofibers were developed by the vapor phase deposition technique (Nair, Natarajan, and Kim, 2005). A 10.7% solution of PEO and $FeCl_3$ (1:2.5) was electrospun into a water/ethanol solvent system. The fibers formed were exposed to PPy vapor at ambient condition for 1–14 days when the vapor phase polymerization occurred. Blend fibers had an average diameter of 96 nm and a conductivity of 10^{-3} S/cm (Nair, Natarajan, and Kim, 2005). Since the conductivity of the PEO-PPy nanofibers was low, attempts have been made to develop composite nanofibers using polystyrene as the template (Nair, Hsiao, and Kim, 2008). Polystyrene having a molecular weight of 90,000 along with $FeCl_3$ or ferric p-toluenesulfonate (FeTS) as oxidants were dissolved using tetrahydrofuran (THF) and acetone and electrospun into fibers. The nanofibrous mat obtained was exposed to PPy vapors to impart conductivity. Fibers obtained from the two different oxidants have different morphology and performance. The PS-FeCl₃ fibers had diameters of 625 ± 300 nm when compared to 930 ± 150 nm for the PS-FeTS fibers. Morphologically, the PS-FeCl₃ fibers had a worm-like structure on the surface, whereas the PS-FeTS fibers had a serrated and separated appearance due to different doping levels (21% and 28%, respectively; Figure 13.5). In terms of conductivity, the two fibers were also different, with 2×10^{-3} and 5×10^{-3} S/cm for the PS-FeCl₃ and PS-FeTS fibers, respectively. However, the conductivity could be increased to 0.13 S/cm by removing the polystyrene using THF without affecting the morphology of the fibers, which is due to the better interaction between the PPy polymers. It was suggested that using FeTS instead of $FeCl_3$ as the oxidant was preferable since FeTS provides better crystallinity and conductivity (Nair, Hsiao, and Kim, 2008).

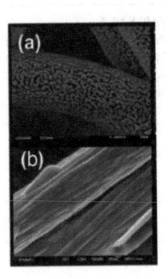

FIGURE 13.5

PPy was also blended with polyurethane (93,000 g/mol) and made into electrospun nanofibers. Polyurethane was dissolved in dimethylformamide (DMF) and tetrahydrofuran (1/1 v/v) ratio and later in Ce(IV) and a known amount of PPy was added to the solution and electrospun into fibers (Yanilmaz, Kalaoglu, Karakas, and Sarac, 2012). Fibers without PPY had an average diameter of 2 μm but decreased to 1.3 μm when 7.5% PPy was added. An increase in tensile strength of the fibers from 54 to 69 MPa was observed due to the addition of PPy. Similar enhancement was also observed in thermal properties and Young's modulus. Conductivity of the polyurethane fibers without PPy was 7×10^{-7} S/cm compared to 1.4×10^{-6} S/cm for the fibers containing 5% PPy. However, the conductivity was also found to be dependent on the frequency of measurement (Yanilmaz, Kalaoglu, Karakas, and Sarac, 2012).

A novel approach of electrospinning and subsequent oxidation was used to develop PPY and poly(vinyl pyrrolidine) (PVP) nanofibers (Tavakkol, Tavanai, Abdolmaleki, and Morshed, 2017). In this study, PPy was made electrospinnable with adequate viscosity by adding poly(vinyl pyrrolidine). The electrospun blends were collected in a solution of $FeCl_3 \cdot 6H_2O$ as dopant in ethanol and oxidized in situ for the PPY to be polymerized. Fibers obtained had an average diameter of 440 nm and a crystallinity of 26%. Although no chemical reaction was observed between PPY and PVP, good conductivity of 5.22×10^{-1} S/cm was obtained. A comparison of the various methods used to develop PPy-based conductive fibers is listed in Table 13.3.

To develop pure and composite PPy fibers, pyrrole was chemically modified using poly(3,4-diethylpyrrole). Modified PPy was soluble in various solvents including DMF. However, it was not possible to produce pure PPy electrospun fibers even while using the solubilized PPy. Hence, PPy solution was combined with

TABLE 13.2

Properties of PPy obtained after polymerization using APS at 0°C (Kang, Lee, Joo, and Lee, 2005). Reproduced with permission from Elsevier

Polymerization time (min)	Apparent yield (%)	Doping level (%)	Actual yield (%)	Conductivity (S/cm)	Solubility in chloroform
5	8.4	14.7	4.6	5.8	Insoluble
10	9.4	16.8	4.8	5.0	Insoluble
30	12.6	17.2	6.3	1.3	Insoluble
60	14.6	17.7	7.2	6.2×10^{-1}	Insoluble
120	15.2	19.0	7.1	3.3×10^{-1}	Soluble
240	14.6	20.5	6.5	2.5×10^{-1}	Soluble
480	14.5	19.8	6.6	2.1×10^{-1}	Soluble
1440	15.2	20.8	6.7	2.0×10^{-1}	Soluble

polystyrene solution and made into electrospun composite fibers. To further improve the amount of PPY and hence the conductivity, the composite fibers were treated in DMF under ambient conditions for 24 h. This resulted in selective removal of polystyrene and formation of a hollow PPy fiber (Figure 13.6). However, the conductivity of such pure or hollow PPy fibers was not reported (Sen, Davis, Mitchell, and Robinson, 2009). Conducting polycaprolactone (PCL) fibers were developed by incorporating various percentages of PPy (Číková, Mičušík, Šišková, Procházka, Fedorko, and Omastová, 2018). PCL having a molecular weight of 67×10^3/g was dissolved using a blend of dichloromethane (DCM) and DMF. A blend solution was made into electrospun fibers using a flow rate of 0.6 mL/h and a voltage of 15.6 kV and 15 cm distance

TABLE 13.3

Properties of PPy-based conductive fibers obtained using different techniques (Tavakkol, Tavanai, Abdolmaleki, and Morshed, 2017). Reproduced with permission from Tavakkol

Process	Additive polymer	Dopant	Oxidant	Conductivity (S/cm)	Fiber diameter (nm)
Electrospinning	–	DBSA	APS	0.5	3×10^3
Coating	–PEO	NaDEHS	–	2.7×10^{-2}	70
Electrospinning,+ oxidation	S-SEBS	NaDEHS	–	4.9×10^{-8}	200–300
	P(An-co-Vac)	DBSA	$FeCl_3$	0.52	300
	PCL/gelatin	–	Ce(IV)	10^{-7}	200–400
	PU	–	–	1.3×10^{-5}	216
	PVDF	–	Ce(IV)	1.4×10^{-6}	2×10^3
	PEO	DBSA	$FeCl_3$	–	460
	PVP	–	$FeCl_3$	10^{-3}	96
	–	–	$FeCl_3$	10^{-3}	100–350
	PS	AOT	$FeCl_3$	–	5×10^3
	PAN	CI	$FeCl_3$	2×10^{-3}	265, 930
	PS	–	$AgNO_3$	1.3×10^{-3}	180–292
	PA-6	–	$FeCl_3$	–	300–400
	PCL and PLA	$FeCl_3$	$FeCl_3$	–	260
	PAN	PTSA or CA	APS or Fe^{3+}	–	300
	PVP	AQSA	$FeCl_3$	–	986
	PVP	–	$FeCl_3$	3.05×10^{-4}	730
	PVP	PTSA	$FeCl_3$	3.59×10^{-2}	438
	PVP	AQSA	$FeCl_3$	5.22×10^{-1}	833
	PVP	–	$FeCl_3$	2.29×10^{-6}	1153
	PVP	PTSA	$FeCl_3$	2.26×10^{-5}	496
		AQSA	$FeCl_3$	1.98×10^{-1}	961

between the collector and needle tip. To make the fibers conductive, PPy was mixed with 2-napthalensulfonate and FeCl$_3$.

Electrospun PCL fibers were added into the PPy mixture and stirred for 8 h and later washed and dried to obtain a dark colored electrospun membrane (Figure 13.7). Physical properties of the fibers and the conductivity varied considerably with the addition of PPy. In addition, the extent of change was dependent on the ratio of PCl and PPy (Table 13.4). A considerably good conductivity of up to 9 S/cm was achievable when an optimum level of PPy was used. The PCL-PPy blend fibers were considered suitable for use in bioelectronics due to their good conductivity and nontoxicity of PCL.

Conductive PCL/gelatin blend fibers incorporated with PPy were developed for bone tissue engineering. The PCL/gelatin was combined with PPy dissolved in HFIP to obtain a PPy ratio of up to 30%. Electrospun fibers obtained were characterized for their tensile properties, conductivity, and degradation. Furthermore, the fibers were used as substrates to culture rabbit cardiomyocytes (Kai, Prabhakaran, Jin, and Ramakrishna, 2011). Unlike other studies, the diameter of the fiber decreased with an increase in PPY content, whereas modulus and conductivity increased (Table 13.5). Cardiomyocytes seeded on the fibers showed a considerably higher attachment and also an increased proliferation

(Figure 13.8) on the 15% PPy containing fibers. Furthermore, cardiac-specific proteins were expressed, indicating that the scaffolds could be ideal for cardiac tissue generation, particularly those containing 15% PPy (Kai, Prabhakaran, Jin, and Ramakrishna, 2011).

In what was reported as a novel approach, blends of pure and blends of pyrrole and multiwalled carbon nanotubes (MWCNTs) were developed (Han, 2007). Pure pyrrole fibers were produced by first synthesizing an organic salt FeAOT by chemical interaction between 1,4-bis(2-ethylhexyl)sulfosuccinate (AOT) and ferric chloride (Fe). The organic salt was made into fibers by mechanical means. Pyrrole fibers were formed by vapor deposition onto the salt fiber. In addition to the pure pyrrole, blends of the fiber with various ratios of MWCNTs were also developed. Fibers containing the nanofibers had good conductivity (10–15 S/cm) and tensile strength (12–43 MPa). In terms of electrical conductivity, the fibers showed strong and broad anodic and cathodic peaks, suggesting redox potential in the range of −2 to +3 V at a scan rate of 20 mV/s (Han, 2007).

13.3 PEDOT-BASED CONDUCTIVE FIBERS

Poly(3,4-ethylenedixoythiophene) (PEDOT) has also been considered an ideal polymer for developing conductive

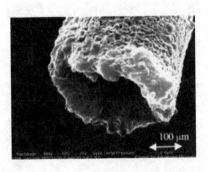

FIGURE 13.6

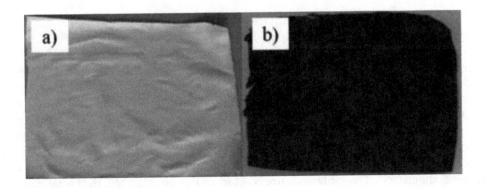

FIGURE 13.7

TABLE 13.4

Comparison of the properties and conductivity of PCL fibers before and after treating (polymerization) with various levels of PPy (Číková, Mičušík, Šišková, Procházka, Fedorko, and Omastová, 2018). Reproduced with permission from Elsevier

Sample	Fiber diameter (nm)		Contact angle (°)		Conductivity (S/cm)	
	Before	After	Before	After	Before	After
PCL	103 ± 56	620 ± 23	121.9 ± 2.4	–	4×10^{-8}	–
PCL/PPy (85:15)	363 ± 44	517 ± 39	90.9 ± 1.2	70.6 ± 2.3	4×10^{-8}	0.03
PCL/PPy (75:25)	430 ± 29	383 ± 59	105.0 ± 1.6	96.2 ± 2.3	6×10^{-8}	0.8
PCL/PPy (60:40)	276 ± 23	508 ± 26	112.7 ± 2.5	101.1 ± 2.1	9×10^{-8}	1
PCL/PPy (50:50)	240 ± 50	600 ± 20	114.9 ± 3.6	111.0 ± 2.9	5×10^{-8}	3
PCL/PPy (40:60)	395 ± 25		120.2 ± 1.8	117.8 ± 1.2	3×10^{-8}	9

TABLE 13.5

Properties of PCL–glycerol electrospun fibers without and with 15% and 30% PPy (Kai, Prabhakaran, Jin, and Ramakrishna, 2011). Reproduced with permission from John Wiley and Sons

Nanofibers	Fiber diameter (nm)	Contact angle (°)	Young's modulus (MPa)	Elongation (%)	Conductivity (mS/cm)
PCL–glycerol	239 ± 37	24.3 ± 1.8	7.9 ± 1.6	61.1 ± 17.3	0
PCL–glycerol + 15% PPy	216 ± 36	46.9 ± 2.0	16.8 ± 1.9	13.6 ± 3.2	0.013
PCL–glycerol + 30% PPy	191 ± 45	63.5 ± 2.8	50.3 ± 3.3	3.7 ± 1.4	0.37

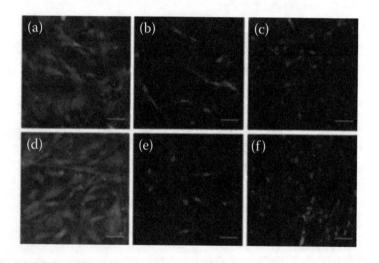

FIGURE 13.8

materials. However, PEDOT is insoluble in water and other solvents and hence is grafted with polymers such as poly(4-styrene sulfonate) (PSS) to make it soluble. A blend of PEDOT and PSS was combined with 3,4-ethylenedioxytiophene and polymerized. PEDOT/PSS fibers were produced by extruding the solution into a acetone coagulation bath using a single-hole spinneret of different diameters. The obtained fibers had diameter of 10 μm and tensile strength, elongation, and modulus of 17 MPa, 1.1 GPa, and 4.3%, respectively. Conductivity of the fibers was stable and was about 10^{-1} S/cm (Okuzaki, 2003). The PEDOT–PSS system was also used to coat polyamide fabrics for use in humidity and strain sensors (Daoud, Xin, and Szeto, 2005) by immersing the fabrics in the solution and later drying at 80°C for 10 h. Although the conductivity of the fabrics was not reported, they were suggested to be suitable for direct sensing of humidity, temperature, and applied

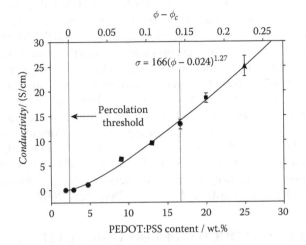

FIGURE 13.9

strain. PEDOT–PSS coating was also used to develop sensors with polyurethane as the substrate. Polyurethane was dissolved using DMF, DMSO, and THF. Solution containing the substrate and conductive polymers was extruded into a coagulation bath at a rate of 5 mL/h (Seyedin, 2014). The addition of PEDOT–PSS did not result in major changes in the properties of the fibers. However, a 30-fold increase in modulus from 7 to 247 MPa occurred when 25% PEDOT–PSS was used. Similarly, the conductivity of fibers increased nearly linearly (Figure 13.9) with an increase in PEDOT–PSS and passed the percolation threshold (Seyedin, 2014). The composite fibers were considered suitable for strain-sensing applications due to the ability of the PEDOT–PSS polymer to deform and rearrange within the PU matrix. Highly conductive PEDOT–PSS fibers were manufactured using a simple one-step wet-spinning approach (Jalili, 2011). PEDOT–PSS was dispersed in an aqueous solution to which ethylene glycol and polyethylene glycol were added. Fibers were extruded from the solution using a custom-built equipment at a speed between 0.8 and 2.0 mL/h into a coagulation bath consisting of acetone and isopropyl alcohol. The obtained fibers were treated with ethylene glycol,

drawn and annealed at 120°C for 30 min. Fibers with diameters between 10 and 15 μm were obtained and had a maximum stress of 125 MPa, tensile modulus of 3.3 GPa, and 15.8% elongation when isopropanol was used for coagulation. Fibers developed from PEDOT–PSS–EG with 10% PEG had a conductivity of 264 S/cm, considerably higher than previous reports on developing conductive fibers. Some of the processing conditions and properties of the fibers obtained are listed in Table 13.6. It was suggested that including PEG in the spinning dope was necessary to achieve high conductivity and with this approach, the need for posttreatment can also be eliminated (Jalili, 2011).

Cellulose pulp made from wood was carboxymethylated to obtain a degree of substitution of 0.065 (Wistrand, Lingström, and Wågberg, 2007). To make the cellulose conductive, polyelectrolyte multilayers were built using a sequential process. Single cellulose fibers were fixed using tapes on a dynamic contact angle analyzer. In addition, cellulose fibers in solution were treated with poly(3,4-ethylenedioxythiophene): poly(styrene sulfonate) (PEDOT:PSS) and poly(allyl amine). The treated fibers were made into sheets and the properties were measured. Conductivity of the fibers was dependent (Figure 13.10) on the initial fiber charge density achieved after fiber modification and the type of conducting polymer and number of layers coated on the fibers. For instance, untreated pulp had a charge density of 54 μeq/g compared to 350 μeq/g for the carboxymethylated fibers. Similarly, surface charge density was 3.5 μeq/g against 56 μeq/g. The conductivity of the cellulose sheets made using carboxymethylated pulp was considerably higher than that of the untreated fibers. Also, the conductivity was dependent on the number of layers of PEDOT:PSS present on the paper sheets (Wistrand, Lingström, and Wågberg, 2007).

Most conductive polymers have high molecular weights, do not dissolve in electrospinnable solvents, and hence are difficult to be made into nano- and microfibers. However, researchers have used several approaches to either modify the polymers or blend them with other polymers and developed conducting nanofibers and membranes. A combination of electrospinning and oxidative polymerization was used to convert poly(methyl

TABLE 13.6

Processing conditions used and properties of PEDOT:PSS fibers (Jalili, 2011). EG is ethylene glycol. Reproduced xwith permission from John Wiley and Sons

Processing	Coagulation bath	Modulus, GPa	Strength, MPa	Strain at break, %	Conductivity, S/cm
As-spun	Acetone	2.4 ± 0.3	86 ± 8.3	12.5 ± 1.0	8 ± 2
EG treated	Acetone	3.7 ± 0.4	98 ± 8	11 ± 0.7	201 ± 12
As spun	Isopropanol	3.3 ± 0.3	125 ± 6.7	15.8 ± 1.2	9 ± 2
EG treated	Isopropanol	5 ± 0.3	130 ± 5	12.1 ± 0.5	223 ± 10
EG treated under tension	Isopropanol	5.7 ± 0.4	134 ± 6	12 ± 0.6	351 ± 15
PEG treated	Isopropanol	2.5 ± 0.2	97 ± 4	13.5 ± 0.6	264 ± 17

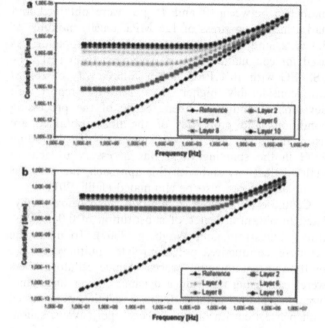

FIGURE 13.10

methacrylate) and PEDOT into nanofibers. In this approach, PMMA was dissolved in DMF and the EDOT monomers were added to the PMMA solution in different ratios. Later the blend solution was electrospun into a $FeCl_3$ solution for the polymers to oxidize and form the fibers (Figure 13.11). An average diameter of the fibers obtained was 500 nm, with the PEDOT forming a 50 nm layer on the surface of the PMMA fibers. Conductivity of the fibers was reliant on the ratio of the two polymers and $FeCl_3$ concentration during oxidation. A conductivity of up to 8 S/cm could be achieved under the optimum conditions (Park, Lee, Kim, Ryu, Lee, Choi, Cheong, and Kim, 2013).

PEDOT was also considered to be suitable for stimulating cell growth (Bolin, Svennersten, Wang, Chronakis, Richter-Dahlfors, Jager, and Berggren, 2009). In a study by Bolin et al., 10% PET solution was obtained by dissolving in a 1:1 mixture of trifluoroacetic acid and DCM. The solution was electrospun using a 0.9 mm tip needle at a voltage of 25 kV. The fibers obtained were exposed to vapors of EDOT at 60°C for 6 h for the PEDOT to be polymerized. In addition to the vapor phase polymerization, the PET fibers were spin coated with PEDOT and later washed with butanol, isopropanol, and distilled water to remove unreacted conductive polymers (Bolin, Svennersten, Wang, Chronakis, Richter-Dahlfors, Jager, and Berggren, 2009). The PET fibers coated (thickness of 100 nm) with PEDOT had diameters of 200–400 nm and were porous with a pore size between 5 and 10 μm. Coating of PEDOT onto PET was also found to have substantially increased the hydrophilicity of the fibers. The resistance of the fiber

was between 1000 and 20,000 Ω/sq and but decreased inversely with an increase in the deposition time. Neuroblastoma cells seeded onto the PET–PEDOT nanofibers showed development of neurites and actin, indicating good biocompatibility. The voltage-operated calcium channels were activated by applying a potential of −3.0 V on the nanofibers and the Ca^{2+} response was measured. Cells were found to respond to the stimulation depending on the amount of PEDOT on the fibers. It was suggested that PET coated with PEDOT could be used as electrodes for simulation of cells (Bolin, Svennersten, Wang, Chronakis, Richter-Dahlfors, Jager, and Berggren, 2009).

13.4 CONDUCTING FIBERS USING CARBON NANOTUBES

Carbon nanotubes (single and multiwalled) have been extensively studied as potential materials to develop conductive fibers. An in situ polymerization approach was used to develop single carbon nanotube (length of 1.5 μm and diameter of 1.4 nm) polymer composite fibers and films using three different polymers (poly(3-hexyl thiophene)), pyrrole, or EDOT. A schematic representation of the process used to develop the composite fibers and films is shown in Figure 13.12. Fibers with diameters ranging from 22 to 70 μm were obtained. A maximum conductivity of 170 S/cm, considerably higher than that reported in previous reports, was possible under the optimized conditions (Table 13.7). However, the amount of conducting polymer in the fibers was crucial to obtain the desired conductivity (Allen, Pan, Fuller, and Bao, 2014).

Highly conducting polyethyleneamine–carbon nanotube composite fibers were prepared with a conductivity as high as 100–200 S/cm. In this approach, aqueous single-walled carbon nanotube dispersions were injected into PEI methanol solution to form gel-like flat fibers (Muñoz et al., 2005). Subsequent removal of methanol from the fibers resulted in strong uniform fibers with a width between 15 and 50 μm. In terms of mechanical properties, the composite fibers had a tensile strength of 50 MPa cm^3/g, toughness of 2 J/g, and high modulus between 110 and 140 GPa cm^3/g.

Silk fibroin obtained from *Bombyx mori* was dissolved in 9.3 M LiBr at 60°C to obtain a 20% solution. This solution was electrospun into membranes that were later dipped into a dispersion of MWCNTs. Fibers obtained had an average diameter of 460 ± 40 nm (Kang, 2007). A substantial absorption of the CNTs on the surface of the fibers was evident from the morphological images (Figure 13.13). The conductivity of the fibers between 10^{-6} and 10^{-4} S/cm was considered suitable for application in electromagnetic shielding and electrostatic dissipation (Kang, 2007). In

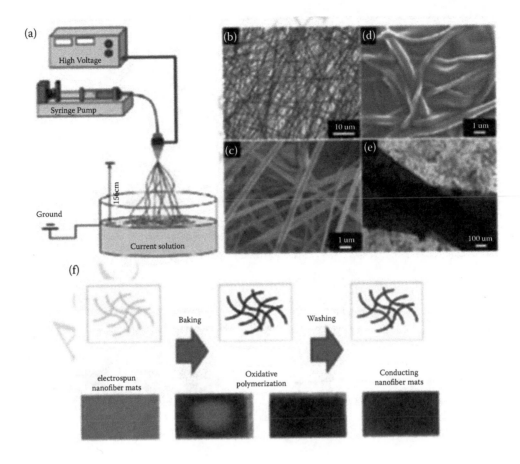

FIGURE 13.11

a similar approach, electrospun poly(lactic acid) fibers were made conductive by embedding MWCNTs. Initially, the MWCNTs were prepared by acid-oxidation and made soluble in N,N-DMF with the addition of pluornic F27 and sodium dodecyl sulfate as the catalyst. The CNT solution was combined with the PLA solution to obtain up to 5% CNTs by weight of PLA and electrospun into aligned and random fibers. The conducting PLA electrospun nanofibers were used to study the response of osteoblasts to electrical stimulation. The average length and diameter of the fibers was about 425 and 402 nm, respectively, with an increase in CNT concentration substantially increasing the diameter of the fibers (Shao, Zhou, Li, Li, Luo, Wang, Li, and Weng, 2011). The mechanical properties of the PLA fibers showed significant variations with an increase in the amount of CNTs in the fibers. Randomly oriented fibers provided better strength than the aligned ones at all concentrations of CNTs (Figure 13.14). Similarly, osteoblast cells on the random fibers responded positively to the electrical signals, much better than those on the aligned fibers. It was suggested that the addition of CNTs into PLA not only assisted in obtaining better quality

fibers but could also be used for specific medical applications such as fracture healing (Figure 13.15).

13.5 CONCLUSIONS

Conductive fibers with potential for use in electronics and medical applications have been developed using various techniques and polymers. Electrospinning appears to be most convenient to obtain nanofibers when compared to the other approaches used. PPy, PANI, and PEDOT are the primary conductive polymers used for developing the fibers. In addition to pure conductive fibers, conventional fibers, such as cotton, silk, PET, and PVA, have also been made conductive by adding conductive polymers. Chemical modifications of the conductive polymers have also helped to increase their solubility and interaction between the polymers and substrates. In addition to increased conductivity, the presence of conducting polymers has also improved mechanical properties and crystallinity of the fiber, under specific conditions. Further research into adapting the fibers for newer applications and development of conducting polymers are necessary to increase the use of conductive fibers.

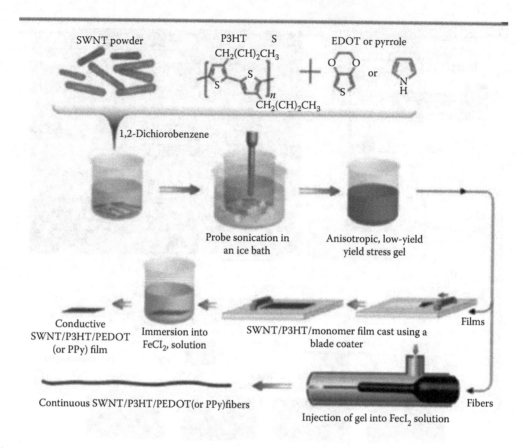

FIGURE 13.12

TABLE 13.7

Comparison of the properties of conducting fibers produced using various combinations of SWNTs and conducting polymers (Allen, Pan, Fuller, and Bao, 2014). Reproduced with permission from American Chemical Society

CNT dispersion	Oxidant solution	Fiber formation	Max. conductivity (S/cm)
SWNT/P3HT/PPy(1/14)	0.5 M FeCl$_3$ in ethanol	None	–
SWNT/P3HT/PPy(1/1/400)	0.5 M FeCl$_3$ in ethanol	None	–
SWNT/P3HT/PPy (1/1/40)	0.5 M FeCl$_3$ in ethanol	None	–
SWNT/P3HT/EDOT(1/140)	0.5 M FeCl$_3$ in ethanol	Semicontinuous	170
	0.5 M FeCl$_3$ in ethanol	Discontinuous	3
	0.5 M FeCl$_3$ in ethanol+phytic acid in ethanol	Semicontinuous	19
		Discontinuous	7
		Semicontinuous	64

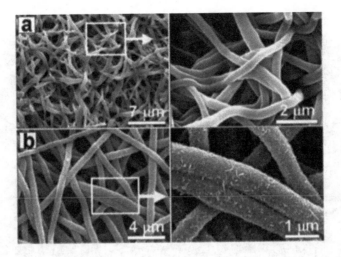

FIGURE 13.13

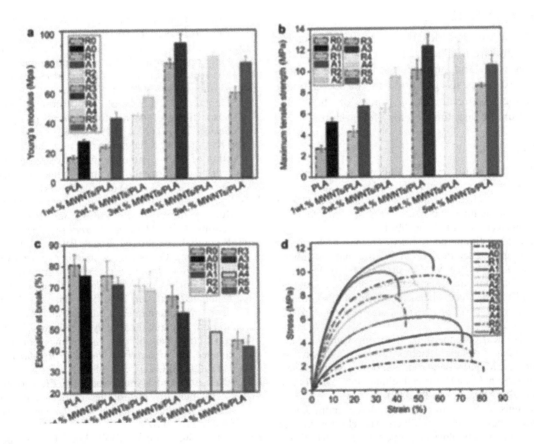

FIGURE 13.14

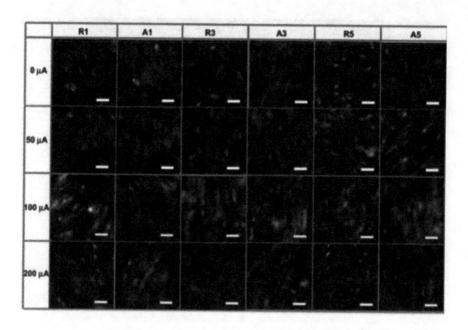

FIGURE 13.15

ACKNOWLEDGMENTS

Narendra Reddy expresses his thanks to the Center for Incubation, Innovation, Research and Consultancy for their encouragement and support to complete this chapter. NR also expresses his gratitude to the Department of Biotechnology, Government of India, for their support by the Ramalingaswami Re-entry fellowship.

REFERENCES

Allen, Ranulfo, Lijia Pan, Gerald G. Fuller, and Zhenan Bao. Using in-situ polymerization of conductive polymers to enhance the electrical properties of solution-processed carbon nanotube films and fibers. *ACS Applied Materials & Interfaces* 6 (13) (2014): 9966–9974.

Bhadani, Suraj N., Sumanta K. Sen Gupta, and Manoj Kumar Gupta. Electrically conducting natural fibers. *Indian Journal Fibre Textile Research* 18 (1993): 46–47.

Bhadani, Suraj N., Sumanta K. Sen Gupta, Guru C. Sahu, and Madhuri Kumari. Electrochemical formation of some conducting fibers. *Journal of Applied Polymer Science* 61 (2) (1996): 207–212.

Bolin, Maria H., Karl Svennersten, Xiangjun Wang, Ioannis S. Chronakis, Agneta Richter-Dahlfors, Edwin W. H. Jager, and Magnus Berggren. Nano-fiber scaffold electrodes based on PEDOT for cell stimulation. *Sensors and Actuators B: Chemical* 142 (2) (2009): 451–456.

Chronakis, Ioannis S., Sven Grapenson, and Alexandra Jakob. Conductive polypyrrole nanofibers via electrospinning: Electrical and morphological properties. *Polymer* 47 (5) (2006): 1597–1603.

Číková, Eliška, Matej Mičušík, Alena Šišková, Michal Procházka, Pavol Fedorko, and Mária Omastová. Conducting electrospun polycaprolactone/polypyrrole fibers. *Synthetic Metals* 235 (2018): 80–88.

Daoud, Walid A., John H. Xin, and Yau S. Szeto. Polyethylene-dioxythiophene coatings for humidity, temperature and strain sensing polyamide fibers. *Sensors and Actuators B: Chemical* 109 (2) (2005): 329–333.

Granato, Flavio, Andrea Bianco, Chiara Bertarelli, and Giuseppe Zerbi. Composite polyamide 6/polypyrrole conductive nanofibers. *Macromolecular Rapid Communications* 30 (6) (2009): 453–458.

Han, Gaoyi, and Gaoquan Shi. Novel route to pure and composite fibers of polypyrrole. *Journal of Applied Polymer Science* 103 (3) (2007): 1490–1494.

Hong, Kyung Hwa, Kyung Wha Oh, and Tae Jin Kang. Preparation of conducting nylon-6 electrospun fiber webs by the in situ polymerization of polyaniline. *Journal of Applied Polymer Science* 96 (4) (2005): 983–991.

Hosseini, Seyed Hossein, and Ali Pairovi Preparation of conducting fibres from cellulose and silk by polypyrrole coating. *Iranian Polymer Journal* 14 (11) (2005): 934–940.

Jalili, Rouhollah, Joselito M. Razal, Peter C. Innis, and Gordon G. Wallace. One-step wet-spinning process of poly (3, 4-ethylenedioxythiophene): Poly (styrenesulfonate) fibers and the origin of higher electrical conductivity. *Advanced Functional Materials* 21 (17) (2011): 3363–3370.

Kai, Dan, Molamma P. Prabhakaran, Guorui Jin, and Seeram Ramakrishna. Polypyrrole-contained electrospun conductive nanofibrous membranes for cardiac tissue engineering. *Journal of Biomedical Materials Research Part A* 99 (3) (2011): 376–385.

Kang, Minsung, and Hyoung-Joon Jin. Electrically conducting electrospun silk membranes fabricated by adsorption of carbon nanotubes. *Colloid and Polymer Science* 285 (10) (2007): 1163–1167.

Kang, Tae Su, Soong Wook Lee, Jinsoo Joo, and Jun Young Lee. Electrically conducting polypyrrole fibers spun by electrospinning. *Synthetic Metals* 153 (1–3) (2005): 61–64.

Khan, Saima Naz, Aurangzeb Khan, and Martin E. Kordesch. Conducting polymer fibers of polyaniline doped with camphorsulfonic acid. *MRS Online Proceedings Library Archive* 948 (2006): B04–B10.

Kim, Bohwon, Vladan Koncar, and Claude Dufour. Polyaniline-coated PET conductive yarns: Study of electrical, mechanical, and electro-mechanical properties. *Journal of Applied Polymer Science* 101 (3) (2006): 1252–1256.

Mirmohseni, Abdolreza, Dariush Salari, and Reza Nabavi. Preparation of conducting polyaniline/Nylon 6 blend fibre by wet spinning technique. *Iranian Polymer Journal* 15 (3) (2006): 259–264.

Muñoz, Edgar, Dong-Seok Suh, Steve Collins, Miles Selvidge, Alan B. Dalton, Bog G. Kim, Joselito M. Razal et al. Highly conducting carbon nanotube/polyethyleneimine composite fibers. *Advanced Materials* 17 (8) (2005): 1064–1067.

Nair, Sujith, Erik Hsiao, and Seong H. Kim. Fabrication of electrically-conducting nonwoven porous mats of polystyrene–polypyrrole core–shell nanofibers via electrospinning and vapor phase polymerization. *Journal of Materials Chemistry* 18 (42) (2008): 5155–5161.

Nair, Sujith, Sudarshan Natarajan, and Seong H. Kim. Fabrication of electrically conducting polypyrrole-poly (ethylene oxide) composite nanofibers. *Macromolecular Rapid Communications* 26 (20) (2005): 1599–1603.

Norris, Ian D., Manal M. Shaker, Frank K. Ko, and Alan G. MacDiarmid. Electrostatic fabrication of ultrafine conducting fibers: Polyaniline/polyethylene oxide blends. *Synthetic Metals* 114 (2) (2000): 109–114.

Okuzaki, Hidenori, and Masayoshi Ishihara. Spinning and characterization of conducting microfibers. *Macromolecular Rapid Communications* 24 (3) (2003): 261–264.

Park, Hongkwan, Sun Jong Lee, Seyul Kim, Hyun Woog Ryu, Seung Hwan Lee, Hyang Hee Choi, In Woo Cheong, and Jung-Hyun Kim. Conducting polymer nanofiber mats via combination of electrospinning and oxidative polymerization. *Polymer* 54 (16) (2013): 4155–4160.

Picciani, Paulo H. S., Eliton S. Medeiros, Zhongli Pan, William J. Orts, Luiz H. C. Mattoso, and Bluma G. Soares. Development of conducting polyaniline/poly (lactic acid) nanofibers by electrospinning. *Journal of Applied Polymer Science* 112 (2) (2009): 744–753.

Sen, S., Frederick John Davis, Geoffrey Robert Mitchell, and E. Robinson. Conducting nanofibres produced by electrospinning. *Journal of Physics: Conference Series* 183 (1) (2009): 012020.

Seyedin, Mohammad Ziabari, Joselito M. Razal, Peter C. Innis, and Gordon G. Wallace. "Strain-responsive polyurethane/PEDOT: PSS elastomeric composite Fibers with high electrical conductivity." *Advanced Functional Materials* 24, no. 20 (2014): 2957-2966.

Shao, Shijun, Shaobing Zhou, Long Li, Jinrong Li, Chao Luo, Jianxin Wang, Xiaohong Li, and Jie Weng. Osteoblast function on electrically conductive electrospun PLA/MWCNTs nanofibers. *Biomaterials* 32 (11) (2011): 2821–2833.

Sujith, Kalluri, Anish Madhavan Asha, P. Anjali, Niniya Sivakumar, K. R. V. Subramanian, Shantikumar V. Nair, and Avinash Balakrishnan. Fabrication of highly porous conducting PANI-C composite fiber mats via electrospinning. *Materials Letters* 67 (1) (2012): 376–378.

Tavakkol, Elham, Hossein Tavanai, Amir Abdolmaleki, and Mohammad Morshed. Production of conductive electrospun polypyrrole/poly (vinyl pyrrolidone) nanofibers. *Synthetic Metals* 231 (2017): 95–106.

Wang, Yu, S. Serrano, and J. J. Santiago-Aviles. Conductivity measurement of electrospun PAN-based carbon nanofiber. *Journal of Materials Science Letters* 21 (13) (2002): 1055–1057.

Wistrand, Ingemar, Rikard Lingström, and Lars Wågberg. Preparation of electrically conducting cellulose fibres utilizing polyelectrolyte multilayers of poly (3, 4-ethylenedioxythiophene): Poly (styrene sulphonate) and poly (allyl amine). *European Polymer Journal* 43 (10) (2007): 4075–4091.

Yanilmaz, Meltem, Fatma Kalaoglu, Hale Karakas, and A. Sezai Sarac. Preparation and characterization of electrospun polyurethane–polypyrrole nanofibers and films. *Journal of Applied Polymer Science* 125 (5) (2012): 4100–4108.

Yu, Qiao-Zhen, Min-Min Shi, Meng Deng, Mang Wang, and Hong-Zheng Chen. Morphology and conductivity of polyaniline sub-micron fibers prepared by electrospinning. *Materials Science and Engineering: B* 150 (1) (2008): 70–76.

Zhou, Yangxin, Marcus Freitag, James Hone, Cristian Staii, A. T. Johnson Jr, Nicholas J. Pinto, and A. G. MacDiarmid. Fabrication and electrical characterization of polyaniline-based nanofibers with diameter below 30 nm. *Applied Physics Letters* 83 (18) (2003): 3800–3802.

14 Conducting Polymer-Metal-Based Binary Composites for Supercapacitor Applications

A.B.V. Kiran Kumar
Nano world India, Buchireddy Plaem
Nellore, India

Amity Institute of Nanotechnology
Amity University
Noida, India

Swati Chaudhary
Amity Institute of Nanotechnology
Amity University
Noida, India

C.H.V.V. Ramana
Department of Electrical and Electronics Engineering Science
University of Johannesburg, Auckland Park Campus
Johannesburg , South Africa

CONTENTS

14.1   INTRODUCTION

14.1.1   PRINCIPLES OF SUPERCAPACITORS

The rapid industrialization and urbanization increased the burden on the fossil fuels. In addition, the consumption of these fossil fuels gives rise to major environmental problems, including their deteriorating impact on environment such as global warming, emission of greenhouse gases, lack of afforestation, increasing deforestation, and also the depletion of fossil fuels. These factors necessitate the utilization of renewable environment friendly energy sources. But these renewable sources also rely on the regional climatic conditions. Therefore, to overcome the ongoing environmental and practical energy demands, the best solution is to fabricate/develop the efficient energy-storage systems. In this regard, supercapacitors (SCs), rechargeable batteries, and fuel cells are designed as efficient energy-storage devices to meet the present and future energy requirements (Kim et al. 2018).

SCs, basically refer to those energy-storage systems that particularly bridge the gaps between the conventional batteries and the capacitors. However, there is a fundamental difference between these two devices (Figure 14.1); the conventional batteries have high energy density, whereas SCs have high power density. In particular, SCs are especially designed for applications with high energy and power density. These devices offer many advantages such as long cycle stability and wide operating thermal range (Zhang et al. 2018). Some basic principles of SCs are given below:

$$C \propto \frac{A}{d} \ or \ C = \varepsilon_0 \varepsilon_r \frac{A}{d} \qquad (14.1)$$

$$\text{Energy} \ (E) = \frac{1}{2} C V^2 \qquad (14.2)$$

$$\text{Power} (P) = \frac{V^2}{4R} \qquad (14.3)$$

where C = capacitance; ε_0 = permittivity of free space; ε_r = permeability; A = area; d = distance; V = voltage; R = resistance.

The capacitance of a SC increases with the area of an electrode and decreases with the distance between the electrodes. Energy and power can be calculated using Equations 14.2 and 14.3. The energy of a SC depends on the capacitance and operating voltage. According to Equation 14.3, power of a SC also depends on the internal resistance of a device, which is caused by the components of that device, such as electrode material, active material, and electrolyte. The resistance can be measured by equivalent series resistance.

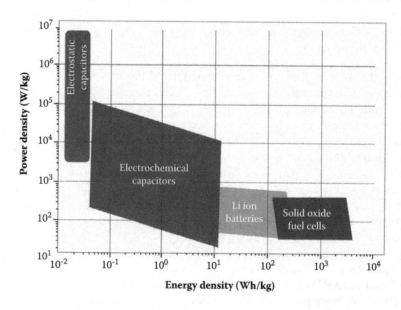

FIGURE 14.1   Ragone plot of energy-storage devices. (reproduced with permission from Raza et al. (2018.))

SCs, primarily consist of three components, which are electrode, electrolyte, and separator (Figure 14.2). The performance of a SC is mainly controlled by the storage property of the electrode material. On this basis, it can be classified into two types of charge storage mechanisms. (Zhang et al. 2015; Iro 2016). One such mechanism follows nonfaradic process in which charge separation and accumulation take place at the active material and electrolyte interface called electrical double-layer capacitance (EDLC) mechanism. (Candelaria et al. 2012). Another mechanism is pseudocapacitance type of mechanism, where oxidation and reduction reactions take place between electrode and electrolyte, resulting in the charge transportation. This type of mechanism is faradic in nature.

Since the active materials decide the performance of the SC, it is important to select an electrode material for better energy storage. In another words, altering the properties such as morphology and chemical properties of the active material plays a vital role in the performance of the SCs.

Therefore as shown in Figure 14.3, a SC can be classified into three types based on the combination of various active materials, namely EDLC, pseudo, and hybrid capacitors.

14.1.2 ELECTRICAL DOUBLE-LAYER CAPACITORS

EDLC SCs consist of solid, gel, or liquid electrolyte between two electrodes. EDLC SC stores energy at the electrode and electrolyte interface by forming a double

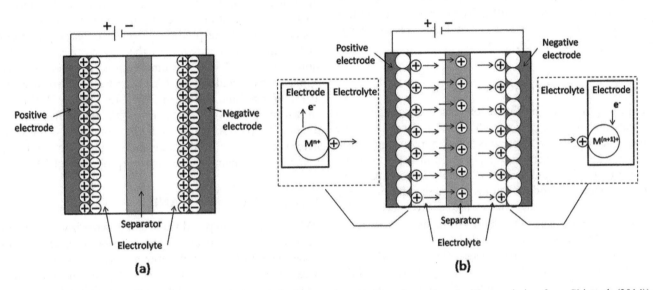

FIGURE 14.2 Charge-storage mechanisms of (a) EDLCs (b) pseudocapacitors. (reproduced with permission from Shi et al. (2014)).

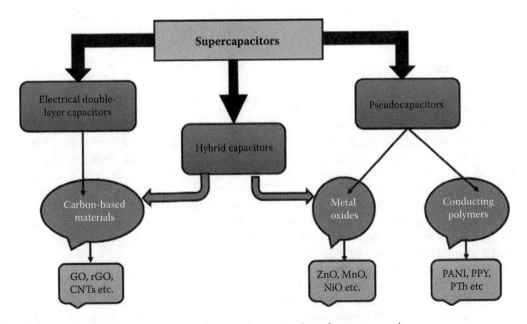

FIGURE 14.3 Classification of electrode materials based on the mechanisms for supercapacitors.

layer. The thickness of the EDLC is about 1 nm and includes an electrode space charging layer, an electrolyte diffusion layer, and a thin Helmholtz layer. The amount of energy storage is proportional to the surface area, pore size, and distribution. The process of charging and discharging mainly depends on the process of sorption and desorption of ions at the interface. This process is very fast in EDLC SC and is stable for over 100,000 charge and discharge cycles. This class of capacitors includes all the carbon-based materials, such as graphene oxide (GO) (Pawar, Shukla, and Saxena 2016), reduced graphene oxide (rGO) (Mondal, Rana, and Malik 2017), carbon nanotubes (CNTs) (Wang et al. 2016), carbon quantum dots (CQDs) (Dang et al. n.d.), and charcoal, as active materials in the functioning of SCs. These active materials undergo EDLC type of mechanism in which charge separation takes place to store the charge (Su and Schlögl 2010). These capacitors produce high power due to fast cycles, but lower specific energy because of the limited electrode area.

14.1.3 Pseudocapacitors

Because of the reversible fast redox reactions, pseudocapacitors are able to store energy both on the surface and in the bulk. Therefore, these capacitors store more energy, which leads to the higher capacitance per gram of active material than EDLC. CPs and transition metal oxides (MOs) are very common electrode materials for these capacitors. For example, pseudocapacitors makes use of MOs such as zinc oxide (Haldorai, Voit, and Shim 2014), manganese oxide (Rusi and Majid 2015), nickel oxide (Ji et al. 2015), titanium oxide (Li et al. 2011), cobalt oxide (Zhu et al. 2017), and conducting polymers (CPs) such as polypyrrole (PPy) (Adhikari et al. 2017), polyaniline (PANI) (Poyraz et al. 2014), and polythiophene (PTh) (Rawat, Ghosal, and Ahmad 2014) as the electrode materials. These materials perform faradic redox reaction to meet the energy requirements (Wei et al. 2011; Huang et al. 2015).

14.1.4 Hybrid Capacitors

Hybrid capacitors (as the name suggest) refer to those types of SCs which makes use of both, that is, carbon and MO/CP materials for device fabrication. They consist of both EDLC and redox reaction mechanisms, which enhances the power density as well as energy density. These capacitors produce high power density, but much lower energy density than batteries and fuel cells. Therefore, these storage devices are considered as the bridge for conventional capacitors and batteries. Furthermore, hybrid capacitors are divided into three types: first, asymmetric pseudo type, where both the electrodes are coated with different active materials such as anode coated with carbon material and cathode with either MO or CP. Second type is composites type, in which both the electrodes are coated with the same active material. This active material is a combination of both carbon and MO/

CPs. Third type is rechargeable battery type, in which lithium-based composites are used as the cathode electrode material and this type resembles the battery-type energy-storage devices (Etxeberria et al. 2010; Abdul Bashid et al. 2017; Mondal, Rana, and Malik 2017).

14.1.5 Brief Introduction of CP and Their Properties for Energy-storage Applications

CPs, basically referred to as the organic polymers that have conjugated double bonds. These organic polymers have high charge density, and cyclic, thermal, and mechanical stability (Megha et al. 2018). In addition, CPs give faster redox reactions, can be easily synthesized, are highly flexible, and are affordable. This can be attributed to the faster movement of electrons along the long polymeric chains of CPs. Upon oxidation, an electron is removed from the π system of backbone producing a cation. These properties of CPs make them suitable for the energy-storage applications. The CPs that are commonly used in energy-storage applications includes PPy, PANI, and PTh (Shown et al. 2015; Wang, Lin, and Shen 2016; Yang et al. 2017). As shown in Figure 14.5. these three CPs are similar in conducting properties. PANI and PPy contains Nitrogen as hetero atom and PTh contains sulfur. These hetero atoms influences the chemical and physical properties of CPs.

14.1.6 Polyaniline (PANI)

PANI, generally prepared via *in situ* polymerization or electrochemical polymerization, proves to be an efficient electrode material for SC applications (Chaudhary et al. 2018) because of its facile doping chemistry, excellent thermal, cyclic, mechanical, and environmental stability. In addition, PANI also has the stability to survive in various oxidation states (leucoemaraldine, emeraldine, and pernigraniline). Moreover, the morphology, synthetic method, nature and amount of the dopant and solvent added determine the efficiency of PANI in energy-storage applications (Weerakoon et al. 2012; Camurlu 2014).

14.1.7 Polypyrrole (PPy)

PPy being a conductive polymer, is highly conductive, shows fast charging–discharging mechanism, have cyclic stability, and offers high degree of flexibility. PPy can be synthesized by either chemical polymerization or by electrochemical polymerization. However, electrochemical synthesis is more preferable because of its simplicity and command over the various parameters of the materials such as density, orientation, and facile doping (Camurlu 2014).

14.1.8 Polythiophene (PTh)

PTh is the potent candidate for energy-storage applications, as it is thermally, electrically, and environmentally

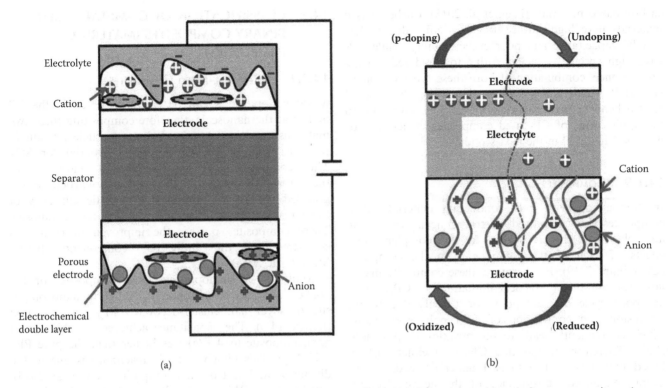

FIGURE 14.4 The general charge-storage mechanisms of (a) EDLC (b) conducting polymer electrode for supercapacitors. (reproduced with permission from Meng et al. (2017)).

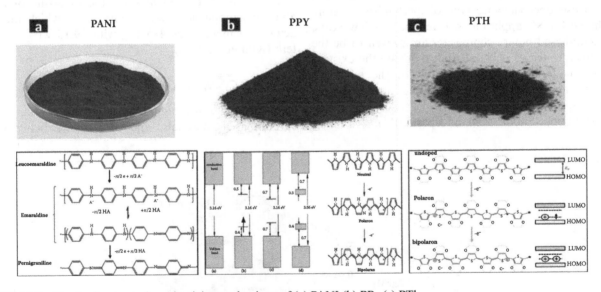

FIGURE 14.5 Powder images and conductivity mechanisms of (a) PANI (b) PPy (c) PTh.

stable. It also shows faster kinetics due to large surface area. Its electrochemical performance depends on its morphology, method of synthesis, and superficial properties. However, its lower electrochemical performance in comparison with PPy and PANI lead to its limited

practicability. This can be ascribed to the low inherited specific capacitance and low power density of PTh (Valderrama-García et al. 2016).

Over the years, CPs have gained worldwide attention due to their ability to counteract the drawbacks of carbon-based materials (Peng et al. 2015). Carbon-based materials have high surface area but suffer due to low specific capacitance and poor cyclic stability, while CPs have high specific capacitance due to rapid redox reactions. Hence, combination of both these types of materials (CPs and carbon-based) improves the overall electrochemical performance for SCs. This gives rise to the applications of CP-based composites, which will be discussed in detail in the next section.

14.1.9 COMPOSITES

In comparison with bulk conducting materials, their composites show improved electrochemical, mechanical, and electrical properties due to the complementary effects of each of the components in the composite (Chaudhary 2019). Furthermore, these composite materials also counteract practical limitations of CPs, which are poor cycle life and poor conductivity due to the expansion and contraction of CP's polymeric chains, which consequently leads to the breaking of polymeric chain. Therefore, compositing CPs with either carbon-based materials or metal-based materials leads to the overall enhanced performance of the active electrode material for SC applications (Shown et al. 2015).

Metal hydroxides, metal cobaltites, metal molybdates, metal-organic frameworks (MOFs), and sulphides come under the metal-based electrode materials. The binary composites of CPs and metal-based electrode materials undergo pseudocapacitance type of mechanism to store the charge for SC applications. Over the last few years, CPs-metal-based binary composites are proven to be the most supercapacitive materials that motivate the existing electrode materials to pave way toward the advanced energy-storage devices for meeting the future energy-storage requirements, which will be discussed in detail in the next section (Snook, Kao, and Best 2011).

14.2 CLASSIFICATION OF CP-METAL-BASED BINARY COMPOSITES (MATERIAL PROPERTIES)

14.2.1 CP-METAL-BASED COMPOSITES

Metal nanoparticles can be easily dispersed in the CP matrix at the nanoscale, therefore compositing these two materials (namely CP and metal nanoparticles) results in the formation of efficient electrode material for SCs. These composites have high stability and conductivity as the uniform dissipation of the metal nanoparticles in CP grid led toward the improved synergistic efficiency of these two components in the resulting nanocomposite. These composites possess the improved properties as compared to the individual components (Ćirić-Marjanovic 2013).

Many CP-M nanoparticle-based composite for SC applications have been reported. A hybrid nanocomposite of silver nanocluster-decorated PPy was prepared (Figure 14.6). The capacitance achieved for this hybrid nanocomposite (414 F/g) was higher than the pure PPy (273 F/g). This high value of capacitance is due to the distinctive architect of nanocomposite (Gan et al. 2015). Furthermore, PANI being highly electroactive formed the nanocomposites with the noble metals, including Pd, Pt, Au, and Ag, as these noble metals have novel optical and electrical properties. Also, the content of metal nanoparticles present in the composite plays a major role. In this regard, an *in situ* nanocomposite of PANI and Au nanoparticles was reported, for which the capacitance achieved was 485 F/g with 19.15 wt% Au content (Kim et al. 2012).

14.2.2 CP-METAL OXIDE-BASED COMPOSITES

MOs usually gives higher energy density values than carbon-based materials, and also shows better cyclic

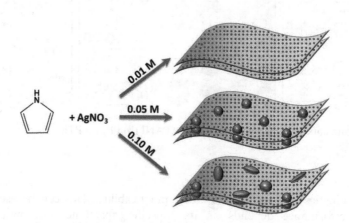

FIGURE 14.6 Schematic representation of development of various Ag/PPy nanocomposites. (reproduced with permission from Gan et al. (2015)).

stability than the conventional CPs. These electrochemical properties of MOs make them suitable for energy-storage applications. This superior performance of MOs can be ascribed to the electrochemical activities occurred in interior and exterior of MO-based active materials. Moreover, CPs in the composites prevent the clump formation of MO nanoparticles, which consequently allows the uniform dispersion of MO nanoparticles onto the CP matrix (Kumar et al. 2018; Ghosh et al. 2019). There are a number of research works reported with the high capacitance values that makes use of CP-MO-based binary composites as the electrode materials (Shi et al. 2014; Wang, Lin, and Shen 2016).

The research works performed so far suggested that among the CP-MO-based binary composites, one is a binary nanocomposite of PPy and cobalt oxide (CoO), with capacitance value 2223 F/g (Figure 14.7a). This value can be attributed to fast electron mobility of PPy; the porous structure of CoO increases the rate of ion diffusion (Zhou et al. 2013). Among the MOs, titanium dioxide (TiO_2) has large surface area and electronic conductivity, which makes it more suitable for energy applications. TiO_2 and PANI nanocomposite is prepared electrochemically by the electropolymerization of aniline onto an anode titania nanotube (Figure 14.7b). The reported capacitance value for this binary nanocomposite was 732 F/g at 1 A/g (Li et al. 2011).

14.2.3 CP-Metal Hydroxides

Apart from CP-MO-based composites, binary composites of CPs and M-OH can also be used as efficient active materials in SC applications. Furthermore, among the metal hydroxides, nickel hydroxide $Ni(OH)_2$ is easy to prepare, has fast redox reactions, is economical, and has high conductivity, these properties make $Ni(OH)_2$ a potent candidate for SCs. Therefore, compositing $Ni(OH)_2$ with CPs results in high electrochemical performance for SCs (Wang, Lin, and Shen 2016; Shendkar et al. 2018).

The reversibility and stability for a binary nanocomposite of PANI and $Ni(OH)_2$ was found to be better than when they were used individually as a single-component (PANI or $Ni(OH)_2$ electrode. The capacitance achieved for the composite was 113.8 F/g, while for PANI it was 0.59 F/g and for $Ni(OH)_2$ it was 39.96 F/g. The achieved electrochemical performance showed the high efficiency of composites over the individual components for SCs (Shendkar et al. 2018).

14.2.4 CP-Metal Cobaltites

Over the last few years, the spinel-structured transition MOs such as zinc cobaltite ($ZnCo_2O_4$) and copper cobaltite ($CuCo_2O_4$) have also emerged as the suitable active material candidates in energy-storage applications. In addition, high electrochemical activity of these metal cobaltites can be ascribed to the two MOs present in one molecule of cobaltite. Furthermore, they lead to better physical and chemical properties, including high conductivity, stability, and redox behavior of the cobaltites in comparison to the single MOs. Also, these MOs are environment friendly and economic. However, they have narrow operating range and low inherited specific capacitance. These problems can be overcome by compositing the cobaltites with CPs.

A $CuCo_2O_4$ composite with PANI has been reported with the capacitance of 403 F/g (Figure 14.8). This enhanced performance of $CuCo_2O_4$ is achieved after the incorporation of PANI. This observation clearly interprets the improved performance of metal cobaltites-CP-based composites (Omar et al. 2017b). PANI nanocomposite with Zn-based cobaltite has also been reported where the

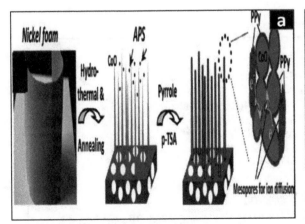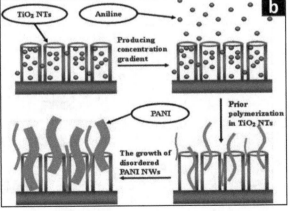

FIGURE 14.7 Schematic representation of preparation of (a) CoO/PPy nanocomposite and (b) TiO_2/PANI nanocomposite. ((a) reproduced with permission from Zhou et al. (2013), published by American Chemical Society; (b) image reproduced with permission from Li et al. (2011)).

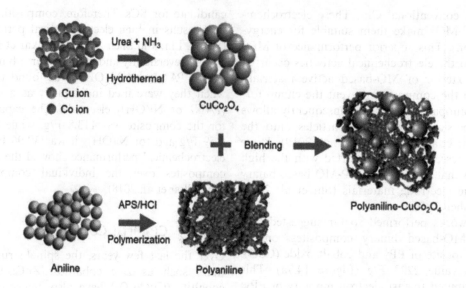

FIGURE 14.8 Schematic representation of development of $CuCo_2O_4$/PANI nanocomposite. (reproduced with permission from Omar et al. (2017b)).

$ZnCo_2O_4$ was intercalated in the PANI matrix by physical blending method. Owing to this interaction between metal and CP, the capacitance achieved was 398 F/g (Omar et al. 2017a).

14.2.5 CP-METAL MOLYBDATES

Metal molybdates such as $MnMoO_4$ and $CoMoO_4$, because of their superior electrochemical performance in comparison to single-transition MOs, are used as active materials in SC applications. This can be ascribed to the different oxidation states and high electronic conductance of metal molybdates. Furthermore, these metal molybdates also improve the swelling and shrinkage property of the CPs by the introduction of metal

molybdate ions into the matrix of CP by the process of intercalation. Therefore, a composite formation between CP and metal molybdates results in better electrochemical performance than the individual components (Segal et al. 2012; Zhou et al. 2016).

A binary nanocomposite of PPy and $MnMoO_4$ was reported where $MnMoO_4$ was used as the template to modify the morphology of the nanocomposite and the nanocomposite also form a huge conducting network for facilitating the electron transportation (Figure 14.9). The capacitance achieved for nanocomposite was 462.9 F/g (Zhou et al. 2016). Furthermore, $ZnCoMoO_4$- and PANI-based binary nanocomposite synthesized via *in situ* polymerization with capacitance 246 F/g has also been reported, whereas for individual component, PANI,

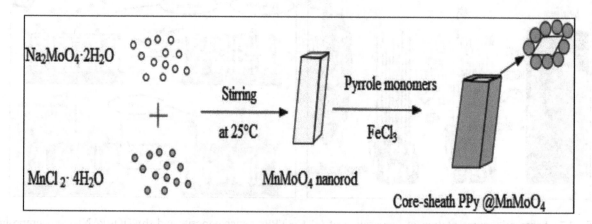

FIGURE 14.9 Schematic representation of synthesis of $MnMoO_4$/PPy nanocomposite. (reproduced with permission from Zhou et al. (2016)).

the capacitance achieved was only 160 F/g. This clearly indicates the synergistic effect of PANI and ZnCoMoO$_4$ in the nanocomposite (Veerasubramani et al. 2016).

14.2.6   CP-MOFs

Over the recent few years, MOFs have attracted a considerable recognition for an effective performance in the arena of SCs due to their various advantages such as functionality, diversity, tailorability, and versatility. Moreover, they have large surface area, controllable size and morphology, high conductivity, and long cyclic stability. Furthermore, when utilized as supporters with CPs in the composite, MOFs can overcome the problems of poor cyclic stability of CPs. The MOF- and CP-based nanocomposites give rise to the physicochemical properties that are not present in the individual components (Segal et al. 2012).

A PPy- and Zr-based MOF was reported where a facile one pot strategy was used to obtain uniform PPy/Zr-MOF hybrid coated on fibrous substrates (Figure 14.10). The capacitance of the hybrid obtained was 206 F/g (Qi et al. 2018).

14.2.7   CP-Sulfides

Metal sulfides such as molybdenum disulfide (MoS$_2$) and nickel disulfide (Ni$_3$S$_2$) have gained a lot of attraction for SC applications. This can be ascribed to large surface area, fast intrinsic ionic conductivity, and high theoretical capacity of these metal sulfides in comparison to MOs. However, these MS$_2$ suffer from poor electrical conductivity and specific capacitance. These discrepancies can be overcome by combining them with the CPs to form the composites. Similarly, the swelling/shrinkage and poor cyclic stability of CPs can be countered by the metal sulfides. During the composite formation, metal sulfide provides the large surface area and this area is

considered as matrix for stopping compaction of CPs during charging and discharging (Xia et al. 2012; Zhang et al. 2016).

A nanocomposite of mechanically exfoliated MoS$_2$ and PANI was synthesized where the few layers of MoS$_2$ were obtained via mechanical separation of layers where surface was decorated with PANI through polymerization method (Figure 14.11). The capacitance achieved for this nanocomposite was 510.12 F/g (Cho et al. 2017). Similarly, a binary nanocomposite of PPy and Ni$_3$S$_2$ in which the honey comb-shaped Ni$_3$S$_2$ was *in situ* grown on nickel foam while the PPy was *in situ*-electropolymerized and the capacitance achieved was 1.13 F/cm^2 (Zhang et al. 2016).

14.3   PREPARATION METHODS

With the advancement in experimental accessibilities such as instrumentations and scientific technologies, a number of advanced synthetic techniques are used nowadays to synthesize the CP-M-based binary composites, as these composites pave the way for future energy requirements.

14.3.1   In Situ Polymerization

This method has been considered as the most prime method for the synthesis of CP-based composites. This can be attributed to the low cost and ease of synthesis in this method. In addition, bulk amount of polymers can be prepared by this synthetic method. However, the excess of oxidizing agent and high ionic concentration of medium leads to the formation of uncontrolled material.

A number of research works in which CP-M-based composites were prepared by the *in situ* polymerization have been reported. A binary nanocomposite of PPy and MoO$_3$ was synthesized via the process of *in situ* polymerization by coating PPy on MoO$_3$. The capacitance

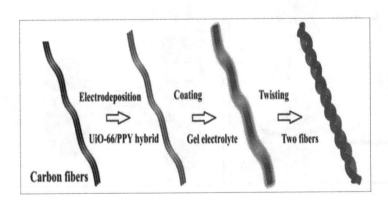

FIGURE 14.10   Illustration of fabrication of Zr-MOF (UiO-66)/PPy flexible electrode. (reproduced with permission from Qi et al. (2018)).

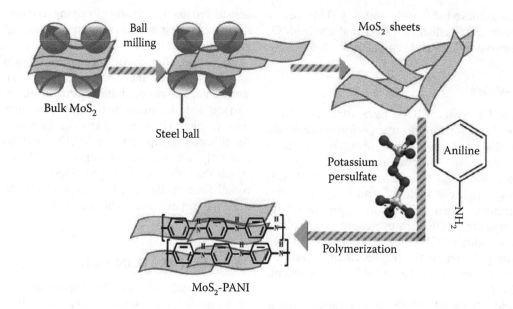

FIGURE 14.11 Schematic illustration of development of MoS₂/PANI nanocomposite. (reproduced with permission from Cho et al. (2017)).

achieved for this nanocomposite was 129 F/g (Xia et al. 2012). *In situ* encapsulation method has been also evoked as one of the major synthetic strategies to form the CP-MO-based composites for energy-storage applications. As shown in Figure 14.12a, PPy encapsulated flower like NiO binary nanocomposite was also prepared through this method and the capacitance obtained was 595 F/g at 1 A/g current density (Ji et al. 2015). In addition, solvothermal in combination with this method also gained attention in the synthetic field. In this regard, a binary nanocomposite of PPy and CuS was reported where solvothermal approach (temperature 150°C) was

used to prepare CuS and then CuS was used as the polymerization substrate for the polymerization of pyrrole monomer in aqueous medium (Figure 14.12b). The reported capacitance was 427 F/g (Peng et al. 2014).

14.3.2 ELECTROPOLYMERIZATION

The electropolymerization method is a simple single-step method giving rise to the formation of CP films. As it can be carried out using either one of the three techniques, namely constant current (galvanostatic), constant voltage (potentiostatic), and variable current and voltage

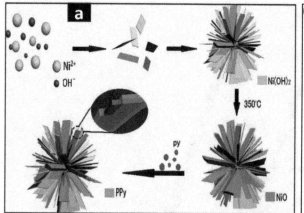
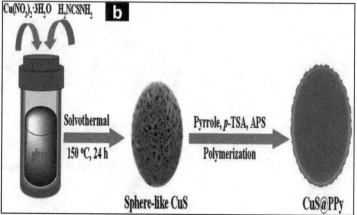

FIGURE 14.12 Schematic illustration of preparation of (a) NiO/PPy and (b) CuS/PPy nanocomposites synthesized via *in situ* polymerization mechanism. (a) reproduced with permission from Ji et al. (2015); (b) from Peng et al. (2014).

(potentiodynamic); it involves versatility, which allows efficient production. Also the thickness of the films produced by this method can be controlled. However, it does not give much yield as compared to the *in situ* oxidative polymerization method.

Many research papers of CP-M-based composites using electropolymerization synthetic method have been reported. A composite electrode of PANI and TiO_2 was synthesized, in which PANI was electropolymerized onto the titania anode. The capacitance value achieved was quite high, that is, 732 F/g at 1 A/g, and such a considerable value of capacitance may be attributed to the novelty of the nanocomposite, which consists of disordered PANI nanoarrays encapsulated inside the titania nanotube (Li et al. 2011). A core shell nanocomposite of α-Fe_2O_3 and PANI was synthesized via electrodeposition of PANI on α-Fe_2O_3 nanowire-coated carbon cloth (Figure 14.13a). The obtained composite displayed a high volumetric capacitance of 2.02 mF/cm^3 (Tong et al. 2015). Hydrothermal-assisted electrodeposition technique has also been identified as one of the efficient techniques to synthesize the CP-M-based composites. A binary nanocomposite of PPy and $ZnCoO_4$ was reported where the $ZnCoO_4$ nanostructures were prepared hydrothermally and then pyrrole monomers were electrodeposited onto the $ZnCoO_4$ nanostructures (Figure 14.13b). The capacitance achieved was very high 1559 F/g at 2 mA/cm^2. The reason behind getting this much capacitance is due to the high stability of nanocomposite, as $ZnCoO_4$ nanostructures grown on Ni foam strengthen the bonding between the active material and the substrate. Also the synthesized nanocomposite has the high surface area for the efficient electron transportation (Chen et al. 2015). A binary

nanocomposite thin film of PANI and Ag nanoparticles was prepared by the process of depositing on Ag nanoparticles over the stainless steel substrate. Owing to the synergistic effect of two components, the capacitance obtained was larger for Ag/PANI (512 F/g) as compared to pure PANI (285 F/g) (Kim et al. 2012).

14.3.3 BLENDING METHOD

Blending method is a kind of conventional method used for the synthesis of composites. The prime advantage of this methodology is that the ratios of individual components can be controlled in the resulting composite and also the required morphology can be selected (Chaudhary 2019). Sonication-assisted blending process also improves the dispersion properties of CP-based composites (Camargo, Satyanarayana, and Wypych 2009).

Earlier, blending method was considered as the standard method for preparing nanocomposites. PANI-implanted copper cobaltite ($CuCo_2O_4$) nanocomposite was synthesized via the physical blending process. The PANI matrix was intercalated with $CuCo_2O_4$ and capacitance reported for the nanocomposite was 403 F/g. This value of capacitance can be ascribed to the improved performance of $CuCo_2O_4$ after the manifestation of PANI (Omar et al. 2017b). Hydrothermal treatment on combining with blending has also been used for synthesizing CP-based composites. A nanocomposite of PANI (synthesized by oxidative polymerization) and $ZnCo_2O_4$ (synthesized hydrothermally) was prepared by the blending process (Figure 14.14). The optimum capacitance (398 F/g) was achieved (Omar et al. 2017a).

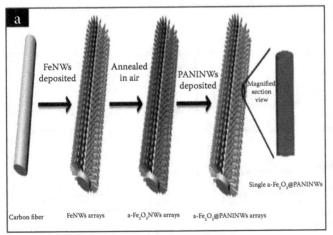
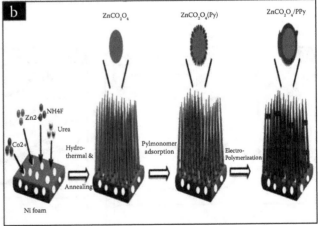

FIGURE 14.13 Schematic illustration of fabrication of (a) Fe_2O_3/PANI, (b) $ZnCo_2O_4$/PPy nanocomposite electrodes prepared via electropolymerization method. ((a) reproduced with permission from Tong et al. (2015); (b) reproduced with permission from Chen et al. (2015)).

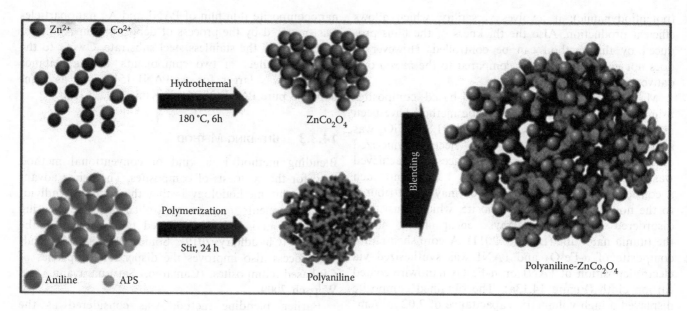

FIGURE 14.14 Schematic representation of preparation of ZnCo$_2$O$_4$/PANI nanocomposite synthesized via blending method. (reproduced with permission from Omar et al. (2017a)).

14.4 CHARACTERIZATION METHODS

14.4.1 X-RAY DIFFRACTION (XRD)

XRD is the preliminary analysis to study the crystalline properties of the materials. The XRD spectra of PANI mainly consists of three significant peaks at $2\theta = 14.7°$, $20.2°$, and $25.42°$ with the corresponding d-values 6.0, 4.3, and 3.5 Å, while the peaks at $2\theta = 15.21°$, $25.42°$, and $2\theta = 20.64°$ correspond to the perpendicular and parallel arrangement of polymer chains, respectively. These parallel and perpendicular periodicities basically refer to the arrangement of monomeric (aniline) units in the long polymeric chain (Figure 14.15a). In addition,

the broad appearance of the peaks clearly indicates the semicrystalline nature of PANI. Similarly, a significant broad peak obtained at $2\theta = 23.6$ with d-spacing = 0.38 nm initially for PPy nanowires. But on annealing the PPy samples, significant change has been observed in peak position as well as in peak intensity (Figure 14.15b). On increasing the annealing temperature from 400, 600 to 800°C, the corresponding peak position ($2\theta = 24.1$, 24.6, and 25.5°) shifts to the higher diffraction angle side with the increasing intensity and corresponding d-spacings (0.37, 0.36, and 0.35 nm). The observed values of d-spacing close to 0.34 nm of graphite clearly indicate that there are chances that on

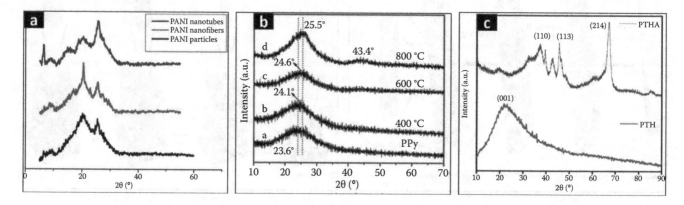

FIGURE 14.15 The XRD patterns of (a) PANI nanotubes, nanofibers, and nanoparticles (b) PPy nanowires at different annealing temperatures after 2 h and (c) PTh and PTh/Al$_2$O$_3$ nanocomposite. (reproduced with permission from Khalid et al. (2012b); (b) reproduced with permission from Ma et al. (2009); (c) reproduced with permission from Vijeth et al. (2018)).

annealing at high temperature, complete transformation of PPy sample to any kind of carbon sample takes place (Ma et al. 2009). Like PANI and PPy, PTh also shows the broad XRD pattern owing to the semi-crystalline nature of PTh (Figure 14.15 c). The main peak was obtained at $2\theta = 21.92°$. This peak can be attributed to the π–π stacking of thiophene monomeric units in the amorphous structure of PTh (Vijeth et al. 2018).

Compositing CPs with any metal or its compounds results in the significant change in the appearance as well as in the intensity of peaks in the XRD spectra. Generally, the sharp and well-defined peaks have been obtained for MO-based composites owing to their crystalline nature. For instance, the PPy/MoO₃ composite shows a sharp XRD spectra, clearly indicating the proper incorporation of the two components in the resulting binary nanocomposite. The obtained pattern also resembles to the MoO₃ XRD spectra (Figure 14.16a). Also it indicates that the PPy does not restrict the formation of crystalline structure of MoO₃ (Xia et al. 2012). The amount of the metal content present in the composite also plays a significant role in the appearance of XRD spectra. Such as the XRD spectra of gold nanoparticles (AuNP) and PANI binary nanocomposites with different contents of AuNP, depict a gradual increase in intensity with the increasing Au content. The characteristics' peaks of PANI ($2\theta = 25°$) and AuNP ($2\theta = 38.2°$, $44.6°$, $64.9°$, and $77.8°$) have been obtained for all the nanocomposites (Ran et al. 2018). Intercalation of components while preparing the composites also had an impact on the XRD pattern. In addition, PANI/CuCo₂O₄ binary nanocomposite displayed spectra in which the peaks obtained appeared weaker and broader (Figure 14.16b). For CuCo₂O₄, all the characteristics' peaks were obtained at $2\theta = 19.07°$, $31.36°$, $36.96°$, $38.96°$, $45.07°$, $56.03°$, $59.59°$, and $65.70°$, suggesting high purity of CuCo₂O₄ and similarly for PANI, the semi-crystalline nature was observed and the peaks at $2\theta = 14.67°$, $20.72°$, and $25.59°$ attributed to parallel and perpendicular periodicities of monomeric units in the polymer chain. The bad appearance obtained for PANI/CuCo₂O₄ nanocomposite can be ascribed to the intercalation of CuCo₂O₄ in the PANI matrix to form composite. Depending on the synthetic method, sometimes the presence of one component dominates the other one in the resulting nanocomposite. This can also be clearly depicted from XRD analysis (Omar et al. 2017b). For instance, in the XRD of PPy and MnMoO₄ composite, the characteristics obtained belonged to PPy only indicating the proper coating of PPy on MnMoO₄ and also indicating the low content of MnMoO₄ present in the binary nanocomposite (Zhou et al. 2016).

14.4.2 Infrared (IR) Spectroscopy

Infrared (IR) spectroscopy has been performed to identify the characteristic functional group vibrations of the materials. Therefore, it is done to identify the chemical structure of the individual components and their nanocomposites. The characteristic IR patterns of PANI consist of stretching of quinoid- and benzenoid-type rings (Figure 14.17a). In this regard, for PANI, stretching of the characteristic quinoid and benzenoid rings has been observed at 1585 and 1490 cm^{-1}, respectively. In addition, the C–N stretching peaks obtained at 1225, 1160, and 1035 cm^{-1} corresponds to N–H, N=C=N (Khalid et al. 2012a). Similarly, the corresponding C–C stretching vibrations of pyrrole ring have been observed at 1552 and 1473 cm^{-1} (Figure 14.17b). Furthermore, the corresponding peaks obtained at 1310 and 1090 cm^{-1} belongs

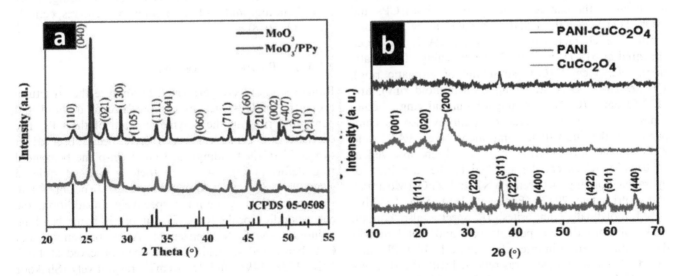

FIGURE 14.16 The XRD spectra of (a) MoO₃/PPy and (b) CuCo₂O₄/PANI. ((a) reproduced with permission from Login (2015); (b) reproduced with permission from Omar et al. (2017b)).

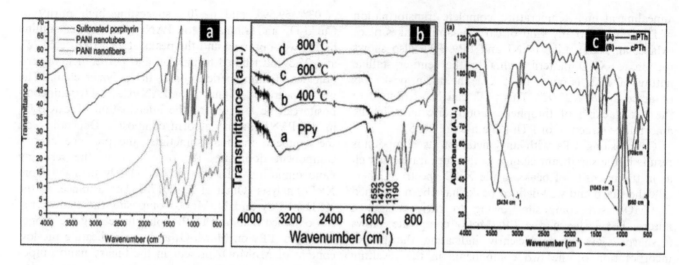

FIGURE 14.17 The IR spectra of (a) PANI nanotubes and nanofibers, (b) PPy nanowires at different annealing temperatures after 2 h, and (c) PTh prepared by microwave technique (mPTh) and conventional emulsion polymerization technique (cPTh). ((a) reproduced with permission from Khalid et al. (2012b); (b) reproduced with permission from Ma et al. (2009); (c) reproduced with permission from Rawat, Ghosal, and Ahmad (2014)).

to =C–H in-plane deformation (Ma et al. 2009). Like PPy and PANI, the characteristic peaks for PTh are observed at 2957, 1622, 724, and 672 cm^{-1}, which corresponds to C–H, C=C, C–S, and C–S–C stretching, respectively (Rawat, Ghosal, and Ahmad 2014) (Figure 14.17c).

The IR pattern of the CP-M composites primarily reveals the same trend as that of the pristine CP. In relation to this, a binary nanocomposite of Ag and PANI was reported (Figure 14.18a). Here the peaks obtained for IR include N–H and C–H stretchings at 3340 and 3000 cm^{-1}, respectively. The doping states of PANI were represented by the N–H in-plane deformation and C–H out-of-plane vibration, respectively (Kim et al. 2012). However, the IR pattern of the CP-MO-based composites includes the characteristics peaks of both CPs and MOs. That is, for the PANI and ZnO binary nanocomposite, the characteristics' peaks for PANI has been appeared at 515.71 cm^{-1} (C–N–C stretching of aromatic ring), 592.85 cm^{-1} (C–C stretching of aromatic ring), 1503.09 cm^{-1} (C–N stretching of benzenoid ring), and 1572.52 cm^{-1} (C=N stretching of quinoid ring). Sometimes it can depict the reaction pathway also as compared to the individual components (i.e., PANI and ZnO), an additional broad peak has been observed at 3470 cm^{-1}, which can be ascribed to the formation of hydrogen bonding between PANI and ZnO (Mostafaei and Zolriasatein 2012) (Figure 14.18b). Like CP-MO, CP-M molybdate composites' IR spectra also contain peaks of both the components (CP and metal molybdate) with variation in intensity (Figure 14.18c). PPy and MnMoO$_4$ nanocomposite was reported with the IR spectra, which consist of characteristic MnMoO$_4$ peaks at 750 cm^{-1} (Mo–O stretching) and 876 cm^{-1} (Mo–O–Mo

stretching). Similarly, the characteristic peaks of PPy were also obtained in the nanocomposite. In addition, a peak at 907 cm^{-1} that belongs to =C–H out-of-plane vibration showed decreased intensity, which can be ascribed to the mutual interaction between the PPy and MnMoO$_4$ (Zhou et al. 2016). The MOF-based composites also depict the individual components in the IR pattern (Figure 14.18d). Furthermore, a binary nanocomposite of PPy- and Zr-based MOF was synthesized, and the IR spectra of the composite consists of the peaks at 1400, 1500, and 1575 cm^{-1} (C–O, C=C, and C=O of functional groups binded on Zr-based MOF, respectively). The peaks 920, 1170, 1620, and 2920 cm^{-1} were related to C–H in-plane bending, C=C, N–H bending, and C–H stretching of PPy, respectively (Qi et al. 2018).

14.4.3 Raman Spectroscopy

Raman studies were performed to analyze the structural properties of CPs such as the presence of quinoid and benzenoid rings in the polymeric chain of the CPs. The characteristic peaks for PANI have been observed at 416 cm^{-1} (C–N bending), 573 cm^{-1} (in-plane benzenoid ring deformation), 748 cm^{-1} (deformation of quinoid group), 778 cm^{-1} (quinoid group bending), and 840 cm^{-1} (C–H wagging), respectively. In addition, the corresponding peaks for C–H quinoid in-plane bending, quinoid C–N, C–N$^+$ benzenoid, Ar–N benzenoid, and C–C benzenoid stretchings have been observed at 1162, 1220, 1336, 1403, and 1560 cm^{-1}, respectively (Shakoor and Rizvi 2010). Similarly, for PPy, the characteristics' peaks have been observed at 950, 992, 1053, 1326, and

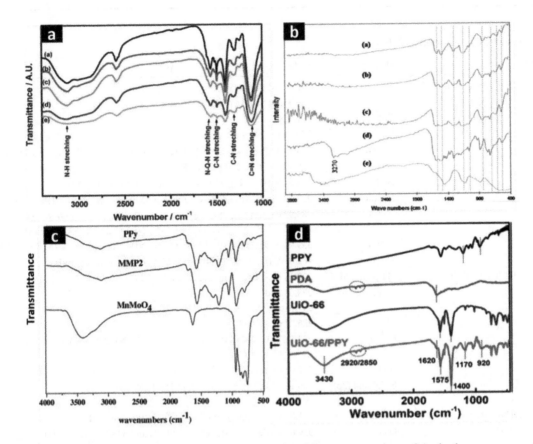

FIGURE 14.18 The IR spectra of (a) Ag/PANI nanocomposites with different proportions of Ag in the nanocomposites, (b) ZnO/PANI with different weight percentages of ZnO in the nanocomposites, (c) PPy, MnMoO$_4$, MnMoO$_4$/PPy (MMP2) nanocomposite, and (d) UiO-66/PPy nanocomposite, PPy, PDA, and UiO-66. ((a) reproduced with permission from Kim et al. (2012); (b) reproduced with permission from Mostafaei and Zolriasatein (2012); (c) reproduced with permission from Zhou et al. (2016); (d) reproduced with permission from Qi et al. (2018).

1529 cm^{-1}, which belongs to ring deformation due to dication and radical cations, C–H in-plane bending, asymmetrical C–N stretching, and C–C backbone stretching, respectively (Minh and Quoc 2013). Likewise, for PTh characteristic peaks have been obtained at 1500 cm^{-1} corresponding to aromatic C=C stretching (Wong et al. 2017) (Figure 14.19a). Raman spectra of the CP-M composites reveal the same trend as of the

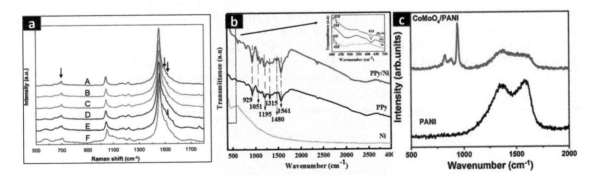

FIGURE 14.19 Raman spectra of (a) PTh films deposited under various conditions. (b) PPy/Ni nanocomposite, PPy, and Ni, and (c) CoMoO$_4$/PANI nanocomposite and PANI. ((a) reproduced with permission from Borrelli, Lee, and Gleason (2014); (b) reproduced with permission from Muhamad et al. (2017); (c) reproduced with permission from Veerasubramani et al. (2016).

TABLE 14.1

XPS analysis of various nanocomposites

S. no.	Composite	Element	State
1	NiCo$_2$O$_4$@PPy (Kong et al. 2015)	Ni^{2+} (2p)	855.1 and 872.8 eV
		Ni^{3+}	856.3 and 874.3 eV
		Co^{2+} (2p)	781.2 and 796.1 eV
		Co^{3+}	779.3 and 794.9 eV
		C (1s)	284.1 eV (β-carbon), 285.1 eV (α-carbon), 286.2 eV (C=N), and 287.8 eV (–C=N+)
		O (1s)	529.1 eV (O^{2-}), 530.8 eV (OH$^-$)
		N (1s)	398.3 eV (–C=N–), 399.3 eV (–NH–), and 401 eV (–NH$^+$–)
2	MoS$_2$-PANI (Cho et al. 2017)	Mo (3d)	229.17 eV (C-MoS2-PANI) and 229.55 eV (M-MoS$_2$-PANI)
		C (1s)	283.92 eV (PANI), 284.33eV (C-MoS$_2$-PANI), and 284.66 eV (M-MoS$_2$-PANI)
			283.92 eV (PANI), 284.33 eV (C-MoS$_2$-PANI), and 284.66 eV (M-MoS$_2$-PANI)
		N (1s)	399.41 eV (PANI), 399.12 eV (C-MoS$_2$-PANI), 399.37 eV (M-MoS$_2$-PANI)
		S (2p)	162.0 eV (C-MoS$_2$-PANI) and 162.44 eV (M-MoS$_2$-PANI)
3	MoS$_2$/PANI (Nam et al. 2016)	2H-phase, Mo^{4+}	229.38 eV (3d5/2) and 232.72 eV (3d3/2).
		1T phase Mo	228.63 and 231.94 eV (MoS2)
		Mo^{6+}	226.22 eV (3d5/2)
		C (1s)	283.5 eV (C–C/C–H), 284.75 eV (C–N/C=N), 285.82 eV (C–O), and 288 eV (C=O) eV
		N (1s)	394.63 eV (benzenoid-amine –NH–), 397.92 eV (the quinoid-imine –NH$^+$)
		S (2s) and (2p)	226 eV (2S), 161.8 (2p3/2), and 163.1 eV (2p1/2), (S^{2-})
4	α-Fe$_2$O$_3$@PANI core-shell nanowire (Tong et al. 2015)	Fe (2p)	711.0 eV (Fe 2p3/2) and 724.6 eV (Fe 2p1/2) for α-Fe$_2$O$_3$
		O (1s)	529.8 eV (Fe–O), 531.3 eV (Fe–O–H) and 533.1 eV (H–O–H)
		N (1s)	400.04 eV (amine, –NH–) in benzenoid amine. 399.3 eV (imine, =N–) and 401.08 eV (N+)
5	MnO$_2$ nanorods GO/polyaniline ternary composites (Zhang et al. 2014)	Mn (2p)	643.0 eV (2p3/2) and 654.5 eV (2p1/2)
		N (1s)	398.6 eV (–N=), 399.7 eV(–NH–), and 400.1 eV (–N$^+$–)
6	Ag/polyaniline (Kim et al. 2012)	Ag (3d)	366.91 eV (3d5/2) and 372.77 eV (3d3/2)
		N (1s)	399.34 eV (quinoid imine) and 400.53 eV (benzenoid amine)
7	Ag@PPy nanocomposite (Gan et al. 2015)	Ag (3d)	368.0 eV (3d5/2) and 374.1 eV (3d3/2) with 6.1 eV spin energy separation
		N (1s)	397.9 eV (imine nitrogen –N=); 399.1 eV (neutral amine nitrogen –N–H); 400.3 eV (–N+)
8	AuNP/PANI (Ran et al. 2018)	Au (4f)	84.0 eV (4f7/2) and 87.6 eV (4f5/2)
		N (1s)	400.5 eV (–NH–) and 402.0 eV (protonated amine)
9	Pure PPy chlorine doped (Cao et al. 2015)	N (1s)	397.6 eV (imine nitrogen –N=); 399.5 eV (neutral amine nitrogen –N–H); 400.8 eV (–N+)
		C (1s)	284.5 eV (C–C), 285.4 eV (C–N), and 288.4 eV (C=O)
10	Pure PANI (Golczak et al. 2008; Smolin, Soroush, and Lau 2017)	C (1s)	284.4 eV (C–C/C–H); 285.0 eV (C–N/C=N); 285.8 eV (C–N+ /C=N+/C–Cl); 286.7 eV (C=O/C–O)
	PANI-XPS_golczak2008	N (1s)	398.5 eV (–N=), 399.6 eV (–NH–); 401.1 (N+)
11	Polythiophene (pPTh) film (Pohoata et al. 2015)	C1s	284.6 eV (C–C, C=C, C–H); 286.00 eV (C–S); 287 eV (C–O), which correspond to the neutral sulfur units and positively charged sulfur atoms, respectively
		S2p	164.00 eV (2p3/2); 165.18 eV(2p1/2)

individual components, only the differences lie in the intensities of the peaks. The binary nanocomposite of PPy and Ni was prepared and the peaks for PPy were observed at 1374 (ring stretching of PPy) and 1590 cm^{-1} (C=C stretching). However, these peaks shifted to 1382 and 1575 cm^{-1}, respectively, with the increased intensity indicating some sort of interaction between PPy and Ni during the polymerization process (Muhamad et al. 2017) (Figure 14.19b). For CP-M molybdate-based composites also, Raman spectroscopy plays a major role. The Raman pattern of PANI and CoMoO$_4$ showed the characteristic peaks of PANI while the peaks in the lower wavenumber range that is below 1000 cm^{-1} indicates the presence of CoMoO$_4$ in the composite (Veera-subramani et al. 2016) (Figure 14.19c). Raman spectroscopy was also performed to analyze the bonding structure of the components with the varying content in the nanocomposite. Nanocomposites of PTh and TiO$_2$ with different TiO$_2$ were prepared. The Raman spectra of pure PTh contain the peaks at 1500 and 1450 cm^{-1} corresponding to C=C stretching of pure PTh. While in the composite, the peak at 630 cm^{-1} indicate the presence of PTh in the composite. Also, the peak at 150 cm^{-1} belongs to TiO$_2$ (Thakur et al. 2017).

14.4.4 X-Ray Photoelectron Spectroscopy (XPS)

X-Ray photoelectron spectroscopy (XPS) is one of the important methods for the analysis of nanomaterials. It is widely used for the surface chemical analysis of up to 1–10 nm depth. This analysis provides a valuable information related to the chemical state of the material from surface quantitatively. When the material interacts with the X-ray photon energy, one electron is ejected from the surface atom or molecule. The kinetic energy of the ejected electron depends on the electron binding energy and incident photon energy. By analyzing the emitted electron kinetic energy, chemical state of the atoms can be determined. XPS analysis is one of the versatile methods used to analyze the different components in the nanocomposite. Table 14.1 summarized the XPS analysis of various nanocomposites.

14.4.5 Morphology: SEM and TEM

Studying the morphology of the nanomaterials is important for nanomaterials because their properties are dependent on their size, shape, and structure. In general, SEM and TEM methods are used to analyze the morphology of nanomaterials. During the growth of the nanomaterials from seeds, morphology can be controlled on

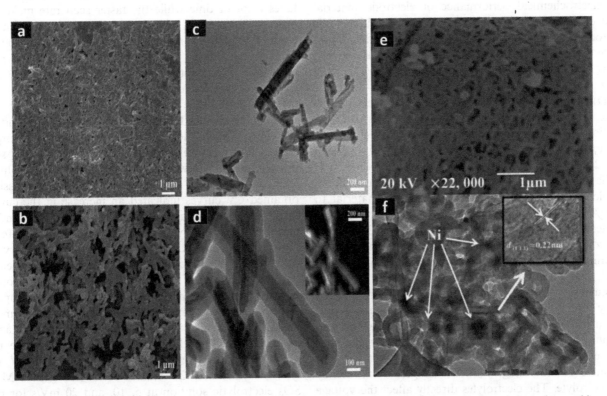

FIGURE 14.20 ((a) and (b) FESEM images of the pristine MoO$_3$ nanobelts and MoO$_3$/PPy nanocomposite are shown. (c) and (d) TEM images of pristine MoO$_3$ nanobelts and the MoO$_3$/PPy nanocomposite are shown. The inset is the dark-field TEM image. (e) SEM image of PPy/Ni nanocomposite, (f) TEM image of PPy/Ni nanocomposite (inset: HRTEM image of the PPy/Ni nanocomposite). ((a–d) are reproduced with permission from Login (2015); (e) and (f) are reproduced with permission from Muhamad et al. (2017).

the basis of the reaction conditions and other components such as capping agents, templates, and solvent systems. Similarly, the properties of nanocomposites are determined by the individual component morphology and also by the type of interactions. In this section, few selected examples have been given to understand the morphology of nanocomposites.

SEM and TEM images of MoO_3 nanobelts and MoO_3/PPy nanocomposite are shown in Figure 14.20. From these results, it can be observed that MoO_3 nanobelts are 50–150 nm wide and 0.3–1.0 μm long (Figure 14.20a and b). At the same time, Figure 14.20c shows that the dimensions for MoO_3/PPy nanocomposite are 1−3 μm long and 200−300 nm wide. The formation of nanocomposite with PPy layer over the MoO_3 is very clear and the thickness of PPy shell is about 60–100 nm. The morphology of the PPy/Ni nanocomposite is shown in Figure 14.20d and e. The SEM results, show only the nanofibers with 30 nm diameter, whereas TEM results confirm the formation of PPy/Ni nanocomposite. These results show that Ni nanoparticles are dispersed into PPy nanofibers. Furthermore, the HRTEM image (inset) shows the interplanar d spacing 0.22 nm, which is attributed to (111) lattice planes of the face-centered cubic phase of Ni.

14.5 ELECTROCHEMICAL STUDIES

The electrochemical performance of electrode material has a crucial role in the working of SCs; the higher the electrochemical performance, higher is the efficiency of device. These studies of the SCs have been analyzed primarily through three major characterizations: (1) cyclic voltammetry (CV), (2) galvanostatic charge–discharge (GCD), and (3) electrochemical impedance spectroscopy (EIS). Electrochemical characterization relies on, the selection of electrode (substrate), current collector, electrolyte, and most importantly, the active material (Zhang et al. 2011).

The electrode substrate or current collector should complement the performance of the active material to increase its conductivity. Basically, it is aimed to transfer the electrode current toward the external load. The current collector should be conductive and strong enough to handle the electrolytic disturbances. Stainless steel, nickel, and aluminum are generally used for collecting the current. The ideal electrolyte for getting higher energy and power densities should be such that it provides a facile passage for the movement of ions and also it should be nonaqueous with low resistance, as the nonaqueous electrolytes survive high operating voltages. The resistance of the voltage window also gets influenced by the electrolyte. The electrolytes directly affect the voltage window and its resistance. Mainly, three kinds of electrolytes are used for SCs. These are aqueous solutions of acids and bases, organic, and ionic liquid electrolytes

(Rafique et al. 2017). As far as the active material is concerned, ideally, it should have high energy and power density, because these properties make the electrode material compatible for energy-storage applications. For EDLC mechanism-based SCs, the active material with controllable porosity and high specific area is preferred. For pseudocapacitance mechanism-based SCs, the active material with high stability is generally preferable (Kim et al. n.d.).

14.5.1 CYCLIC VOLTAMMETRIC (CV) STUDIES

CV studies are performed to analyze or study the oxidizing–reducing (redox) trends of the active electrode materials. CV plot shows the current that flows along the electrochemical cell (along the x axis) and simultaneously the voltage that sweeps over the voltage range (along the y axis). The ideal CV curve should be rectangular in shape, in which the current increases rapidly to give elevated value that lies few millivolts from the applied potential range and shows the speedy charging process of the active material. The EDLC and pseudocapacitance both behave nonideally, owing to the different properties of electrode materials and give irregular rectangular shape. The voltage scan rates in CV also play a crucial role; the CV can be performed at different scan rates. The slower scan rate allows slower process that takes a lot of time while the faster scan rate makes the process faster. We can also calculate the capacitance from CV. However, higher capacitance is generally achieved at lower scan rates. This is because more number of electrolyte ions can penetrate into the electrode pores for longer time at lower scan rates (Arunkumar and Paul 2017).

CV analysis of CP-based composite helps in studying the effect of metal doping, as Ni nanoparticles incorporated in PANI nanofibers were reported and the CV studies were conducted with 1 M H_2SO_4 electrolytic solution at 5 mV/s (Figure 14.21a). Furthermore, successively obtained CV curves indicated an increase in the current displaying the conductive behavior of the PANI and Ni nanoparticles' binary nanocomposite. Furthermore, to understand the effect of Ni doping, the CV curve of pure Ni was also plotted and the obtained increment in current response of nanocomposite in comparison of pure PANI showed beneficial effects of metal nanoparticles with CP. The capacitance reported for this composite was 850 F/g at 0.86 A/g current density (Kazemi et al. 2014). CP-MO-based binary nanocomposite usually displays the pseudocapacitive behavior in CV analysis. For instance, CV studies conducted with 1 M H_2SO_4 electrolytic solution at 5, 10, and 20 mV/s for pure SnO_2 and PANI/SnO_2 nanocomposite (Figure 14.21b). The CV curves obtained for both the samples displayed almost ideal (rectangular) behavior. As for PANI, the

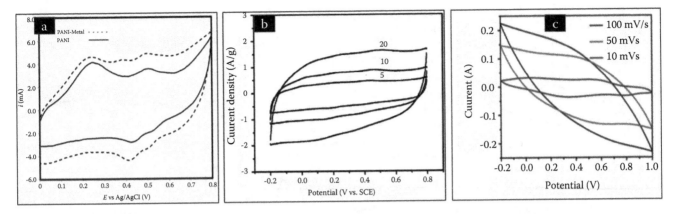

FIGURE 14.21 The cyclic voltammograms of (a) PANI and PANI/Ni composite at 5 mV/s scan rate. (b) PANI/SnO₂ composite at the scan rates of 5, 10, and 20 mV/s. (c) PANI/cobalt-based MOF (ZIF-67) at the scan rates of 10, 50, and 100 mV/s. ((a) reproduced with permission from Kazemi et al. (2014); (b) reproduced with permission from Hu et al. (2009); (c) reproduced with permission from Wang et al. (2015)).

redox processes occur simultaneously with doping and undoping of counter ions. Therefore, the CV curve of pure PANI shows the dominance of pseudocapacitance mechanism. The PANI/SnO₂ nanocomposite also depicted the similar redox mechanism for charge storage and the capacitance achieved was 305.3 F/g (Hu et al. 2009). Recently, MOFs have gained great attention for SC applications as the electrode material. Although MOFs have poor conductivity but their internal resistance can be reduced by intertwining MOF crystal with the CPs that are further electrodeposited on to the surface of MOF. In this regard, a flexible, conductive, and porous electrode was prepared in which the PANI was electrochemically polymerized and electrodeposited onto the cobalt-based MOF crystals (ZIF-67). The CV studies have been performed with 3 M KCl electrolytic solution. The maximum capacitance values, 2146, 1466, and 901 mF/cm², were achieved at 10, 50, and 100 mV/s, respectively. These higher values of the CP-MOF-based electrode attributed to the optimistic effect resulted from the EDLC mechanism of MOF and the pseudocapacitance of PANI (Lu et al. 2014).

Apart from the scan rate (as discussed in the first para of this section), the polymerization duration and also the extent of composite formation also play an efficient role in the CV analysis of CP-based composites (Figure 14.22). A nanocomposite of nickel cobaltite (NiCo₂O₄) and PPy was reported in which the synergistic effect of NiCo₂O₄ and PPy was utilized to report the high value of 2244 F/g. The aforementioned three factors were properly analyzed here. Four nanocomposites with the abovementioned same composition and different polymerization durations were prepared. From the CV results, the higher MW of the polymer higher the current response, which consequently led to the higher capacitance. Similarly, with an increment from 2 to 30 mV/s, the corresponding current response and nonlinear or

asymmetric behavior also increased. This clearly indicated the pseudocapacitance mechanism of the prepared nanocomposite. All of the individual components (NiCo₂O₄, PPy, and carbon cloth) along with the composite were analyzed with CV at 10 mV/s (Figure 14.22). The higher current response was obtained for the composite, unlike the individual components (Kong et al. 2015).

The electrolyte also acts as a crucial factor in improving the functioning of electrode material. For instance, two composites of PPy and CoMoO₄ were prepared: PPy/CoMoO₄ and CoMoO₄/PPy. As CoMoO₄ can easily loose or gain electrons in alkaline electrolyte, therefore 1 M KOH electrolytic solution was used for PPy/CoMoO₄ and 0.5 M Na₂SO₄ neutral electrolytic solution was used for CoMoO₄/PPy, as PPy structure gets destroyed in the alkaline electrolyte. The CV curve of PPy and CoMoO₄/PPy were plotted at 5 mV/S and it was found out that higher current response was obtained for CoMoO₄/PPy as compared to PPy attributed to doping/dedoping of electrolyte ion (sodium) to improve the performance of composite. Also the integrated CV area obtained for CoMoO₄/PPy is higher than that of PPy (Chen et al. 2016).

14.5.2 GALVANOSTATIC CHARGE–DISCHARGE (GCD) STUDIES

GCD studies were performed to study charging–discharging stability of the active material, which means that examination of GCD studies determines the cyclic behavior of active material. Capacitance, energy density, and power density can also be calculated from this study. This study uses constant current density and evaluates the corresponding potential in time reference. Like CV, EDLC- and pseudocapacitance-based materials also display different charging–discharging behavior

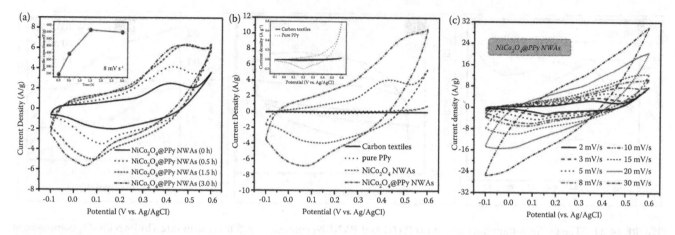

FIGURE 14.22 The cyclic voltammograms of (a) $NiCo_2O_4$/PPy nanocomposites with different polymerization duration at the scan rate of 8 mV/s. (b) Carbon textiles, PPy, $NiCo_2O_4$, and $NiCo_2O_4$/PPy nanocomposite at the scan rate of 10 mV/s. (c) $NiCo_2O_4$/PPy nanocomposite at different scan rates. (reproduced with permission from Kong et al. (2015)).

based on their different charge storage mechanisms. Although symmetrical-shaped charge–discharge curve is considered as the ideal one for SCs, the EDLC-based materials show a linear curve in charging–discharging and pseudocapacitive materials because redox reactions display nonlinear cyclic behavior (Ban et al. 2013; Hiray and Kushare 2014; Y. Chen et al. 2016).

A binary nanocomposite of PPy and MoS_2 was prepared and the GCD studies were performed with 1 M KCl electrolytic solution at 1, 2, 3, 5, and 10 A/g current density values (Figures 14.23a and b). Almost symmetric-shaped GCD curves are obtained in simulation with low internal resistance (IR) drop. Furthermore, discharge capacitance has been plotted which indicated that at high current density (10 A/g), capacitance remained stable at 450 F/g. This observation clearly indicated that high capacitance can be achieved at higher current density for the PPy/MoS_2 nanocomposite. The maximum capacitance, energy, and power densities calculated were 553.7 F/g, 5700 W/kg, and 49.2 W/kg, respectively (Mu et al. 2012). The composition of nanocomposites also plays a major role in achieving higher capacitance value. In this regard, a hybrid nanocomposite of 1D PANI and 2D MoS_2 was reported with three different MoS_2:PANI ratios as 1:1, 1:2, and 1:3. The GCD studies for nanocomposites were conducted in 1 M Na_2SO_4 electrolyte (Figures 14.23c and d). To compare all the nanocomposites, GCD curves were plotted at 1 mA/cm² for all the three composites. The nonlinear-shaped curves obtained for all the nanocomposites indicated the pseudocapacitance mechanism and also the higher discharging time obtained for 1:2 nanocomposite, which clearly indicated that 1:2 is the most suitable composition among all for SCs. The highest values of capacitance (812 F/g), energy density (112 Wh/kg), and power density (0.6 kW/kg) were achieved for 1:2 compositions. Furthermore, 1:2 nanocomposite was

plotted at 1 to 6 mA/cm² current densities and it was found that capacitance decreases with increase in current density, owing to the low access of active material at higher scan rates (Nam et al. 2016). As discussed in the previous section, two binary nanocomposites PPy/$CoMoO_4$ and $CoMoO_4$/PPy were prepared and their electrochemical properties were studied with different electrolytes based on the properties of the active material. The asymmetric-shaped charge–discharge curves obtained for both the nanocomposites indicate the dominance of pseudocapacitance mechanism in the nanocomposite. Furthermore, capacitance values were calculated for PPy, $CoMoO_4$/PPy, $CoMoO_4$, and PPy/$CoMoO_4$, and they resulted in 182, 232, 219, and 230 F/g, respectively (Chen et al. 2016).

14.5.3 CYCLIC STABILITY

The cyclic stability also occupies a dominant participation in assessing electrochemical performance of active material for SC applications. Cyclic stability, basically refers to the number of cycles up to which the given electrode material is stable in charging and discharging processes. One cycle indicates the corresponding charging then discharging. Therefore, the higher the number of cycles, the higher is the electrochemical strength of active electrode material.

The cyclic stability of PANI/TiO_2 nanocomposite was analyzed at 5 A/g current density with 1 M HCl electrolytic solution for 2000 cycles by plotting specific capacitance and Coulombic efficiency (y axis) vs. number of cycles (x axis) (Figure 14.24a). It was found that ≈100% capacitance retention was achieved during the 2000 long cycles, indicating the high potential of electrode material for SC applications (Li et al. 2011). An *in situ* PPy/Ni binary nanocomposite was also cyclically analyzed by plotting capacitance retention vs. number of cycles at 5 mA/g current density (Figure 14.24b). It can be observed that

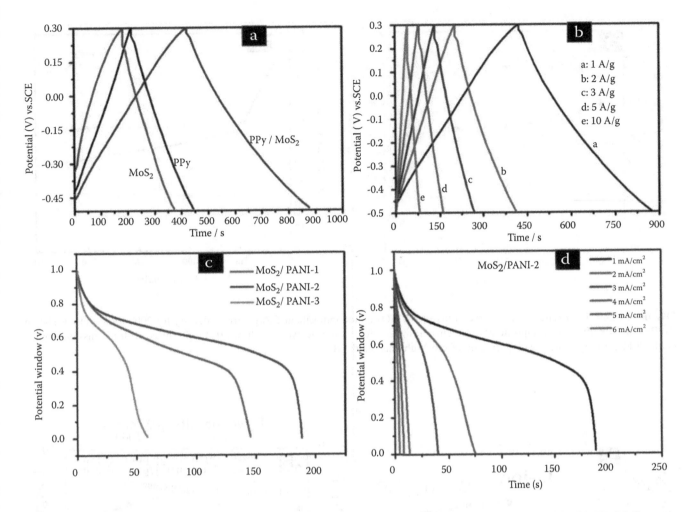

FIGURE 14.23 The GCD curves of (a) MoS₂, PPy, and PPy/MoS₂ composite at the current density of 1 A/g. (b) PPy/MoS₂ composite at the different current densities. (c) MoS₂/PANI nanocomposites with three different ratios of PANI. (d) MoS₂/PANI nanocomposite at the different charging currents. (a and b reproduced with permission from Mu et al. (2012); c and d reproduced with permission from Nam et al. (2016)).

capacitance drastically decreased until 200 cycles and reached 58%, there after there is no considerable changes upto 1000 cycles. However, decreasing to 58% within 200 cycles is indicates poor performance of the nanocomposite. (Muhamad et al. 2017).

14.5.4 RAGONE PLOT

The evaluation of the electrode materials for the performance of SC can be done from their charge density values. That is, higher the energy and power densities, higher is the performance of SC. Ragone plot is the plot that is drawn between power density (x axis) and energy density (y axis), and it also helps in inferring the charge densities' range of the given electrode material (Figure 14.25a and b). An asymmetric SC device comprising CoO and PPy had displayed a power density (5500 W/kg) and energy density (11.8 Wh/kg) (Zhu et al. 2017). Also a hybrid nanocomposite of PPy- and Zr-based MOF demonstrated high power and energy

densities of 2102 μW/cm² and 0.8 μWh/cm², respectively (Qi et al. 2018).

14.5.5 ELECTROCHEMICAL IMPEDANCE SPECTROSCOPIC (EIS) STUDIES

This study has been performed for analyzing resistance that occurs due to charge transportation and diffusion of active electrode material. In EIS, the alternating current was supplied with the smaller amplitude (in mV) from 0.01 Hz to 1 MHz. The Nyquist graph was plotted between real (x axis) and imaginary (y axis) components of resistance. The ideal plot shows an incomplete semi-circle at higher frequency and almost a straight line in lower frequency range. Furthermore, semi-circle obtained is called the charge transfer resistance (R_{CT}). R_{CT} indicates charge transfer barrier between electrode and electrolyte. Therefore, low R_{CT} is preferable for increase in the efficiency of SC. The linear part of EIS at low

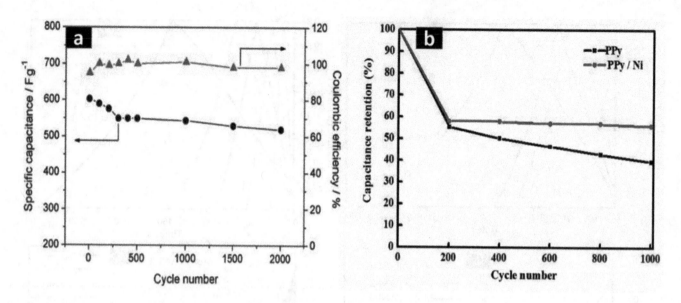

FIGURE 14.24 The cyclic stability curve of (a) PANI/TiO$_2$ nanocomposite at 5 A/g current density for 2000 long cycles is shown. (b) PPy and PPy/Ni nanocomposite at 5 A/g current density for 1000 long cycles is shown ((a) reproduced with permission from Li et al. (2011); (b) reproduced with permission from Muhamad et al. (2017).

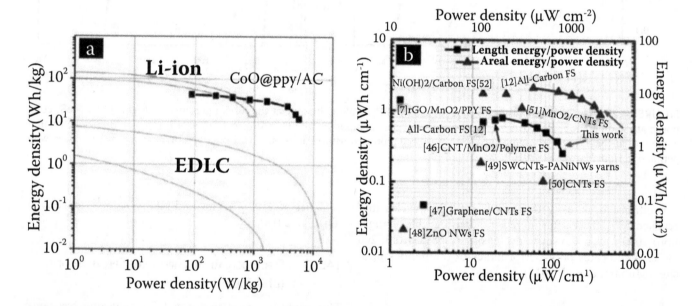

FIGURE 14.25 Ragone plot of (a) CoO/PPy nanocomposite and (b) area and length, energy and power densities compared with some of the fiber supercapacitors. (a) reproduced with permission from Zhou et al. (2013); (b) reproduced with permission from Qi et al. (2018).

frequency range corresponds to ion diffusion. A binary nanocomposite of PANI and RuO$_2$ was synthesized, in which RuO$_2$ nanoparticles were embedded on a thin film of PANI. The EIS studies were performed for RuO$_2$, RuO$_2$/PANI, and PANI with 0.5 M H$_2$SO$_4$ electrolytic solution. All the three electrodes showed ideal electrochemical impedance behavior. Also, the lower values of resistance obtained for RuO$_2$ (0.85Ω), RuO$_2$/PANI

(2.24Ω), and PANI (7.42Ω) indicated the high potential of the nanocomposite electrode for future energy-storage applications (Li et al. 2012).

A binary nanocomposite of PANI and Ni nanoparticles was prepared for enhancing the conductivity and capacitance of PANI with the incorporation of metal nanoparticles onto it (also analyzed in 'Cyclic Voltammetric (CV) Studies' section). EIS studies were conducted in

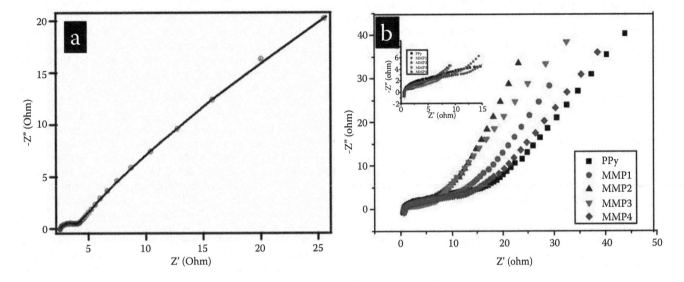

FIGURE 14.26 The Nyquist plots of (a) Ni/PANI nanocomposite and (b) PPy and PPy/MnMoO₄ nanocomposites with different monomer ratios. (a) reproduced with permission from Kazemi et al. (2014); (b) reproduced with permission from Zhou et al. (2016).

1 M H_2SO_4 electrolytic solution and the nanocomposite exhibited the ideal-shaped impedance spectra with a smaller semi-circle in higher range (Figure 14.26a). The diameter of the semi-circle (R_{CT}) was found to be 1.5Ω (Kazemi et al. 2014). Furthermore, to simulate the electrical parameters of EIS with CV and GCD studies, the obtained results have been fitted in an equivalent circuit, which consisted of two RC loop circuits in parallel and electrolytic resistance in series. EIS studies were also performed to analyze the nanocomposites with same composition but with different ratio. A binary *in situ* nanocomposite comprising $MnMoO_4$ and PPy was prepared in monomeric five different ratios of 0:1, 1:99, 2:98, 3:97, and 4:96, the EIS studies were done for all the compositions in 2 M KCl electrolytic solution and in frequency region of 0.05 to 100 kHz (Figure 14.26b). All the Nyquist plots obtained indicate that the ideal pattern and the lower value (2.31Ω) of R_{CT} was obtained for 2:98 monomeric ratio nanocomposite (Zhou et al. 2016).

REFERENCES

Abdul, Bashid, Hamra Assyaima, Hong Ngee Lim, Sazlinda Kamaruzaman, Suraya Abdul Rashid, Robiah Yunus, Nay Ming Huang, et al. 2017. "Electrodeposition of Polypyrrole and Reduced Graphene Oxide onto Carbon Bundle Fibre as Electrode for Supercapacitor." *Nanoscale Research Letters* 12 (1): 246. doi:10.1186/s11671-017-2010-3.

Adhikari, Amrita De, Ramesh Oraon, Santosh Kumar Tiwari, Joong Hee Lee, Nam Hoon Kim, and Ganesh Chandra Nayak. 2017. "A V2O5 Nanorod Decorated Graphene/ Polypyrrole Hybrid Electrode: A Potential Candidate for Supercapacitors." *New Journal of Chemistry* 41 (4): 1704–13. doi:10.1039/c6nj03580a.

Arunkumar, M., and Amit Paul. 2017. "Importance of Electrode Preparation Methodologies in Supercapacitor Applications." 8039–50. doi:10.1021/acsomega.7b01275.

Ban, Shuai, Jiujun Zhang, Lei Zhang, Ken Tsay, Datong Song, and Xinfu Zou. 2013. "Electrochimica Acta Charging and Discharging Electrochemical Supercapacitors in the Presence of Both Parallel Leakage Process and Electrochemical Decomposition of Solvent." *Electrochimica Acta* 90: 542–49. doi:10.1016/j.electacta. 2012.12.056.

Borrelli, David C., Sunghwan Lee, and Karen K. Gleason. 2014. "Optoelectronic Properties of Polythiophene Thin Films and Organic TFTs Fabricated by Oxidative Chemical Vapor Deposition." *Journal of Materials Chemistry C* 2 (35): 7223–31. doi:10.1039/c4tc00881b.

Camargo, Pedro Henrique Cury, Kestur Gundappa Satyanarayana, and Fernando Wypych. 2009. "Nanocomposites: Synthesis, Structure, Properties and New Application Opportunities." *Materials Research* 12 (1): 1–39. doi:10.1590/S1516-14392009000100002.

Camurlu, Pinar. 2014. "Polypyrrole Derivatives for Electrochromic Applications." *RSC Advances* 4 (99): 55832–45. doi:10.1039/c4ra11827h.

Candelaria, Stephanie L., Yuyan Shao, Wei Zhou, Xiaolin Li, Jie Xiao, Ji Guang Zhang, Yong Wang, Jun Liu, Jinghong Li, and Guozhong Cao. 2012. "Nanostructured Carbon for Energy Storage and Conversion." *Nano Energy* 1 (2): 195–220. doi:10.1016/j.nanoen.2011.11.006.

Cao, Jianyun, Yaming Wang, Junchen Chen, Xiaohong Li, Frank C. Walsh, Jia Hu Ouyang, Dechang Jia, and Yu Zhou. 2015. "Three-Dimensional Graphene Oxide/Polypyrrole Composite Electrodes Fabricated by One-Step Electrodeposition for High Performance Supercapacitors." *Journal of Materials Chemistry A* 3 (27): 14445–57. doi:10.1039/c5ta02920a.

Chaudhary, Swati, P.S. Bhashyam, and A.B.V. Kiran Kumar. 2018. "Polyaniline and Charcoal Binary Nanocomposite as an Electrode Material for Supercapacitor Applications." IEEE proceedings 2018.

Chaudhary, Swati, A.B.V. Kiran Kumar, Sharma Nita Dilawar, and Gupta Mukul. 2019. "Cauliflower Shaped Ternary Nanocomposites with Enhanced Power and Energy Density for Supercapacitors." *International Journal of Energy Research* 43 (8): 3446–3460. doi:10.1002/er.4486.

Chen, Tingting, Yong Fan, Guangning Wang, Qing Yang, and Ruixiao Yang. 2015. "Rationally Designed Hierarchical ZnCo2O4/Polypyrrole Nanostructures for High-Performance Supercapacitor Electrodes." *RSC Advances* 5 (91): 74523–30. doi:10.1039/c5ra14808a.

Chen, Yong, Guiying Kang, Hui Xu, and Long Kang. 2016. "Two Composites Based on CoMoO4 Nanorods and PPy Nanoparticles: Fabrication, Structure and Electrochemical Properties." *Synthetic Metals* 215: 50–55. doi:10.1016/j.synthmet.2016.02.006.

Cho, Moo Hwan, S. G. Ansari, Md Palashuddin Sk, Sajid Ali Ansari, and Hassan Fouad. 2017. "Mechanically Exfoliated MoS2 Sheet Coupled with Conductive Polyaniline as a Superior Supercapacitor Electrode Material." *Journal of Colloid and Interface Science* 504: 276–82. doi:10.1016/j.jcis.2017.05.064.

Ćirić-Marjanovic, Gordana. 2013. "Recent Advances in Polyaniline Composites with Metals, Metalloids and Nonmetals." *Synthetic Metals* 170 (1): 31–56. doi:10.1016/j.synthmet.2013.02.028.

Dang, Yong-Qiang, Shao-Zhao Ren, Guoyang Liu, Jiangtao Cai, Yating Zhang, Jieshan Qiu, and Bingqing Wei. 2016. "Electrochemical and Capacitive Properties of Carbon Dots/Reduced Graphene Oxide Supercapacitors." *Nanomaterials (Basel)* 6 (11): 212. doi:10.3390/nano6110212.

Etxeberria, Aitor, Ionel Vechiu, Haritza Camblong, and Jean-Michel Vinassa. 2010. "Hybrid Energy Storage Systems for Renewable Energy Sources Integration in Microgrids: A Review." *2010 9th International Power and Energy Conference, IPEC 2010*, 532–37. doi:10.1109/IPECON.2010.5697053.

Gan, John Kevin, Yee Seng Lim, Nay Ming Huang, and Hong Ngee Lim. 2015. "Hybrid Silver Nanoparticle/Nanocluster-Decorated Polypyrrole for High-Performance Supercapacitors." *RSC Advances* 5 (92): 75442–50. doi:10.1039/c5ra14941j.

Ghosh, Srijayee, Billa Sanjeev, Mukul Gupta, and A. B. V. Kiran Kumar. 2019. "XAS Studies of Brain-Sponge CNClZnO Nanostructures Using Polyaniline as Dual Source for Solar Light Photocatalysis." *Ceramics International* 45 (1): 1314–21. doi:10.1016/j.ceramint.2018.10.019.

Golczak, Sebastian, Anna Kanciurzewska, Mats Fahlman, Krzysztof Langer, and Jerzy J. Langer. 2008. "Comparative XPS Surface Study of Polyaniline Thin Films." *Solid State Ionics* 179 (39): 2234–39. doi:10.1016/j.ssi.2008.08.004.

Haldorai, Yuvaraj, Walter Voit, and Jae-jin Shim. 2014. "Electrochimica Acta Nano ZnO @ Reduced Graphene Oxide Composite for High Performance Supercapacitor: Green Synthesis in Supercritical Fluid." *Electrochimica Acta* 120: 65–72. doi:10.1016/j.electacta.2013.12.063.

Hiray, Pankaj G., and Bansidhar E. Kushare. 2014. "Design a Controller for Discharging and Charging of Supercapacitor as Energy Storage in Medium Voltage AC System." *International Journal of Research Studies in Science, Engineering and Technology* 1 (4): 49–56.

Hu, Zhong Ai, Yu Long Xie, Yao Xian Wang, Li Ping Mo, Yu Ying Yang, and Zi Yu Zhang. 2009. "Polyaniline/SnO2 Nanocomposite for Supercapacitor Applications." *Materials Chemistry and Physics* 114 (2–3): 990–95. doi:10.1016/j.matchemphys.2008.11.005.

Huang, Ming, Fei Li, Fan Dong, Yu Xin Zhang, and Li Li Zhang. 2015. "MnO2-Based Nanostructures for High-Performance Supercapacitors." *Journal of Materials Chemistry A* 3 (43): 21380–423. doi:10.1039/c5ta05523g.

Iro, Zaharaddeen S. 2016. "A Brief Review on Electrode Materials for Supercapacitor." *International Journal of Electrochemical Science* 11: 10628–43. doi:10.20964/2016.12.50.

Ji, Wenjing, Junyi Ji, Xinghong Cui, Jianjun Chen, Daijun Liu, Hua Deng, and Qiang Fu. 2015. "Polypyrrole Encapsulation on Flower-Like Porous NiO for Advanced High-Performance Supercapacitors." *Chemical Communications* 51 (36): 7669–72. doi:10.1039/c5cc00965k.

Kazemi, S. H., M. A. Kiani, R. Mohamadi, and L. Eskandarian. 2014. "Metal-Polyaniline Nanofibre Composite for Supercapacitor Applications." *Bulletin of Materials Science* 37 (5): 1001–6. doi:10.1007/s12034-014-0037-y.

Kim, Brian Kihun, Serubbable Sy, Aiping Yu, and Jinjun Zhang. n.d. "Electrochemical Supercapacitors for Energy Storage and Conversion." doi:10.1002/9781118991978.hces112.

Kim, J. H., R. S. Devan, C. J. Park, Dipali S. Patil, Y. R. Ma, P. S. Patil, S. A. Pawar, J. S. Shaikh, R. S. Kalubarme, and A. V. Moholkar. 2012. "Investigations on Silver/Polyaniline Electrodes for Electrochemical Supercapacitors." *Physical Chemistry Chemical Physics* 14 (34): 11886. doi:10.1039/c2cp41757j.

Kim, Ki-Hyun, Waseem Raza, Eilhann E. Kwon, Jianhua Yang, Sandeep Kumar, Andleeb Mehmood, Nadeem Raza, et al. 2018. "Recent Advancements in Supercapacitor Technology." *Nano Energy* 52 (August): 441–73. doi:10.1016/j.nanoen.2018.08.013.

Kong, Dezhi, Weina Ren, Chuanwei Cheng, Ye Wang, Zhixiang Huang, and Hui Ying Yang. 2015. "Three-Dimensional NiCo2O4@Polypyrrole Coaxial Nanowire Arrays on Carbon Textiles for High-Performance Flexible Asymmetric Solid-State Supercapacitor." *ACS Applied Materials and Interfaces* 7 (38): 21334–46. doi:10.1021/acsami.5b05908.

Kumar, Akash, Sanjeev Billa, Swati Chaudhary, A. B. V. Kiran Kumar, Ch V. V. Ramana, and D. Kim. 2018. "Ternary Nanocomposite for Solar Light Photoatalytic Degradation of Methyl Orange." *Inorganic Chemistry Communications* 97 (August): 191–95. doi:10.1016/j.inoche.2018.09.038.

Li, Jie, Zhi'an Zhang, Yanqing Lai, Yexiang Liu, Haitao Huang, Keyu Xie, and Guoge Zhang. 2011. "Polyaniline Nanowire Array Encapsulated in Titania Nanotubes as a Superior Electrode for Supercapacitors." *Nanoscale* 3 (5): 2202. doi:10.1039/c0nr00899k.

Li, Xiang, Weiping Gan, Feng Zheng, Lulu Li, Nina Zhu, and Xiaoqing Huang. 2012. "Preparation and Electrochemical Properties of RuO2/Polyaniline Electrodes for Supercapacitors." *Synthetic Metals* 162 (11–12): 953–57. doi:10.1016/j.synthmet.2012.04.002.

Login, Register. 2015. "Investigation of a Branchlike MoO3/Polypyrrole Hybrid with Enhanced Electrochemical Performance Used as an Electrode In." 1–6. doi:10.1016/S0167.

Lu, Xihong, Minghao Yu, Gongming Wang, Yexiang Tong, and Yat Li. 2014. "Flexible Solid-State Supercapacitors: Design, Fabrication and Applications." *Energy & Environmental Science* 7 (7): 2160–81. doi:10.1039/C4EE00960F.

Ma, Yanwen, Shujuan Jiang, Guoqiang Jian, Haisheng Tao, Leshu Yu, Xuebin Wang, Xizhang Wang, Jianmin Zhu, Zheng Hu, and Yi Chen. 2009. "CNx Nanofibers Converted from Polypyrrole Nanowires as Platinum Support for Methanol Oxidation." *Energy & Environmental Science* 2 (2): 224–29. doi:10.1039/b807213m.

Megha, R., Farida A. Ali, Y. T. Ravikiran, C. H. V. V. Ramana, and A. B. V. Kiran Kumar. 2018. "Conducting Polymer Nanocomposite Based Temperature Sensors: A Review." *Inorganic Chemistry Communications* 98 (June): 11–28. doi:10.1016/j.inoche.2018.09.040.

Meng, Qiufeng, Kefeng Cai, Yuanxun Chen, and Lidong Chen. 2017. "Research Progress on Conducting Polymer Based Supercapacitor Electrode Materials." *Nano Energy* 36 (February): 268–85. doi:10.1016/j.nanoen.2017.04.040.

Minh, Le, and Vu Quoc. 2013. "Layers of Inhibitor Anion – Doped Polypyrrole for Corrosion Protection of Mild Steel." *Materials Science – Advanced Topics* no. June 2013. doi:10.5772/54573.

Mondal, Sanjoy, Utpal Rana, and Sudip Malik. 2017. "Reduced Graphene Oxide/Fe3O4/Polyaniline Nanostructures as Electrode Materials for an All-Solid-State Hybrid Super-capacitor." *Journal of Physical Chemistry C* 121 (14): 7573–83. doi:10.1021/acs.jpcc.6b10978.

Mostafaei, Amir, and Ashkan Zolriasatein. 2012. "Synthesis and Characterization of Conducting Polyaniline Nano-composites Containing ZnO Nanorods." *Progress in Natural Science: Materials International* 22 (4): 273–80. doi:10.1016/j.pnsc.2012.07.002.

Mu, Jingjing, Ziqiang Lei, Hui Peng, Haohao Huang, Guofu Ma, and Xiaozhong Zhou. 2012. "In Situ Intercalative Polymerization of Pyrrole in Graphene Analogue of MoS2 as Advanced Electrode Material in Supercapacitor." *Journal of Power Sources* 229: 72–78. doi:10.1016/j.jpowsour.2012.11.088.

Muhamad, Sarah Umeera, Nurul Hayati Idris, Hanis Mohd Yusoff, Muhamad Faiz Md Din, and S. R. Majid. 2017. "In-Situ Encapsulation of Nickel Nanoparticles in Poly-pyrrole Nanofibres with Enhanced Performance for Super-capacitor." *Electrochimica Acta* 249: 9–15. doi:10.1016/j.electacta.2017.07.174.

Nam, Min Sik, Umakant Patil, Byeongho Park, Heung Bo Sim, and Seong Chan Jun. 2016. "A Binder Free Synthesis of 1D PANI and 2D MoS2 Nanostructured Hybrid Compos-ite Electrodes by the Electrophoretic Deposition (EPD) Method for Supercapacitor Application." *RSC Advances* 6 (103): 101592–601. doi:10.1039/c6ra16078f.

Omar, Fatin Saiha, Arshid Numan, Navaneethan Duraisamy, Shahid Bashir, K. Ramesh, and S. Ramesh. 2017a. "A Promising Binary Nanocomposite of Zinc Cobaltite Inter-calated with Polyaniline for Supercapacitor and Hydrazine Sensor." *Journal of Alloys and Compounds* 716: 96–105. doi:10.1016/j.jallcom.2017.05.039.

Omar, Fatin Saiha, Arshid Numan, Navaneethan Duraisamy, Mohammad Mukhlis Ramly, R. Ramesh, and R. Ramesh. 2017b. "Binary Composite of Polyaniline/Copper Cobal-tite for High Performance Asymmetric Supercapacitor Application." *Electrochimica Acta* 227: 41–48. doi:10.1016/j.electacta.2017.01.006.

Patil, Dipali S., J. S. Shaikh, S. A. Pawar, R. S. Devan, Y. R. Ma, A. V. Moholkar, J. H. Kim, R. S. Kalubarme, C. J. Parkc, and P. S. Patil. 2012. "Investigations on Silver/Polyaniline Electrodes for Electrochemical Supercapacitors." *Physical Chemistry Chemical Physics* 14 (34): 11886. doi:10.1039/c2cp41757j.

Pawar, Pranav Bhagwan, Shobha Shukla, and Sumit Saxena. 2016. "Graphene Oxide – Polyvinyl Alcohol Nanocomposite Based Electrode Material for Supercapacitors." *Journal of Power Sources* 321 (May): 102–5. doi:10.1016/j.jpowsour.2016.04.127.

Peng, Hui, Guofu Ma, Kanjun Sun, Jingjing Mu, Hui Wang, and Ziqiang Lei. 2014. "High-Performance Supercapacitor Based on Multi-Structural CuS@polypyrrole Composites Prepared by in Situ Oxidative Polymerization." *Journal of Materials Chemistry A* 2 (10): 3303–7. doi:10.1039/c3ta13859c.

Peng, Lele, Yu Zhao, Guihua Yu, Yu Ding, and Ye Shi. 2015. "Nanostructured Conductive Polymers for Advanced Energy Storage." *Chemical Society Reviews* 44 (19): 6684–96. doi:10.1039/c5cs00362h.

Pohoata, V., T. Teslaru, N. Dumitrascu, M. Dobromir, L. Cure-cheriu, and I. Topala. 2015. "Polythiophene Films Obtained by Polymerization under Atmospheric Pressure Plasma Conditions." *Materials Chemistry and Physics* 169: 120–27. doi:10.1016/j.matchemphys.2015.11.038.

Poyraz, Selcuk, Idris Cerkez, Tung Shi Huang, Zhen Liu, Litao Kang, Jujie Luo, and Xinyu Zhang. 2014. "One-Step Syn-thesis and Characterization of Polyaniline Nanofiber/ Silver Nanoparticle Composite Networks as Antibacterial Agents." *ACS Applied Materials and Interfaces* 6 (22): 20025–34. doi:10.1021/am505571m.

Qi, Kai, Ruizuo Hou, Shahid Zaman, Yubing Qiu, Bao Yu Xia, and Hongwei Duan. 2018. "Construction of Metal-Organic Framework/Conductive Polymer Hybrid for All-Solid-State Fabric Supercapacitor." *ACS Applied Materials and Inter-faces* 10 (21): 18021–28. doi:10.1021/acsami.8b05802.

Rafique, Amjid, Stefano Bianco, Marco Fontana, Candido F. Pirri, and Andrea Lamberti. 2017. "Flexible Wire-Based Electrodes Exploiting Carbon/ZnO Nanocomposite for Wearable Supercapacitors." *Ionics* 23 (7): 1839–47. doi:10.1007/s11581-017-2003-3.

Ran, Fen, Yongtao Tan, Wenju Dong, Zhen Liu, Lingbin Kong, and Long Kang. 2018. "In Situ Polymerization and Reduc-tion to Fabricate Gold Nanoparticle-Incorporated Polya-niline as Supercapacitor Electrode Materials." *Polymers for Advanced Technologies* 29 (6): 1697–705. doi:10.1002/pat.4273.

Rawat, Neha Kanwar, Anujit Ghosal, and Sharif Ahmad. 2014. "Influence of Microwave Irradiation on Various Properties of Nanopolythiophene and Their Anticorrosive Nanocom-posite Coatings." *RSC Advances* 4 (92): 50594–605. doi:10.1039/c4ra06679k.

Raza, Waseem, Faizan Ali, Nadeem Raza, Yiwei Luo, Ki-hyun Kim, and Jianhua Yang. 2018. "Recent Advancements in Supercapacitor Technology Nano Energy Recent Advance-ments in Supercapacitor Technology." *Nano Energy* 52 (August): 441–73. doi:10.1016/j.nanoen.2018.08.013.

Rusi, and S. R. Majid. 2015. "Green Synthesis of In Situ Elec-trodeposited RGO/MnO2 Nanocomposite for High Energy Density Supercapacitors." *Scientific Reports* 5 (November): 1–13. doi:10.1038/srep16195.

Segal, David, Richa Srivastava, Sa Jamal, Michelangelo Moroni, Daniele Borrini, Luca Calamai, Luigi Dei. 2012. "Ceramic Nanomaterials from Aqueous and 1,2-Ethanediol Supersat-urated Solutions at High Temperature." *Chemical Sciences Journal* 2013 (1): 543–50. doi:10.1039/a700881c.

Shakoor, Abdul, and Tasneem Zahra Rizvi. 2010. "Raman Spec-troscopy of Conducting Poly (Methyl Methacrylate)/Polya-niline Dodecylbenzenesulfonate Blends." *Journal of Raman Spectroscopy* 41 (2): 237–40. doi:10.1002/jrs.2414.

Shendkar, Janardhan H., Vijaykumar V. Jadhav, Pritamkumar V. Shinde, Rajaram S. Mane, and Colm O'Dwyer. 2018. "Hybrid Composite Polyaniline-Nickel Hydroxide Elec-trode Materials for Supercapacitor Applications." *Heliyon* 4 (9): e00801. doi:10.1016/j.heliyon.2018.e00801.

Shi, Fan, Lu Li, Xiu Li Wang, Chang Dong Gu, and Jiang Ping Tu. 2014. "Metal Oxide/Hydroxide-Based Materials for Supercapacitors." *RSC Advances* 4 (79): 41910–21. doi:10.1039/c4ra06136e.

Shown, Indrajit, Abhijit Ganguly, Li Chyong Chen, and Kuei Hsien Chen. 2015. "Conducting Polymer-Based Flexible

Supercapacitor." *Energy Science and Engineering* 3 (1): 1–25. doi:10.1002/ese3.50.

Smolin, Yuriy Y., Masoud Soroush, and Kenneth K. S. Lau. 2017. "Oxidative Chemical Vapor Deposition of Polyaniline Thin Films." *Beilstein Journal of Nanotechnology* 8 (1): 1266–76. doi:10.3762/bjnano.8.128.

Snook, Graeme A., Pon Kao, and Adam S. Best. 2011. "Conducting-Polymer-Based Supercapacitor Devices and Electrodes." *Journal of Power Sources* 196 (1): 1–12. doi:10.1016/j.jpowsour.2010.06.084.

Su, Dang Sheng, and Robert Schlögl. 2010. "Nanostructured Carbon and Carbon Nanocomposites for Electrochemical Energy Storage Applications." *ChemSusChem* 3 (2): 136–68. doi:10.1002/cssc.200900182.

Thakur, Anukul K., Ashvini B. Deshmukh, Ram Bilash, Indrapal Karbhal, Mandira Majumder, and Manjusha V. Shelke. 2017. "Facile Synthesis and Electrochemical Evaluation of PANI/CNT/MoS 2 Ternary Composite as an Electrode Material for High Performance Supercapacitor." *Materials Science & Engineering B* 223: 24–34. doi:10.1016/j.mseb.2017.05.001.

Tong, Ye-Xiang, Xiao-Yan Chen, Gao-Ren Li, Xue-Feng Lu, and Wen Zhou. 2015. "α-Fe 2 O 3 @PANI Core–Shell Nanowire Arrays as Negative Electrodes for Asymmetric Supercapacitors." *ACS Applied Materials and Interfaces* 7 (27): 14843–50. doi:10.1021/acsami.5b03126.

Valderrama-García, Bianca X., Efraín Rodríguez-Alba, Eric G. Morales-Espinoza, Kathleen Moineau Chane-Ching, Ernesto Rivera, Scott Reed, and Marino Resendiz. 2016. "Synthesis and Characterization of Novel Polythiophenes Containing Pyrene Chromophores: Thermal, Optical and Electrochemical Properties." *Molecules* 21 (2). doi:10.3390/molecules21020172.

Veerasubramani, Ganesh Kumar, Karthikeyan Krishnamoorthy, Sivaprakasam Radhakrishnan, Nam Jin Kim, and Sang Jae Kim. 2016. "In-Situ Chemical Oxidative Polymerization of Aniline Monomer in the Presence of Cobalt Molybdate for Supercapacitor Applications." *Journal of Industrial and Engineering Chemistry* 36: 163–68. doi:10.1016/j.jiec.2016.01.031.

Vijeth, H., M. Niranjana, L. Yesappa, S. P. Ashokkumar, and H. Devendrappa. 2018. "Polythiophene Nanocomposites as High Performance Electrode Material for Supercapacitor Application." *AIP Conference Proceedings* 1942 (April). doi:10.1063/1.5029148.

Wang, Huanhuan, Jianyi Lin, and Ze Xiang Shen. 2016. "Polyaniline (PANi) Based Electrode Materials for Energy Storage and Conversion." *Journal of Science: Advanced Materials and Devices* 1 (3): 225–55. doi:10.1016/j.jsamd.2016.08.001.

Wang, Junchuan, Jiahui Tian, Xueqin Zhang, Baoping Lin, Yidong Zhang, Ying Sun, and Hong Yang. 2016. "All-Solid-State Asymmetric Supercapacitors Based on ZnO Quantum Dots/Carbon/CNT and Porous N-Doped Carbon/CNT Electrodes Derived from a Single ZIF-8/CNT Template." *Journal of Materials Chemistry A* 4 (26): 10282–93. doi:10.1039/c6ta03633c.

Wang, Lu, Xiao Feng, Lantian Ren, Qiuhan Piao, Jieqiang Zhong, Yuanbo Wang, Haiwei Li, Yifa Chen, and Bo Wang. 2015. "Flexible Solid-State Supercapacitor Based on a Metal-Organic Framework Interwoven by Electrochemically-Deposited PANI." *Journal of the American Chemical Society* 137 (15): 4920–23. doi:10.1021/jacs.5b01613.

Weerakoon, K. A., J. H. Shu, Mi-Kyung Park, and B. A. Chin. 2012. "Polyaniline Sensors for Early Detection of Insect Infestation." *ECS Journal of Solid State Science and Technology* 1 (5): Q100–5. doi:10.1149/2.014205jss.

Wei, Weifeng, Xinwei Cui, Weixing Chen, and Douglas G. Ivey. 2011. "Manganese Oxide-Based Materials as Electrochemical Supercapacitor Electrodes." *Chemical Society Reviews* 40 (3): 1697–721. doi:10.1039/c0cs00127a.

Wong, Y. T. Angel, Johannes Landmann, Maik Finze, and David L. Bryce. 2017. "Dynamic Disorder and Electronic Structures of Electron-Precise Dianionic Diboranes: Insights from Solid-State Multinuclear Magnetic Resonance Spectroscopy." *Journal of the American Chemical Society* May. doi:10.1021/jacs.7b01783.

Xia, Xinhui, Jiangping Tu, Yongqi Zhang, Xiuli Wang, Changdong Gu, Xin Bing Zhao, and Hong Jin Fan. 2012. "High-Quality Metal Oxide Core/Shell Nanowire Arrays on Conductive Substrates for Electrochemical Energy Storage." *ACS Nano* 6 (6): 5531–38. doi:10.1021/nn301454q.

Yang, Jing, Ying Liu, Siliang Liu, Le Li, Chao Zhang, and Tianxi Liu. 2017. "Conducting Polymer Composites: Material Synthesis and Applications in Electrochemical Capacitive Energy Storage." *Materials Chemistry Frontiers* 1 (2): 251–68. doi:10.1039/c6qm00150e.

Zhang, Guanggao, Yadong Yao, Hongjing Wang, Guangfu Yin, Lu Long, Minglei Yan, Menglai Kong, et al. 2016. "Ni3S2@polypyrrole Composite Supported on Nickel Foam with Improved Rate Capability and Cycling Durability for Asymmetric Supercapacitor Device Applications." *Journal of Materials Science* 52 (7): 3642–56. doi:10.1007/s10853-016-0529-9.

Zhang, Jiujun, Cheng Zhong, Yida Deng, Wenbin Hu, Jinli Qiao, and Lei Zhang. 2015. "A Review of Electrolyte Materials and Compositions for Electrochemical Supercapacitors." *Chemical Society Reviews* 44 (21): 7484–539. doi:10.1039/c5cs00303b.

Zhang, Lei, Xiaosong Hu, Zhenpo Wang, Fengchun Sun, and David G. Dorrell. 2018. "A Review of Supercapacitor Modeling, Estimation, and Applications: A Control/Management Perspective." *Renewable and Sustainable Energy Reviews* 81 (February 2017): 1868–78. doi:10.1016/j.rser.2017.05.283.

Zhang, Li Li, Zhibin Lei, Jintao Zhang, Xiaoning Tian, and Xiu Song Zhao. 2011. "Supercapacitors: Electrode Materials Aspects." doi:10.1002/0470862106.ia816.

Zhang, Lingling, Weihua Tang, Erjun Kan, Shaopeng Zhang, Guangqiang Han, Jian Tang, and Yun Liu. 2014. "MnO2 Nanorods Intercalating Graphene Oxide/Polyaniline Ternary Composites for Robust High-Performance Supercapacitors." *Scientific Reports* 4 (1): 1–7. doi:10.1038/srep04824.

Zhou, Cheng, Yangwei Zhang, Yuanyuan Li, and Jinping Liu. 2013. "Construction of High-Capacitance 3D CoO@Polypyrrole Nanowire Array Electrode for Aqueous Asymmetric Supercapacitor." *Nano Letters* 13 (5): 2078–85. doi:10.1021/nl400378j.

Zhou, Jingkuo, Ruinan Xue, Jianping Gao, Yahui Song, Jing Yan, Xiaoyang Xu, Yu Liu, et al. 2016. "High-Performance Supercapacitor Materials Based on Polypyrrole Composites Embedded with Core-Sheath Polypyrrole@MnMoO 4 Nanorods." *Electrochimica Acta* 212: 775–83. doi:10.1016/j.electacta.2016.07.035.

Zhu, Yuanyuan, Shuang Cheng, Weijia Zhou, Jin Jia, Lufeng Yang, Minghai Yao, Mengkun Wang, et al. 2017. "Construction and Performance Characterization of α-Fe2O3/RGO Composite for Long-Cycling-Life Supercapacitor Anode." *ACS Sustainable Chemistry and Engineering* 5 (6): 5067–74. doi:10.1021/acssuschemeng.7b00445.

15 Polyethylenedioxythiophene (PEDOT)-Based Supercapacitor Applications

Manoj Karakoti, Sandeep Pandey and Nanda Gopal Sahoo
Prof. R. S. Nanoscience and Nanotechnology Centre
Department of Chemistry DSB Campus, Kumaun University
Nainital, India

Anirban Dandapat
Department of Chemistry
DSB Campus, Kumaun University
Nainital, India

Suman Mahendia
Department of Physics
Kurukshetra University
Kurukshetra, India

CONTENTS

15.1 INTRODUCTION

A global demand for energy is rapidly increasing with the growing population due to the dependency of human on the energy-consumption appliances. To fulfill this ever-growing demand of energy, we are in urgent need of some portable and flexible electronic devices having convenient and empirical energy-storage capacities with superior energy and power density (Nishide and Oyaizu 2008; Rogers, Someya, and Huang 2010; Lipomi and Bao 2011; Tarascon and Armand 2011).

Presently, supercapacitors have been emerged as suitable materials to be applied in energy-storage devices attributable to their extraordinary characteristics, which include higher energy storage and power density, and faster charge–discharge cycles as compared to those of the Li-ion batteries (Miller and Simon 2008; Jennings et al. 2009). Supercapacitors are the crucial energy-storing devices that generate the higher energy-storage abilities than the conventional batteries and possess higher power supply abilities than the conventional capacitors. Thus, a supercapacitor could bridge between the capacitor and the battery, and possesses extensive life cycles and the ecofriendly mechanism as shown in Figure 15.1 (Shown et al. 2015).

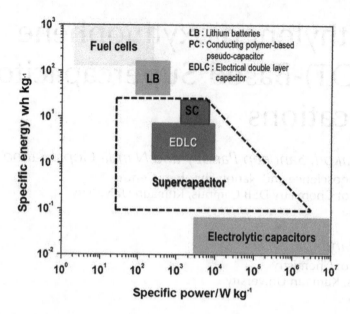

FIGURE 15.1 Comparison of energy density and power density of a supercapacitor with respect to other energy devices displayed by Ragone plot (Shown et al. 2015).

Supercapacitors are the energy-storage devices along with higher capacitive properties and lower internal resistance that enable to store and transmit energy at a greater rate than that of the conventional batteries due to the energy-storage mechanism and simple charge partition at the interfaces between the electrolyte and electrodes (Stoller et al. 2008; Wu et al. 2010). A supercapacitor can play a vital role in the energy-consuming appliances as compared to other energy devices because of its many obvious advantages, including a high power density, long lifecycles, broad thermal range, light weight, low maintenance cost, and flexible packaging (Wang et al. 2009). The supercapacitors can be used for the energy-recapture sources (e.g., load cranes, forklifts, fuel cell vehicles, electrical vehicles, and low-emission hybrid vehicles) and for the power quality enhancement (Wu 2002; Miller and Simon 2008). A typical structure of a supercapacitor comprises two electrodes, one electrolyte, and a separator which electrically take apart the electrodes from each other (Iro, Subramani, and Dash 2016). Among these three components, the electrode material plays the crucial role in the electrochemical functioning of supercapacitors. The supercapacitors are basically divided into the three groups, viz. electrochemical double-layer capacitors (EDLCs), pseudocapacitors, and hybrid supercapacitors. Primarily, the EDLCs, which are being preferably used for real applications, utilize the porous carbon materials with larger assessable surface area as an electrode material (Salanne et al. 2016). EDLCs are capable of storing charge electrostatically or through non-faradic process without any transfer of charge between the electrodes and electrolyte (Jayalakshmi and Balasubramanian 2008; Kiamahalleh et al. 2012). This process allows holding the capacitance properties and delivering high

power along with outstanding life cycles. The capacitance properties of EDLCs mainly depend on the two major factors, viz. specific capacitance and operational voltage. The overall performance of EDLCs mainly depends upon the electrical conductivity and surface area of the electrode materials. Highly porous materials are the prime choice for the electrode materials for EDLCs as the porosity results in high surface area. By applying the voltage, a potential difference has been created onto the electrodes and the charge is accumulated over its surfaces, which enforces the resultant electrolyte ions to be diffused over the separator toward the oppositely charged electrodes (Halper and Ellenbogen 2006; Choi and Yoon 2015). Besides the aforementioned factors, proper selection of the electrolyte also plays a crucial role for gaining high power and energy in EDLCs. In EDLCs, mainly carbonaceous materials (i.e., activated carbon (AC), carbon nanotubes (CNTs), graphene, and CNTs–graphene hybrid materials) are utilized for the fabrication of the electrode materials (Gamby et al. 2001; Stoller et al. 2008; Izadi-Najafabadi et al. 2010; Liu et al. 2010; Yoo et al. 2011; Zheng et al. 2012).

Second group (i.e., pseudocapacitors), wherein the faradic charge transfer (also known as the redox reaction) occurs to generate the capacitance with high specific capacitance and energy density, suffers from the low power density and life cycle as compared to EDLCs. The power density is directly related to the rate of charging of the supercapacitors, whereas the energy density is related with the energy-storage capacities of the supercapacitors. The performance of this type of capacitors is mainly responsible for how much transfer of charge occurred in the active electrode materials, and the diffusion of ions between the electrode and electrolyte interface. Due to this faradic reaction, the charge transfer takes much longer time than EDLCs and suffers from the lower power density. On the other hand, this reaction favors to store relatively higher energy than EDLCs. The materials used in psuedocapacitors are different from EDLCs and show very fast faradic charge transfer. For example, metal oxides (such as ruthenium dioxide, nickel oxide, and manganese oxide) and conducting polymers (such as polypyrrole, polyaniline, and poly(3,4-ethylenedioxythiophene)) are commonly used as active materials for electrodes in psuedocapacitors (Hashmi and Upadhyaya 2002; Park et al. 2002; Gujar et al. 2007; Cheng et al. 2011; Wu and Wang 2012; Cai et al. 2013; Jalili, Razal, and Wallace 2013).

The third group, that is, hybrid supercapacitor, which is also known as the asymmetric supercapacitor, consists of two different electrodes in which one shows capacitive behaviors and the other shows faradic responses (Simon and Gogotsi 2010; Yu et al. 2013). This type of structural arrangement shows battery-like electrode to offer the higher energy density and EDLC-like electrode to facilitate the high power capacity within a single device. The hybrid nature of supercapacitor improves the overall cell potential, energy, and power densities than the EDLCs (Zhang, Yedlapalli, and Lee 2009; Wang, Zhang, and Zhang 2012; Yu et al. 2013). For achieving the high performance from the hybrid supercapacitor, we need to develop the potential materials for electrode with balanced design of composite material, particle size, and morphology with appropriate electrolyte system (Yu et al. 2013).

Over the past few years, many materials were developed and used for the high-performance supercapacitors, but among those the conducting polymer shows great potential for the efficient supercapacitor devices. The conducting polymers show the good pseudocapacitive properties due to their fast charging–discharging rate, suitable structure, low cost, and fast doping and undoping process (Ryu et al. 2004). According to the literature, various conducting polymers such as polyaniline, polythiophene, and poly (3,4-ethylenedioxythiophene) (PEDOT) are used for the supercapacitor application. Among these, PEDOT is the most popular polymer for the energy application devices, especially for the supercapacitor, because of its stability and higher electrical conductivity (>500 S cm^{-1}) with potential window of 1.2 V (Snook, Kao, and Best 2011).

PEDOT, a derivative of polythiophene, is initially synthesized by the Bayer AG Research Laboratory, Germany, in the 1980s, having a structure shown in Scheme 15.1 (Jonas and Schrader 1991; Heywang and Jonas 1992; Winter et al. 1995).

PEDOT has several advantages such as excellent optical transparency within conducting state, significantly stable oxidized state attributed to the lower oxidation potential, suitable bandgap, and most importantly very good conductivity at ambient condition and raised temperature (Jonas and Heywang 1994). Also, it is a conjugated polymer that belongs to the polymer family that has both p- and n-type dopable nature. The p-doping (n-doping) conducting polymer creates oxidation (reduction) in the polymer resulting in increase in its electrical conductivity (Damlin, Kvarnström, and Ivaska 2004). It has many applications such as supercapacitors (Liu et al. 2015), electrochromic displays, antistatics, photovoltaics, electroluminescent displays, printed wiring, and sensors (Groenendaal et al. 2000; Kirchmeyer and Reuter 2005). For the supercapacitor applications, there are mainly two types of PEDOT available: one is the synthesized form of the polymerization of 3,4-ethylenedioxythiophene (EDOT) via various methods (Chu, Tsai, and Sun 2012; Lee et al. 2012) and the other one is the aqueous phase (PEDOT:PSS) that has the surfactant to suspend PEDOT in water. Previous study for the PEDOT-based supercapacitor showed the electrochemical capacitance with ranges from 80 to 180 F g^{-1} (Carlberg and Inganäs 1997; Liu and Lee 2008; Liu and Reynolds 2010; Anothumakkool et al. 2013) in contrast to its theoretical capacitance of 210 F g^{-1} (Snook, Kao, and Best 2011).

Recently, PEDOT has captured the scientific attention due to the aforementioned properties and its application for the development of high capacitive supercapacitors. This chapter discusses the synthesis methods for pure

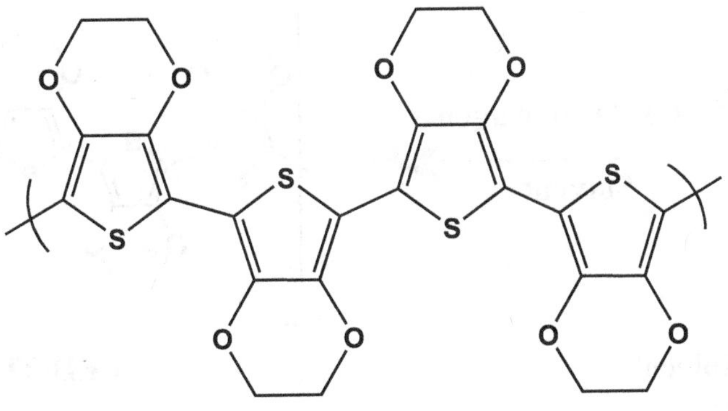

SCHEME 15.1 Chemical structure of PEDOT.

PEDOT and PEDOT-based composites with various carbonaceous materials and metal oxides for the high-performance supercapacitor applications.

15.2 SYNTHESIS OF PEDOT VIA VARIOUS METHODS

PEDOT is a sky blue color conducting polymer, commonly synthesized by the chemical and electrochemical methods via polymerization of EDOT monomer (Dietrich et al. 1994; Jonas and Heywang 1994; Pei et al. 1994). In addition, vapor-phase polymerization and hydrothermal methods are also used for the synthesis of PEDOT. In these methods, mostly one-dimensional and two-dimensional nanostructures (such as nanorods, nanospheres, nanotubes, and nanofibers) were prepared in the past decades (Martin 1996; Kim et al. 2005; Jang 2006; Wu, Li and Feng 2007; Du, Zhou, and Mai 2009). However, some recent reports show the more complicated and hierarchical structure of PEDOT (Zhou et al. 2011). The synthesis methods are discussed below.

15.2.1 CHEMICAL SYNTHESIS

The general procedure (Scheme 15.2) for the development of PEDOT involves the polymerization of EDOTs in the presence of various solvents and oxidative agents. In this way, various morphologies of PEDOT objects were developed with size ranges from nano- to microregion having narrow size distribution. Corradi et al. synthesized PEDOT with the conductivity of 20 Ω^{-1} cm^{-1} via chemical polymerization of EDOT in an aqueous medium using FeCl$_3$, Ce(SO$_4$)$_2$, and (NH$_4$)$_2$Ce(NO$_3$)$_6$ as oxidizing agents (Corradi and Armes 1997).

Many research groups already reported numerous techniques for the synthesis of PEDOT nano-objects and upgrading of its processability (Khan and Armes 1999; Henderson et al. 2001; Oh and Im 2002; Han and Armes 2003; Han and Foulger 2004; Müller, Klapper, and Müllen 2006; Zhang et al. 2006). In this order, Sun

et al. reported the PEDOT nanowire synthesis method in the presence of poly(acrylic acid) (Sun and Hagner 2007). Mumtaz et al. reported a synthetic method for PEDOT nano-objects along with a variety of morphologies such as donuts and particles by using Py-grafted poly(vinyl alcohol) (PVA-g-Py) with iron(III) p-toluenesulfonate hexahydrate [Fe(III)(OTs)$_3$·6(H$_2$O)] and ammonium persulfate as an oxidizing agent and stabilizer, respectively (Mumtaz et al. 2008). Zhang et al. reported a unique one-step bulk synthesis of electrically conductive powder PEDOT nonofibers (100–180 nm diameter) at room temperature by using the V$_2$O$_5$ seeding approach (Zhang, MacDiarmid, and Manohar 2005). Oh et al. showed the preparation method for PEDOT particles in aqueous media by using the dodecylbenzene sulfonic acid (DBSA) as a surfactant in the presence of FeCl$_3$/Na$_2$S$_2$O$_8$ as an oxidative reagent. As-prepared PEDOT particles showed the aggregation with poor defined structure (Oh and Im 2002). Again, polystyrene and silica-coated particles were developed by Armes and colleagues (Khan and Armes 1999) and Han and Foulger (2004), respectively. One report was published by Müller et al. for the preparation of PEDOT nanoparticles through the polymerization emulsion techniques. In these techniques, poly(isoprene)-b-poly(methyl methacrylate) (PI-b-PMMA) and iron(III) chloride are used as a stabilizer and oxidizing agent, respectively, in cyclohexane (Müller, Klapper, and Müllen 2006). Cloutet et al. synthesized the well-defined spherical PEDOT latexes with the help of poly(ethylene oxide) end-functionalized with a (3,4-ethylenedioxythiophene) moiety in the presence of ammonium persulfate or iron(III) p-toluenesulfonate hexahydrate as oxidizing agent in alcoholic medium (Cloutet, Mumtaz, and Cramail 2009).

15.2.2 ELECTROCHEMICAL METHOD

This method is very handy for the polymerization of EDOT into the PEDOT, basically a film obtained over the anode surface. This reaction can be performed in both aqueous and organic media. Initially, the research

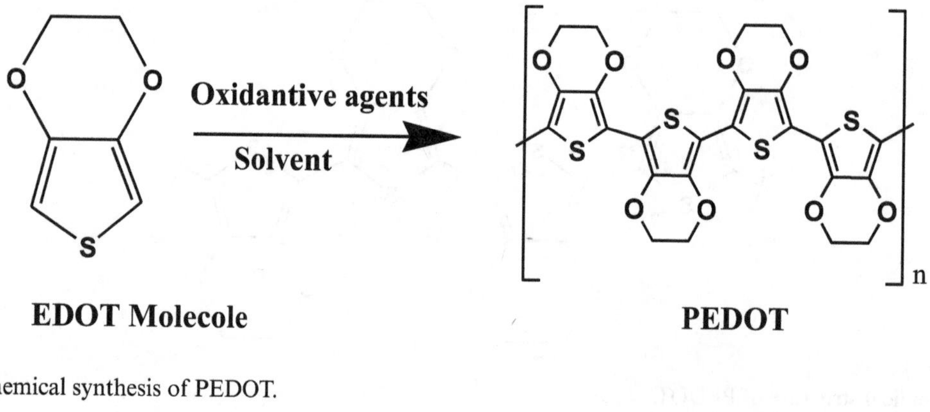

EDOT Molecole   **PEDOT**

SCHEME 15.2 Chemical synthesis of PEDOT.

was focused toward the aqueous-phase synthesis (Du and Wang 2003; Gao et al. 2006; Han et al. 2007; Bhandari et al. 2008; Patra, Barai, and Munichandraiah 2008; Tamburri et al. 2009; Wen et al. 2009; Zhou et al. 2010). However, some limitations of EDOT such as low solubility in water at room temperature have shifted most of the research to be performed in any organic solvents such as propylene carbonate and acetonitrile (Sakmeche et al. 1999; Melato, Viana, and Abrantes 2008). Besides, several surfactants are being used during the electrochemical synthesis of PEDOT to control their structures, since the conducting properties of a polymer are directly related to their chemical and physical structures. The sodium dodecylsulfate (SDS) can be used as the surfactant for the synthesis of PEDOT with globular and fibrous morphologies (Patra, Barai, and Munichandraiah 2008). Generally, conductive polymer is synthesized in an electrochemical method in their oxidation form with the help of doping with different ions and their concentrations. In the same way, electrical properties of PEDOT are greatly influenced and tuned with the help of doping ions. For example, the Cl^- ion-doped film is less conductive than the NO_3^- ion (Han et al. 2007; Patra, Barai, and Munichandraiah 2008). Zhou et al. showed the doping effect of Cl^- and NO_3^- ions in the PEDOT polymer film toward electro-polymerization behavior, morphologies, and electrochemical properties (Zhou et al. 2011).

Xiao et al. showed the polymerization of EDOT into the PEDOT via the electrochemical method at various potential rates ranging from 1.0 to 1.8 V vs. Ag/AgCl by utilizing the counter-electrode of Pt foil performed under 0.1 M LiClO₄ in acetonitrile solution at 25°C (Xiao et al. 2007). Tsai et al. synthesized the poly(3,4-ethylenedioxythiophene) film-modified glassy electrode in 1-butyl-3-methylimidazolium tetrafluoroborate (BMT) ionic liquid at pH 1.2. This experiment was performed at –0.5 to 1.1 V (scan rate = 100 mV s^{-1}) in CV with 0.01 M of EDOT (Tsai, Thiagarajan, and Chen 2011). Most researchers used cyclic voltammetry and chronoamperometry for the synthesis of PEDOT (Du and Wang 2003). The high charge capacity (114 F g^{-1}) of PEDOT was synthesized via pulse polymerization technique on glassy carbon substrate. In this process, the pulse rate and time play crucial role in the morphologies, size, and chain defects, while pulse off time is responsible for the polymer conjugation and orientation (El 2009).

15.2.3 VAPOR-PHASE POLYMERIZATION

This method is widely used to synthesize conducting polymer and PEDOT. Initially, this method was performed by using different oxidizing agents such as FeCl₃ and H₂O₂ for the polymerization of polypyrrole films via chemical vapor deposition (CVD) process (Mohammadi et al. 1986). Recently, this method is used for the polymerization of PEDOT via EDOT, using FeCl₃ as oxidizing agent (Kim

et al. 2003). Winther-Jensen et al. developed a new synthetic pathway for PEDOT with a conductivity exceeding 1000 S cm^{-1}. This process is mainly based on volatile base-inhibited (pyridine) vapor-phase polymerization of EDOT, with ferric *p*-toluenesulfonate as oxidizing agent. Here, vapor-phase polymerization was performed within a single chamber, which was connected with air, nitrogen, or argon gases and a heater that provided the temperature (30–50°C) to the monomer to accelerate the polymerization (Figure 15.2). The samples were enclosed with PEDOT that was coated initially by oxidizing agents such as ethanol or butanol solution of Fe(III) tosylate and then drying at 60°C in air followed by the transfer of the sample into a polymerization chamber. This chamber has a quartz crystal microbalance (QCM) that is used for checking the growth and conversion of monomer into the polymer during the process. This process was operated with the 10-MHz crystal that was also coated with the similar oxidizing agent as the substrates prior to contact in EDOT vapor. During the progress of reaction, a weight change was found on the surface film over the crystal by observing the change in resonating frequency of crystal. This chamber also has an oscillator that was coupled with the frequency counter for excitation of the crystal and the observed change in frequency by recording the analog from the frequency counter. The change in the frequency and weight of crystal at the time of polymerization indicated the growth of PEDOT from the EDOT monomer (Winther-Jensen and West 2004).

15.2.4 HYDROTHERMAL SYNTHESIS

The hydrothermal method is a single-step facile method for the synthesis of conducting polymers. In addition, this method is also employed for the synthesis of PEDOT and its nanostructure. Rajesh et al. developed the PEDOT nanostructures on the carbon fiber cloth (CFC) (in-situ polymerization of EDOT) via this method for the high-performance supercapacitor (Rajesh et al. 2017). Later, Ahmed et al. prepared PEDOT from EDOT by applying the hydrothermal method. EDOT monomer (0.5 M) was first dispersed in 50 ml of ethanol as a solvent. In another beaker, ammonium persulfate (1 g) was dissolved in the DI water (10 ml). After mixing both solutions, 3 ml of sulfuric acid was added to it and stirred for 12 h and subsequently kept at the autoclave at 150°C for another 12 h for hydrothermal treatment. The finally obtained hydrothermally treated product was filtered followed by washing with DI water, dried, and finally stored for further use, that is, the fabrication of the electrode for supercapacitor (Ahmed and Rafat 2018).

Although the solution-phase chemical polymerization of PEDOT has shown potential for upscale production, this method produced PEDOT with lower conductivity with lesser porosity, which hinders to achieve the high performance from the supercapacitor. On the other hand, electro-polymerization in-situ techniques can

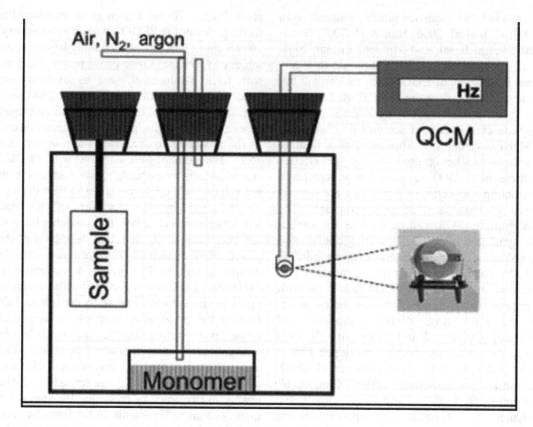

FIGURE 15.2 Systematic representation of vapor-phase polymerization of PEDOT (Winther-Jensen and West 2004).

produce conducting polymers with a high conjugation length, thin, uniform, and adherent polymer films. Synthesis conditions were chosen properly by the deposition of films on a limited surface area having the high degree of geometrical conformity with desirable thickness using definite numbers of growth cycles in potentiodynamic cycling. Mostly peoples used the vapor-phase or electrochemical-deposition methods for the fabrication of PEDOT-based electrode (D'Arcy et al. 2014). However, the problem associated with these methods is that it requires hours or even a day, which can be comparatively costly and need higher temperature for the completion of the process. Moreover, PEDOT layers have the thickness in the order of 10 μm to several hundreds of nanometers (Yang et al. 2013; Kurra, Park, and Alshareef 2014; Pandey, Rastogi, and Westgate 2014), which show the very low capacitance and render its practical application and commercial production. Also, the amount of active materials obtained from these methods is usually too low because of the formation of a thin layer of PEDOT with low surface area or a thick layer of PEDOT which is unable to give the adequate and fast electrochemical performances. On the other hand, hydrothermal synthesis is a one-step and facile method for the fabrication of electrode, but it is beneficial only for the laboratory scale. To achieve a PEDOT-based supercapacitor with high electrochemical performance, electrode martials play

a vital role with sufficiently thin PEDOT layers and large surface areas.

15.3 PEDOT AND PEDOT-BASED COMPOSITES FOR SUPERCAPACITOR APPLICATION

15.3.1 PEDOT-BASED SUPERCAPACITOR

Many electrically conductive polymers are used for the high-performance supercapacitor application. Among these conducting polymers, researchers paid significant attention towards PEDOT due to its unique features, such as high redox capacitance, narrow bandgap, wide potential window, good optical transparency, structural stability along with high electrical conductivity (Kirchmeyer and Reuter 2005; Anothumakkool et al. 2015; Xia et al. 2015), which make it suitable to be used as high-performance supercapacitors. In this context, Rajesh et al. performed the in-situ hydrothermal (at 150°C) polymerization of EDOT into PEDOT nanostructure deposited on CFC (Figure 15.3).

It was observed that a homogenous orientation of PEDOT nanostructure was grown over the CFC substrate (Figure 15.4) that played an important role for increasing the interfacial contact area throughout electrochemical process and enhanced the electrical conductivity. (Rajesh et al. 2017)

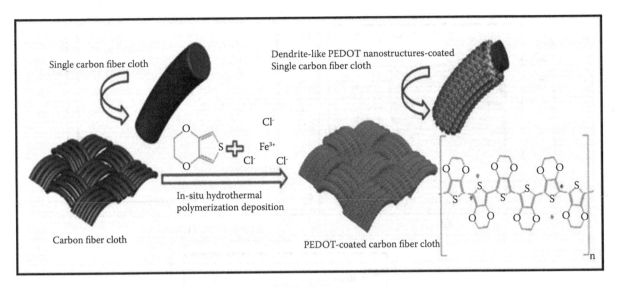

FIGURE 15.3 Schematic representation for the growth of PEDOT nanostructures over the carbon fiber cloth by in-situ hydrothermal polymerization (Rajesh et al. 2017).

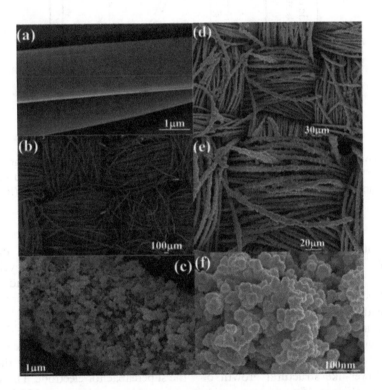

FIGURE 15.4 SEM images of (a, b) bare CFC and (c–f) PEDOT nanostructures over the CFC with different magnifications (Rajesh et al. 2017).

The developed CFC/PEDOT displayed the higher specific capacitance of 203 F g^{-1} at a scan rate of 5 mV s^{-1} with higher power and energy density of 40.25 kW kg^{-1} and 4.4 Wh kg^{-1} (Figure 15.5(a) and (b)), respectively, in 1M H$_2$SO$_4$. Specific capacitance of CFC/PEDOT supercapacitor was evaluated to be 198 F g^{-1} from galvanostatic charge/discharge curve at 0.5 A g^{-1}. It also

exhibited the excellent cyclic stability after 12,000 charging/discharging cycles along with the retention capacity of 86% from its original capacitance (Figure 15.5c).

D'Arcy et al. showed the vapor-phase polymerization technique for the development of nanofibrillar PEDOT for supercapacitors application without using any organic binder and other additives. This showed the

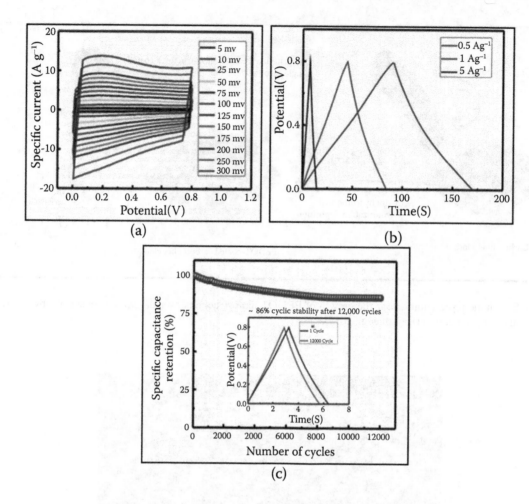

FIGURE 15.5 (a) CV curves of CFC/PEDOT at various scan rates (5–300 mV s^{-1}); (b) charge/discharge curves (selected 0.5– 5 A g^{-1}) of CFC/PEDOT; and (c) cycling stability showed by CFC/PEDOT nanostructure-based supercapacitor at 10 A g^{-1} for 12,000 cycles (inset figure shows the charge/discharge curves of 1st and 12,000th cycles) (Rajesh et al. 2017)

excellent gravimetric capacitance of 175 F g^{-1} along with the retention of its 94% efficiency after 1000 cycles (D'Arcy et al. 2014).

Anothumakkool et al. showed the electro-deposition of PEDOT over the individual carbon fiber substrate at 10 mA cm^{-2}, and resultant composite exhibited the supercapacitive behavior with polymer gel electrolyte of polyvinyl alcohol–sulfuric acid (PVA–H$_2$SO$_4$) (Anothumakkool et al. 2014). Flower-like structural growth of PEDOT over the substrate (Figure 15.6) is responsible for enhancement of the material properties such as surface area and electrical conductivity.

Performance (Figure 15.7) of the developed solid-state device shows very low internal resistance with a higher specific capacitance of 181 F g^{-1} at a discharging current density of 0.5 A g^{-1}. This device also shows a very high areal capacitance of 836 mF cm^{-2} and a volumetric capacitance of 28 F cm^{-3} with a mass-specific capacitance of 111 F g^{-1} which is in close agreement with liquid-state device (112 F g^{-1} in 0.5 M H$_2$SO$_4$) of the same material.

This device also exhibits the outstanding charging/discharging stability for 12,000 cycles at 5 A g^{-1} with constant capacitive behavior and stability under a wide range of humidity and temperature of 30–80% and –10 to 80°C, respectively.

Template-free direct synthesis of PEDOT nanowires (PEDOT-NWs) was also carried out via electro-polymerization of EDOT over the carbon cloth (CC) for the supercapacitor electrode. This device showed the higher gravimetric capacitance of 256.1 F g^{-1} at 0.8 A g^{-1} in neutral aqueous electrolyte of Na$_2$SO$_4$ along with specific energy and specific power of 182.1 Wh kg^{-1} and 13.1 kW kg^{-1}, respectively, at 1.6 V operating potential window (Hsu et al. 2013). Du et al. exhibited the capacitive properties of PEDOT via electro-deposition by using a unipolar-pulsed method on the different substrate such as indium tin oxide (ITO), stainless steel (SS), stainless steel mesh (SSM), and carbon paper (CP) (Du et al. 2018). Among these, PEDOT/CP electrode exhibited the highest mass-specific capacitance of 126.24 F g^{-1} at 1 mA cm^{-2}, which

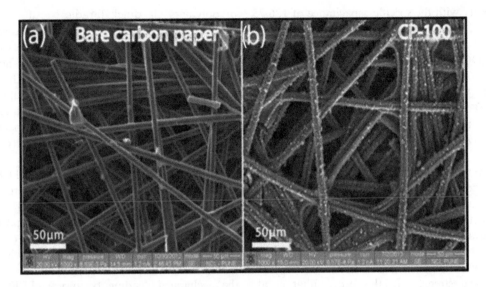

FIGURE 15.6 Deposition of PEDOT over carbon substrate developed via electro-deposition technique shown by SEM images: (a) bare carbon paper and (b) electro-deposited PEDOT on carbon substrate (Anothumakkool et al. 2014). Reproduced by permission of The Royal Society of Chemistry.

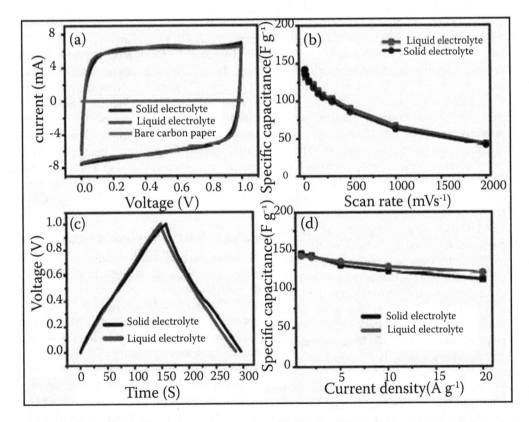

FIGURE 15.7 Comparison between the electrochemical performance of the solid- and liquid-state devices: (a) CV of the both devices investigated at 50 mV s^{-1} (b) show the investigation of both PEDOT-based devices at various scan rates; (c) charging/discharging curve for the both devices measured at 0.5 A g^{-1}. (d) Both PEDOT-based devices show the specific capacitance measured at different current densities (Anothumakkool et al. 2013). Reproduced by permission of The Royal Society of Chemistry.

shows the reliability with the specific capacitance of 126.75 $F g^{-1}$ at $5 mV s^{-1}$. It was also observed that the PEDOT/ITO, PEDOT/CP showed better capacitance behavior as compared to PEDOT/SSM, PEDOT/SS with superior performance rate of 76.78% at $5mA cm^{-2}$ and charging–discharging stability with preservation of 80.57% capacitive behavior for 5000 cycles. Among the aforementioned types of substrate or current collector, ITO was widely used in previous reports (Zhou et al. 2010; Aradilla et al. 2012; Jacob et al. 2014). However, the commercial production of ITO substrate is very difficult due to its heavy weight, fragile nature, and poor adhesion toward the conducting polymers. Therefore, the researchers were the looking forward from the ITO to the metal substrates such as SS (Bhat and Selva Kumar 2007; Patra and Munichandraiah 2007; Mukkabla, Deepa, and Srivastava 2015) and SSM (Vadiyar et al. 2016; Jian et al. 2017), which possess the low sheet resistance with excellent mechanical strength. In this order, Bhat et al. showed the synthesis of n- and p-doped PEDOT deposited over SS substrate that exhibited the specific capacitance of $121 F g^{-1}$ at $10 mV s^{-1}$ (Bhat and Selva Kumar 2007).

Carbon paper (CP) (Anothumakkool et al. 2014; Pandey, Rastogi, and Westgate 2014) is prepared by carbon fiber with porous and rough surface, which not only benefited the PEDOT for wrapping the carbon fiber but also enhanced the electrode specific capacitance, electronic and ionic conductivity, and mechanical integrity, along with the superior stability in charging–discharging cycles (Chu, Tsai, and Sun 2012). Pandey et al. showed the current pulse polymerization techniques for the preparation of PEDOT over the carbon fiber papers, which displayed the specific capacitance of $154.5 F g^{-1}$ with 1–butyl–3–methylimidazolium tetrafluoroborate ($BMIBF_4$) with poly(vinylidene fluoride-hexafluoropropylene) as ionic liquid-based gel polymer electrolyte (Pandey, Rastogi, and Westgate 2014). Among these materials, carbon fiber paper is considered to be the best material as a current collecting substrate for the PEDOT-based supercapacitor attributed to its porous nature and lower cost. Porosity of the substrate will help facilitate the faster ion diffusion rate into the electrodes materials (PEDOT/CP) and also boost the performance of the supercapacitors. Apart from the valuable properties of PEDOT as discussed above, it suffers from downside such as poor cyclic stability (Yang et al. 2015) and low charge-storage capacity (Suppes, Deore, and Freund 2008). This drawback limits its potential application for the supercapacitor as an electrode material. Therefore, to overcome these issues, the current research is oriented toward the addition of conductive materials to produce the high electrochemical performing electrode materials for supercapacitor.

15.3.2 PEDOT AND CARBONACEOUS MATERIALS

Generally conducting polymers show three-dimensional charge-storage mechanisms. They store charge through the electrical double layer as well as from the framework of polymers via quick faradic transfer of charge results in show the high capacitive behavior. However, such type of faradic charge transfer occurred in pseudo-capacitor is slower as compared to EDLCs and hence lowers the charge/discharge cycles. To overcome this issue, recently carbonaceous materials such as graphene, activated carbon, and CNTs are being incorporated into the conducting polymers to enhance their capacitive properties and maintain their high level of power density (Mastragostino, Paraventi, and Zanelli 2000).

Previous reports showed that the PEDOT/carbon composite displayed the high power and energy density for the supercapacitors (Liu and Lee 2008; Kelly, Yano, and Wolf 2009). The PEDOT/carbon-based composite was mainly developed via the in-situ polymerization of EDOT over the carbon materials and through the mechanical blending. However, in-situ polymerization has the advantages over the mechanical blending, because the PEDOT/carbon composite electrode showed the higher conductivity which help to enhance the capacitive properties of supercapacitor (Lei, Wilson, and Lekakou 2011). Liu et al. reported the electrochemical polymerization of EDOT into PEDOT nanotubes for the supercapacitor application. This cell exhibited the specific capacitance of $120 F g^{-1}$ with high power density of $25 kW kg^{-1}$, at the same time retaining its 80% energy density ($5.6 Wh kg^{-1}$) (Liu and Lee 2008).

Moreover, PEDOT has the capability to take a place over the PVDF as an effective binder and overcome the discharge of harmful solvent during electrode processing. The PEDOT is an efficient binder for the activated carbon–carbon black composite as electrode materials and exhibited higher specific capacitance over the use of PVDF binder for the same composite. Moreover, PEDOT has the advantages of acting as a binder for carbonaceous materials and increasing the electrochemical performance of the supercapacitor as well.

15.3.2.1 PEDOT-Activated Carbon Composite

Activated carbon is a frequently used material for the fabrication of electrode in supercapacitor devices because of its ecofriendly nature and lower cost, in comparison to other forms of carbon (Gu and Yushin 2014). It has the pore volume of $>0.5 cm^3 g^{-1}$ with large surface area of $>1000 m^2 g^{-1}$ that is suitable for the supercapacitor applications (Sevilla and Mokaya 2014). Lei et al. reported the activated carbon with carbon black and PEDOT-based electrode for the supercapacitor in an organic electrolyte of $1M TEABF_4/PC$. Loading of PEDOT could enhance the capacitance of electrode and cell with increasing the energy density via its pseudo-capacitive effect (Lei, Wilson, and Lekakou 2011). In this study, three systems were selected: activated carbon/carbon black (ACB), activated carbon/carbon black/5% PEDOT:PSS (ACBP5), and activated carbon/carbon black/10% PEDOT:PSS (ACBP10). For these systems the capacitance value initially enhanced with the addition of PEDOT:PSS (most with addition of 5%) as observed in their CV analysis and galvanostatic charge–discharge curves (Figure 15.8 (a) and (b)). However, further increase in the concentration of PEDOT:PSS to 10% would decrease the capacitance value due to slower ion diffusion at high PEDOT concentration. Thereafter, the results for ACBP5 were converted into the Ragone plot for this supercapacitor (Figure 15.8 (c)), and it was observed that the increasing of capacitance value also enhanced the energy density of the cell which approached 200 Wh kg^{-1}, whereas the power density reached 10 kW kg^{-1}.

15.3.2.2 PEDOT–Graphene-Based Composite

Graphene and graphene-based composites are very demanding for the supercapacitor application due to their multifunctional properties. The graphene is a one-atom thick sheet of carbon that is arranged in a hexagonal pattern has paid the attention toward the energy storage and conversion devices (Novoselov et al. 2004). The graphene has several unique features, such as

very high electrical conductivity, interesting optical, catalytic, and mechanical properties along with the enormously high surface area (2630 m^2 g^{-1}), which made it suitable for the high-performance supercapacitor application and have the theoretical capacitance of about 550 F g^{-1} (Xia et al. 2009). Despite of having this high theoretical capacitance, the observed capacitance is limited due to high surface energy of graphene nanosheet which induces to agglomerate quickly; thus, we are not able to use the entire surface area. For this reason, it is hard to reach the intrinsic capacitance of graphene sheets.

Hence, to utilize maximum surface area and increase the distance between graphene sheets, we need to modify its surface via covalent and noncovalent methods (Karakoti et al. 2018). Many materials, such as polymers, small organic molecules, metal nanoparticles, and other carbon nanomaterials (Bai, Li, and Shi 2011), are used for enhancing the properties of graphene-based composite and help use the maximum surface area. Among these materials the PEDOT has drawn much attention for tailoring the properties of graphene such as excellent conductivity, transparency,

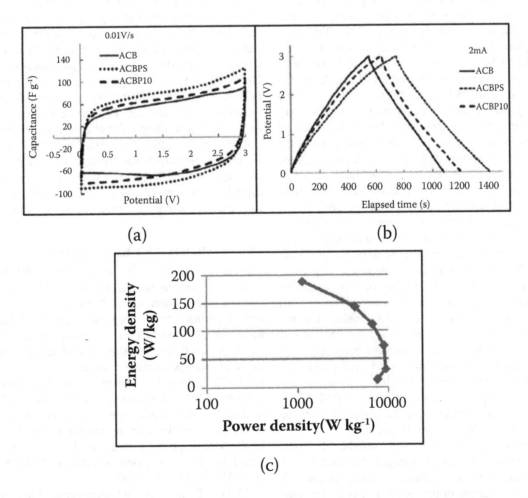

(a)   (b)

(c)

FIGURE 15.8 (a) CV analysis of ACB, ACBP5, and ACBP10 electrodes in 1M TEABF$_4$–PC electrolyte at the scan rate of 0.01 V s^{-1}; (b) charge/discharge curves for ACB, ACBP5, and ACBP10 electrodes at the current density of 2 mA; (c) Ragone plot for the ACBP5-based supercapacitor (Lei, Wilson, and Lekakou 2011).

and environmental stability (Groenendaal et al. 2000; Liu and Lee 2008, Liu et al. 2010). Basically, three common methods are used for the development of graphene/PEDOT composites—direct mixing of PEDOT with grapheme (Hong et al. 2008); polymerization of EDOT into PEDOT via in-situ chemical method that is performed under the pre-existence of graphene (Wang et al. 2009); and the reduction of graphene oxide (GO) via chemical method within PEDOT:PSS that is used as a dispersing agent (Jo et al. 2011). Among these methods the chemical method of the polymerization of PEDOT is useful for the development of high-performing graphene/PEDOT composite, but still it has many challenges in this field. To overcome this issue, Lee and coworkers proposed the layer-by-layer stacking of G/PEDOT composite via electrochemical method and achieved the specific capacitance of 139 F g^{-1} at the current density of 1 A g^{-1} (Lee et al. 2012). Reduced graphene oxide (rGO), a derivative of graphene, becomes a promising material for the energy-storage devices because of its good electrical conductivity and higher surface area (Stoller et al. 2008; Wang et al. 2009). Many conducting polymers such as polyaniline (PANi), poly(3,4-ethylenedioxythiophene) (PEDOT), and polypyrrole (PPy) are used to provide composite with rGO for the development of electrode in supercapacitor applications (Stankovich et al. 2006; Saha et al. 2010; Wu et al. 2010). To explore this study, Zhang et al. showed the direct coating of PEDOT over the rGO sheet for the high-performance supercapacitive behavior which displayed the specific capacitance of 108 F g^{-1} at the current density of 0.3 A g^{-1} (Zhang and Zhao 2012). Although rGO is very a promising material for energy-storage devices, it has several limitations, including poor water solubility that restricts the direct coating of conducting polymers over it, especially for the low water-soluble polymers such as PEDOT. To overcome this issue, Xu et al. proved that the sulfonated rGO has the better water solubility used for the synthesis of sulfonated graphene and PEDOT composites (Xu et al. 2009).

Alvi et al. presented the oxidative chemical polymerization of EDOT and made a composite with the graphene in 1:1 ratio for the supercapacitor application by using 2M HCl and 2M H$_2$SO$_4$. CV analysis of this composite was performed Figure 15.9 (a) and (b) at a different scan rate and showed the different capacitive value, but the highest value was found to be 304 F g^{-1} for 2M HCl at 10 mV s. On the other hand, charge–discharge curve of PEDOT and graphene composite in 2M HCl performed under a different scan rate (Figure 15.10) and has the capacitance of 374 F g^{-1} at the current density of 0.01 A g^{-1}, which is in close agreement with the capacitance value evaluated by CV analysis (Alvi et al. 2011).

Urgent demand of the flexible and wearable electronic devices in various fields has encouraged the invention of more flexible and sophisticated energy-storing devices with higher capacitance values. In this order, Liu et al. reported the fabrication of highly flexible and conducting composite of reduced graphene oxide/PEDOT:PSS through the simple bar-coating method (Liu et al. 2015). This device is bendable and rolled up without sacrificing the electrochemical performance of the devices. The device exhibited the moderately high areal electrochemical performance of 448 mF cm^{-2} at a scan rate of 10 mV s^{-1} in PVA–H$_3$PO$_4$ gel as electrolyte with mass loading of 8.49 mg cm^{-2} of composite materials for electrode fabrication. However, this device showed a very low gravimetric capacitance of 81 F g^{-1} at the lower scan rate of 5 mV s^{-1}, which is lower than the previous reports on graphene–PEDOT-based supercapacitors (100–300 F g^{-1}). Sun et al. reported the microwave-assisted in-situ method for the synthesis of graphene/PEDOT hybrid which exhibited the specific capacitance of 270 F/g at the current density of 1 A/g by using 1M H$_2$SO$_4$ electrolyte In addition, it has 34 Wh kg^{-1} energy density at 25 kW kg^{-1} power density with 90% retention capacity after the 10,000 cycles at 5 A g^{-1} current density (Sun et al. 2013).

Recently, Ahmed et al. synthesized the PEDOT/rGO composite as electrode materials for supercapacitor through hydrothermal method. The result exhibited the specific capacitance of 102.8 F g^{-1} at 10 mV s^{-1} using 6M KOH as electrolyte solution and also provided the specific power of 4.6 kW kg^{-1} and specific energy of 1.8 Wh kg^{-1} (Ahmed and Rafat 2018). Kumar et al. reported the large-area, all-solid-state, and flexible supercapacitors through spray deposition techniques over the CC (carbon cloth)/PEDOT:PSS–rGO (20 wt.%) that exhibited specific capacitances of 170 F g^{-1} at 10 mV s^{-1} and also displayed the energy and power densities of 11.0 Wh kg^{-1} and 4460 W kg^{-1}, respectively, in a PVA/H$_3$PO$_4$-based gel electrolyte. This device showed the outstanding cyclic and bending stability with 100% retention of capacitive behavior after 2000 cycles and 95% stability under 0.5 mm of bending radius. Three devices of the CC (carbon cloth)/PEDOT:PSS–rGO (20 wt.%) linked in a series and capable to glow white and green LEDs for ~45 s when it is fully charged (Kumar, Ginting, and Kang 2018).

15.3.2.3 PEDOT–CNT-Based Composite

The discovery of CNTs by Iijima (1991) has opened a new window in nanotechnology and material science due to their novel characteristics, such as electronic, optoelectronic, high chemical stability, larger surface area, and electrochemical properties (Lu et al. 2011), which make them potential candidates for the development of high-performance supercapacitor. At molecular level, CNTs look like the graphene sheet rolled up into cylindrical tube at a nanoscale dimension, and the CNTs obtained from a single layer of graphene sheet are known as the single-walled carbon nanotubes (SWCNTs). If further addition of more than one layers around SWCNTs called multiwalled carbon nanotubes (MWCNTs) (Iijima 1991; Harris 2004; Dai 2006).

MWCNT network interconnected high mesoporous structure with diameter ranging from 2 to 50 nm that allows unbroken charge distribution and provides the effortless diffusion path for the extensive range of electrolyte ion on

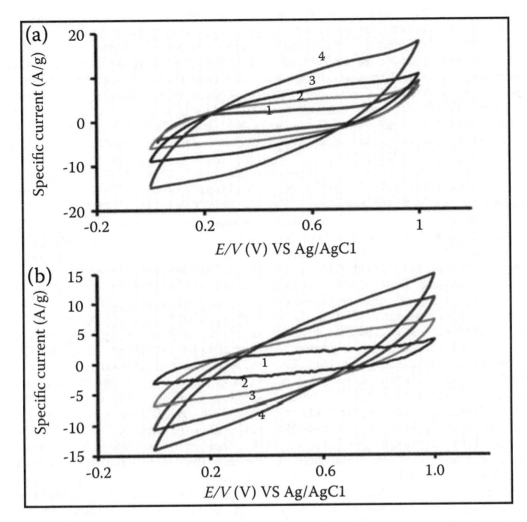

FIGURE 15.9 (a) CV analysis of G-PEDOT nanocomposite-based electrodes in 2M H_2SO_4 at different scan rates: (1) 10 mV s^{-1}, (2) 20 mV s^{-1}, (3) 50 mV s^{-1}, and (4) 100 mV s^{-1}; (b) CV of G-PEDOT nanocomposite-based electrodes in 2M HCl at different scan rates: (1) 10 mV s^{-1}, (2) 20 mV s^{-1}, (3) 50 mV s^{-1}, and (4) 100 mV s^{-1} (Alvi et al. 2011).

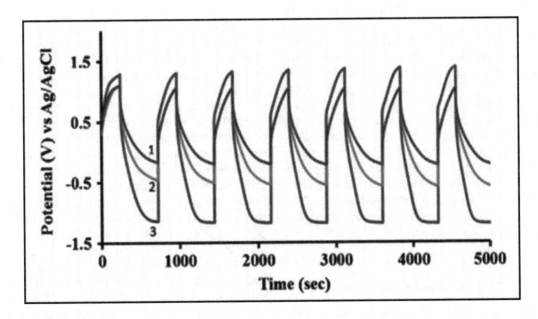

FIGURE 15.10 Charging/discharging curve of G-PEDOT-based nanocomposite in 2M HCl at various discharging currents: (1) 1 mA, (2) 10 mA, and (3) 20 mA (Alvi et al. 2011).

electrode surface (Boyea et al. 2007). Due to these intrinsic properties, it exhibits the specific capacitance comparable to that of activated carbon, although it has lower surface area (Zhang et al. 2011). Conway et al. reported the MWCNT-based supercapacitor with achieving the specific capacitance of 108 F g^{-1} at a scan rate of 10 mV s^{-1} (Conway and Pell 2003). In this order, the SWCNT-based electrode also reported in KOH electrolyte that exhibits the promising power density and energy density of 20 kW kg^{-1} and 10 Wh kg^{-1}, respectively (Lu and Dai 2010).

According to the literature, CNT can act as the potential candidate and synergistic host for polymer/MWCNT composite materials as electrode for high-performing supercapacitors (Li et al. 2009). Chen et al. published a report on core-shell PEDOT/PSS- modified MWCNT nanocomposite synthesized under hydrothermal process via in-situ technique of polymerization with high specific capacitive behavior of 198.2 F g^{-1} at the current density of 0.5 A g^{-1}, and the degradation of capacitance observed after 2000 cycles was 26.9% from the original capacitive value (Chen et al. 2009). Lota et al. described the chemical polymerization of PEDOT and also prepare its composite with the CNTs by the mechanical mixing techniques. The composite PEDOT (15 wt.%) exhibits the good specific capacitance of 95 F g^{-1} with the three-electrode system. On the other hand, capacitance values of pure CNTs and PPEDOT were evaluated to be 10–15 F g^{-1} and 80–100 F g^{-1}, respectively (Lota, Khomenko, and Frackowiak 2004). Antiohos et al. showed the compositional effects of PEDOT–PSS/SWNT (20% SWNT, 30% SWNT, 40% SWNT, and 50% SWNT) films on electrochemical performance of supercapacitor (Antiohos et al. 2011). According to this, dispersed CNTs on PEDOT matrix boosted the capacitance value by 65% and the energy density by a factor of 3 while showing retention of capacitive behavior over 1000 cycles. Here, 40% SWCNT-based composite exhibited the best performance in terms of capacitance and energy density, and maintained power density compared to 20% SWCNT-, 30% SWCNT-, and 50% SWCNT-based composites. The device with 40% SWCNT-based composite showed the specific capacitance of 104 F g^{-1} at 1 A g^{-1} with maximum energy and power density of 7 Wh kg^{-1} and 825 W kg^{-1}, respectively, and displayed retention capacitive value greater than 90% over the 1000 cycles. In this order, Zhao et al. reported the cellulose-mediated PEDOT:PSS/MWCNT composite film as a flexible and conductive electrode for supercapacitor application. This composite showed the excellent electrochemical performance (higher specific capacitance of 485 F g^{-1} at 1 A g^{-1}) and the lower resistance of 0.45 Ω with retention of the capacitive value to 95% after 2000 cycles at 2 A g^{-1} (Zhao et al. 2017).

15.3.3 PEDOT–METAL OXIDE-BASED COMPOSITES

Several metal oxides, especially some transition metal oxides, have shown the potential to be used as the electrode

materials for supercapacitors. These metal oxides should have some favorable properties such as electronically conductive, existence of multiple oxidation state without any phase. Besides, it also provides the path to the protons for freely intercalation into oxide lattice which is suitable for the supercapacitor application (Wang, Zhang, and Zhang 2012). Among a series of transition metal oxides, only a few metal oxides are used as electrode fabrication for supercapacitors such as MnO_2, Co_3O_4, NiO, Fe_3O_4, RuO_x, SnO_2, and V_2O_5. According to the previous reports, RuO_x (showed the 1358 F g^{-1} theoretical specific capacitance)-based supercapacitor has remarkable capacitive properties. But its cost and toxic nature limit its application. Other low-cost metal oxides such as MnO_2, Co_3O_4, and NiO are also used, but they suffer from the poor stability and poor electrical conductivity, which result in lower specific capacitance. Accordingly, many efforts have been made by the researchers for improving and optimizing the overall performance of the low-cost metal oxide as electrode for the supercapacitor by the development of composite with carbon materials, secondary transition metal oxide, and conducting polymer (Wang, Zhang, and Zhang 2012). On the other hand, pure PEDOT-based supercapacitors suffer from serious problem of typical volumetric swelling and shrinking at the time of insertion and ejection of counterions, which significantly reduce the overall electrochemical performance. To overcome this issue, we will discuss about the metal oxide composite with the conductive polymer, especially PEDOT, which is able to enhance the stability, conductivity, and flexibility in the electrodes along with increasing the overall performance of the supercapacitors.

In this order, Sen et al. showed the PEDOT/MnO_2-based nanocomposite for high-performance supercapacitor with 315 F g^{-1} in comparison with MnO_2 which showed lower capacitance of 158 F g^{-1}. The device performance was evaluated at the operating voltage of 1.2 V with 1M $LiClO_4$ in acetonitrile, a nonaqueous electrolyte solution. The capacitance was originated by the intercalation of metal ions at the time of reduction, and de-intercalation upon oxidation process predominates over the surface adsorption and desorption of metal ions into electrode materials (Sen et al. 2013). Ranjusha et al. reported that the freeze-dried MnO_2-embedded PEDOT:PSS hybrid sponge can be used as a electrode for the supercapacitor. The hybrid sponge exhibits the mass-specific capacitance value of 10,688 F g^{-1} with KOH electrolyte, which was 35% greater than PEDOT:PSS sponges as evaluated from Wei bull statistics. This further explores the fully functional asymmetric coin cell that shows the energy density of 200 mWh kg^{-1} and power density of 6.4 kW kg^{-1} (Ranjusha et al. 2014). Tang et al. exhibited the controlled fabrication of MnO_2@PEDOT@MnO_2 hierarchical nanocomposites as positive electrode material for high-performing supercapacitor on nickel foam (NF) substrate (Tang et al. 2015). This electrode showed the maximum specific capacitance of 449 F g^{-1} with aqueous solution of Na_2SO_4 as the electrolyte that was evaluated through galvanostatic discharge at the current density of 0.5 A g^{-1}. Here, the prepared MnO$_2$@PEDOT@MnO$_2$ acted as positive electrode and activated carbon as negative electrode operating at the voltage window of 0–1.8 V, with high energy density and power density of 47.8 Wh k g^{-1} and 180 W k g^{-1}, respectively, with 91.3% retention of capacitive value after 5000 cycles. The performance of this supercapacitor is superior over the AAO (anodic aluminum oxide)-templated MnO$_2$ nanofibril/nanowire arrays (Duay et al. 2013) and MnO$_2$/PEDOT coaxial nanowires (270 F g^{-1} at 250 mV s^{-1}) (Liu and Lee 2008; Liu, Duay, and Lee 2010, 2011). Lee et al. used the surface-modified Ni(OH)$_2$ nanosheets with PEDOT:PSS (P–Ni(OH)$_2$) as an electrode for high capacitance supercapacitor. Here, the composite of P–Ni(OH)$_2$ displayed higher capacitance value of 502 F g^{-1} than pure Ni(OH)$_2$ of 448 F g^{-1} at the current density of 1 A g^{-1}, and the result clearly indicates that the PEDOT enhances the performance of the pure Ni (OH)$_2$ (Lee et al. 2019). Lee and Bae reported the fabric-based PEDOT:PSS hybridized SnO$_2$ nanoparticles for the high-performing supercapacitor. Herein, composite of PEDOT/SnO$_2$ was synthesized by the dispersion of SnO$_2$ nanopowder in aqueous PEDOT: PSS solution. This composite-based electrode exhibits the excellent specific capacitance of 126 F g^{-1} at the scan rate of 20 mV s^{-1} using 2M H$_2$SO$_4$ as an electrolyte, whereas the value was 6 F g^{-1} for bare PEDOT:PSS at the same scan rate (Lee and Bae 2015). Huang et al. reported the electrochemical loading of hydrous RuO$_2$ particles into the PEDOT:PSS matrix via cyclic voltammetry at various potential cycles for the fabrication of PEDOT–PSS–RuO$_2 \cdot x$H$_2$O electrode to be used as supercapacitor. Here, this electrode exhibits the maximum specific capacitance of 653 F g^{-1} using 0.5M H$_2$SO$_4$ as an electrolyte (Huang et al. 2006).

15.4 CONCLUSION AND REMARKS

The urgent need of flexible energy devices with the growing world population has drawn the attention toward the efficient conducting polymer materials and related nanocomposites for the wider scope of material science and technology to fulfill the demand of the today's energy crises. This chapter briefly discussed the class of conducting polymers especially dedicated to supercapacitor based on the PEDOT and its composites for tailoring their properties along with the methods for synthesizing PEDOT, and the recent developments in the field, which covers various flourishing and potential works for achieving the high electrochemical performance. We hope this chapter acts as a stimulant to generate new ideas for the development of new techniques and new materials in the area of PEDOT-based materials for more-efficient and high-performance supercapacitors. Nonetheless, many issues need to be overcome related to PEDOT and its materials-based supercapacitors such as energy density,

power density, and charge–discharge cycle stability, suitable for the real applications. In addition, we need the strong and dedicated efforts for promoting the large-scale commercial applications. This chapter also helps in better understanding the effects of doping and modification on the pristine and hybrids conducting PEDOT, and these findings can be utilized as a pathway in future for the development of new PEDOT-based materials to be used in supercapacitor application.

PEDOT is a class of conducting polymer that has greatly attracted toward the supercapacitor application. PEDOT has high conductivity, narrow bandgap, low redox potential, and good stability at ambient conditions. Besides, it has some other advantages such as readily available and simple process for the fabrication of its film or 3D structure. Due to having these properties it is a suitable candidate for the supercapacitor.

In spite of these advantages, PEDOT has some major drawbacks, which lead to a poor electrochemical performance of the supercapacitor. PEDOT has the difficulties to be used directly as the electrode materials for the supercapacitor, because it swells and shrinks at the time of intercalation/de-intercalation mechanism, which results in the lowering of its electrochemical cyclic performance (Hu et al. 2010). In addition, the polymeric nature of PEDOT hinders to maintain the agreeable cyclic stability after many cycles which lead to short life of supercapacitor.

Another disadvantage of PEDOT is the poor solubility in common solvents and thus shows relatively poor performance, which deters its application in flexible supercapacitor. Therefore, highly efficient and scalable pathways for the fabrication of PEDOT-based flexible conducting films which are capable to show high capacitance value with long life are significantly essential (Wang et al. 2016). To make it soluble, it may be combined with suitable insulating materials such as PSS (polystyrene sulfonate) moiety that ultimately enhances its solubility; however, it reduces the conductivity as well. Besides, the acidic (pH between 1 and 2) and hygroscopic nature of PEDOT:PSS is very harmful as those factors corrode the electrodes and thus result in the degradation of the device and reduce the electrochemical performance. Nonetheless, mostly PEDOT:PSS is being used as the electrode material for the supercapacitor. To overcome these problems and modify its properties, several approaches have been employed by adding dopant/additives and synthesizing the nanocomposites with some other materials such as carbon nanomaterials and metal oxide.

ACKNOWLEDGMENT

Authors acknowledge the financial support from National Mission of Himalayan Studies, Kosi Kataramal, Almora, India (Ref No.: NMHS/MG-2016/002/8503-7) and

Department of Science and Technology INSPIRE division (Ref No. IF150750 & IFA16-MS81), New Delhi, India.

REFERENCES

Ahmed, Sultan, and M Rafat. 2018. "Hydrothermal synthesis of PEDOT/rGO composite for supercapacitor applications." *Materials Research Express* no. 5 (1):015507.

Alvi, Farah, Manoj K Ram, Punya A Basnayaka, Elias Stefanakos, Yogi Goswami, and Ashok Kumar. 2011. "Graphene–polyethylenedioxythiophene conducting polymer nanocomposite based supercapacitor." *Electrochimica Acta* no. 56 (25):9406–9412.

Anothumakkool, Bihag, Siddheshwar N Bhange, Manohar V Badiger, and Sreekumar Kurungot. 2014. "Electrodeposited polyethylenedioxythiophene with infiltrated gel electrolyte interface: a close contest of an all-solid-state supercapacitor with its liquid-state counterpart." *Nanoscale* no. 6 (11):5944–5952.

Anothumakkool, Bihag, Siddheshwar N Bhange, Sreekuttan M Unni, and Sreekumar Kurungot. 2013. "1-Dimensional confinement of porous polyethylenedioxythiophene using carbon nanofibers as a solid template: an efficient charge storage material with improved capacitance retention and cycle stability." *RSC Advances* no. 3 (29):11877–11887.

Anothumakkool, Bihag, Roby Soni, Siddheshwar N Bhange, and Sreekumar Kurungot. 2015. "Novel scalable synthesis of highly conducting and robust PEDOT paper for a high performance flexible solid supercapacitor." *Energy & Environmental Science* no. 8 (4):1339–1347.

Antiohos, Dennis, Glenn Folkes, Peter Sherrell, Syed Ashraf, Gordon G Wallace, Phil Aitchison, Andrew T Harris, Jun Chen, and Andrew I Minett. 2011. "Compositional effects of PEDOT-PSS/single walled carbon nanotube films on supercapacitor device performance." *Journal of Materials Chemistry* no. 21 (40):15987–15994.

Aradilla, David, Francesc Estrany, Elaine Armelin, and Carlos Alemán. 2012. "Ultraporous poly (3, 4-ethylenedioxythiophene) for nanometric electrochemical supercapacitor." *Thin Solid Films* no. 520 (13):4402–4409.

Bai, Hua, Chun Li, and Gaoquan Shi. 2011. "Functional composite materials based on chemically converted graphene." *Advanced Materials* no. 23 (9):1089–1115.

Bhandari, Shweta M Deepa S Singh, Govind Gupta, and Rama Kant. 2008. "Redox behavior and optical response of nanostructured poly(3,4-ethylenedioxythiophene) films grown in a camphorsulfonic acid based micellar solution." *Electrochimica Acta* no. 53 (7):3189–3199. doi: 10.1016/j.electacta.2007.11.018.

Bhat, D Krishna, and M Selva Kumar. 2007. "N and p doped poly(3,4-ethylenedioxythiophene) electrode materials for symmetric redox supercapacitors." *Journal of Materials Science* no. 42 (19):8158–8162. doi: 10.1007/s10853-007-1704-9.

Boyea, JM, RE Camacho, SP Sturano, and WJ Ready. 2007. "Carbon nanotube-based supercapacitors: technologies and markets." *Nanotechnology Law & Business* no. 4:19.

Cai, Zhenbo, Li Li, Jing Ren, Longbin Qiu, Huijuan Lin, and Huisheng Peng. 2013. "Flexible, weavable and efficient microsupercapacitor wires based on polyaniline composite fibers incorporated with aligned carbon nanotubes." *Journal of Materials Chemistry A* no. 1 (2):258–261.

Carlberg, JC, and O Inganäs. 1997. "Poly (3, 4-ethylenedioxythiophene) as electrode material in electrochemical capacitors." *Journal of the Electrochemical Society* no. 144 (4):L61–L64.

Chen, Li, Changzhou Yuan, Hui Dou, Bo Gao, Shengyao Chen, and Xiaogang Zhang. 2009. "Synthesis and electrochemical capacitance of core–shell poly (3, 4-ethylenedioxythiophene)/poly (sodium 4-styrenesulfonate)-modified multi-walled carbon nanotube nanocomposites." *Electrochimica Acta* no. 54 (8):2335–2341.

Cheng, Qian, Jie Tang, Jun Ma, Han Zhang, Norio Shinya, and Lu-Chang Qin. 2011. "Polyaniline-coated electro-etched carbon fiber cloth electrodes for supercapacitors." *The Journal of Physical Chemistry C* no. 115 (47):23584–23590.

Choi, Hojin, and Hyeonseok Yoon. 2015. "Nanostructured electrode materials for electrochemical capacitor applications." *Nanomaterials* no. 5 (2):906–936.

Chu, Chun-Yu, Jin-Ting Tsai, and Chia-Liang Sun. 2012. "Synthesis of PEDOT-modified graphene composite materials as flexible electrodes for energy storage and conversion applications." *International Journal of Hydrogen Energy* no. 37 (18):13880–13886.

Cloutet, Eric, Muhammad Mumtaz, and Henri Cramail. 2009. "Synthesis of PEDOT latexes by dispersion polymerization in aqueous media." *Materials Science and Engineering: C* no. 29 (2):377–382.

Conway, BE, and WG Pell. 2003. "Double-layer and pseudocapacitance types of electrochemical capacitors and their applications to the development of hybrid devices." *Journal of Solid State Electrochemistry* no. 7 (9):637–644.

Corradi, R, and SP Armes. 1997. "Chemical synthesis of poly (3, 4-ethylenedioxythiophene)." *Synthetic Metals* no. 84 (1–3):453–454.

Dai, Liming. 2006. *Carbon Nanotechnology: Recent Developments in Chemistry, Physics, Materials Science and Device Applications.* Elsevier.

Damlin, Paul, C Kvarnström, and Alex Ivaska. 2004. "Electrochemical synthesis and in situ spectroelectrochemical characterization of poly (3, 4-ethylenedioxythiophene)(PEDOT) in room temperature ionic liquids." *Journal of Electroanalytical Chemistry* no. 570 (1):113–122.

D'Arcy, Julio M, Maher F El-Kady, Pwint P Khine, Linghong Zhang, Sun Hwa Lee, Nicole R Davis, David S Liu, Michael T Yeung, Sung Yeol Kim, and Christopher L Turner. 2014. "Vapor-phase polymerization of nanofibrillar poly (3, 4-ethylenedioxythiophene) for supercapacitors." *ACS Nano* no. 8 (2):1500–1510.

Dietrich, Michael, Jürgen Heinze, Gerhard Heywang, and Friedrich Jonas. 1994. "Electrochemical and spectroscopic characterization of polyalkylenedioxythiophenes." *Journal of Electroanalytical Chemistry* no. 369 (1–2):87–92.

Du, Hai Yan, Xiao Xiao Liu, Zhe Ren, and Pan Pan Liu. 2018. "Capacitance characteristic of PEDOT electrodeposited on different substrates." *Journal of Solid State Electrochemistry* no. 22 (12):3947–3954.

Du, X, and Z Wang. 2003. "Effects of polymerization potential on the properties of electrosynthesized PEDOT films." *Electrochimica Acta* no. 48 (12):1713–1717.

Du, Xu-Sheng, Cui-Feng Zhou, and Yiu-Wing Mai. 2009. "Novel synthesis of poly (3, 4-ethylenedioxythiophene) nanotubes and hollow micro-spheres." *Materials Letters* no. 63 (18–19):1590–1593.

Duay, Jonathon, Stefanie A Sherrill, Zhe Gui, Eleanor Gillette, and Sang Bok Lee. 2013. "Self-limiting electrodeposition of hierarchical MnO2 and M (OH) 2/MnO2 nanofibril/nanowires: mechanism and supercapacitor properties." *ACS Nano* no. 7 (2):1200–1214.

El, Enany GM. 2009. "Electrochemical synthesis of high charge capacity poly (3, 4-ethylenedioxythiophene) using pulse polymerization technique." *Zaštita materijala* no. 50 (4):193–196.

Gamby, J, PL Taberna, P Simon, JF Fauvarque, and M Chesneau. 2001. "Studies and characterisations of various activated carbons used for carbon/carbon supercapacitors." *Journal of Power Sources* no. 101 (1):109–116.

Gao, Yan, Gilles P Robertson, Michael D Guiver, Serguei D Mikhailenko, Xiang Li, and Serge Kaliaguine. 2006. "Low-swelling proton-conducting copoly (aryl ether nitrile) s containing naphthalene structure with sulfonic acid groups meta to the ether linkage." *Polymer* no. 47 (3):808–816.

Groenendaal, Lambertus, Friedrich Jonas, Dieter Freitag, Harald Pielartzik, and John R Reynolds. 2000. "Poly (3, 4-ethylenedioxythiophene) and its derivatives: past, present, and future." *Advanced Materials* no. 12 (7):481–494.

Gu, Wentian, and Gleb Yushin. 2014. "Review of nanostructured carbon materials for electrochemical capacitor applications: advantages and limitations of activated carbon, carbide-derived carbon, zeolite-templated carbon, carbon aerogels, carbon nanotubes, onion-like carbon, and graphene." *Wiley Interdisciplinary Reviews: Energy and Environment* no. 3 (5):424–473.

Gujar, TP, Woo-Young Kim, Indra Puspitasari, Kwang-Deog Jung, and Oh-Shim Joo. 2007. "Electrochemically deposited nanograin ruthenium oxide as a pseudocapacitive electrode." *International Journal of Electrochemical Science* no. 2:666–673.

Halper, Marin S, and James C Ellenbogen. 2006. *Supercapacitors: A Brief Overview*. The MITRE Corporation, McLean, VA. 1–34.

Han, Dongxue, Guifu Yang, Jixia Song, Li Niu, and Ari Ivaska. 2007. "Morphology of electrodeposited poly (3, 4-ethylenedioxythiophene)/poly (4-styrene sulfonate) films." *Journal of Electroanalytical Chemistry* no. 602 (1):24–28.

Han, Moon Gyu, and Steven P Armes. 2003. "Synthesis of poly (3, 4-ethylenedioxythiophene)/silica colloidal nanocomposites." *Langmuir* no. 19 (11):4523–4526.

Han, Moon Gyu, and Stephen H Foulger. 2004. "Preparation of poly (3, 4-ethylenedioxythiophene)(PEDOT) coated silica core–shell particles and PEDOT hollow particles." *Chemical Communications* no. 19:2154–2155.

Harris, Peter JF. 2004. *Carbon Nanotubes and Related Structures: New Materials for the Twenty-First Century*. AAPT.

Hashmi, SA, and HM Upadhyaya. 2002. "Polypyrrole and poly (3-methyl thiophene)-based solid state redox supercapacitors using ion conducting polymer electrolyte." *Solid State Ionics* no. 152:883–889.

Henderson, Allister MJ, Jennifer M Saunders, James Mrkic, Paul Kent, Jeff Gore, and Brian R Saunders. 2001. "A new method for stabilising conducting polymer lattices using short chain alcohol ethoxylate surfactants." *Journal of Materials Chemistry* no. 11 (12):3037–3042.

Heywang, Gerhard, and Friedrich Jonas. 1992. "Poly (alkylenedioxythiophene) s—new, very stable conducting polymers." *Advanced Materials* no. 4 (2):116–118.

Hong, Wenjing, Yuxi Xu, Gewu Lu, Chun Li, and Gaoquan Shi. 2008. "Transparent graphene/PEDOT–PSS composite films as counter electrodes of dye-sensitized solar cells." *Electrochemistry Communications* no. 10 (10):1555–1558.

Hsu, Yu-Kuei, Ying-Chu Chen, Yan-Gu Lin, Li-Chyong Chen, and Kuei-Hsien Chen. 2013. "Direct-growth of poly (3, 4-ethylenedioxythiophene) nanowires/carbon cloth as hierarchical supercapacitor electrode in neutral aqueous solution." *Journal of Power Sources* no. 242:718–724.

Hu, Liangbing, Mauro Pasta, Fabio La Mantia, LiFeng Cui, Sangmoo Jeong, Heather Dawn Deshazer, Jang Wook Choi, Seung Min Han, and Yi Cui. 2010. "Stretchable, porous, and conductive energy textiles." *Nano Letters* no. 10 (2):708–714.

Huang, Li-Ming, Hong-Ze Lin, Ten-Chin Wen, and A Gopalan. 2006. "Highly dispersed hydrous ruthenium oxide in poly (3, 4-ethylenedioxythiophene)-poly (styrene sulfonic acid) for supercapacitor electrode." *Electrochimica Acta* no. 52 (3):1058–1063.

Iijima, Sumio. 1991. "Helical microtubules of graphitic carbon." *Nature* no. 354 (6348):56.

Iro, Zaharaddeen S, C Subramani, and SS Dash. 2016. "A brief review on electrode materials for supercapacitor." *International Journal of Electrochemical Science* no. 11 (12):10628–10643.

Izadi-Najafabadi, Ali, Satoshi Yasuda, Kazufumi Kobashi, Takeo Yamada, Don N Futaba, Hiroaki Hatori, Motoo Yumura, Sumio Iijima, and Kenji Hata. 2010. "Extracting the full potential of single-walled carbon nanotubes as durable supercapacitor electrodes operable at 4 V with high power and energy density." *Advanced Materials* no. 22 (35):E235–E241.

Jacob, Dona, PA Mini, Avinash Balakrishnan, SV Nair, and KRV Subramanian. 2014. "Electrochemical behaviour of graphene–poly (3, 4-ethylenedioxythiophene)(PEDOT) composite electrodes for supercapacitor applications." *Bulletin of Materials Science* no. 37 (1):61–69.

Jalili, Rouhollah, Joselito M Razal, and Gordon G Wallace. 2013. "Wet-spinning of PEDOT: PSS/functionalized-SWNTs composite: a facile route toward production of strong and highly conducting multifunctional fibers." *Scientific Reports* no. 3:3438.

Jang, Jyongsik. 2006. "Conducting polymer nanomaterials and their applications." In *Emissive Materials Nanomaterials*, 189–260. Springer.

Jayalakshmi, Mandapati, and K Balasubramanian. 2008. "Simple capacitors to supercapacitors-an overview." *International Journal of Electrochemical Science* no. 3 (11):1196–1217.

Jennings, Aaron A, Sara Hise, Bryant Kiedrowski, and Caleb Krouse. 2009. "Urban battery litter." *Journal of Environmental Engineering* no. 135 (1):46–57.

Jian, Xuan, Hui-min Yang, Jia-gang Li, Er-hui Zhang, and Zhen-hai Liang. 2017. "Flexible all-solid-state high-performance supercapacitor based on electrochemically synthesized carbon quantum dots/polypyrrole composite electrode." *Electrochimica Acta* no. 228:483–493.

Jo, Kiyoung, Taemin Lee, Hyun Jung Choi, Ju Hyun Park, Dong Jun Lee, Dong Wook Lee, and Byeong-Su Kim. 2011. "Stable aqueous dispersion of reduced graphene nanosheets via non-covalent functionalization with conducting polymers and application in transparent electrodes." *Langmuir* no. 27 (5):2014–2018.

Jonas, Friedrich, and Gerhard Heywang. 1994. "Technical applications for conductive polymers." *Electrochimica Acta* no. 39 (8–9):1345–1347.

Jonas, Friedrich, and L Schrader. 1991. "Conductive modifications of polymers with polypyrroles and polythiophenes." *Synthetic Metals* no. 41 (3):831–836.

Karakoti, Manoj, Sunil Dhali, Sravendra Rana, Sanka Rama V Siva Prasanna, SPS Mehta, and Nanda G Sahoo. 2018. "Surface modification of carbon-based nanomaterials for polymer nanocomposites." In Ahmad Fauzi Ismail and Pei Sean Goh (eds.), *Carbon-Based Polymer Nanocomposites for Environmental and Energy Applications*, 27–56. Elsevier.

Kelly, Timothy L, Kazuhisa Yano, and Michael O Wolf. 2009. "Supercapacitive properties of PEDOT and carbon colloidal microspheres." *ACS Applied Materials & Interfaces* no. 1 (11):2536–2543.

Khan, MA, and SP Armes. 1999. "Synthesis and characterization of micrometer-sized poly (3, 4-ethylenedioxythiophene)-coated polystyrene latexes." *Langmuir* no. 15 (10):3469–3475.

Kiamahalleh, Meisam Valizadeh, Sharif Hussein Sharif Zein, Ghasem Najafpour, SUHAIRI ABD SATA, and Surani Buniran. 2012. "Multiwalled carbon nanotubes based nanocomposites for supercapacitors: a review of electrode materials." *Nano* no. 7 (02):1230002.

Kim, BH, DH Park, J Joo, SG Yu, and SH Lee. 2005. "Synthesis, characteristics, and field emission of doped and de-doped polypyrrole, polyaniline, poly (3, 4-ethylenedioxythiophene) nanotubes and nanowires." *Synthetic Metals* no. 150 (3):279–284.

Kim, Jinyeol, Eungryul Kim, Youngsoon Won, Haeseong Lee, and Kwangsuck Suh. 2003. "The preparation and characteristics of conductive poly (3, 4-ethylenedioxythiophene) thin film by vapor-phase polymerization." *Synthetic Metals* no. 139 (2):485–489.

Kirchmeyer, Stephan, and Knud Reuter. 2005. "Scientific importance, properties and growing applications of poly (3, 4-ethylenedioxythiophene)." *Journal of Materials Chemistry* no. 15 (21):2077–2088.

Kumar, Neetesh, Riski Titian Ginting, and Jae-Wook Kang. 2018. "Flexible, large-area, all-solid-state supercapacitors using spray deposited PEDOT: PSS/reduced-graphene oxide." *Electrochimica Acta* no. 270:37–47.

Kurra, Narendra, Jihoon Park, and Husam N Alshareef. 2014. "A conducting polymer nucleation scheme for efficient solid-state supercapacitors on paper." *Journal of Materials Chemistry A* no. 2 (40):17058–17065.

Lee, Kuen-Chan, Cai-Wan Chang-Jian, Er-Chieh Cho, Jen-Hsien Huang, Wei-Ting Lin, Bo-Cheng Ho, Jia-An Chou, and Yu-Sheng Hsiao. 2019. "Surface modification of Ni (OH) 2 nanosheets with PEDOT: PSS for supercapacitor and bendable electrochromic applications." *Solar Energy Materials and Solar Cells* no. 195:1–11.

Lee, Minbaek, and Joonho Bae. 2015. "High-performance fabric-based electrochemical capacitors utilizing the enhanced electrochemistry of PEDOT: PSS hybridized with SnO2 nanoparticles." *Bulletin of the Korean Chemical Society* no. 36 (8):2101–2106.

Lee, Soojeong, Mi Suk Cho, Hyuck Lee, Jae-Do Nam, and Youngkwan Lee. 2012. "A facile synthetic route for well defined multilayer films of graphene and PEDOT via an electrochemical method." *Journal of Materials Chemistry* no. 22 (5):1899–1903.

Lei, Chunhong, Peter Wilson, and Constantina Lekakou. 2011. "Effect of poly (3, 4-ethylenedioxythiophene)(PEDOT) in carbon-based composite electrodes for electrochemical supercapacitors." *Journal of Power Sources* no. 196 (18):7823–7827.

Li, Jiao, Juncheng Liu, Congjie Gao, Jinling Zhang, and Hanbin Sun. 2009. "Influence of MWCNTs doping on the structure and properties of PEDOT: PSS films." *International Journal of Photoenergy* no. 2009.

Lipomi, Darren J, and Zhenan Bao. 2011. "Stretchable, elastic materials and devices for solar energy conversion." *Energy & Environmental Science* no. 4 (9):3314–3328.

Liu, Chenguang, Zhenning Yu, David Neff, Aruna Zhamu, and Bor Z Jang. 2010. "Graphene-based supercapacitor with an ultrahigh energy density." *Nano Letters* no. 10 (12):4863–4868.

Liu, David Y, and John R Reynolds. 2010. "Dioxythiophene-based polymer electrodes for supercapacitor modules." *ACS Applied Materials & Interfaces* no. 2 (12):3586–3593.

Liu, Ran, Jonathon Duay, and Sang Bok Lee. 2010. "Redox exchange induced MnO2 nanoparticle enrichment in poly (3, 4-ethylenedioxythiophene) nanowires for electrochemical energy storage." *ACS Nano* no. 4 (7):4299–4307.

Liu, Ran, Jonathon Duay, and Sang Bok Lee. 2011. "Electrochemical formation mechanism for the controlled synthesis of heterogeneous MnO2/Poly (3, 4-ethylenedioxythiophene) nanowires." *ACS Nano* no. 5 (7):5608–5619.

Liu, Ran, and Sang Bok Lee. 2008. "MnO2/poly (3, 4-ethylenedioxythiophene) coaxial nanowires by one-step coelectrodeposition for electrochemical energy storage." *Journal of the American Chemical Society* no. 130 (10):2942–2943.

Liu, Yuqing, Bo Weng, Joselito M Razal, Qun Xu, Chen Zhao, Yuyang Hou, Shayan Seyedin, Rouhollah Jalili, Gordon G Wallace, and Jun Chen. 2015. "High-performance flexible all-solid-state supercapacitor from large free-standing graphene-PEDOT/PSS films." *Scientific Reports* no. 5:17045.

Lota, K, V Khomenko, and E Frackowiak. 2004. "Capacitance properties of poly (3, 4-ethylenedioxythiophene)/carbon nanotubes composites." *Journal of Physics and Chemistry of Solids* no. 65 (2–3):295–301.

Lu, Wen, and Liming Dai. 2010. "Carbon nanotube supercapacitors." In Jose Mauricio Marulanda (ed.), *Carbon Nanotubes*, 561–589. IntechOpen.

Lu, Xiaofeng, Wanjin Zhang, Ce Wang, Ten-Chin Wen, and Yen Wei. 2011. "One-dimensional conducting polymer nanocomposites: synthesis, properties and applications." *Progress in Polymer Science* no. 36 (5):671–712.

Martin, Charles R. 1996. "Membrane-based synthesis of nanomaterials." *Chemistry of Materials* no. 8 (8):1739–1746.

Mastragostino, M, R Paraventi, and A Zanelli. 2000. "Supercapacitors based on composite polymer electrodes." *Journal of the Electrochemical Society* no. 147 (9):3167–3170.

Melato, AI, AS Viana, and LM Abrantes. 2008. "Different steps in the electrosynthesis of poly (3, 4-ethylenedioxythiophene) on platinum." *Electrochimica Acta* no. 54 (2):590–597.

Miller, John R, and Patrice Simon. 2008. "Electrochemical capacitors for energy management." *Science* no. 321 (5889):651–652.

Mohammadi, A, M-A Hasan, B Liedberg, I Lundström, and WR Salaneck. 1986. "Chemical vapour deposition (CVD) of conducting polymers: polypyrrole." *Synthetic Metals* no. 14 (3):189–197.

Mukkabla, Radha, Melepurath Deepa, and Avanish Kumar Srivastava. 2015. "Poly (3, 4-ethylenedioxythiophene)/nickel disulfide microspheres hybrid in energy storage and conversion cells." *RSC Advances* no. 5 (120):99164–99178.

Müller, Kevin, Markus Klapper, and Klaus Müllen. 2006. "Synthesis of conjugated polymer nanoparticles in non-aqueous emulsions." *Macromolecular Rapid Communications* no. 27 (8):586–593.

Mumtaz, Muhammad, Emmanuel Ibarboure, Christine Labrugère, Eric Cloutet, and Henri Cramail. 2008. "Synthesis of PEDOT nano-objects using poly (vinyl alcohol)-based reactive stabilizers in aqueous dispersion." *Macromolecules* no. 41 (23):8964–8970.

Nishide, Hiroyuki, and Kenichi Oyaizu. 2008. "Toward flexible batteries." *Science* no. 319 (5864):737–738.

Novoselov, Kostya S, Andre K Geim, Sergei V Morozov, D Jiang, Y Zhang, Sergey V Dubonos, Irina V Grigorieva, and

Alexandr A Firsov. 2004. "Electric field effect in atomically thin carbon films." *Science* no. 306 (5696):666–669.

Oh, Seong-Geun, and Seung-Soon Im. 2002. "Electroconductive polymer nanoparticles preparation and characterization of PANI and PEDOT nanoparticles." *Current Applied Physics* no. 2 (4):273–277.

Pandey, Gaind P., Alok C. Rastogi, and Charles R Westgate. 2014. "All-solid-state supercapacitors with poly (3, 4-ethylenedioxythiophene)-coated carbon fiber paper electrodes and ionic liquid gel polymer electrolyte." *Journal of Power Sources* no. 245:857–865.

Park, Jong Hyeok, Jang Myoun Ko, O Ok Park, and Dong-Wen Kim. 2002. "Capacitance properties of graphite/polypyrrole composite electrode prepared by chemical polymerization of pyrrole on graphite fiber." *Journal of Power Sources* no. 105 (1):20–25.

Patra, S, K Barai, and N Munichandraiah. 2008. "Scanning electron microscopy studies of PEDOT prepared by various electrochemical routes." *Synthetic Metals* no. 158 (10):430–435.

Patra, S, and N Munichandraiah. 2007. "Supercapacitor studies of electrochemically deposited PEDOT on stainless steel substrate." *Journal of Applied Polymer Science* no. 106 (2):1160–1171.

Pei, Qibing, Guido Zuccarello, Markus Ahlskog, and Olle Inganäs. 1994. "Electrochromic and highly stable poly (3, 4-ethylenedioxythiophene) switches between opaque blue-black and transparent sky blue." *Polymer* no. 35 (7):1347–1351.

Rajesh, Murugesan, C Justin Raj, Ramu Manikandan, Byung Chul Kim, Sang Yeup Park, and Kook Hyun Yu. 2017. "A high performance PEDOT/PEDOT symmetric supercapacitor by facile in-situ hydrothermal polymerization of PEDOT nanostructures on flexible carbon fibre cloth electrodes." *Materials Today Energy* no. 6:96–104.

Ranjusha, R, KM Sajesh, S Roshny, V Lakshmi, P Anjali, TS Sonia, A Sreekumaran Nair, KRV Subramanian, Shantikumar V Nair, and KP Chennazhi. 2014. "Supercapacitors based on freeze dried MnO2 embedded PEDOT: PSS hybrid sponges." *Microporous and Mesoporous Materials* no. 186:30–36.

Rogers, John A, Takao Someya, and Yonggang Huang. 2010. "Materials and mechanics for stretchable electronics." *Science* no. 327 (5973):1603–1607.

Ryu, Kwang Sun, Young-Gi Lee, Young-Sik Hong, Yong Joon Park, Xianlan Wu, Kwang Man Kim, Man Gu Kang, Nam-Gyu Park, and Soon Ho Chang. 2004. "Poly (ethylenedioxythiophene)(PEDOT) as polymer electrode in redox supercapacitor." *Electrochimica Acta* no. 50 (2–3):843–847.

Saha, Arindam, SK Basiruddin, Sekhar Chandra Ray, SS Roy, and Nikhil R Jana. 2010. "Functionalized graphene and graphene oxide solution via polyacrylate coating." *Nanoscale* no. 2 (12):2777–2782.

Sakmeche, Nacer, Salah Aeiyach, Jean-Jacques Aaron, Mohamed Jouini, Jean Christophe Lacroix, and Pierre-Camille Lacaze. 1999. "Improvement of the electrosynthesis and physicochemical properties of poly (3, 4-ethylenedioxythiophene) using a sodium dodecyl sulfate micellar aqueous medium." *Langmuir* no. 15 (7):2566–2574.

Salanne, Mathieu, Benjamin Rotenberg, Katsuhiko Naoi, Katsumi Kaneko, P-L Taberna, Clare P Grey, Bruce Dunn, and Patrice Simon. 2016. "Efficient storage mechanisms for building better supercapacitors." *Nature Energy* no. 1 (6):16070.

Sen, Pintu, Amitabha De, Ankan Dutta Chowdhury, SK Bandyopadhyay, Nidhi Agnihotri, and M Mukherjee. 2013. "Conducting polymer based manganese dioxide nanocomposite as supercapacitor." *Electrochimica Acta* no. 108:265–273.

Sevilla, Marta, and Robert Mokaya. 2014. "Energy storage applications of activated carbons: supercapacitors and hydrogen storage." *Energy & Environmental Science* no. 7 (4):1250–1280.

Shown, Indrajit, Abhijit Ganguly, Li-ChyongChen, and Kuei-Hsien Chen. 2015. "Conducting polymer-based flexible supercapacitor." *Energy Science & Engineering* no. 3 (1):2–26.

Simon, Patrice, and Yury Gogotsi. 2010. "Materials for electrochemical capacitors." In Peter Rodgers (ed.), *Nanoscience and Technology: A Collection of Reviews from Nature Journals*, 320–329. World Scientific.

Snook, Graeme A, Pon Kao, and Adam S Best. 2011. "Conducting-polymer-based supercapacitor devices and electrodes." *Journal of Power Sources* no. 196 (1):1–12.

Stankovich, Sasha, Dmitriy A Dikin, Geoffrey HB Dommett, Kevin M Kohlhaas, Eric J Zimney, Eric A Stach, Richard D Piner, SonBinh T Nguyen, and Rodney S Ruoff. 2006. "Graphene-based composite materials." *Nature* no. 442 (7100):282.

Stoller, Meryl D, Sungjin Park, Yanwu Zhu, Jinho An, and Rodney S Ruoff. 2008. "Graphene-based ultracapacitors." *Nano Letters* no. 8 (10):3498–3502.

Sun, Dong, Li Jin, Yun Chen, Jian-Rong Zhang, and Jun-Jie Zhu. 2013. "Microwave-assisted in situ synthesis of graphene/PEDOT hybrid and its application in supercapacitors." *ChemPlusChem* no. 78 (3):227–234.

Sun, Xuping, and Matthias Hagner. 2007. "Novel poly (acrylic acid)-mediated formation of composited, poly (3, 4-ethylenedioxythiophene)-based conducting polymer nanowires." *Macromolecules* no. 40 (24):8537–8539.

Suppes, Graeme M, Bhavana A Deore, and Michael S Freund. 2008. "Porous conducting polymer/heteropolyoxometalate hybrid material for electrochemical supercapacitor applications." *Langmuir* no. 24 (3):1064–1069.

Tamburri, Emanuela, Silvia Orlanducci, Francesco Toschi, Maria Letizia Terranova, and Daniele Passeri. 2009. "Growth mechanisms, morphology, and electroactivity of PEDOT layers produced by electrochemical routes in aqueous medium." *Synthetic Metals* no. 159 (5–6):406–414.

Tang, Pengyi, Lijuan Han, Lin Zhang, Shijie Wang, Wei Feng, Guoqing Xu, and Li Zhang. 2015. "Controlled construction of hierarchical nanocomposites consisting of MnO2 and PEDOT for high-performance supercapacitor applications." *ChemElectroChem* no. 2 (7):949–957.

Tarascon, J-M, and Michel Armand. 2011. "Issues and challenges facing rechargeable lithium batteries." In Vincent Dusastre (ed.), *Materials for Sustainable Energy: A Collection of Peer-Reviewed Research and Review Articles from Nature Publishing Group*, 171–179. World Scientific.

Tsai, Tsung Hsuan, Soundappan Thiagarajan, and Shen-Ming Chen. 2011. *International Journal of Electrochemical Science* no. 6:3878–3889. www.spm.com.cn.

Vadiyar, Madagonda M, Sanjay S Kolekar, Jia-Yaw Chang, Anil A Kashale, and Anil V Ghule. 2016. "Reflux condensation mediated deposition of Co3O4 nanosheets and ZnFe2O4 nanoflakes electrodes for flexible asymmetric supercapacitor." *Electrochimica Acta* no. 222:1604–1615.

Wang, Guoping, Lei Zhang, and Jiujun Zhang. 2012. "A review of electrode materials for electrochemical supercapacitors." *Chemical Society Reviews* no. 41 (2):797–828.

Wang, Yan, Zhiqiang Shi, Yi Huang, Yanfeng Ma, Chengyang Wang, Mingming Chen, and Yongsheng Chen. 2009. "Supercapacitor devices based on graphene materials." *The Journal of Physical Chemistry C* no. 113 (30):13103–13107.

Wang, Zhaohui, Petter Tammela, Jinxing Huo, Peng Zhang, Maria Strømme, and Leif Nyholm. 2016. "Solution-processed poly (3, 4-ethylenedioxythiophene) nanocomposite paper electrodes for high-capacitance flexible supercapacitors." *Journal of Materials Chemistry A* no. 4 (5):1714–1722.

Wen, Yangping, Jingkun Xu, Haohua He, Baoyang Lu, Yuzhen Li, and Bin Dong. 2009. "Electrochemical polymerization of 3, 4-ethylenedioxythiophene in aqueous micellar solution containing biocompatible amino acid-based surfactant." *Journal of Electroanalytical Chemistry* no. 634 (1):49–58.

Winter, I, C Reese, J Hormes, G Heywang, and F Jonas. 1995. "The thermal ageing of poly (3, 4-ethylenedioxythiophene). An investigation by X-ray absorption and X-ray photoelectron spectroscopy." *Chemical Physics* no. 194 (1):207–213.

Winther-Jensen, Bjørn, and Keld West. 2004. "Vapor-phase polymerization of 3, 4-ethylenedioxythiophene: a route to highly conducting polymer surface layers." *Macromolecules* no. 37 (12):4538–4543.

Wu, Hong-Ying, and Huan-Wen Wang. 2012. "Electrochemical synthesis of nickel oxide nanoparticulate films on nickel foils for high-performance electrode materials of supercapacitors." *International Journal of Electrochemical Science* no. 7:4405–4417.

Wu, HP, DW He, YS Wang, M Fu, ZL Liu, JG Wang, and HT Wang. 2010. Graphene as the electrode material in supercapacitors. Paper read at 2010 8th International Vacuum Electron Sources Conference and Nanocarbon.

Wu, Jun, Yu Li, and Wei Feng. 2007. "A novel method to form hollow spheres of poly (3, 4-ethylenedioxythiophene): growth from a self-assemble membrane synthesized by aqueous chemical polymerization." *Synthetic Metals* no. 157 (22–23):1013–1018.

Wu, Nae-Lih. 2002. "Nanocrystalline oxide supercapacitors." *Materials Chemistry and Physics* no. 75 (1–3):6–11.

Xia, Chuan, Wei Chen, Xianbin Wang, Mohamed N Hedhili, Nini Wei, and Husam N Alshareef. 2015. "Highly stable supercapacitors with conducting polymer core-shell electrodes for energy storage applications." *Advanced Energy Materials* no. 5 (8):1401805.

Xia, Jilin, Fang Chen, Jinghong Li, and Nongjian Tao. 2009. "Measurement of the quantum capacitance of graphene." *Nature Nanotechnology* no. 4 (8):505.

Xiao, Rui, Seung Il Cho, Ran Liu, and Sang Bok Lee. 2007. "Controlled electrochemical synthesis of conductive polymer nanotube structures." *Journal of the American Chemical Society* no. 129 (14):4483–4489.

Xu, Yanfei, Yan Wang, Jiajie Liang, Yi Huang, Yanfeng Ma, Xiangjian Wan, and Yongsheng Chen. 2009. "A hybrid material of graphene and poly (3, 4-ethyldioxythiophene) with high conductivity, flexibility, and transparency." *Nano Research* no. 2(4): 343–348.

Yang, Wenyao, Yuetao Zhao, Xin He, Yan Chen, Jianhua Xu, Shibin Li, Yajie Yang, and Yadong Jiang. 2015. "Flexible conducting polymer/reduced graphene oxide films: synthesis, characterization, and electrochemical performance." *Nanoscale Research Letters* no. 10 (1):222.

Yang, Yajie, Luning Zhang, Shibin Li, Wenyao Yang, Jianhua Xu, Yadong Jiang, and Junfeng Wen. 2013. "Electrochemical performance of conducting polymer and its nanocomposites prepared by chemical vapor phase polymerization method." *Journal of Materials Science: Materials in Electronics* no. 24 (7):2245–2253.

Yoo, Jung Joon, Kaushik Balakrishnan, Jingsong Huang, Vincent Meunier, Bobby G Sumpter, Anchal Srivastava, Michelle Conway, Arava Leela Mohana Reddy, Jin Yu, and Robert Vajtai. 2011. "Ultrathin planar graphene supercapacitors." *Nano Letters* no. 11 (4):1423–1427.

Yu, Guihua, Xing Xie, Lijia Pan, Zhenan Bao, and Yi Cui. 2013. "Hybrid nanostructured materials for high-performance electrochemical capacitors." *Nano Energy* no. 2 (2):213–234.

Zhang, Bin, Yiting Xu, Yifang Zheng, Lizong Dai, Mingqiu Zhang, Jin Yang, Yujie Chen, Xudong Chen, and Juying Zhou. 2011. "A facile synthesis of polypyrrole/carbon nanotube composites with ultrathin, uniform and thickness-tunable polypyrrole shells." *Nanoscale Research Letters* no. 6 (1):431.

Zhang, Jintao, and XS Zhao. 2012. "Conducting polymers directly coated on reduced graphene oxide sheets as high-performance supercapacitor electrodes." *The Journal of Physical Chemistry C* no. 116 (9): 5420–5426.

Zhang, Junshe, Prasad Yedlapalli, and Jae W Lee. 2009. "Thermodynamic analysis of hydrate-based pre-combustion capture of CO2." *Chemical Engineering Science* no. 64 (22):4732–4736.

Zhang, Xinyu, Jeong-Soo Lee, Gil S Lee, Dong-Kyu Cha, Moon J Kim, Duck J Yang, and Sanjeev K Manohar. 2006. "Chemical synthesis of PEDOT nanotubes." *Macromolecules* no. 39 (2):470–472.

Zhang, Xinyu, Alan G MacDiarmid, and Sanjeev K Manohar. 2005. "Chemical synthesis of PEDOT nanofibers." *Chemical Communications* no. 42:5328–5330.

Zhao, Dawei, Qi Zhang, Wenshuai Chen, Xin Yi, Shouxin Liu, Qingwen Wang, Yixing Liu, Jian Li, Xianfeng Li, and Haipeng Yu. 2017. "Highly flexible and conductive cellulose-mediated PEDOT: PSS/MWCNT composite films for supercapacitor electrodes." *ACS Applied Materials & Interfaces* no. 9 (15):13213–13222.

Zheng, Chao, Weizhong Qian, Chaojie Cui, Qiang Zhang, Yuguang Jin, Mengqiang Zhao, Pinghen Tan, and Fei Wei. 2012. "Hierarchical carbon nanotube membrane with high packing density and tunable porous structure for high voltage supercapacitors." *Carbon* no. 50 (14):5167–5175.

Zhou, Cuifeng, Zongwen Liu, Xusheng Du, and Simon P Ringer. 2010. "Electrodeposited PEDOT films on ITO with a flower-like hierarchical structure." *Synthetic Metals* no. 160 (15–16):1636–1641.

Zhou, Cuifeng, Zongwen Liu, Yushan Yan, Xusheng Du, Yiu-Wing Mai, and Simon Ringer. 2011. "Electro-synthesis of novel nanostructured PEDOT films and their application as catalyst support." *Nanoscale Research Letters* no. 6 (1):364.

16 Electrospun Composite Separator for Li-Ion Batteries: A Short Review

A. Dennyson Savariraj
Advanced Ceramics and Nanotechnology Laboratory,
Department of Materials Engineering, Faculty of Engineering
University of Concepcion
Concepcion, Chile

R.V. Mangalaraja
Advanced Ceramics and Nanotechnology Laboratory,
Department of Materials Engineering, Faculty of Engineering
University of Concepcion
Concepcion, Chile

Technological Development Unit (UDT)
University of Concepcion
Coronel, Chile

G. Bharath and Fawzi Banat
Department of Chemical Engineering
Khalifa University
Abu Dhabi, United Arab Emirates

N. Pugazhenthiran
Laboratorio de Tecnologías Limpias, Faculty of Engineering
University of Concepcion
Concepcion, Chile

CONTENTS

16.1 INTRODUCTION

The rapid population growth and the eventual ever-growing need for energy pose challenges to researchers to go for high-profile energy harvesting and energy-storage devices. Ample efforts have been dedicated to date to bring out cost-effective devices with novel electrolyte materials, high ionic conductivity with large electrochemical window, and high ionic conductivity keeping in mind safety and eco-friendliness (Tarascon and Armand 2001, 359; Goodenough and Kim 2010, 587–603; Yang, Zheng, and Cui 2013, 3018–3032; Freitag et al. 2017, 2100–2107).

Batteries are fabricated for commercial use in several configurations and shapes, including flat, button, cylindrical, and prismatic (rectangular), depending on the kind of application. Researchers around the globe contributed much toward improving the ambient temperature lithium battery technology. Lithium-ion batteries (LIBs) became an inevitable power source, because they yield higher power energy density, longer cycle life, and higher operational voltage.

It consists of positive and negative electrodes. The positive electrodes usually consist of a metal oxide ($LiCoO_2$) mounted on aluminum foil; the negative electrode is made of a thin layer of carbon material (graphite) coated on copper foil, with a porous plastic film sandwiched in between, which serves as the separator. The separator film is usually soaked in lithium hexafluorophosphate ($LiPF_6$) that is dissolved in a mixture of organic solvents. The lithium ions are inserted and extracted from the interstitial space between atomic layers within the vicinity of the active materials (Arora and Zhang 2004, 4419–4462). Figure 16.1 shows the schematic of the electrospinning device.

A separator is usually made of porous membrane permitting ionic flow, positioned in between the electrodes of opposite polarity, however preventing electric contact. Most separators have been manufactured from cellophane, cellulosic papers, nonwoven fabrics, ion-exchange membranes, foams, and microporous flat-sheet membranes made of polymeric materials. The role of the separator is more demanding and complex, because battery research is more complicated (Arora and Zhang 2004, 4419–4462), even though separators play a more important role than

other components in terms of obtaining high power density and safe operation of batteries (Jung et al. 2016, 703–750). Figure 16.2 shows the process for the preparation of the cell using electrospun nanofiber-coated membranes.

Based on the shape and design of the cell, the different cell components, including separators, are engineered. The separators are either wound together along the electrode or placed in between the electrodes to make jellyroll and sandwich-like structures, respectively. A good separator is expected not to undergo changes after they are stacked and bound in the cell. To retain the intactness of the separator, it is coated with an additional layer with reduced interfacial resistance. Usually, the pressure from the cell container holds the stacked cells, and in the case of lithium-ion gel polymer stacked cells, the layers of electrodes and separators are either bound or laminated (Arora and Zhang 2004, 4419–4462). The main drawbacks in using the commercially available polyolefin are small porosities, small pore size, and poor wetting capability which all reduce ionic conductivity (Arora and Zhang 2004, 4419–4462), and its thermal instability at high temperature leads to internal short circuit (Uchida et al. 2003, 821–825; Wu et al. 2004, 1803–1812). The above-mentioned bottlenecks led the researchers to investigate electrospun nanofibers of enhanced ionic conductivity and thermal stability as a suitable alternative to polyolefin separators (Cao et al. 2014, 224–229).

Electrospinning is a powerful solution-based technique to fabricate one-dimensional polymer-based fibers and

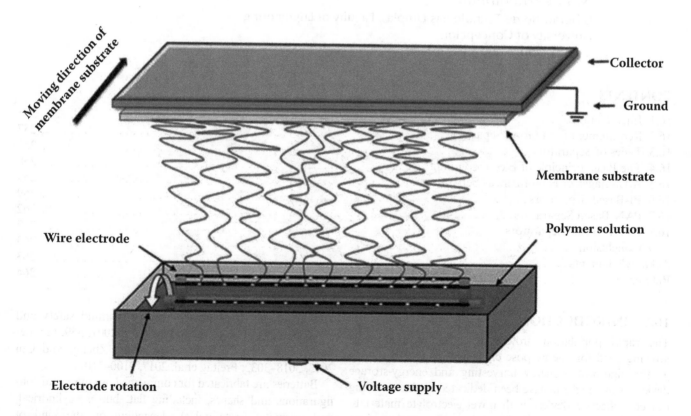

FIGURE 16.1 Schematic of the electrospinning device. Reproduced from Lee et al. 2014a, with permission from Springer Nature.

Membrane preparation   **_Cell assembly_**

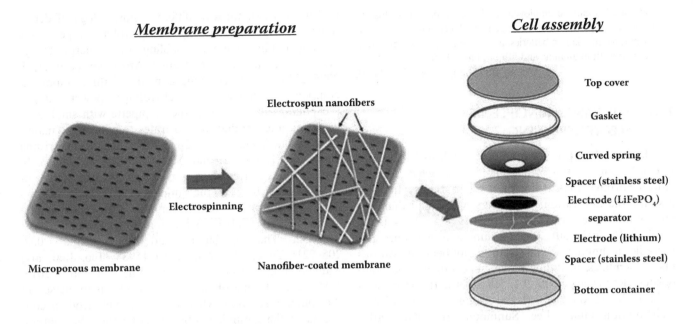

FIGURE 16.2 Process for the preparation of the cell using electrospun nanofiber-coated membranes. Reproduced from Lee et al. 2014a, with permission from Springer Nature.

membranes in the nanoscale (Choi et al. 2007, 104–115; Persano et al. 2013, 504–520; Wang et al. 2014, 538–544; Aravindan et al. 2015, 2225–2234). This technique can be adopted for large production and extended to industrial production ranging from medicine (Grafahrend et al. 2010, 67) to water filtration (Gibson et al. 2001, 469–481) which are already commercialized.

Electrospun polymer fibers such as poly(vinylidene fluoride) (PVDF) or polyacrylonitrile (PAN), PVDF copolymers, or PVDF-inert oxide blends (Jung et al. 2016, 703–750) soaked with liquid electrolytes have already been employed in batteries; however, solvent-free, electrolyte-free solid-state PEO/SN/LiBF$_4$-based separators are also reported (Freitag et al. 2017, 2100–2107). In this chapter, we present a detailed report on different types of electrospun composite separators for LIBs and the advantages of using them over the commercially available counterparts.

16.2 REQUIREMENT FOR A GOOD SEPARATOR

The selection of separators is based on the requirement for the type of battery to be fabricated. Therefore, the choice of the separator always depends on the need to be fulfilled. The electrical resistance of the separator is highly influenced by its total volume, pore structure, the tortuosity of the pores, and the design of the separator (Weighall 1995, 273–282). A good separator should have the following features:

1. uniform thickness,
2. short wetting time in order to reduce the electrolyte filling time,
3. electronically insulating,
4. minimum electronic resistance,
5. good mechanical and dimensional stabilities,
6. enough chemical resistance against degradation by impurities and physical strength for easy handling, and
7. capable of avoiding migration of particles or colloidal or soluble species in between the electrodes.

The type of separator and the method of incorporation depend on the kind of cell, and the criteria vary for each kind; some compromise has to be made keeping in mind cost, performance, and safety. The required properties of separators differ from each type. For example, the separators used in sealed nickel–cadmium (NiCd) and nickel–metal hydride (NiMH) batteries should have high permeability to allow gas molecules to pass through so as to have overcharge protection, while the separator used in for an SLI (starting, lighting, and ignition) battery should be strong enough to bear mechanical shock. The separators used in alkaline batteries should have enough flexibility for wrapping around the electrodes, whereas the separators employed in lithium-ion cells must possess a proper shutdown feature for safety considerations

(Arora and Zhang 2004, 4419–4462).

16.3 TYPES OF SEPARATORS

The classification of separators is made based on their physical and chemical properties. The materials used are molded, nonwoven, woven, bonded, microporous-bonded papers or laminates; however, single component comprises solid or gel electrolytes and separator.

Separators made of organic materials like polymers and cellulosic papers, inorganic materials that include glass wool, SiO$_2$, and asbestos can be employed for batteries operated close to ambient temperatures while regenerated cellulose is used as a separator in alkaline batteries, and microporous polymer films are used both in alkaline and lithium-based batteries

(Arora and Zhang 2004, 4419–4462).

16.4 THE BASIC PRINCIPLE OF ELECTROSPINNING

The history of electro spinning dates back to 1902 and J. F Cooley and W.J. Morton obtained patented however Anton Formhal took it to commercialization of the process for the fabrication nano fibers with small diameter and obtained several patents from 1934 to 1944 (Jung et al. 2016, 703–750). Electrospinning is a simple and versatile technique to fabricate nanofibers of various fiber assemblies, including polymers, ceramics, and composites (Raghavan et al. 2012, 915–930). Its basic principle is based on the uniaxial elongation of polymer melt/solution (Jung, Lee, Sunmoon, and Kim 2016, 703–750). Electrospinning has a significant advantage of obtaining fibers of high surface-to-volume ratios providing numerous active sites, controllable fiber diameter,

and fibrous structures of different morphology of dense, hollow, and porous allowing the electrolytes to penetrate, leading to high potential enabling better energy-storage system (Ji et al. 2011, 2682–2699; Aravindan et al. 2015, 2225–2234). Electrospinning consists of three important components, namely a high-voltage power supply, a capillary tube with a needle or pipette with small-bore diameter, and conductive substrates to collect the material (Ramakrishna et al. 2005; Teo and Ramakrishna 2006, R89–R106; Aravindan et al. 2015, 2225–2234). In order to induce free charge an electric field of high voltage is applied between the sample collector and the polymer source kept in solution or in a melt (Ramaseshan et al. 2007, 111101; Schiffman and Schauer 2008, 317–352; Thavasi, Singh, and Ramakrishna 2008, 205–221; Yu et al. 2009, 15984–15985; Teo, Inai, and Ramakrishna 2011, 013002; Xue et al. 2013, 13807–13813). One of the two electrodes is immersed in the spinning solution, whereas the other electrode is connected to the grounded substrate collector. The solution fluid is kept inside the capillary tube and held by its surface tension. When an electric field is applied to the end of this capillary tube during the elongation of the

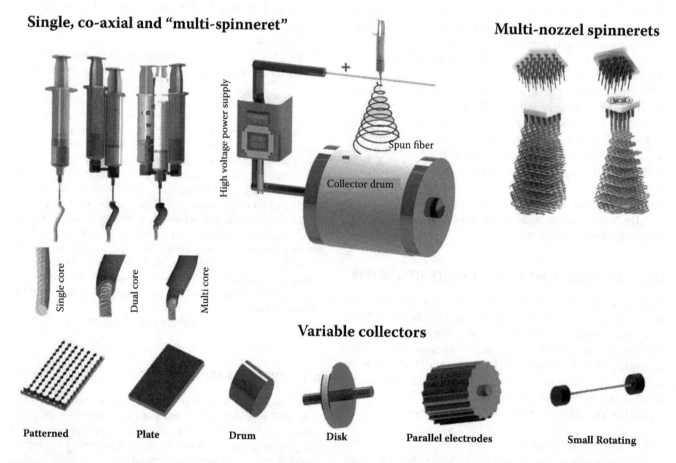

FIGURE 16.3 Schematic representation of the electrospinning setup with variable needles, collecting substrates, and syringes. Reproduced from Vanchiappan et al. 2015, with permission from The Royal Society of Chemistry.

viscous electrospinning solution, the electrostatic force produces continuous nanofibers. The charges carried in the organic solvent usually travels longer-distance of the liquid since their mobility in organic solvent is less than that of the aqueous media (Raghavan et al. 2012, 915–930). As the intensity of the applied electric field is increased until the surface tension of the hemispherical fluid is lower than the repulsive force, the droplet at the end of the capillary tube elongates to form an electrically charged conical shape called the Taylor cone. On increasing the electric field further the repulsive electrostatic force between the charged surface and evaporation of the solvent exceeds the surface tension attaining a critical point whereby the electrically charged stream of fluids is emitted from the tip of the Taylor cone in whipping motion though the stream is stable near the Taylor cone and collected on the collector. The emitted polymer solution is obtained as a continuous network of film due to instability and elongation (Sundaramurthy et al. 2012, 8201–8208; Zhang et al. 2013, 5973–5980; Jayaraman et al. 2013, 6677–6679; Sundaramurthy et al. 2014, 16776–16781). Figure 16.3 shows the schematic illustration of electrospinning setup with variable needles, collecting substrates, and syringes.

The characteristics of the electrospun nanofibers are influenced by: (i) operating parameters or process parameters such as electric potential, needle gauge, nozzle type, flow rate, tip to collector screen distance, polymer concentration; (ii) ambient parameters such as humidity, air viscosity in the chamber, and temperature; (iii) system parameters such as molecular weight, distribution of molecular weight, and architecture of the polymer (linear or branched); and (iv) polymer solution parameters such as viscosity, solubility of the polymer, dielectric constant, and surface tension of the solvents (Huang et al. 2003, 2223–2253; Ramakrishna et al. 2005; Teo and Ramakrishna 2006, R89–R106). The spinning solution consists of precursors (usually inorganic and carbon precursors), sacrificial polymer templates such as polystyrene (PS), poly(vinyl pyrrolidone) (PVP), poly (methyl methacrylate) (PMMA), poly(ethylene oxide) (PEO), PAN, poly(vinyl alcohol) (PVA), poly(vinyl acetate) (PVAc), styrene-acrylonitrile (SAN), PVDF, poly (vinyl chloride) (PVC), and polyimide (PI). Both the precursors and the sacrificial polymer templates are dissolved in one or more mixture of suitable solvents (Jung et al. 2016, 703–750).

16.5 ADVANTAGES OF NANOFIBERS AS SEPARATORS

Polymer-based flexible separators with electrical insulation and ionic conduction remain as an inevitable class of separators in LIBs as they have several advantages over others in terms of solving leakage issue and forming in to any desired shape (Meyer 1998, 439–448; Arora and Zhang 2004, 4419–4462; Arico et al. 2005, 366–377; Fergus 2010, 4554–4569; Aravindan et al. 2011, 14326–14346; Baskakova, Yarmolenko, and Efimov 2012, 367–380; Hallinan and Balsara 2013, 503–525). Although a list of polymers, copolymers, and polymer mix is employed in LIB with a dual role of separator and electrolytes, polyvinylidene fluoride–co-hexafluoropropylene (PVDF–HFP) is of paramount importance due to extreme anodic stability and high dielectric constant to supply ample charge carriers resulting in very high electrocatalytic activity. The high anodic stability is attributed to the presence of electron-withdrawing group such as –C–F while the chemical and mechanical stabilities of the membranes are due to the crystallinity of VDF and amorphous nature of HFB, respectively (Aravindan and Vickraman 2007, 5121–5127; Aravindan et al. 2010, 033105; Baskakova, Yarmolenko, and Efimov 2012, 367–380; Costa, Silva, and Lanceros-Mendez 2013, 11404–11417). Interfacial stability of the separator with the lithium electrode is very important in exhibiting the compatibility with the electrode. When compared to its counterparts prepared by polymer dissolution and conventional phase inversion method, electrospun PVDF–HFP as separator cum electrolytes shows superior compatibility and high performance (Kim et al. 2006, 5636–5644; Subramania et al. 2007, 8–15).

The electrospun PVDF–HFP membranes obey the Vogel–Fulcher–Tammann model that is evident from the nonlinear increase in the ionic conductivity under ambient conditions ($\sim$3.2 mS cm^{-2}). Moreover, as the temperature is increased, the conductivity proportionally increases since there is an increase in the number of charge carriers when the activation energy is high (Prasanth, Aravindan, and Srinivasan 2012, 299–307; Aravindan et al. 2013, 308–316; Aravindan et al. 2015, 2225–2234). The above-mentioned advantages make the electrospun PVDF–HFP nanofiber membrane a superior candidate to be employed as the separator cum electrolyte in designing and fabricating high-performance LIBs (Aravindan et al. 2015, 2225–2234; Jung et al. 2016, 703–750).

16.6 PI-BASED SEPARATORS

PI is one of the polymers that can be electrospun and used as a separator in LIBs. The high instability of Celgard at high temperature paved the way to adopt PI as an alternative candidate. PI exhibits thermal stability even up to 500°C, while Celgard membrane undergoes shrinkage at 150°C and melts above 167°C. PI nanofiber-based nonwoven shows superior wettability for polar solvents over the Celgard membrane (Miao et al. 2013, 82–86). Table 16.1 shows different types of electrospun nanofibers and their composites as separators for LIBs and their electrocatalytic parameters. It supports better and faster Li$^+$-ion transport and rate capability due to larger pore volume of the PI nanofiber mat, although PI might yield irreversible increase in cell impedance after a number of cycles (Lee et al. 2014b, 1211–1217).

TABLE 16.1

Electrospun nanofiber separators for LIBs. Reproduced from Jung et al. (2016), with permission from The Royal Society of Chemistry

Materials	Electrospinning solution (precursor/polymer/solvent)	Heat treatment	Electrochemical performance	Ref.
PI separators	–/PMDA, ODA/DMAc	–	~160 mAh g^{-1} after 100 cycles at 0.2 C (Li$_4$Ti$_5$O$_{12}$/Li batteries)	(Miao et al. 2013, 82–86)
PI separators	–/PMDA, ODA/DMAc	–	~146.8 mAh g^{-1} after 200 cycles at 5 C (Li/Pls/Li$_4$Ti$_5$O$_{12}$ cells)	(Wang et al. 2014, 538–544)
PI–Al$_2$O$_3$ PI–SiO$_2$	PDA/ODA/PMDA and TEA	150°C	~163.9 mAh g^{-1} for PI ~169.0 mAh g^{-1} for PI–SiO$_2$ ~167.5 mAh g^{-1} for PI–Al$_2$O$_3$ ~159.9 mAh g^{-1} for SV718 at 0.1 C (LiCoO$_2$ batteries)	(Shayapat, Chung, and Park 2015, 110–121)
Al$_2$O$_3$-coated PI separators	–/PMDA, ODA/DMF	–	119.3 mAh g^{-1} after 200 cycles at 1 C (graphite/Li(Ni$_{0.5}$Co$_{0.2}$Mn$_{0.3}$)O^{2-} LiMn$_2$O$_4$ batteries)	(Lee et al. 2014, 1211–1217)
PAN separators	–/PAN/DMF	–	105.83 mAh g^{-1} after 200 cycles at 0.5 C (Li/LiCoO$_2$batteries)	(Cho et al. 2007, A159–A162)
PAN separators	–/PAN/DMF	–	102.7 mAh g^{-1} after 250 cycles at 0.2 C (graphite/LiCoO$_2$ batteries)	(Cho et al. 2008, 155–160)
Lignin/PAN nonwoven composite	PAN/DMF	150°C for 15 min	148.9 mAh g^{-1} after 50 cycles at 0.2 C (Li/LiFePO$_4$)	(Zhao et al. 2015, 101115–101120)
(Agarose/PAA)/PAN separators	–/Agarose, PAA/DMF –/PAN/DMF	–	~100 mAh g^{-1} after 100 cycles at 1 C, 60°C (Li/LiMn$_2$O$_4$ batteries)	(Kim et al. 2015, 10687–10692)
PVDF membrane	–/PVDF/DMF, acetone	160°C for 2 h in air	119 mAh g^{-1} after 50 cycles at 0.5 C (Li/LiMn$_2$O$_4$ batteries)	(Gao et al. 2006, 100–105)
PVDF–CTFE coated on the PP or PE substrate	–/PVDF–CTFE/DMF, acetone	–	155 mAh g^{-1} after 50 cycles at 0.2 C (Li/LiFePO$_4$ batteries)	(Lee et al. 2014b, 2451–2458)
PSA@PVDF–HFP separators	(Core) –/PSA/DMAc (Shell) –/PVDF–HFP/DMAc		105 mAh g^{-1} after 100 cycles at 0.5 C (graphite/LiCoO$_2$ batteries)	(Zhou et al. 2013, A1341–A1347)
PI@PVDF–HFP separators	(Core) –/PDA, ODA/DMAc (Shell) –/PVDF–HFP/DMF	–	105 mAh g^{-1} after 100 cycles at 0.5 C (graphite/LiCoO$_2$ batteries)	(Liu et al. 2013, 806–813)
Cellulose @ PVDF–HFP separators	(Core) –/Cellulose acetate/ DMAc, acetone (Shell) –/PVDF–HFP/ DMAc, Acetone	–	105.8 mAh g^{-1} after 100 cycles at 0.2 C (Li/LiCoO$_2$ batteries)	(Huang et al. 2015, 932–940)
PVDF	PVDF)/N,N-dimethylaceta- mide (DMAC)		101.3 mAh g^{-1} at first cycle at 0.5 C (LiMn$_2$O$_4$ /Li batteries)	(Cao et al. 2014, 224–229)
PDA-coated PVDF	PDA/PVDF/DMAC DMAC, acetone, and 3,4-dihydroxy phenethylamine	–	104.7 mAh g^{-1} at first cycle at 0.5 C LiMn$_2$O$_4$ /Li batteries)	(Cao et al. 2014, 224–229)
PVDF/SiO$_2$ separators	–/PVDF/DMAc, acetone	–	132 mAh g^{-1} after 100 cycles at 1 C (Li/LiFePO$_4$ batteries)	(Yanilmaz et al. 2014, 57–65)
PVDF/SiO$_2$ separators	TEOS/PVDF/DMAc, acetone, EtOH	–	159 mAh g^{-1} after 50 cycles at 0.2 C (Li/LiFePO$_4$ batteries)	(Zhang et al. 2014, 423–431)
PVDF/PMIA/PVDF separators	–/PMIA/LiCl, DMAc –/PVDF/DMAc	–	135.29 mAh g^{-1} after 100 cycles at 0.2 C (Li/LiCoO$_2$ batteries)	(Zhai et al. 2014, 14511–14518)
PVDF–HFP/TiO$_2$@ Li$^+$ filler separators	TiO2@Li$^+$ fillers/PVDF– HFP/ DMF	–	135.29 mAh g^{-1} after 100 cycles at 0.2 C (Li/LiCoO$_2$ batteries)	(Cao et al. 2015, 8258–8262)

Yue et al. made a comparative study using Celgard and electrospun PI as separators for LIBs and found that the latter performed far superior to the former both in terms of thermal stability and electrochemical performance (higher capacity, lower resistance, and higher rate capability) in a wide current range that certifies PI for high-rate applications (Miao et al. 2013, 82–86). Qiujun et al. too fabricated PI membranes through electrospinning using pyromellitic dianhydride (PDMA) along with oxydianiline (ODA) dimethylacetamide (DMAc). They tested the as-electrospun nonwoven membranes of 800-nm diameter for thermal stability, including flame test, and compared them with the counterpart Celgard. PI showed high porosity and high uptake of electrolytes (when activated) in comparison to Celgard. The batteries fabricated with PI separators showed significantly superior charge/discharge performance and retained appreciable capacity even after 200 cycles (99.3%) at the high current rate (5 C) (Wang et al. 2014, 538–544).

Jaritphun et al. went one step further to fabricate PI-based composite along with Al_2O_3 and SiO_2 and reported their suitability as separators for LIBs based on their properties and compared it with the commercial separator SV718. The composites were fabricated by first electrospinning a mixture containing pre-polymer, poly(amic acid) ammonium salt (PAAS), and inorganic nanoparticles of SiO_2 or Al_2O_3 followed by imidizing the electrospun nanofibers of PAAS composited at a high temperature of 350°C under nitrogen atmosphere. The thermal properties examined through the thermal gravimetric analyzer (TGA) showed that the commercial separator SV718 failed to exhibit good thermal stability as it melted even at 137°C and decomposed at 300°C while PI and its composites (PI–Al_2O_3 and PI–SiO_2). Withstood hot-oven test up to 200°C without shrinking, whereas SV718 was stable only below 100°C. In addition to that, SV718 exhibited poor wettability with a high contact angle of 54° against 10° for PI and its composites. The PI and its composites showed high uptake of electrolytes due to high porosity (the surface morphology of PI, and the composites of PI and SV718 are shown in Figure 16.4). The average discharge capacities of PI, PI–SiO_2, PI–Al_2O_3, and SV718 separators are reported to be 163.9, 169.0, 167.5, and 159.9 mAh g^{-1}, respectively, at 0.1 C. From the results, it is evident that PI–SiO_2 and PI–Al_2O_3 separators stand out to be better candidates than the commercial SV718 (Shayapat, Chung, and Park 2015, 110–121). Juneun et al. too made a detailed study using PI–Al_2O_3 composite and compared it with commercial polypropylene (PP) separator. They fabricated PI nanofibers with an average diameter of 300 nm by electrospinning and sandwiched between thin Al_2O_3 overlayers

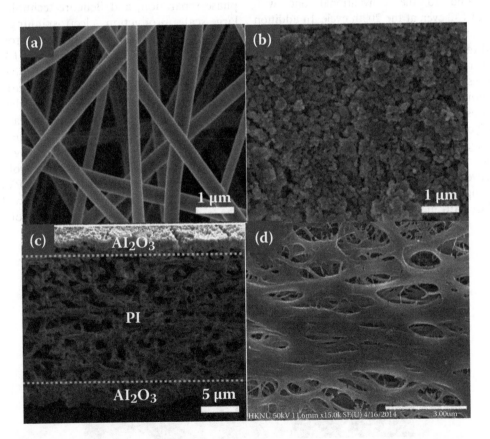

FIGURE 16.4 (a) FE-SEM images of the PI separator; (b-c) surface and cross-sectional images of the Al_2O_3–PI separator; (d) FE-SEM image of the commercial porous separator SV718. Reproduced from Lee et al. (2014); Shayapat, Chung, and Park (2015), with permission from Elsevier.

that were also electrospun and then dip-coated with Al_2O_3 nanopowders. Dip-coating Al_2O_3 on both the sides of PI was adopted to avoid irreversible increase in cell impedance due to the large pore size of PI nanofiber that also caters faster Li^+-ion transport and greater rate. The Al_2O_3-coated PI separators proved to be a better material than commercial PP with better capacity and cyclability (Lee et al. 2014b, 1211–1217).

16.7 PAN-BASED SEPARATORS

PAN stands out to be one among the several alternative candidates to replace polyolefin as a separator with better wettability, lower Gurley, high porosities, and better cycle performance. T.H. Cho et al. made two attempts to investigate the role of PAN as a separator. In the first one, they fabricated PAN nanofiber-based separators with close thickness and similar pore size to the conventional microporous separators and in the second one with similar porosity and air permeability like that of the conventional nonwoven separator.

The electrospun PAN nanofibers showed homogenous thickness and uniform diameter of 350 nm and exhibited electrochemical stability in cyclic voltammetry study between 3 and 4.5 V vs. Li. The cells fabricated using the above separator performed with higher retained capacities as compared to the conventional one with a moderate rate of C/2 even at the 200th cycle. In addition to the above, the superior rate capability of the PAN-based device attributed to smaller diffusion resistance of the separator. In another study, T.H. Cho et al. attempted to develop two types of PAN-based nonwoven separators with a homogenous diameter of 380 and 250 nm. The PAN-based devices performed better than the ones with conventional polyolefin and did not exhibit any internal short circuit during charge–discharge cycles. The improved ionic conductivity of PAN-based device is due to increased porosities. Moreover, the PAN nonwoven separators show thermal stability up to 120°C but fail after 150°C (digital photographs of Celgard membrane (Figure 16.5a)) and

PAN nonwoven after hot oven tests (Figure 16.5b) (Cho et al. 2008, 155–160).

In order to improve the devices' performance, few attempts were made for making PAN-based composites to serve as separators. Man Zhao et al. reported lignin/polyacrylonitrile (L–PAN) composite fiber-based nonwoven membranes by electrospinning. They varied the lignin:polyacrylonitrile weight ratios (1:9, 3:7, and 5:5 by weight) and carried out the electrospinning. It was found that the discharge capacity of the cell with L–PAN (3:7 by weight) retained 148.9 mAh g^{-1} after 50 cycles at 0.2 C with 95% of the capacity retention. The enhanced electrocatalytic activity of the device is due to the higher porosity (74%) of L–PAN as compared to that of the commercial PP separator (42%) (Zhao et al. 2015, 101115–101120).

16.8 PVDF-BASED SEPARATORS

The hydrophilic surface of the conventional polyolefin along with its low surface energy does not have considerable affinities toward polar organic electrolytes. This increases the resistance and brings down the battery performance by seriously hindering its cycle life. PVDF is yet another choice of researchers to replace polyolefin. Although microporous PVDF membranes can be fabricated by conventional casting method, phase separation, and Bellcore technology method, for large-scale production electrospinning is more preferred since it enhanced improved porosity (Gao et al. 2006, 100–105). Gao et al. fabricated PVDF separators by electrospinning with varied applied voltage and improved the crystallinity by 5–6% with thermal treatment. The thickness of the PVDF fibers can be controlled by varying the applied voltage, that is, the higher the voltage thinner will be the fiber thickness, and when all the other variables are made constant there is a more jumbled distribution of fiber diameter. They incorporated the electrospun PVDF into LIBs and compared its performance with commercial Celgard$_{TM}$ 2400 (PP separator) and observed that the

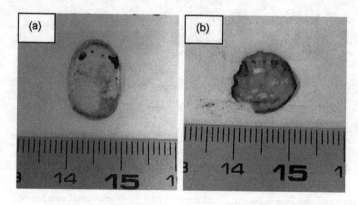

FIGURE 16.5 Photographs of the (a) Celgard membrane and the (b) PAN nonwoven after the hot oven test at 150°C. Reproduced from Cho et al. (2008) with permission from Elsevier.

former performed better than the latter with enhanced charge–discharge performance with very little loss even after 50 cycles at C/2 rate. The higher temperature yielded soft PVDF fibers with improved properties and this PVDF fiber as a separator exhibited better affinity toward Li metal yielding very less inter resistance and retained the ability of the electrolyte (Gao et al. 2006, 100–105). Lee et al. prepared a separator by electrospinning polyvinylidene fluoride–cochlorotrifluoroethylene (PVDF–co-CTFE) nanofibers over three different microporous membrane substrates. The fibers thus produced exhibited an average diameter ranging from 129 to 134 nm, and different properties such as ionic conductivities, electrolyte uptake, and interfacial resistances were studied and incorporated in LIB as a separator. The nanofiber-coated membranes showed better performance than the uncoated membrane (Lee et al. 2014, 2451–2458). Core-shell structured membranes were also investigated for their suitability as separators in LIBs by Zhou et al. and Liu et al. Polysulfonamide (PSA) and PVDF–HFP had been used as the inner (core) layer and outer (shell) layer, respectively, by Zhou et al. to form PSA@PVDF–HFP composite nonwoven separators. They compared the performance of PSA@PVDF–HFP with PVDF–HFP and PP as a separator in LIBs. PSA@PVDF–HFP showed superior behavior in the electrolyte uptake (350%), higher percentage of porosity (75%), low Gurley value (22 s/100cc), better electrochemical stability, and higher electronic conductivity when compared to the other two candidates (Zhou et al. 2013, A1341–A1347). Liu et al. fabricated core @ sheath-based nanofibrous separator with the following composition: PI@PVDF–HFP via coaxial electrospinning. It was found that PI@PVDF–HFP had a very high thermal stability up to 300°C and tensile strength of 23 MPa. This also behaved as a robust gel polymer electrolyte that can be used for high-temperature batteries. Both PI@PVDF–HFP and PI@PVDF–HFP exhibited discharged capacity of 105 mAh g^{-1} after 100 cycles at 0.5 C rate (Liu et al. 2013, 806–813; Zhou et al. 2013, A1341–A1347).

Huang and co-workers made an innovative attempt of extracting cellulose from cigarette filter and fabricating cellulose (core)/PVDF–HFP (shell) composite to be used as a separator in LIB. The above-said composite possessed excellent properties such as good tensile strength (34.1 MPa), thermal stability up to 200°C high porosity (66%), more ionic conductivity (6.16 mS cm^{-1}), low interfacial resistance (98.5 Ω), and high electrolyte compatibility (355%) which all make them a better choice as a separator for LIBs over others. The LIBs fabricated with cellulose/PVDF–HFP (105.8 mAh g^{-1} after 100 cycles) performed better than the ones fabricated using conventional separators some other previous reports (Huang et al. 2015, 932–940). Cao et al. attempted to use the bare PVDF and PDA (polydopamine)-coated PVDF as separators and found that PDA coating over the hydrophilic surface of PVDF improved several properties that reflected in the LIB performance (Cao et al. 2014, 224–229).

Zhang et al. and Yanilmaz et al. studied the electrical and mechanical properties of PVDF/SiO$_2$ composites and used them as separators in LIBs and compared their performance with that of Celgard and polyolefin-based separators, respectively. Zhang and coworkers made PVDF/SiO$_2$ in two types, typically by varying the ratio as 10:1 and 9:1, and reported that PVDF/SiO$_2$ with 9:1 ratio had more porosity in comparison to 10:1 ratio; subsequently, it was exhibited in the charge–discharge behavior of 159 mAh g^{-1} after 50 cycles at 0.2 C; however, the PVDF/SiO$_2$ composite fabricated by Yanilmaz et al. showed a charge–discharge of 132 mAh g^{-1} after 100 cycles at 1 C

(Yanilmaz et al 2014, 57–65;
Zhang et al. 2014, 423–431).

Cao et al. and Zhai et al. fabricated LIBs of Li/LiCoO$_2$ composition with PVDF–HFP fibers with TiO$_2$@Li$^+$ as fillers and PVDF/PMIA(poly(m-phenylene isophthalamide))/PVDF fibers as separators, respectively, and in both the cases the electrochemical performance was found to be 135.29 mAh g^{-1} after 100 cycles at 0.2 C. In the first case, TiO$_2$@Li$^+$ exhibited improved mechanical and electrochemical properties and in the second case the as-prepared PMIA membranes sandwiched between the two PVDF showed robust tensile strength of 13.96 MPa and ionic conductivity of 0.81 mS/ cm at 20°C. (Zhai et al. 2014, 14511–14518; Cao et al. 2015, 8258–8262).

16.9 CONCLUSION

In this chapter, we have summarized and presented the role of a separator, its requirements, different types of one-dimensional (1D) electrospun nanofibers and their composites as separators for the next-generation LIBs. The electrospun composites as separators gained much attention due to nano-sized pores, high porosity, various surface morphologies, increased surface area to volume ratio, impressive thermal stability, mechanical properties, better electrolyte uptake ability or wettability, lower interfacial resistance, and enhanced electrocatalytic activity in terms of cyclability, energy, and power density. The continuous efforts of researchers in designing cost-effective electrospun nanofiber composites with tailor-made properties will lead to the commercialization of these products and give a significant phase lift to LIBs shortly.

ACKNOWLEDGMENTS

A. Dennyson Savariraj and R.V. Mangalaraja gratefully acknowledge FONDECYT Post-doctoral Project No. 3170640, Government of Chile, Santiago, for financial assistance.

REFERENCES

Aravindan, Vanchiappan, Joe Gnanaraj, Srinivasan Madhavi, and Hua-Kun Liu. "Lithium-Ion Conducting Electrolyte Salts for Lithium Batteries." *Chemistry – A European Journal* 17, no. 51 (2011): 14326–14346.

Aravindan, Vanchiappan, Velusamy Senthilkumar, P. Nithiananthi, and P. Vickraman. "Characterization of Poly(Vinylidene-fluoride-Co-Hexafluoroprolylene) Membranes Containing Nanoscopic AlO(OH)n Filler with Li/LiFePO4 Cell." *Journal of Renewable and Sustainable Energy* 2, no. 3 (05/01; 2019/03, 2010): 033105.

Aravindan, Vanchiappan, Nageswaran Shubha, Yan Ling Cheah, Raghavan Prasanth, W. Chuiling, Rajiv Ramanujam Prabhakar, and Srinivasan Madhavi. "Extraordinary Long-Term Cycleability of TiO2-B Nanorods as Anodes in Full-Cell Assembly with Electrospun PVDF-HFP Membranes." *Journal of Materials Chemistry A* 1, no. 2 (2013): 308–316.

Aravindan, Vanchiappan, Jayaraman Sundaramurthy, Palaniswamy Suresh Kumar, Yun-Sung Lee, Seeram Ramakrishna, and Srinivasan Madhavi. "Electrospun Nanofibers: A Prospective Electro-Active Material for Constructing High Performance Li-Ion Batteries." *Chemical Communications* 51, no. 12 (2015): 2225–2234.

Aravindan, Vanchiappan and P. Vickraman. *Polyvinylidenefluoride–Hexafluoropropylene Based Nanocomposite Polymer Electrolytes (NCPE) Complexed with LiPF3(CF3CF2)3.* Vol. 43, 2007. doi:10.1016/j.eurpolymj.2007.10.003.

Arico, Antonino Salvatore, Peter Bruce, Bruno Scrosati, Jean-Marie Tarascon, and Walter van Schalkwijk. "Nanostructured Materials for Advanced Energy Conversion and Storage Devices." *Nature Materials* 4, no. 5 (05/01, 2005): 366–377.

Arora, Pankaj and Zhengming (John) Zhang. "Battery Separators." *Chemical Reviews* 104, no. 10 (10/01, 2004): 4419–4462.

Baskakova, Yu V., Ol'ga V. Yarmolenko, and Oleg N. Efimov. "Polymer Gel Electrolytes for Lithium Batteries." *Russian Chemical Reviews* 81, no. 4 (04/30, 2012): 367–380.

Cao, Chengying, Lei Tan, Weiwei Liu, Jiquan Ma, and Lei Li. *Polydopamine Coated Electrospun Poly(Vinyldiene Fluoride) Nanofibrous Membrane as Separator for Lithium-Ion Batteries.* Vol. 248, 2014. doi:10.1016/j.jpowsour.2013.09.027.

Cao, Jiang, Yuming Shang, Li Wang, Xiangming He, Lingfeng Deng, and Hong Chen. "Composite Electrospun Membranes Containing a Monodispersed Nano-Sized TiO2@Li+ Single Ionic Conductor for Li-Ion Batteries." *RSC Advances* 5, no. 11 (2015): 8258–8262.

Cho, Tae-Hyung, T. Sakai, S. Tanase, K. Kimura, Y. Kondo, T. Tarao, and M. Tanaka. "Electrochemical Performances of Polyacrylonitrile Nanofiber-Based Nonwoven Separator for Lithium-Ion Battery." *Electrochemical and Solid-State Letters* 10, no. 7 (07/01, 2007): A159–A162.

Cho, Tae-Hyung, Masanao Tanaka, Hiroshi Onishi, Yuka Kondo, Tatsuo Nakamura, Hiroaki Yamazaki, Shigeo Tanase, and Tetsuo Sakai. *Battery Performances and Thermal Stability of Polyacrylonitrile Nano-Fiber-Based Nonwoven Separators for Li-Ion Battery.* Special Section Selected Papers from the 1st Polish Forum on Fuel Cells and Hydrogen. Vol. 181, 2008. doi:10.1016/j.jpowsour.2008.03.010.

Choi, Sung Won, Jeong Rae Kim, Young Rack Ahn, Seong Mu Jo, and Elton J. Cairns. "Characterization of Electrospun PVDF Fiber-Based Polymer Electrolytes." *Chemistry of Materials* 19, no. 1 (01/01, 2007): 104–115.

Costa, Carlos M., Maria M. Silva, and S. Lanceros-Mendez. "Battery Separators Based on Vinylidene Fluoride (VDF) Polymers and Copolymers for Lithium Ion Battery Applications." *RSC Advances* 3, no. 29 (2013): 11404–11417.

Fergus, Jeffrey W. *Ceramic and Polymeric Solid Electrolytes for Lithium-Ion Batteries.* Vol. 195, 2010. doi:10.1016/j.jpowsour.2010.01.076.

Freitag, Katharina M., Holger Kirchhain, Leo van Wullen, and Tom Nilges. "Enhancement of Li Ion Conductivity by Electrospun Polymer Fibers and Direct Fabrication of Solvent-Free Separator Membranes for Li Ion Batteries." *Inorganic Chemistry* 56, no. 4 (02/20, 2017): 2100–2107.

Gao, Kun, Xinguo Hu, Chongsong Dai, and Tingfeng Yi. *Crystal Structures of Electrospun PVDF Membranes and Its Separator Application for Rechargeable Lithium Metal Cells.* Vol. 131, 2006. doi:10.1016/j.mseb.2006.03.035.

Gibson, Phillip, Heidi Schreuder-Gibson, and Donald Rivin. *Transport Properties of Porous Membranes Based on Electrospun Nanofibers.* Vol. 187–188, 2001. doi:10.1016/S0927-7757(01)00616-1.

Goodenough, John B. and Youngsik Kim. "Challenges for Rechargeable Li Batteries." *Chemistry of Materials* 22, no. 3 (02/09, 2010): 587–603.

Grafahrend, Dirk, Meike V. Karl-Heinz Heffels, Peter Gasteier Beer, Martin Moller, Paul D. Gabriele Boehm, and Jurgen Groll. "Degradable Polyester Scaffolds with Controlled Surface Chemistry Combining Minimal Protein Adsorption with Specific Bioactivation." *Nature Materials* 10, (12/12, 2010): 67.

Hallinan, Daniel T. and Nitash P. Balsara. "Polymer Electrolytes." *Annual Review of Materials Research* 43, no. 1 (07/01; 2019/03, 2013): 503–525.

Huang, Fenglin, Yunfei Xu, Bin Peng, Yangfen Su, Feng Jiang, You-Lo Hsieh, and Qufu Wei. "Coaxial Electrospun Cellulose-Core Fluoropolymer-Shell Fibrous Membrane from Recycled Cigarette Filter as Separator for High Performance Lithium-Ion Battery." *ACS Sustainable Chemistry & Engineering* 3, no. 5 (05/04, 2015): 932–940.

Huang, Zheng-Ming, Y. -Z. Zhang, M. Kotaki, and S. Ramakrishna. *A Review on Polymer Nanofibers by Electrospinning and Their Applications in Nanocomposites.* Vol. 63, 2003. doi:10.1016/S0266-3538(03)00178-7.

Jayaraman, Sundaramurthy, Vanchiappan Aravindan, Palaniswamy Suresh Kumar, Wong Chui Ling, Seeram Ramakrishna, and Srinivasan Madhavi. "Synthesis of Porous LiMn2O4 Hollow Nanofibers by Electrospinning with Extraordinary Lithium Storage Properties." *Chemical Communications* 49, no. 59 (2013): 6677–6679.

Ji, Liwen, Zhan Lin, Mataz Alcoutlabi, and Xiangwu Zhang. "Recent Developments in Nanostructured Anode Materials for Rechargeable Lithium-Ion Batteries." *Energy & Environmental Science* 4, no. 8 (2011): 2682–2699.

Jung, Ji-Won, Cho-Long Lee, Sunmoon Yu, and Il-Doo Kim. "Electrospun Nanofibers as a Platform for Advanced Secondary Batteries: A Comprehensive Review." *Journal of Materials Chemistry A* 4, no. 3 (2016): 703–750.

Kim, Ju-Myung, Chanhoon Kim, Seungmin Yoo, Jeong-Hoon Kim, Jung-Hwan Kim, Jun-Muk Lim, Soojin Park, and Sang-Young Lee. "Agarose-Biofunctionalized, Dual-Electrospun Heteronanofiber Mats: Toward Metal-Ion Chelating Battery Separator Membranes." *Journal of Materials Chemistry A* 3, no. 20 (2015): 10687–10692.

Kim, Kwang Man, Nam-Gyu Park, Kwang Sun Ryu, and Soon Ho Chang. *Characteristics of PVDF-HFP/TiO2 Composite Membrane Electrolytes Prepared by Phase Inversion and Conventional Casting Methods*. Vol. 51, 2006. doi:10.1016/j.electacta.2006.02.038.

Lee, Hun, Mataz Alcoutlabi, Ozan Toprakci, Guanjie Xu, Jill V. Watson, and Xiangwu Zhang. "Preparation and Characterization of Electrospun Nanofiber-Coated Membrane Separators for Lithium-Ion Batteries." *Journal of Solid State Electrochemistry* 18, no. 9 (09/01, 2014): 2451–2458.

Lee, Juneun, Cho-Long Lee, Kyusung Park, and Il-Doo Kim. *Synthesis of an Al2O3-Coated Polyimide Nanofiber Mat and Its Electrochemical Characteristics as a Separator for Lithium Ion Batteries*. Vol. 248, 2014. doi:10.1016/j.jpowsour.2013.10.056.

Liu, Zhihong, Wen Jiang, Qingshan Kong, Chuanjian Zhang, Pengxian Han, Xuejiang Wang, Jianhua Yao, and Guanglei Cui. "A Core@sheath Nanofibrous Separator for Lithium Ion Batteries obtained by Coaxial Electrospinning." *Macromolecular Materials and Engineering* 298, no. 7 (07/01; 2019/03, 2013): 806–813.

Meyer, Wolfgang H. "Polymer Electrolytes for Lithium-Ion Batteries." *Advanced Materials* 10, no. 6 (04/01; 2019/03, 1998): 439–448.

Miao, Yue-E, Guan-Nan Zhu, Haoqing Hou, Yong-Yao Xia, and Tianxi Liu. *Electrospun Polyimide Nanofiber-Based Nonwoven Separators for Lithium-Ion Batteries*. Vol. 226, 2013. doi:10.1016/j.jpowsour.2012.10.027.

Persano, Luana, Andrea Camposeo, Cagri Tekmen, and Dario Pisignano. "Industrial Upscaling of Electrospinning and Applications of Polymer Nanofibers: A Review." *Macromolecular Materials and Engineering* 298, no. 5 (05/01; 2019/03, 2013): 504–520.

Prasanth, Raghavan, Vanchiappan Aravindan, and Madhavi Srinivasan. *Novel Polymer Electrolyte Based on Cob-Web Electrospun Multi Component Polymer Blend of Polyacrylonitrile/poly(Methyl Methacrylate)/polystyrene for Lithium Ion Batteries—Preparation and Electrochemical Characterization*. Vol. 202 2012. doi:10.1016/j.jpowsour.2011.11.057.

Raghavan, Prasanth, Du-Hyun Lim, Jou-Hyeon Ahn, David C. Changwoon Nah Ho-Suk Ryu Sherrington, and Hyo-Jun Ahn. *Electrospun Polymer Nanofibers: The Booming Cutting Edge Technology*. Special Issue in Honour of Prof. David C. Sherrington FRS. Functional Polymers between Chemistry and Applications. Vol. 72, 2012. doi:10.1016/j.reactfunctpolym.2012.08.018.

Ramakrishna, Seeram, Kazutoshi Fujihara, Wee-Eong Teo, Teik-Cheng Lim, and Zuwei Ma. *An Introduction to Electrospinning and Nanofibers*. World Scientific, 2005.

Ramaseshan, Ramakrishnan, Subramanian Sundarrajan, Rajan Jose, and S. Ramakrishna. "Nanostructured Ceramics by Electrospinning." *Journal of Applied Physics* 102, no. 11 (12/01; 2019/03, 2007): 111101.

Schiffman, Jessica D. and Caroline L. Schauer. "A Review: Electrospinning of Biopolymer Nanofibers and their Applications." *Polymer Reviews* 48, no. 2 (05/01, 2008): 317–352.

Shayapat, Jaritphun, Ok Hee Chung, and Jun Seo Park. *Electrospun Polyimide-Composite Separator for Lithium-Ion Batteries*. Vol. 170, 2015. doi:10.1016/j.electacta.2015.04.142.

Subramania, A., N. T. Kalyana Sundaram, A. R. Sathiya Priya, and G. Vijaya Kumar. *Preparation of a Novel Composite Micro-Porous Polymer Electrolyte Membrane for High Performance Li-Ion Battery*. Vol. 294, 2007. doi:10.1016/j.memsci.2007.01.025.

Sundaramurthy, Jayaraman, Vanchiappan Aravindan, Palaniswamy Suresh Kumar, Srinivasan Madhavi, and Seeram Ramakrishna. "Electrospun TiO2-δ Nanofibers as Insertion Anode for Li-Ion Battery Applications." *The Journal of Physical Chemistry C* 118, no. 30 (07/31, 2014): 16776–16781.

Sundaramurthy, Jayaraman, P. S. Kumar, M. Kalaivani, V. Thavasi, S. G. Mhaisalkar, and S. Ramakrishna. "Superior Photocatalytic Behaviour of Novel 1D Nanobraid and Nanoporous α-Fe2O3 Structures." *RSC Advances* 2, no. 21 (2012): 8201–8208.

Tarascon, J. M. and M. Armand. "Issues and Challenges Facing Rechargeable Lithium Batteries." *Nature* 414, (11/15, 2001): 359.

Teo, Wee-Eong, Ryuji Inai, and Seeram Ramakrishna. "Technological Advances in Electrospinning of Nanofibers." *Science and Technology of Advanced Materials* 12, no. 1 (02/01, 2011): 013002.

Teo, Wee-Eong and Seeram Ramakrishna. "A Review on Electrospinning Design and Nanofibre Aassemblies." *Nanotechnology* 17, no. 14 (06/30, 2006): R89–R106.

Thavasi, V., G. Singh, and S. Ramakrishna. "Electrospun Nanofibers in Energy and Environmental Applications." *Energy & Environmental Science* 1, no. 2 (2008): 205–221.

Uchida, I., H. Ishikawa, M. Mohamedi, and M. Umeda. *AC-Impedance Measurements during Thermal Runaway Process in several Lithium/ Polymer Batteries*. Selected Papers Presented at the 11th International Meeting on Lithium Batteries. Vol. 119–121, 2003. doi:10.1016/S0378-7753(03)00248-9.

Wang, Qiujun, Wei-Li Song, Luning Wang, Yu Song, Qiao Shi, and Li-Zhen Fan. *Electrospun Polyimide-Based Fiber Membranes as Polymer Electrolytes for Lithium-Ion Batteries*. Vol. 132, 2014. doi:10.1016/j.electacta.2014.04.053.

Weighall, M. J. "Battery Separator Design Requirements and Technology Improvements for the Modern Lead/acid Battery." *Journal of Power Sources* 53 1995: 273–282.

Wu, Mao-Sung, Pin-Chi Julia Chiang, Jung-Cheng Lin, and Yih-Song Jan. *Correlation between Electrochemical Characteristics and Thermal Stability of Advanced Lithium-Ion Batteries in Abuse Tests—Short-Circuit Tests*. Vol. 49, 2004. doi:10.1016/j.electacta.2003.12.012.

Xue, Leigang, Xin Xia, Telpriore Tucker, Kun Fu, Shu Zhang, Shuli Li, and Xiangwu Zhang. "A Simple Method to Encapsulate SnSb Nanoparticles into Hollow Carbon Nanofibers with Superior Lithium-Ion Storage Capability." *Journal of Materials Chemistry A* 1, no. 44 (2013): 13807–13813.

Yang, Yuan, Guangyuan Zheng, and Yi Cui. "Nanostructured Sulfur Cathodes." *Chemical Society Reviews* 42, no. 7 (2013): 3018–3032.

Yanilmaz, Meltem, Yao Lu, Mahmut Dirican, Kun Fu, and Xiangwu Zhang. *Nanoparticle-on-Nanofiber Hybrid Membrane Separators for Lithium-Ion Batteries Via Combining Electrospraying and ElectrospinningTechniques*. Vol. 456, 2014. doi:10.1016/j.memsci.2014.01.022.

Yu, Yan, Lin Gu, Changbao Zhu, Peter A. van Aken, and Joachim Maier. "Tin Nanoparticles Encapsulated in Porous Multichannel Carbon Microtubes: Preparation by Single-Nozzle Electrospinning and Application as Anode Material for High-Performance Li-Based Batteries." *Journal of the American Chemical Society* 131, no. 44 (11/11, 2009): 15984–15985.

Zhai, Yunyun, Na Wang, Xue Mao, Yang Si, Jianyong Yu, Salem Al-Deyab, Mohamed El-Newehy, and Bin Ding. "Sandwich-Structured PVDF/PMIA/PVDF Nanofibrous Separators with Robust Mechanical Strength and Thermal Stability for Lithium Ion Batteries." *Journal of Materials Chemistry A* 2, no. 35 (2014): 14511–14518.

Zhang, Feng, Xilan Ma, Chuanbao Cao, Jili Li, and Youqi Zhu. *Poly(Vinylidene Fluoride)/SiO2 Composite Membranes Prepared by Electrospinning and their Excellent Properties for Nonwoven Separators for Lithium-Ion Batteries.* Vol. 251, 2014. doi:10.1016/j.jpowsour.2013.11.079.

Zhang, X., V. Aravindan, P. S. Kumar, H. Liu, J. Sundaramurthy, S. Ramakrishna, and S. Madhavi. "Synthesis of TiO2 Hollow Nanofibers by Co-Axial Electrospinning and Its Superior Lithium Storage Capability in Full-Cell Assembly with Olivine Phosphate." *Nanoscale* 5, no. 13 (2013): 5973–5980.

Zhao, Man, Jing Wang, Chuanbin Chong, Xuewen Yu, Lili Wang, and Zhiqiang Shi. "An Electrospun Lignin/Polyacrylonitrile Nonwoven Composite Separator with High Porosity and Thermal Stability for Lithium-Ion Batteries." *RSC Advances* 5, no. 122 (2015): 101115–101120.

Zhou, Xinhong, Liping Yue, Jianjun Zhang, Qingshan Kong, Zhihong Liu, Jianhua Yao, and Guanglei Cui. "A Core-Shell Structured Polysulfonamide-Based Composite Nonwoven Towards High Power Lithium Ion Battery Separator." *Journal of the Electrochemical Society* 160, no. 9 (01/01, 2013): A1341–A1347.

17 Conducting Polymers
Properties, Synthesis, and Energy Storage Devices

Mohamad Azuwa Mohamed and Nadhratun Naiim Mobarak

Centre for Advanced Materials and Renewable Resources (CAMARR)
Faculty of Science and Technology
Universiti Kebangsaan Malaysia
Selangor, Malaysia

CONTENTS

17.1   INTRODUCTION

Our dependency on energy production and consumption based on fossil fuels has severely affected our lifestyles as well as the environment. In fact, the utilization of fossils fuels as the primary energy sources had led to the emission of greenhouse gases, thereby resulting in catastrophic climate changes. Thus, it has triggered an enormous demand for the utilization of renewable, clean, and green energy technology as well as for maximizing energy efficiency in modern society. Up until today, several renewable energies have been implemented in the contemporary society such as solar energy, wind energy, hydropower, geothermal energy, and biomass energy. The generated energy can be directly used and stored for the backup energy system. In the case of storing the energy, there has been an increasing demand for the development of superior energy storage systems that can improve practicability and energy efficiency in terms of usage, distribution, and transport. It has been suggested that a superior energy storage is required to have excellent energy storage capability, cost-effectiveness, long-term stability, design flexibility, and lightweight (Kim et al. 2016).

Nowadays, the devices such as capacitor, supercapacitor, and battery are promising energy storage devices, and their development for superior storage performances is still on-going. It must be noted the working principle of the aforementioned energy storage devices is quite similar since its relied on the electrochemical. However, as seen in Figure 17.1(a), the specific power and specific energy are different. Nevertheless, they can be implemented with each other and can be configured into one superior energy storage device for maximizing energy efficiency.

Generally, conducting polymers consist of organic polymers that conduct electricity through a conjugated bond system along the polymer chain. As shown in Figure 17.1(b), the conducting polymer has attracted a lot of attention and actively explored the energy storage applications, especially in batteries and supercapacitors. The utilization of conducting polymer in a rechargeable battery also shows gradual increment in the last 18 years. The interest in conducting polymers in energy storage applications has been motivated by their reversible Faradaic redox nature, high charge density, and lower cost when compared with the expensive metal oxides (Nyholm et al. 2011; Shown et al. 2015).

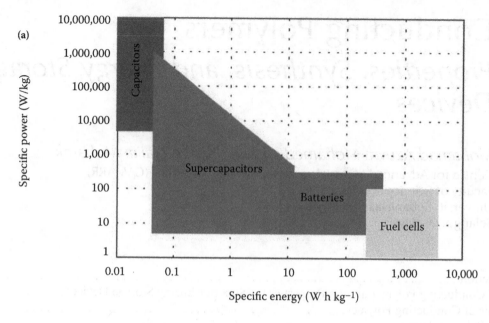

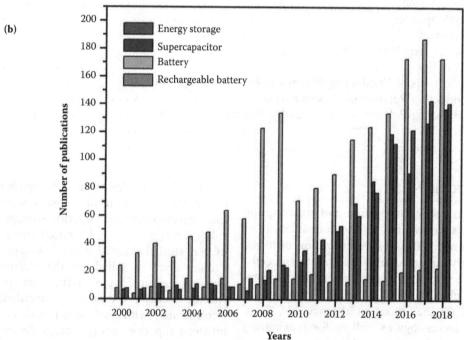

FIGURE 17.1 (a) The Ragone plot shows the energy versus power density comparison of the energy storage device and fuel cell energy conversion device (Saleem, Desmaris, and Enoksson 2016). (b) The evolution of the number of publications in the years 2000–2018 related to conducting polymer for energy storage with two main energy storage devices of supercapacitor, battery, and rechargeable battery. Source of the data: Scopus, access on Mac 2019.

17.2 TYPE OF CONDUCTING POLYMER AND ITS PROPERTIES FOR SUPERIOR ENERGY STORAGE DEVICES

The conducting polymer mostly consists of an organic polymer that exhibits semiconducting or conducting properties. In general, the ability to conduct the electron is associated with its conjugated chemical structure and has a backbone of adjoining sp^2-hybridized orbitals. Hence, delocalized π electrons are formed along their backbone. The conducting polymer can be classified into several categories based on what their main chains contain. The main polymeric chains contain can be categorized into three types that are aromatic cycles (i.e., poly(fluorene)s, polyphenylenes, polystyrenes, polyazulenes, polynaphthalenes), double bonds repeating units (i.e., polyacetylene), and the combination of aromatic and double bonds repeating units (i.e. poly(p-phenylene vinylene)). In addition, there are two main popular heteroatoms that can be present in the conducting polymer's chemical structures; these are nitrogen (i.e., poly(pyrrole)s (PPY), polycarbazole, polyindoles, polyamines, and polyanilines) and sulfur-containing heteroatoms (i.e., poly(thiophene)s (PT), poly(3,4-ethylenedioxythiophene) (PEDOT), and poly (p-phenylene sulfide) (PPS)). In addition, polyselenophenes can also be considered as a good conducting polymer with the Se atom as a heteroatom in its chemical structure (Patra and Bendikov 2010).

Conducting polymers are the focus of research due to their attractive properties, such as a wide range of conductivity, facile production, mechanical stability, light weight, and low cost (Abdelhamid, O'Mullane, and Snook 2015; Kim et al. 2016). The conductive polymers that possess high electrical conductivity are crucial for superior energy storage capability, which leads to high charge/discharge rates. In addition, the conductive polymers that have high surface area and high mesoporosity also contribute to a higher electrical conductivity. The level of conductivity of the conductive polymers can be tuned by introducing doping within its chemical structure (Wang et al. 2017). In general, there are several types of doping that can be used to enhance the conductivity of polymers such as redox doping, nonredox doping, photodoping, charge injection doping, and electrochemical doping (Dai 2004). The presence of carbon-based materials also significantly improves the overall conductivity of the conducting polymers (Abbasi, Antunes, and Velasco 2019).

A simple production of conductive polymers is essential to understand its practical applications. Besides, the low cost and simple production approach that do not compromise its performance efficiency are vital for industrial commercialization production. For instance, the conducting polymer hydrogels that consist of PEDOT-PSS have been prepared by a simple supramolecular self-assembly between polymers and multivalent cations (Dai et al. 2008). The conventional solution casting, which is a straightforward approach, has been employed for the fabrication of porous conductive thermoplastic polyurethane nanocomposite films (Wu and Chen 2017). The resultant conductive polymers exhibited porous structures, large surface area, high porosity, and lightweight; high conductivity was also achieved (2.98×10^{-2} S cm^{-1}) with 40 wt% carbon nanofibers (CNFs).

The long-term stability of conductive polymers is one of the main criteria for better energy storage capability. The mechanical and chemical stability of the conductive polymer can be enhanced by various promising approaches. For instance, the renewable cellulose has the advantages of being lightweight and mechanically robust and of having micro- and

nanostructures with a large surface area; it has also been utilized as robust templates (Kim et al. 2016). The well-defined interconnected network nanoheterostructure resulted in strong synergistic interactions with improved electrical conductivity, reduced junction resistance, increased load transfer efficiency during bending, and enhanced structural stability (Wang et al. 2019). In addition, the excellent cycling stability (93% capacitance retention after 15,000 cycles at 30 mA cm^{-2}) and high specific electrode volumetric energy and power densities of 1.5 mW h cm^{-3} and 1470 mW cm^{-3} have been achieved by the fabrication of robust and highly flexible PEDOT nanocellulose paper composites (Wang et al. 2016). In the case of flexible conductive polymers, the introduction of single-walled carbon nanotubes into a conductive PEDOT:PSS has significantly improved the mechanical and chemical stability, under bending and damp-heat test conditions (Fischer et al. 2018).

17.3 SYNTHESIS OF CONDUCTING POLYMER

Polyacetylene, polythiophenes, polypyrroles, poly(arylene vinylenes), and polyaniline can be categorized as electrically conducting polymers since they contain unsaturated carbon double bond along with their backbone. Each of this polymer can be synthesized by different routes of reactions, either by the chemical oxidative polymerization method where a catalyst will be used or by an electrochemical polymerization method where the potential value plays a vital role. Table 17.1 summarizes synthesis method for different conducting polymers.

17.3.1 POLYACETYLENE

Polyacetylene can be synthesized by either Shirakawa or Durham method. In the Shirakawa method, polyacetylene can be synthesized by using tris(acetylacetonato) titanium (III) and diethylaluminum chloride as a catalyst (Shirakawa and Ikeda 1974). Figure 17.2 shows the scheme routes to synthesize polyacetylene using the Shirakawa method. Based on the research, the reaction of acetylene with Ti(acac)$_2$-AlEt$_2$Cl leads to the formation of benzene, a trace of ethylbenzene, and a small amount of polyacetylene.

Polyacetylene can be synthesized by using a soluble precursor polymer, such as 3,6-bis(trifluoromethyl)pentacyclo[6.2.0.0^{2,4}.0^{3,6}.0^{5,7}]dec-9-en, by a simple thermal elimination reaction using the Durham method, as shown in Figure 17.3 (Feast and Winter 1985). They also claimed that their polyacetylene produced by the Durham method was denser (higher than 1 g cm^{-3}) compared to that obtained by the Shirakawa method (0.4 g cm^{-3}). Besides that, there is only one volatile aromatic leaving group that is formed when the Durham method is used.

TABLE 17.1

Synthesis procedure to produce different types conducting polymers

Types of conducting polymer	Chemical structure	Methods
Polyacetylene	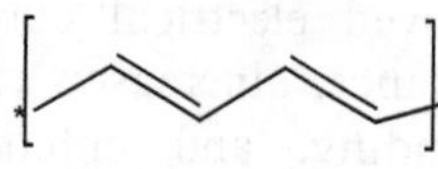	Shirakawa routes (polyacetylene with a low bulk density) Durham routes (polyacetylene with a compact structure with a higher bulk density)
Polythiophenes	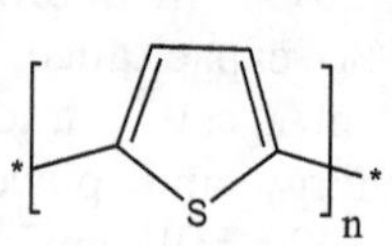	Chemical polymerization (Yamamoto method) Chemical polymerization (Lin method) Electrochemical polymerization
Polypyrroles	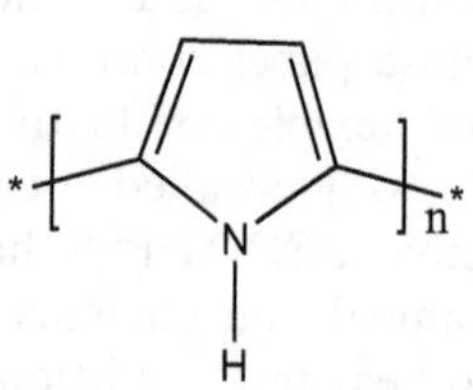	Chemical polymerization Electrochemical polymerization
Poly(arylene vinylenes)	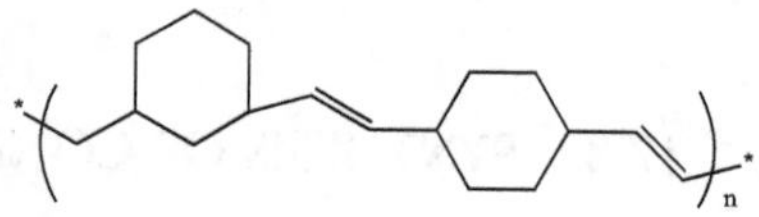	Precursor routes Witting reaction Dehalogenation and dehydrogenation reaction Knoevenagel condensation Heck reaction Ring-opening metathesis polymerization Acyclic diene metathesis (ADMET) polymerization Transition metals catalyzed polycondensation
Polyaniline	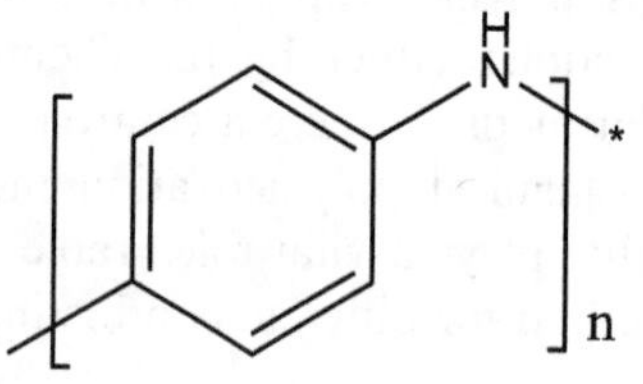	Chemical polymerization

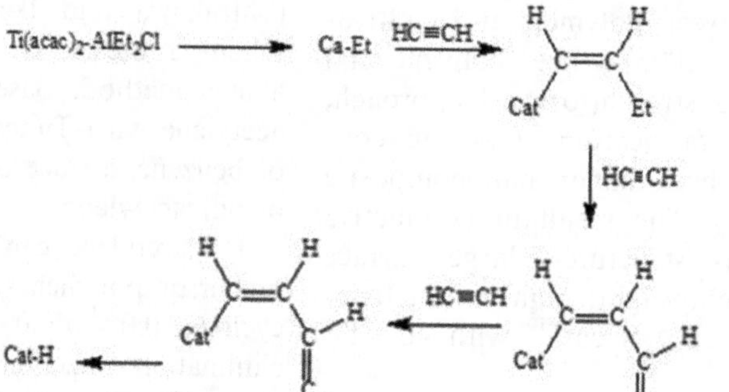

FIGURE 17.2 Routes to synthesize polyacetylene using the Shirakawa method (Shirakawa and Ikeda 1974).

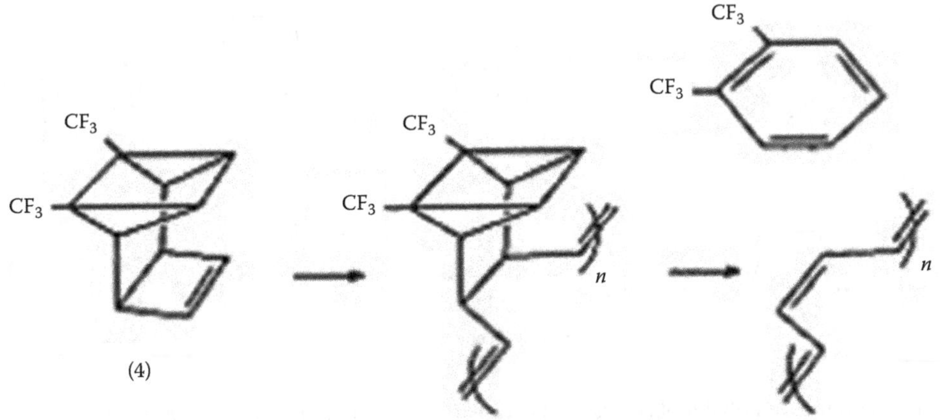

Conducting Polymers

FIGURE 17.3 Routes to synthesize polyacetylene using the Durham method (Feast and Winter 1985).

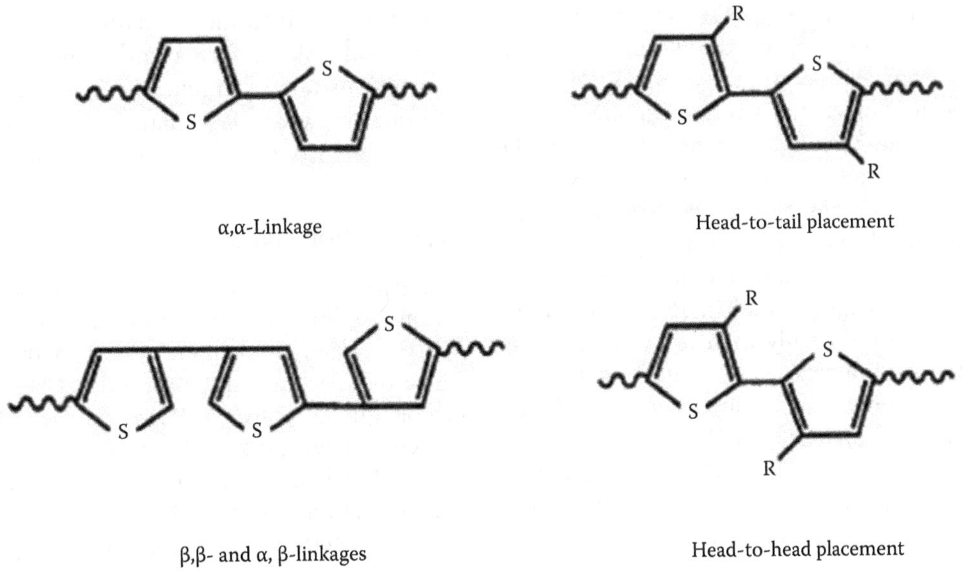

α,α-Linkage

Head-to-tail placement

β,β- and α, β-linkages

Head-to-head placement

FIGURE 17.4 Polymerization of thiophene either at **α** or at **β** position (Wei et al. 1991).

17.3.2 Polythiophenes

Polythiophenes can be prepared by electrochemical or chemical oxidation of thiophene or its derivatives, and the bonding can occur at α or β position Figure 17.4. The polymer chain is dominantly formed by the α,α-linkages due to the high reactivity at the α-position compared to that at the β-position, as proved by other theoretical and experimental studies (Wei et al. 1991).

The first chemical preparation for the production of unsubstituted polythiophene was reported by Yamamoto, Sanechika, and Yamamoto (1980) and Lin et al. (1980). In both the research, 2,5-dibromothiophene and magnesium (Mg) were used as the starting reagents and

tetrahydrofuran as a solvent (Lin and Dudek 1980; Yamamoto, Sanechika, and Yamamoto 1980). However, both of them had used a different type of catalyst. In the research done by Yamamoto, Sanechika, and Yamamoto (1980), they had used nickel(bipyridine) dichloride, while that by Lin and Dudek (1980), they had used palladium(acetylacetonate)$_2$, Ni(acetylacetonate)$_2$, Co(acetylacetonate)$_2$, or Fe(acetylacetonate)$_3$. Their findings showed that their methods could produce a low molecular weight polythiophene; however, it did not dissolve in THF. Figure 17.5 shows the chemical synthesis routes to synthesize polythiophene using the technique proposed by Yamamoto, Sanechika, and Yamamoto (1980) and Lin and Dudek (1980), respectively.

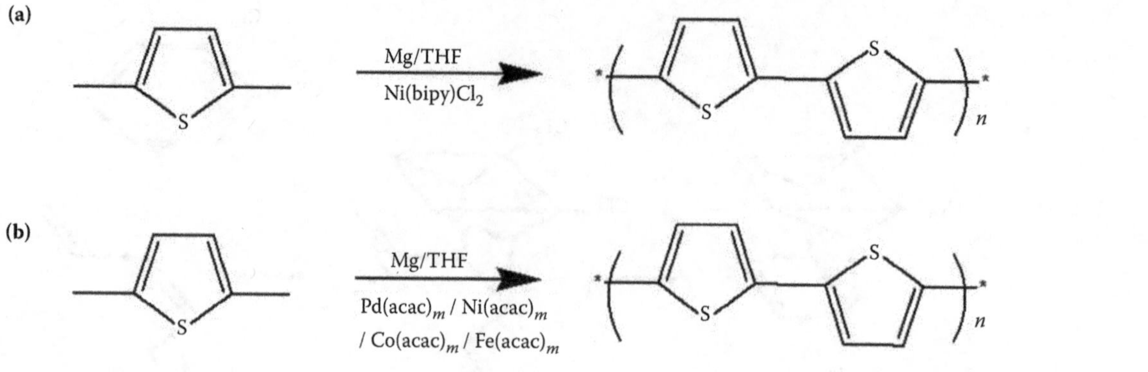

FIGURE 17.5 A route to synthesize polythiophene using the Yamamoto method (Yamamoto, Sanechika, and Yamamoto 1980).

Free-standing films of polythiophene can be obtained by using an electrochemical reaction as the film will be directly fabricated onto the electrodes. Even though during the process the thickness of the films can be controlled, there are several conditions that need to be taken in consideration such as solvent, temperature, the concentration of the monomer, the value of applied oxidation potential, and supporting electrolyte. Based on the previous studies, it can be shown that the supporting electrolyte, especially the anion type, has a strong effect on the electrochemical properties of the polythiophen films (Chrisensen et al. 1988; Tanaka et al. 1988; Krische and Zagorska 1989; Li and Li 2003). Platinum (Pt) or Pt-coated electrodes are typically used to produce a thin uniform film of polythiophene as it is inert against the polymerization reaction at the relatively high oxidation potential of thiophene (Yigit et al. 1996).

In addition, many different alkyl substitutes have been explored to improve the solubility of polythiophene. The usage of alkyl-substituted thiophenes at β-position will lead to the formation of the polythiophene with the regioregular conformation, as shown in Figure 17.6(a), since the polymerization occurs through the 2- and 5-position substitution. However, there is also a possibility of the formation of the regiorandom structure, as illustrated in Figure 17.6(b), because of the steric repulsion between alkyl chains are twisted out of conjugation planarity (Wang, Chen, and Jeng 2014).

17.3.3 POLYPYRROLE

Fine powders of polypyrrole can be obtained by chemical polymerization where using chemical oxidants was necessary. The monomer will undergo oxidative polymerization, where the reaction can be carried out either in aqueous or in nonaqueous solvents. Based on the values of polypyrrole conductivity, iron (III) chloride and water are the best chemical oxidant and solvent, respectively, for chemical polymerization. During the chemical polymerization of pyrrole, electroneutrality of the polymer matrix is

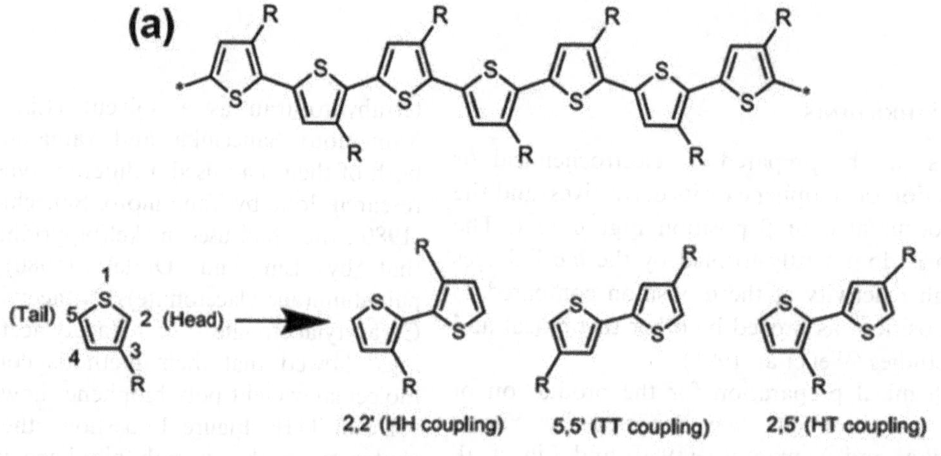

FIGURE 17.6 (a) Poly(3-alkylthiophene) with regioregular conformation. (b) Possible couplings of 3-alkylthiophenes (Wang, Chen, and Jeng 2014).

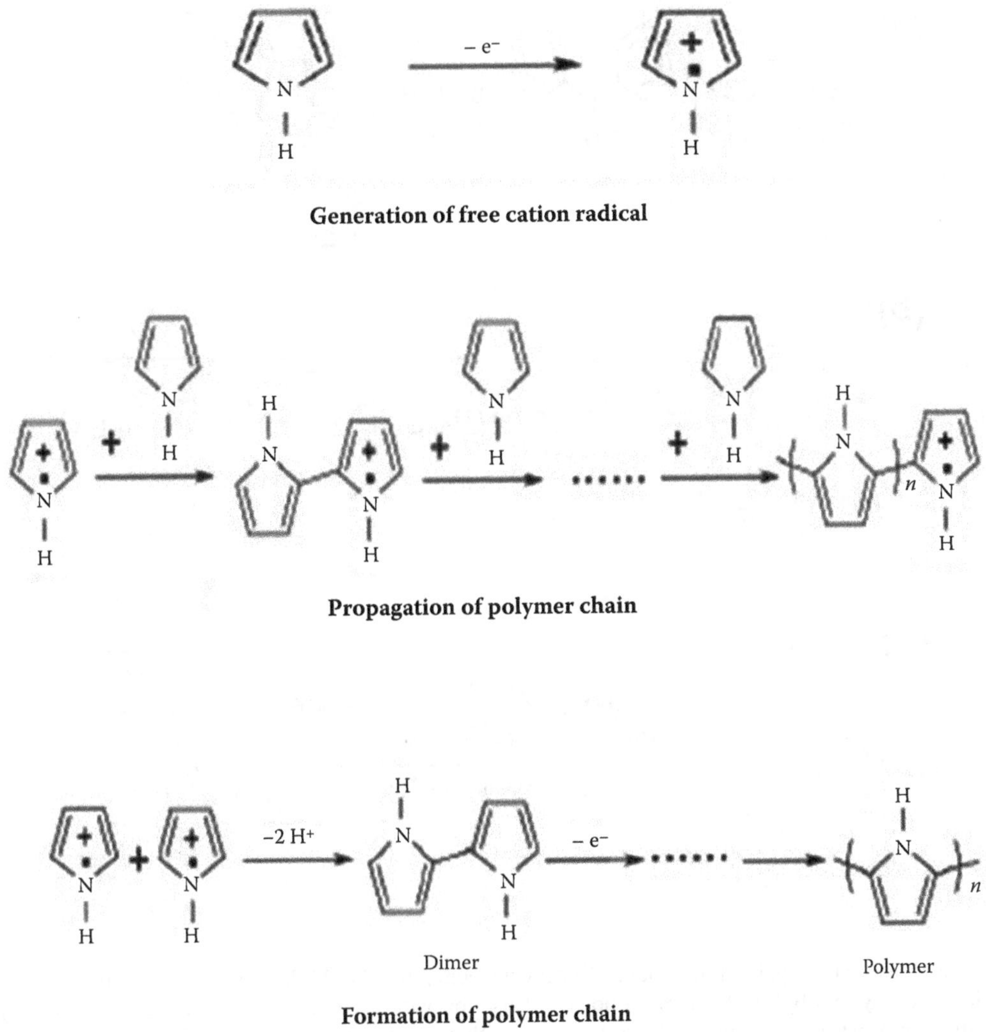

Generation of free cation radical

Propagation of polymer chain

Dimer   Polymer

Formation of polymer chain

FIGURE 17.7 Polymerization step for the pyrrole (Guanggui et al. 2012).

maintained by incorporating anions from the reaction solution. These counterions are usually anions of the chemical oxidant or the reduced product of the oxidant (Ansari 2006). However, only a limited range of polypyrrole can be prepared since a small number of counterions can be incorporated (Armes 1987; Bocchi, Gardini, and Rapi 1987; Mohammadi et al. 1987; Rapi, Bocchi, and Gardini 1988; Chao and March 1988; Machida, Miyata, and Techagumpuch 1989; Ansari 2006).

Polymer films based on polypyrrole can be obtained by an electrochemical reaction. The yield and quality of the resulting film are affected by various factors such as electrode potential, current density, solvent, and electrolyte (Salmon et al. 1982; Otero, Tejada, and Elola 1987; Bradner and Shapiro 1988; Cheung, Bloor, and Stevens 1988; Diaz and Lacroix 1988; Imisides et al. 1991; John and Wallace 1992; Ansari 2006).

Figure 17.7 shows the polymerization scheme for pyrrole where the radical cation was first produced by the initial oxidation of pyrrole dimer, followed by an electrophilic attack on a neutral molecule (Guanggui et al. 2012). However, the coupling of two dimer radical cations is more favored over the electrophilic attack. The coupling occurs between the dimer radical cation and pyrrole radical cation at the electrode surface. The growth of polypyrrole chains was terminated when the ends of the growing chains become sterically blocked (Popa et al. 2008).

17.3.4 Poly(Arylene Vinylenes)

There are several methods to synthesize poly(arylene vinylenes) such as precursor routes, Witting reaction, dehalogenation and dehydrogenation reaction, Knoevenagel condensation, Heck reaction, ring-opening metathesis polymerization, and acyclic diene metathesis

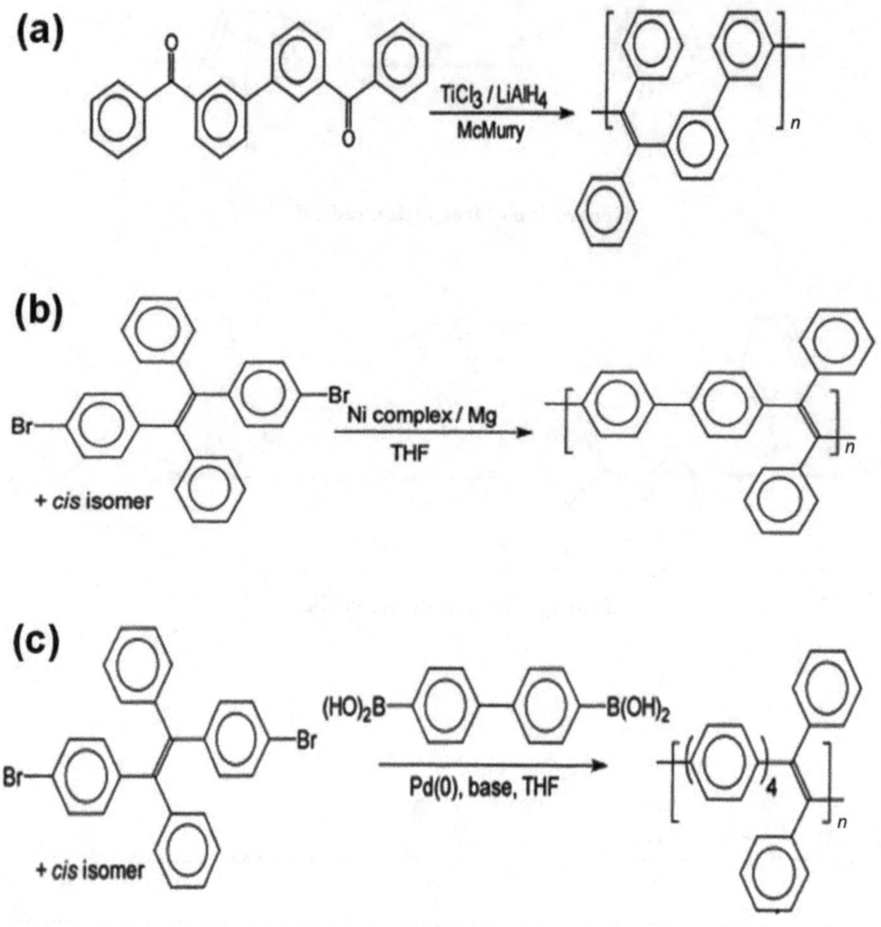

FIGURE 17.8 (a) Synthesis of poly(3,3-diphenylene diphenylvinylene) by the McMurry reaction. (b) The synthesis of poly (4,4-diphenylene diphenylvinylene) by the Yamamoto polycondensation with a control of cis/trans ratio. (c) The synthesis of poly (tetra-*p*-phenylene diphenylvinylene) by Suzuki polycondensation with a control of cis/trans ratio (Cacialli et al. 1999).

polymerization. In addition, transition metal-catalyzed polycondensation methods, such McMurry reaction, Yamamoto, and Suzuki polycondensation, can also be used.

The most common precursor routes to synthesize poly (arylene vinylenes) were based on Wessling and Zimmerman methods. Based on the research done by Lutsen et al. (1999), the precursor route was used in which bis-(sulfoniumhalide) salts of *p*-xylene were treated with an alkali base in water (Lutsen et al. 1999). Meanwhile, the Wittig reaction involves the use of aldehydes and quaternary phosphonium salts for producing poly(arylene vinylenes).

In McMurry condensation methods, poly(*p*-phenylenevinylene) can be obtained by the polymerization of diketone in the presence of TiCl₃ and LiAlH₄, as shown in Figure 17.8(a), while Figure 17.8(b) and (c) shows the production poly(4,4-diphenylene diphenylvinylene) and poly(tetra-*p*-phenylene diphenylvinylene) by Yamamoto and Suzuki polycondensation, respectively (Cacialli et al. 1999). In Yamamoto polycondensation, Ni complex/Mg

in THF solvent is used, while in Suzuki polycondensation Pd is used as the catalyst and THF as the solvent.

Based on the research done by Cacialli et al. (1999), the McMurry route can be used to obtain 1:1 composition between the cis and trans transformation of poly (arylene vinylenes). Yamamoto and Suzuki polymerizations can be used to control the cis/trans-vinylene ratio for producing poly(arylene vinylenes).

17.3.5 POLYANILINE

Polyaniline can exist in leucoemeraldine, emeraldine, and pernigraniline oxidation states (Figure 17.9). However, only polyaniline in the emeraldine states will have electrically conducting properties. As other conducting polymers, the polyaniline can be synthesized by chemical or electrochemical oxidation, where usually chemical polymerization will be used when a large quantity of polymer needs to be produced. Electrochemical polymerization was employed when a film-based polyaniline was

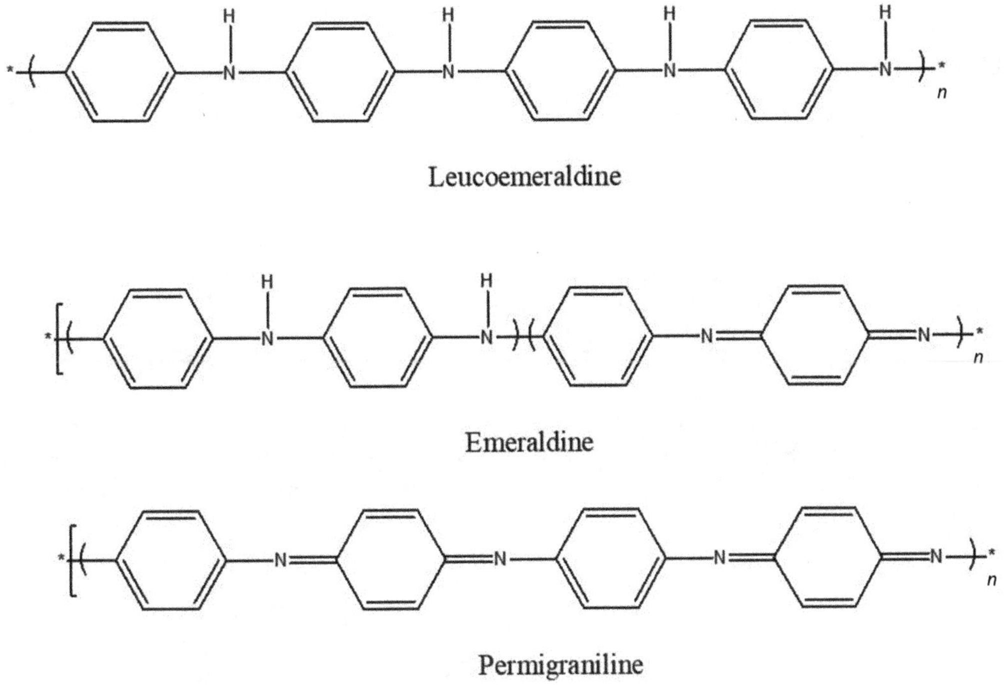

Leucoemeraldine

Emeraldine

Permigraniline

FIGURE 17.9 The structure of leucoemeraldine, emeraldine, and pernigraniline.

required; the mechanism of aniline is shown in Figure 17.10. The experimental conditions, such as electrode material, electrolyte composition, dopant anions, pH of the electrolyte, have a significant influence on the nature of the polymerization process. A low pH is needed for the formation of anilinium radical cation by aniline oxidation on the electrode.

Generally, the process takes place by an anodic oxidation, which involves three steps: initiation, chain propagation, and termination, where aniline radical cations are formed first. The interaction between oligomer radical cations and aniline radical cations takes place in the propagation steps. The reaction is terminated when two radical cations of aniline go through the dimerization process (Beregoi et al. 2016). Overall, the thickness of the polymer and conductivity can be easily controlled by proper design of the electrochemical experiment (Gvozdenović et al. 2011).

17.4 RECENT APPLICATIONS OF CONDUCTING POLYMER IN ENERGY STORAGE MATERIALS

17.4.1 CONDUCTING POLYMER IN SUPERCAPACITOR

A supercapacitor can be defined as an electronic material or the so-called energy storage devices with very high capacity and low internal resistance, which can store and deliver energy. It is also called as electrochemical capacitors. It was suggested that the energy released from a supercapacitor is significantly at a higher rate than that by a regular battery and conventional capacitors; this is due to the mechanism of energy storage that involves a simple charge separation at the interface between the electrode and the electrolyte (Iro 2016). In principle, a superconductor consists of two electrodes (positive and negative electrodes), an electrolyte, and a separator that isolates the two electrodes electrically (Iro 2016). For a superconductor, the amount of energy stored is given by the following equation:

$$E = \frac{1}{2}CV^2$$

where C is the cell capacitance and V is the cell voltage of the supercapacitor (Tammela et al. 2015). Indeed, the electrodes are essential components that warrant a higher performance of the superconductors. Thus, most of the recent researches have been devoted to the modification of the electrodes to improve the energy density and performance efficiency.

In the case of applying a conducting polymer for superconductor applications, it usually suffers from the difficulty to achieve a high specific capacitance and excellent long-term stability simultaneously (Gao et al. 2017). Furthermore, poor ion mobility within conducting polymer electrodes often limits its overall performance (Fong et al. 2017). The capacitance, stability, high rate capability as well as facilitates ion movement within polypyrrole

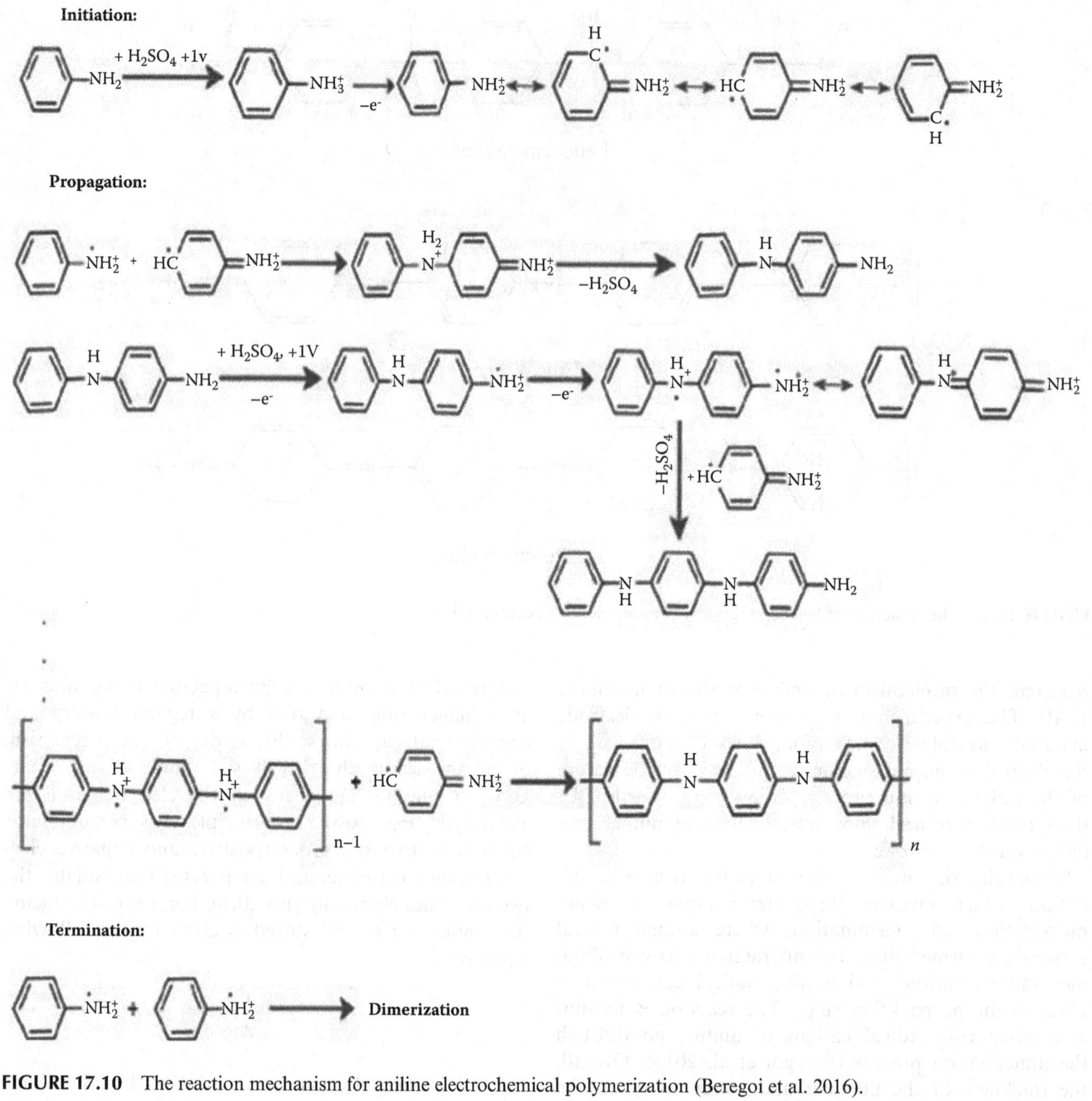

Initiation:

Propagation:

Termination:

FIGURE 17.10 The reaction mechanism for aniline electrochemical polymerization (Beregoi et al. 2016).

(PPy) has significantly improved by composites with cellulosic materials (Tammela et al. 2015). In addition, in an asymmetric superconductor cell, a PPy–cellulose composite is used as a positive electrode, while a CNF obtained from the heat treatment of PPy–cellulose composite is employed as the negative electrode in a 2 M NaCl aqueous electrolyte solution. The resultants asymmetric all-organic superconductor cells exhibit a capacitance of 25 F g^{-1} (or 2.3 F cm^{-2}) at a current density of 20 mA cm^{-2} and a maximum cell voltage of 1.6 V. On the other hand, using other polysaccharide materials other than cellulose, such as lignin, also has

a positive impact on the conductive polymer performance in superconductor applications. For instance, PEDOT incorporated with lignin that consists of a sulfonate group shows double the specific capacitance (170.4 F g^{-1}) when compared to that obtained by the reference PEDOT electrodes (80.4 F g^{-1}) and had relatively superior stability (Ajjan et al. 2016).

It has been suggested that the conducting polymers can store charges not only through pseudocapacitances but also in the electrical double-layer capacitances (EDLCs) (Santino et al. 2016). As a result, the conductive polymer electrode could possess higher specific capacitance in comparison with carbon-based capacitors. However, the pristine conductive polymer electrode suffered poor cycling stability during the long cycle charge/discharge process. One of the approaches to overcome this limitation is by incorporating carbon-based materials in the conductive polymer electrode structures. In 2015, Yang and coworkers successfully fabricated a flexible PEDOT-PSS/RGO (reduced graphene oxide) film with a layered structure by a simple vacuum-filtered method as a high-performance electrochemical electrode (Yang et al. 2015). The resultant materials showed excellent cycling stability, and the capacity retention was 90.6% after 1000 cycles. Later, Hareesh and his team successfully developed an ultrahigh supercapacitance performance of a conducting polymer PEDOT:PSS/RGO coated with MnO_2 nanorods as shown in Figure 17.11 (Hareesh et al. 2017). The electrochemical performance of the prepared materials in acetonitrile-containing lithium perchlorate showed an enhanced specific capacitance of 633 F g^{-1} at a current density of 0.5 A g^{-1} and 100% stability up to 5000 charging–discharging cycles at 1 A g^{-1}. The study suggested that the synergetic effect of the individual components and also the addition of the PEDOT:PSS conducting polymer are important factors that may provide numerous active sites for faradic redox

reactions as well as support the MnO_2 nanorods. Thus, the specific capacitance of the conductive polymer nanocomposite was significantly enhanced.

Furthermore, a new ternary conductive polymer superconductor nanocomposite (SnS_2/N-doped rGO/PANI) developed by Xu et al. revealed a relatively high specific capacitance, specific energy, and specific power values of 1021.67 F g^{-1}, 69.53 W h kg^{-1}, and 575.46 W kg^{-1} (Xu et al. 2018). These results were obtained after 60% polyaniline was deposited onto tin disulfide/nitrogen-doped reduced graphene oxide composites at a current density of 1 A g^{-1}. The illustration of the fabrication method and its morphology and the structure of SnS_2/N-doped rGO/PANI composite are shown in Figure 17.12(a–c). It has been suggested that the surface interaction between PANI and SnS_2/N-doped rGO plays an essential role in ion transport and storage as well as it can withstands the volume change on cycling, leading to a remarkable rate performance and high capacitance behavior. It is important to note that the amount of PANI loading is a crucial element for ensuring the high efficiency of the samples. It was found that the random aggregation of PANI nanofibers on the surface of SnS_2/N-doped GO nanosheets has led to a slight decrement in the specific capacitance. However, an in-depth understanding of this

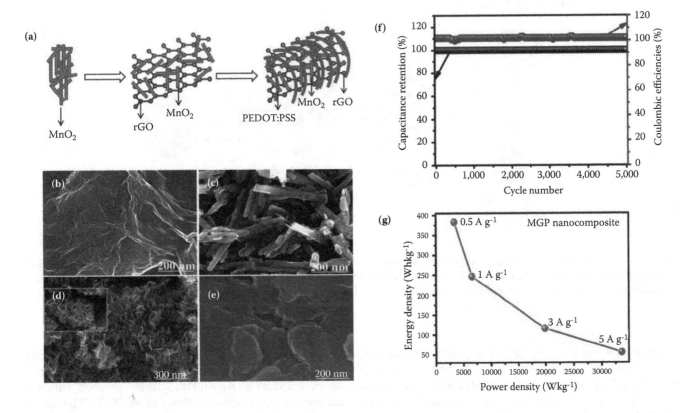

FIGURE 17.11   (a) A schematic diagram of the preparation of the PEDOT:PSS/MnO_2/rGO nanocomposite. FESEM images of (b) GO, (c) MnO_2 nanorods, (d) MnO_2/rGO nanocomposite, and (e) PEDOT:PSS/MnO_2/rGO nanocomposite. (f) Capacitance retention (%) and coulombic efficiency of the PEDOT:PSS/MnO_2/rGO nanocomposite over 5000 cycles at a current density of 1 A g^{-1}; (g) the Ragone plot for the PEDOT:PSS/MnO_2/rGO nanocomposite (Hareesh et al. 2017).

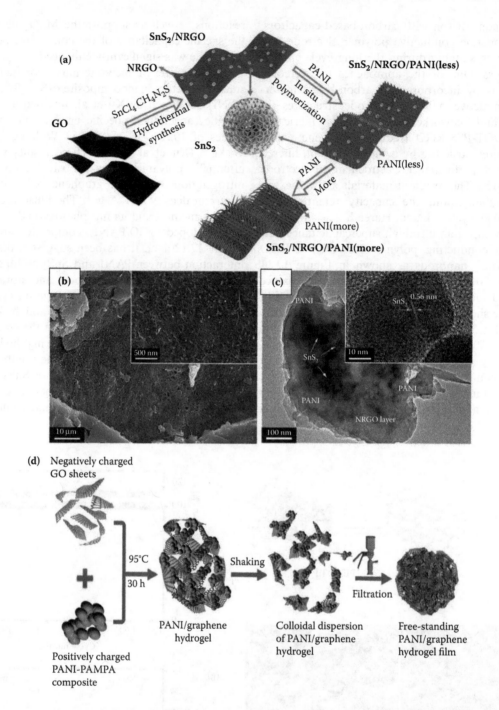

FIGURE 17.12   (a) The synthetic process of SnS₂/N-doped rGO/PANI composite; (b) and (c) FESEM and TEM images of SnS₂ /N-doped rGO/PANI composite (Xu et al. 2018). (d) A schematic diagram of the formation of a polyaniline (PANI)/graphene hydrogel (graphene refers to the reduced graphene oxide) (Moussa et al. 2015).

phenomenon needs to be further explored. In addition, the free-standing composite hydrogel film consists of PANI/graphene hydrogel where PANI played an essential role in the gelation process as shown in Figure 17.12(d), resulting in the volumetric capacitance of 225.42 F cm⁻³ with a two-electrode supercapacitor configuration (Moussa et al. 2015).

Moreover, the transparent, stretchable, and flexible supercapacitor in wearable energy devices has been developed, which consists of Ag/Au/Polypyrrole core-shell nanowire networks as an electrode (Moon et al. 2017). The aim of the core-shell nanowire networks is to overcome the difficulty of applying Ag nanowires as the energy storage devices due to the lower redox potential of Ag than that of

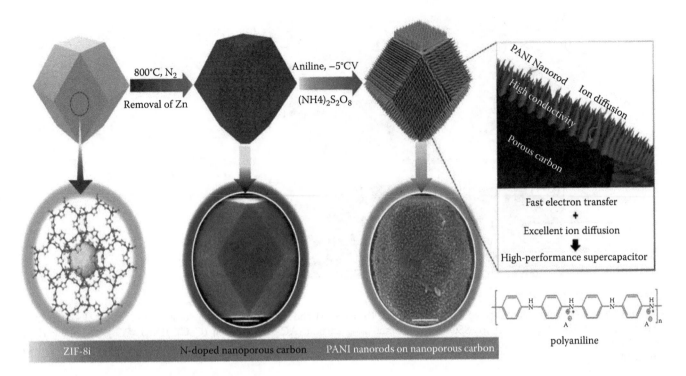

800°C, N₂
Removal of Zn

Aniline, –5°CV
(NH4)₂S₂O₈

PANI Nanorod   Ion diffusion
High conductivity
Porous carbon

Fast electron transfer
+
Excellent ion diffusion
⬇
High-performance supercapacitor

polyaniline

ZIF-8i   N-doped nanoporous carbon   PANI nanorods on nanoporous carbon

FIGURE 17.13 A schematic illustration of the synthetic procedure to attain nanoporous carbon–PANI core-shell nanocomposite materials, starting from the rhombic dodecahedron ZIF-8. For the preparation of nanoporous carbon, ZIF-8 was carbonized, and Zn was removed by washing with HF. The SEM images of the bare carbon and carbon–PANI nanocomposite are shown. The scale bars are 500 nm in length. The PANI nanorods were grown on nanoporous carbon, and the lengths of the nanorods were controlled by the time of polymerization (Salunkhe et al. 2016).

electropolymerizable monomer. A similar concept of core-shell nanoarchitectures has previously been applied to fabricate the ultrahigh supercapacitors. It is believed that a three-dimensional structured electrode comprises of a core-shell architecture consisting of nanoporous carbon as the core with an easy ion diffusion and a conductive redox active material as the shell with high electrical conductivity (Salunkhe et al. 2016). As shown in Figure 17.13, the nanoporous carbon–PANI core-shell nanocomposite electrode can be constructed using the zeolitic imidazolate framework (ZIF) as the porous carbon core, followed by the growth of PANI nanorod by a simple polymerization process.

As can be seen from Figure 17.14(a), the unique, multifaceted nanoarchitecture nanoporous carbon–PANI core-shell nanocomposite is able to avoid the stacking phenomenon that is generally observed in one-dimensional CNTs or two-dimensional graphene. Moreover, this kind of morphology could facilitate easy diffusion of ions deep inside the material by using PANI nanorod arrays, which provide the ions with easy access to the carbon core as well as provide fast conducting channels (electron highways) for electrons to reach the collector surface. The best-fabricated samples (Figure 17.14(b)) indicated that 3 h of PANI polymerization had led to an appropriate length of PANI nanorod arrays on the nanoporous carbon with the highest capacitance value of 1100 F g^{-1}. In addition, the symmetric

supercapacitor (SSC) based on carbon–PANI/carbon–PANI shows the highest specific energy of 21 W h kg^{-1} and the highest specific power of 12 kW kg^{-1} as well as a relatively high capacitance retention of 86% after 20 000 cycles as shown in Figure 17.14(c) and (d). The findings have hugely suppressed the performance of Co_3O_4@polypyrrole (PPy) core/shell nanosheet arrays (Yang et al. 2016) and SiO_2-PPy core-shell nanoparticles (Han and Cho 2018).

Interestingly, in 2019, a low-cost and large-scale manufacturing method to construct a supercapacitor yarn with high power and high energy density has been developed by creating a flexible PDEOT:PSS-PAN (polyacrylonitrile)/Ni cotton (PNF/NiC) capacitor yarn (Sun et al. 2019). The electrode showed a high volumetric capacitance of 26.88 F cm^{-3} (at 0.08 A·cm^{-3}), an energy density of 9.56 mW h·cm^{-3}, and a power density of 830 mW·cm^{-3}. Other exciting features of the novel supercapacitor material are that it is lightweight, highly flexible, resistant to bending fatigue, and can be connected in series or parallel, as shown in Figure 17.15.

More affordable and sustainable supercapacitors can also be developed by using renewable resource-based materials. In this case, previous work has used eggshell membranes for preparing flexible SCs by incorporating carbon nanotubes and by subsequent in situ polymerization of polypyrrole (Alcaraz-Espinoza, de Melo, and de

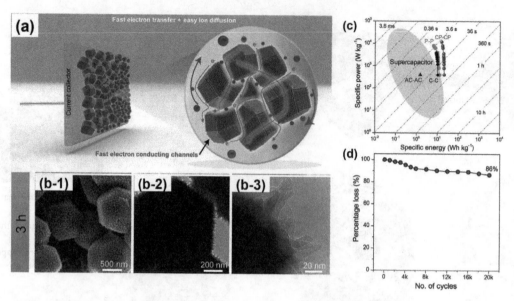

FIGURE 17.14 (a) Unique multifaceted nanoarchitecture; (b) FESEM and TEM images of the best carbon–PANI electrochemical performances. (c) The Ragone plot for SSCs based on activated carbon (AC–AC), carbon (C–C), PANI (P–P), and the carbon–PANI (CP-CP) nanocomposite. (d) A long-term cycling performance for the carbon–PANI core-shell nanocomposite (Salunkhe et al. 2016).

Oliveira 2017). These hybrid devices have shown promising operational characteristics in terms of cyclability, electrochemical behavior, mechanical flexibility, and production costs. These bioinspired supercapacitors exhibited capacitance values of 564.5 mF cm^{-2} (areal capacitance), 24.8 F cm^{-3} (volumetric capacitance), and 357.9 F g^{-1} (gravimetric capacitance). The fabricated devices also showed a capacitance retention of 60% after 4000 cycles of charge/discharge and presented negligible aging even after 500 bending repetitions (at a density of current 5 mA cm^{-2}). Besides, Edberg and coworkers have demonstrated that the introduction of sulfonated lignin into the cellulose-ion-electron-conducting polymer PEDOT:PSS system increases the specific capacitance from 110 to 230 F g^{-1} and the areal capacitance from 160 mF cm^{-2} to 1 F cm^{-2} (Edberg et al. 2018). In the case of solid-state supercapacitors, Pérez-Madrigal and her team have successfully considered the biohydrogel electrolytes system based on k-carrageenan, which is linearly sulfated plant polysaccharides (Pérez-Madrigal et al. 2016). Their work aimed at developing greener and more sustainable energy storage devices by replacing conventional liquid electrolytic systems with biohydrogels in organic electrochemical supercapacitors (OESCs). It was found that the OESCs based on PEDOT electrodes and k-carrageenan hydrogel as an electrolyte exhibit an acceptable specific capacitance, cycling stability, small leakage current, and low self-discharging tendency, which can be considered as a good supercapacitor response.

Even though the aforementioned approaches have successfully enhanced the performance of supercapacitors with facile ion transport properties and high specific capacitance, most of the noticeable improvement in the superconductor based on conductive polymers has been carried out by complicated or costly synthesis methods. The previous study has highlighted the semi-interpenetrating polymer networks (sIPNs) of pseudocapacitive electrically conducting polymer in a cross-linked ionically conducting polymer matrix to improve the ion mobility in supercapacitor electrodes (Fong et al. 2017). The aims are to decrease the ion diffusion length scales and make virtually all the active material accessible for charge storage. The freestanding PEDOT/PEO sIPN film shows a remarkable improvement in specific capacitance, cycling stability, and flexibility in comparison with pristine PEDOT as shown in Figure 17.16.

17.4.2 Conducting Polymer in Rechargeable Battery

A conducting polymer with the emergence of their electrical functionality coupled with their physical and chemical properties has led to electroactive materials for the fabrication of rechargeable secondary batteries. Several researchers have reported using conducting polymers as the electrode materials. The polymers can be either reduced (n-doped) or oxidized (p-doped) to form an electrochemically active conductive material.

Typically, the development of new electrode materials based on conducting polymers often involves incorporating

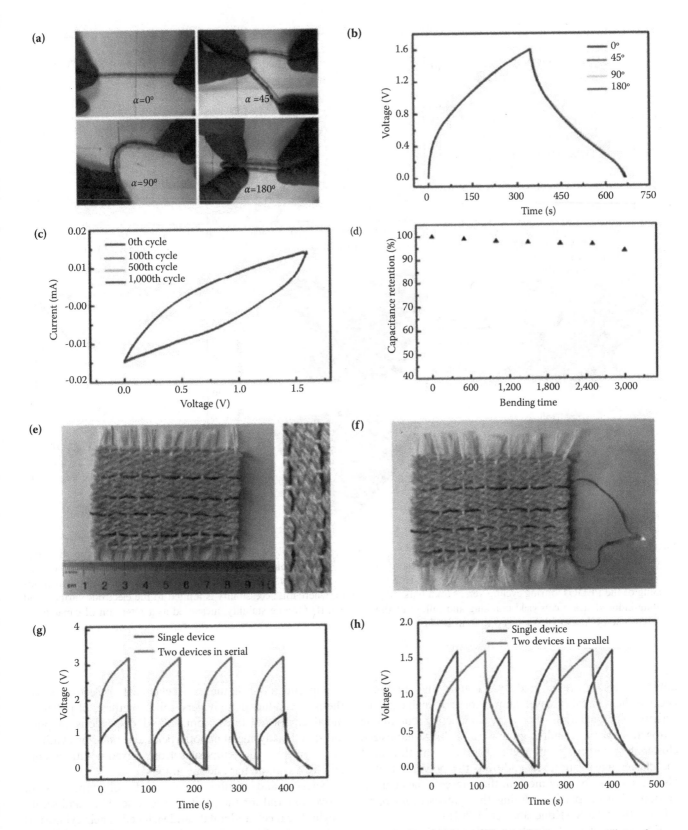

FIGURE 17.15 (a) Silicone tube used for easy bending of PNF/NiC SC yarn at different angles and (b) its corresponding galvanostatic charge and discharge curves. (c) The cyclic voltammetry curves of PNF/NiC SC yarns, which were bent for 100, 500, and 1000 cycles under a 180° bending angle. (d) The capacity retention of SC yarns at 3000 bending cycles. (e) PNF/NiC SC yarns woven into the fabric and (f) lighting a diode. (g) Galvanostatic charge and discharge curves of PNF/NiC SC yarn in series and (h) parallel (Sun et al. 2019).

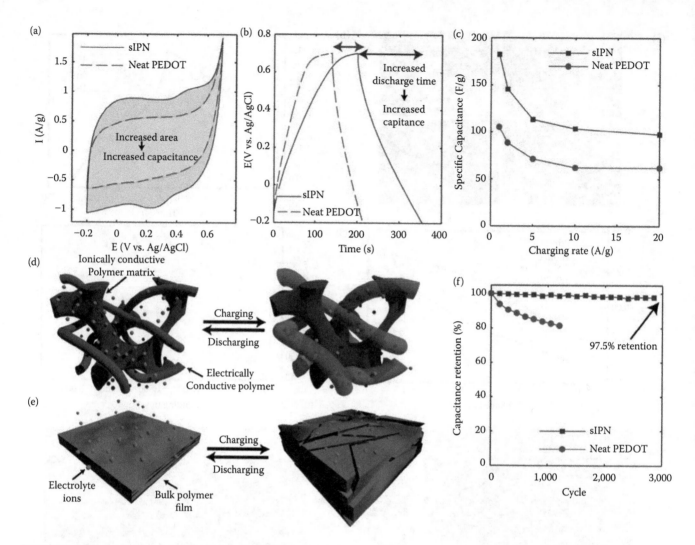

FIGURE 17.16 Performance advantages of the sIPN relative to neat PEDOT. (a) CV curves at 5 mV s^{-1}. (b) Charge–discharge curves at 1 A g^{-1}. (c) Rate capability for charging rates up to 20 A g^{-1}. Morphological origin of the specific capacitance and cycling stability trends of (d) the sIPN, where the cross-linked PEO matrix provides a reservoir of electrolyte ions and locally constrains swelling of the PEDOT during cycling versus (e) a bulk polymer film, where ion accessibility is limited to the electrode surface, and swelling-induced strain can yield cracking and failure of the material. (f) Cycling stability, measured as a retention of capacitance after repeated cycling at 10 A g^{-1} (Fong et al. 2017).

noble metals such as gold, palladium, and platinum because these metals are well known for their high catalytic activity. The materials based on conducting containing nitrogen, oxygen, or sulfur groups can stabilize nanosized clusters by interacting with metal atoms and can thus limit the cluster aggregation. In addition, the polymer matrix also acts as a stabilizing medium that prevents the agglomeration of metal particles during the synthesis and operation of the catalyst (Kondratiev et al. 2016).

Kondratiev et al. (2016) also mention that there are four types of methods that can be used in the preparation of conducting polymer–metal composite films on electrodes: first, the electrochemical or electroless deposition of metal nanoparticles into presynthesized films of conducting polymers; second, incorporation of presynthesized metal

nanoparticles during the electrochemical or electroless synthesis of conducting polymers; third, synthesis of polymer–metal composite films from colloid dispersions of polymers; and lastly, metal nanoparticles and one-step synthesis of polymer–metal composites from mixed solutions containing the monomer and metal ions.

Naegele and Bittihn (1988) used polypyrrole as an active material for the positive electrode. They had set up a cylindrical cell and a flat sandwich cell. Their cylindrical cell consists of two polypyrrole films, one lithium foil, two separators, and a current collector. The cell reaches the energy densities of 15 W. h kg^{-1} at a discharge time of 20 h. The flat sandwich cell consists of a positive electrode that is folded in a rectangular manner between a lithium foil and is enclosed between two separators; they reported that their flat sandwich cell achieved 30 Wh kg at 20 h discharge rate energy densities.

Wei et al. (2018) studied the combination of conducting polymers, such as polystyrene or polyaniline, with bamboo-derived hierarchical porous carbon for a lithium–sulfur battery. Based on their results, a combination of polystyrene and BPDC displays a higher capacity retention during long-term cycles, while the combination of polyaniline with BPDC presents a higher initial capacity. Their studies also showed that the combination of polyaniline with BPDC/S electrode achieved 1320 mA h g^{-1} at 0.1 C, 1195 mA h g^{-1} at 0.2 C, 895 mA h g^{-1} at 0.5 C, 750 mA h g^{-1} at 1 C, and 633 mA h g^{-1} at 2 C, while a combination of polystyrene and BPDC/S electrode achieved 578 mA h g^{-1} at 1 C and 445 mA h g^{-1} at 2 C. In addition, their study also showed that a combination of polyaniline with BPDC/S electrode retained 730 mA h g^{-1} after 100th cycles at 0.1 C with a high Coulombic efficiency (~99%).

Li et al. (2018) studied the potential of polypyrrole as a coating material for Sn nanoparticles, which was further tested as anode-based Li$_x$Sn. They tried to overcome the drawbacks of Sn, such as instability of Li$_x$Sn alloy under ambient atmosphere and enormous volume variation and gradual aggregation of Li$_x$Sn/Sn during the charge and discharge process They reported that by having a 20 nm coating layer, the Li$_x$Sn nanoparticles will have a very high stability in dry air condition as this keeps intact during the thermal lithiation. In addition, 75% of its prelithiated capacity was maintained after exposure to dry air for 5 days. It also delivers a stable reversible capacity of 534 mA h g^{-1} for 300 cycles. Overall, there are a number of other studies that had been done by other researchers who used conducting polymers as a binder. The incorporation of conducting polymer for preparing the electrode is basically with the aim to reduce the volumetric variation of the electrode materials.

17.5 CONCLUSION

Electrically conducting polymers, such as polyacetylene, polythiophenes, polypyrroles, poly(arylene vinylenes), and polyaniline, which have the ability to conduct electrons, are not new for researchers. This is associated with the chemical structures of the conducting polymers that have aromatic cycles, double bonds repeating units, and a combination of aromatic and double bonds repeating units. The conductive polymers that possessed high electrical conductivity and long-term stability are crucial for superior energy storage capability; usually, the electrical conductivity of a conducting polymer can be enhanced by introducing doping within its chemical structure. The properties of the conducting polymer are also influenced by its synthesizing methods. Generally, a conducting polymer can be synthesized either by chemical polymerization or by electrochemical polymerization method. Several studies have proved that the potential of a conducting polymer as electrode materials for supercapacitor application can be improved by adding additives such as cellulose; lignin consists of sulfonate group and a reduced graphene oxide. In addition, a conducting polymer has also been used by other researchers as a binder for preparing electrode (either an anode or a cathode) for applications in batteries. Their studies show that the presence of conducting polymer is not just an alternative to solve the volumetric variation of the electrode materials, but it also helps to increase the capacity retention of the battery.

REFERENCES

Abbasi, H., Antunes, M., and Velasco, J. I. 2019. Recent advances in carbon-based polymer nanocomposites for electromagnetic interference shielding. *Progress in Materials Science* 103: 319–373.

Abdelhamid, M. E., O'Mullane, A. P., and Snook, G. A. 2015. Storing energy in plastics: A review on conducting polymers & their role in electrochemical energy storage. *RSC Advances* 5(15): 11611–11626.

Ajjan, F. N., Casado, N., Rębiś, T., et al. 2016. High performance PEDOT/lignin biopolymer composites for electrochemical supercapacitors. *Journal of Materials Chemistry* 4 (5): 1838–1847.

Alcaraz-Espinoza, J. J., de Melo, C. P., and de Oliveira, H. P. 2017. Fabrication of highly flexible hierarchical polypyrrole/carbon nanotube on eggshell membranes for supercapacitors. *ACS Omega* 2(6): 2866–2877.

Ansari, R. 2006. Polypyrrole conducting electroactive polymers: Synthesis and stability studies. *E-Journal of Chemistry* 3 (4): 186–201.

Armes, S. P. 1987. Optimum reaction conditions for the polymerization of pyrrole by iron(III) chloride in aqueous solution. *Synthetic Metals* 20(3): 365–371.

Beregoi, M., Busuioc, C., Evanghelidis, A., et al. 2016. Electrochromic properties of polyaniline-coated fiber webs for tissue engineering applications. *International Journal of Pharmaceutics* 510(2): 465–473.

Bocchi, V., Gardini, G. P., and Rapi, S. 1987. Highly electroconductive polypyrrole composites. *Journal of Materials Science Letters* 6(11): 1283–1284.

Bradner, F. P., and Shapiro, J. S. 1988. Improvement in the quality of polypyrrole films prepared electrochemically on a mercury anode. *Synthetic Metals* 26(1): 69–77.

Cacialli, F., Daik, R., Feast, W. J., Friend, R. H., and Lartigau, C. 1999. Synthesis and properties of poly(arylene vinylene)s with controlled structures. *Optical Materials* 12 (2–3): 315–319.

Chao, T. H., and March, J. 1988. A study of polypyrrole synthesized with oxidative transition metal ions. *Journal of Polymer Science Part A: Polymer Chemistry* 26(3): 743–753.

Cheung, K. M., Bloor, D., and Stevens, G. 1988. Characterization of polypyrrole electropolymerized on different electrodes. *Polymer* 29(9): 1709–1717.

Christensen, P. A., Hamnett, A., and Hillman, A.R. 1988. An in-situ infra-red study of poly thiophene growth. *Journal of Electroanalytical Chemistry and Interfacial Electrochemistry*, 242(1–2): 47–62.

Dai, L. 2004. Conducting polymer. In *Intelligent Macromolecules for Smart Devices*, edited by Dai, L. 1st ed., 1980:41–80. Engineering Materials and Processes. London: Springer-Verlag.

Dai, T., Jiang, X., Hua, S., Wang, X., and Lu, Y. 2008. Facile fabrication of conducting polymer hydrogels via supramolecular self-assembly. *Chemical Communications* 36: 4279–4281.

Diaz, A. F., and Lacroix, J. C. 1988. Synthesis of electroactive/conductive polymer films: Electrooxidation of hetero-aromatic compounds. *New Journal of Chemistry* 12(4): 171–180.

Edberg, J., Inganäs, O., Engquist, I., and Berggren., M. 2018. Boosting the capacity of all-organic paper supercapacitors using wood derivatives. *Journal of Materials Chemistry* 6 (1): 145–152.

Feast, W. J., and Winter, J. N. 1985. An improved synthesis of polyacetylene. *Journal of the Chemical Society, Chemical Communications* 4: 202–203.

Fischer, R., Gregori, A., Sahakalkan, S. et al. 2018. Stable and highly conductive carbon nanotube enhanced PEDOT:PSS as transparent electrode for flexible electronics. *Organic Electronics: Physics, Materials, Applications* 62: 351–356.

Fong, K. D., Wang, T., Kim, H.-K., Kumar, R. V., and Smoukov, S. K. 2017. Semi-interpenetrating polymer networks for enhanced supercapacitor electrodes. *ACS Energy Letters* 2(9): 2014–2020.

Gao, Z., Yang, J., Huang, J., Xiong, C., and Yang, Q. 2017. A three-dimensional graphene aerogel containing solvent-free polyaniline fluid for high performance supercapacitors. *Nanoscale* 9(45): 17710–17716.

Guanggui, C., Jianning, D., Zhongqiang, Z., Zhiyong, L., and Huasheng, P. 2012. Study on the preparation and multi-properties of the polypyrrole films doped with different ions. *Surface and Interface Analysis* 44(7): 844–850.

Gvozdenović, M. M., Jugović, B. Z., Stevanović, J. S., Trišović, T. L., and Grgur, B. N. 2011. Electrochemical polymerization of aniline. In *Electropolymerization*, edited by Schab-Balcerzak, E. 77–96. London: IntechOpen.

Han, H., and Cho, S. 2018. Ex Situ fabrication of polypyrrole-coated core-shell nanoparticles for high-performance coin cell supercapacitor. *Nanomaterials* 8(9): 726.

Hareesh, K., Shateesh, B., Joshi, R. P., et al. 2017. Ultra high stable supercapacitance performance of conducting polymer coated MnO_2 nanorods/RGO nanocomposites. *RSC Advances* 7(32): 20027–20036.

Imisides, M. D., John, R., Riley, P. J., and Wallace, G. G. 1991. The use of electropolymerization to produce new sensing surfaces: A review emphasizing electrode position of heteroaromatic compounds. *Electroanalysis* 3(9): 879–889.

Iro, Z. S. 2016. A brief review on electrode materials for supercapacitor. *International Journal of Electrochemical Science* 11: 10628–10643.

John, R., and Wallace, G. G. 1992. Factors influencing the rate of the electrochemical oxidation of heterocyclic monomers. *Polymer International* 27(3): 255–260.

Kim, J., Lee, J., You, J., et al. 2016. Conductive polymers for next-generation energy storage systems: Recent progress and new functions. *Materials Horizons* 3(6): 517–535.

Kondratiev, V. V., Malev, V. V., and Eliseeva, S. N. 2016. Composite electrode materials based on conducting polymers loaded with metal nanostructures. *Russian Chemical Reviews* 85(1): 14–37.

Krische, B., and Zagorska, M. 1989. The polythiophene paradox. *Synthetic Metals* 28(1–2): 263–268.

Li, X., and Li, Y. 2003. Electrochemical preparation of polythiophene in acetonitrile solution with boron fluoride-ethyl ether as the electrolyte. *Journal of Applied Polymer Science* 90(4): 940–946.

Lin, J. W.-P., and Dudek, L. P. 1980. Synthesis and properties of poly(2,5-Thienylene). *Journal of Polymer Science: Polymer Chemistry Edition* 18(9): 2869–2873.

Lutsen, L., Adriaensens, P., Becker, H., et al. 1999. New synthesis of a soluble high molecular weight poly(arylene vinylene): Poly[2-methoxy-5-(3,7-dimethyloctyloxy)- p -phenylene vinylene]. polymerization and device properties. *Macromolecules* 32(20): 6517–6525.

Machida, S., Miyata, S., and Techagumpuch, A. 1989. Chemical synthesis of highly electrically conductive polypyrrole. *Synthetic Metals* 31(3): 311–318.

Mohammadi, A., Lundström, I., Salaneck, W. R., and Inganäs, O. 1987. Polypyrrole prepared by chemical vapour deposition using hydrogen peroxide and hydrochloric acid. *Synthetic Metals* 21(1–3): 169–173.

Moon, H., Lee, H., Kwon, J., et al. 2017. Ag/Au/polypyrrole core-shell nanowire network for transparent, stretchable and flexible supercapacitor in wearable energy devices. *Scientific Reports* 7(1): 41981.

Moussa, M., Zhao, Z., El-Kady, M. F., et al. 2015. Free-standing composite hydrogel films for superior volumetric capacitance. *Journal of Materials Chemistry A* 3(30): 15668–15674.

Nyholm, L., Nyström, G., Mihranyan, A., and Strømme, M. 2011. Toward flexible polymer and paper-based energy storage devices. *Advanced Materials* 23(33): 3751–3769.

Otero, T. F., Tejada, R., and Elola, A. S. 1987. Formation and modification of polypyrrole films on platinum electrodes by cyclic voltammetry and anodic polarization. *Polymer* 28(4): 651–658.

Patra, A., and Bendikov, M. 2010. Polyselenophenes. *Journal of Materials Chemistry* 20(3): 422–433.

Pérez-Madrigal, M. M., Estrany, F., Armelin, E., Díaz, D. D., and Alemán, C. 2016. Towards sustainable solid-state supercapacitors: Electroactive conducting polymers combined with biohydrogels. *Journal of Materials Chemistry A* 4(5): 1792–1805.

Popa, C., Turcu, R., Craciunescu, I., et al. 2008. Polypyrrole-porous silicon nanocomposites. *Journal of Optoelectronics and Advanced Materials* 10(9): 2319–2324.

Rapi, S., Bocchi, V., and Gardini, G. P. 1988. Conducting polypyrrole by chemical synthesis in water. *Synthetic Metals* 24 (3): 217–221.

Saleem, A. M., Desmaris, V., and Enoksson, P. 2016. Performance enhancement of carbon nanomaterials for supercapacitors. *Journal of Nanomaterials* 2016: 1–17.

Salmon, M., Diaz, A. F., Logan, A. J., Krounbi, M., and Bargon, J. 1982. Chemical modification of conducting polypyrrole films. *Molecular Crystals and Liquid Crystals* 83(1): 265–276.

Salunkhe, R. R., Tang, J., Kobayashi, N., et al. 2016. Ultrahigh performance supercapacitors utilizing core–Shell nanoarchitectures from a metal–Organic framework-derived nanoporous carbon and a conducting polymer. *Chemical Science* 7(9): 5704–5713.

Santino, L. M., D'Arcy, J. M., Acharya, S., Lu, Y., and Bryan, A. M. 2016. Conducting polymers for pseudocapacitive energy storage. *Chemistry of Materials* 28(17): 5989–5998.

Shirakawa, H., and Ikeda, S. 1974. Cyclotrimerization of acetylene by the tris(acetylacetonato)titanium(III)–Diethylaluminum chloride system. *Journal of Polymer Science* 12(5): 929–937.

Shown, I., Ganguly, A., Chen, L.-C., and Chen, K.-H. 2015. Conducting polymer-based flexible supercapacitor. *Energy Science & Engineering* 3(1): 2–26.

Sun, X., He, J., Qiang, R., et al. 2019. Electrospun conductive nanofiber yarn for a wearable yarn supercapacitor with high volumetric energy density. *Materials* 12(2): 273.

Tammela, P., Wang, Z., Frykstrand, S., et al. 2015. Asymmetric supercapacitors based on carbon nanofibre and polypyrrole/nanocellulose composite electrodes. *RSC Advances* 5 (21): 16405–16413.

Tanaka, K., Shichiri, T., Wang, S., and Yamabe, T. 1988. A study of the electropolymerization of thiophene. *Synthetic Metals* 24(3): 203–215.

Wang, H.-J., Chen, C.-P., and Jeng, R.-J. 2014. Polythiophenes comprising conjugated pendants for polymer solar cells: A review. *Materials* 7(4): 2411–2439.

Wang, J., Wang, J., Kong, Z., Lv, K., Teng, C., and Zhu., Y. 2017. Conducting-polymer-based materials for electrochemical energy conversion and storage. *Advanced Materials* 29(45): 1703044.

Wang, Z., Tammela, P., Huo, J., Zhang, P., Strømme, M., and Nyholm, L. 2016. Solution-processed poly(3,4-ethylenedioxythiophene) nanocomposite paper electrodes for high-capacitance flexible supercapacitors. *Journal of Materials Chemistry A* 4(5): 1714–1722.

Wang, Z., Zhao, S., Huang, A., Zhang, S., and Li, J. 2019. Mussel-inspired codepositing interconnected polypyrrole nanohybrids onto cellulose nanofiber networks for fabricating flexible conductive biobased composites. *Carbohydrate Polymers* 205: 72–82.

Wei, Y., Cheung, C., Tian, C. J., Jang, G. W., and Hsueh, K. F. 1991. Electrochemical polymerization of thiophenes in the presence of bithiophene or terthiophene: Kinetics and mechanism of the polymerization. *Chemistry of Materials* 3(5): 888–897.

Wu, T., and Chen., B. 2017. Facile fabrication of porous conductive thermoplastic polyurethane nanocomposite films via solution casting. *Scientific Reports* 7(1): 1–11.

Xu, Z., Zhang, Z., Gao, L., et al. 2018. Tin disulphide/nitrogen-doped reduced graphene oxide/polyaniline ternary nanocomposites with ultra-high capacitance properties for high rate performance supercapacitor. *RSC Advances* 8 (70): 40252–40260.

Yamamoto, T., Sanechika, K., and Yamamoto, A. 1980. Preparation of thermostable and electric-conducting poly (2,5-thienylene). *Journal of Polymer Science: Polymer Letters Edition* 18(1): 9–12.

Yang, W., Zhao, Y., He, X., et al. 2015. Flexible conducting polymer/reduced graphene oxide films: Synthesis, characterization, and electrochemical performance. *Nanoscale Research Letters* 10(1): 222.

Yang, X., Kaibing, X., Zou, R., and Junqing, H. 2016. A hybrid electrode of Co_3O_4@PPy core/shell nanosheet arrays for high-performance supercapacitors. *Nano-Micro Letters* 8 (2): 143–150.

Yigit, S., Hacaloglu, J., Akbulut, U., and Toppare, L. 1996. Conducting polymer composites of polythiophene with natural and synthetic rubbers. *Synthetic Metals* 79(1): 11–16.

18 Applications of Polyethylenedioxythiophene in Photovoltaics

Jazib Ali
Department of Physics and Astronomy
Shanghai Jiao Tong University
Shanghai, China

Mutayyab Afreen
Department of Physics
University of Agriculture
Faisalabad, Pakistan

Tahir Rasheed
Department of Chemistry and Chemical Engineering
Shanghai Jiao Tong University
Shanghai, China

Muhammad Zeeshan Ashfaq
Department of Physics
University of Agriculture
Faisalabad, Pakistan

Muhammad Bilal
School of Life Science and Food Engineering
Huaiyin Institute of Technology
Huaian China

CONTENTS

18.1 INTRODUCTION

Since the breakthrough of polymeric organic conductors in 1977, the acquaintance in this research field has now full fledged (Chiang et al. 1977). In comparison to the steady rise in publications on "4th generation of polymers" (Heeger 2001), the appearance of a large number of patents represents that the industrial sectors have started to commercialize these conducting polymers. In the last decade, these conducting polymers matured from materials of the laboratory to fully grown industrial products. Polythiophenes are a famous group of conducting polymers currently in the market. Poly(3,4-ethylene dioxythiophene) (PEDOT) plays the most important role in electric, antistatic, and electronic applications (Kirchmeyer and Reuter 2005). Broad applications have been established by employing the conducting properties of the PEDOT composite with polystyrene sulfonic acid along with its in situ polymerized layers of the 3,4-ethylene dioxythiophene monomer.

The PEDOT:poly(styrene sulfonate) (PSS; Figure 18.1) solution exhibits a deep-blue opaque solution in water and forms a continuous and smooth film on both the flexible and hard substrates by using different processing ways like spin coating, doctor blade coating, screen, and inkjet printing. PEDOT:PSS electrical conductivity varies from 10^{-2} to 10^{3} S cm^{-1}, subject to the different synthetic conditions, posttreatment, and processing additives. Thus, high work function (WK) and conductivity can encourage better charge transfer, resulting in PEDOT:PSS with better catalytic properties (Peng et al. 2011). In this chapter, we spotlight on a variety of approaches applied to tailor the properties of PEDOT:PSS hole transporting layer (HTL) in the organic solar cell (OSC) and perovskite solar cell (PSCs). The change in properties of HTL due to tailored PEDOT:PSS and the impact on the crystallization of PVSK and the organic active layer are provided. Finally, the factors that improve the performance after modification of PEDOT:PSS HTL are also discussed.

18.2 APPLICATIONS OF CONDUCTIVE PEDOT FOR THE TRANSPARENT ELECTRODE IN OSCS

The delegated third era of photovoltaic (PV) technology is OSCs, which aims to convert sunlight into electrical energy at very low cost when compared to other sources. So, the objective can be accomplished by using low-price materials as well as manufacturing techniques (Zhao et al. 2012; Xiao et al. 2014). Indium tin oxide (ITO) is present workhorse transparent electrodes (TE) for OSCs, but its high cost and inflexibility do not satisfy the previously stated aim, regardless of its great combination of conductivity as well as transparency. Therefore, the outstanding physical properties as well as good processability of PEDOT:PSS make it a good candidate for TE.

18.2.1 Solution-Processed PEDOT:PSS in Organic Solar Cell

In 2002, Zhang and coworkers made a pioneer effort to exploit PEDOT:PSS as an electrode in OSC. Before that, Cavendish Laboratory investigated its application as an electrode in the organic photodiode in 1999 (Arias et al. 1999). Zhang et al. (1999) also employed "Baytro P" that

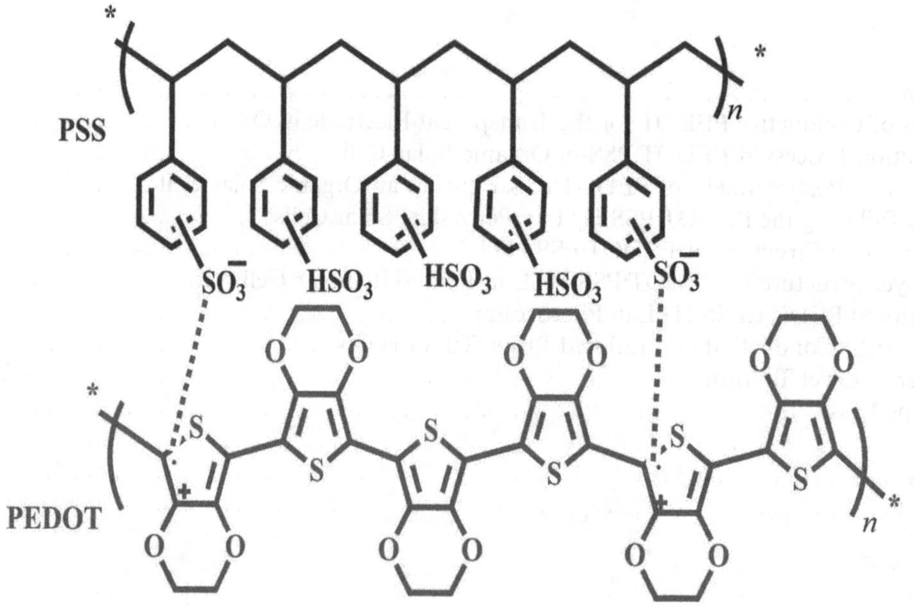

FIGURE 18.1 Chemical structure of PEDOT:PSS.

was the largely conductive grade of PEDOT:PSS at that time. The addition of D-sorbitol and glycerol into the original solution of PEDOT:PSS improved its conductivity to ~10 S cm^{-1} (Arias et al. 1999). Further, Ouyang et al. (2004a) and Ouyang et al. (2005) improved the performance of OSC by the addition of meso-erythritol or ethylene glycol (EG) to PEDOT:PSS solution, which increases its conductivity value to 160 and 155 S cm^{-1} owing to improved inter- and intrachain carrier mobility. Recently, Ishida and coworkers used grazing incidence wide and small-angle X-RAY diffraction to study the ordering of PEDOT nanocrystals in the solid films and crystallinity of EG-doped PEDOT:PSS solution. They found that the charge carrier mobility improves by the addition of EG from 0.045 to 1.7 cm^2 V^{-1} s^{-1} (Wei et al. 2013). Moreover, in another work, it is demonstrated that a 250 nm thick active layer based on MEH-PPV (donor) and PC$_{61}$BM (acceptor) on a PEDOT:PSS as the transparent anode was capable to achieve a performance of 1.5%. Later, it was realized that the incorporation of anionic surfactants might also be the cause of enhancing the conductivity (Fan et al. 2008).

Although the processing additives can enhance the conductivity of two orders of magnitude of PEDOT:PSS, but for meeting the practical application requirements for the electrode, buffer layer grade PEDOT:PSS solution could not provide the required conductivity. So, highly conducive PEDOT:PSS solution was familiarized in 2008 for the consideration of applications of an electrode. Na et al. (2008) studied the comparison of the conductivities of electrode grade PEDOT:PSS (PH500), buffer layer grade PEDOT:PSS (VPAl4083) in the absence and presence of 5% dimethyl sulfoxide (DMSO), on ITO and PET substrate (Na et al. 2010). The average conductivity of PH500 in the absence of solvent additive is less than 1 S cm^{-1} on PET that was three times higher than VPAl4083. Incorporation of 5% DMSO increases the conductivity value to ~500 S cm^{-1} on both PET and glass that was slightly lower in magnitude than ITO/glass (6740 S cm^{-1}) and closed to the ITO on PET (1050 S cm^{-1}). A 100 nm thick film of PEDOT:PSS expresses a transmittance value above 90% in the visible wavelength range, and its sheet resistance reduced (R) to 213 X sq^{-1} by employing poly(3-hexylthiophene-2,5-diyl) (P3HT) as the donor material. Therefore, they achieved a power conversion efficiency (PCE) of 3.27% and 2.8% on glass and PET flexible substrate, respectively, but when compared to the use of ITO/glass as TE (3.66%), these efficiencies were still lower in range (Tait et al. 2013).

Ahlswede et al. (2008) demonstrated that the addition of DMSO into PEDOT:PSS is not suitable for the reproducibility of the device performance, even while using surfynol as a surfactant. A four chemicals-based formulation was devised for solving this problem. Film's transparency and lateral conductivity were improved by DMSO and diethylene glycol (DEG), respectively, whereas surfynol and sorbitol were added for better adhesion and coverage. This work leads to a conductivity of 300 S cm^{-1},

which corresponds to a sheet resistance of 410 X sq^{-1}. A further improvement in the film's performance was attained by the reduction in the sheet resistance by improving the thickness value with additional spray coating layers. So, the preeminent performance of 2.6% was achieved at a maximum T of 74% and R of 80 X sq^{-1} (Ahlswede et al. 2008).

Chen and coworkers incorporated two size-controllable silver nanoparticles; decahedral and icosahedron Ag NPs were embedded into PEDOT:PSS as a hybrid HTL for organic PV applications (Chen et al. 2018). The Ag-decahedral-based nanoparticles exhibited rather red-shifted and larger domain size than that by Ag-icosahedron nanoparticles. Because of the plasmonic and light-harvesting effects of Ag NPs, the PCE increased from 5.8% to 6.5% (decahedral-based) and 6.3% (icosahedron-based) in (PTB7:PC$_{71}$BM) based-OPV on embedding the Ag NPs (Figure 18.2a). External quantum efficiency (EQE) enhancement was also observed due to the plasmonic scattering effect (Figure 18.2b). The morphologies of the HTL films were investigated by atomic force microscopy (AFM), and their relative AFM topographical images are shown in Figure 18.3. The AFM images of the hybrid HTL films show similar morphologies when compared to those of the control PEDOT:PSS HTL film. This indicates that there was no aggregation after introducing the Ag NPs into PEDOT:PSS. The root-mean-square roughness's of the PEDOT:PSS, PEDOT:PSS/Ag-decahedron, and PEDOT:PSS/Ag-icosahedron were 1.5, 2.3, and 1.8 nm, respectively. The PEDOT:PSS/Ag-decahedron hybrid film showed a larger roughness than others, which might enhance the hole collection efficiency.

Zhou and coworkers used an electrode grade PEDOT:PSS during the fabrication of flexible OSC, and realized that the open circuit voltage (V_{oc}) and fill factor (FF) are very low when PH500 was used alone, irrespective of its good conductivity, high WK, and better transmittance. So, the introduction of a PEDOT:PSS buffer layer (VPAl4083) on top of PH500 impressively reduces the leakage current and surface roughness and as a result, we get enhanced V_{oc}, FF, and PCE. Moreover, an electrode grade PEDOT:PSS is not always applied as an anode for the collection of the hole; it also works as a cathode for the collection of the electron after the interface modification. For example, Hau et al. (2009) applied DMSO-doped PEDOT:PSS for both the anode and cathode electrodes in a semitransparent efficient inverted OSC. For this purpose, they used a 50 nm thick layer of ZnO nanoparticles through spin coating, followed by C$_{60}$-SAM (self-assembled monolayer), which helps to block the hole and transport of electron toward the electrode. Unfortunately, a lower transmittance and conductance of transparent electrode decrease the performance of OSC from 4.20% to 3.08% by just replacing the ITO with DMSO-doped PEDOT:PSS (Figure 18.4, architecture 1–2). Furthermore, a layer of PEDOT:PSS coating on the

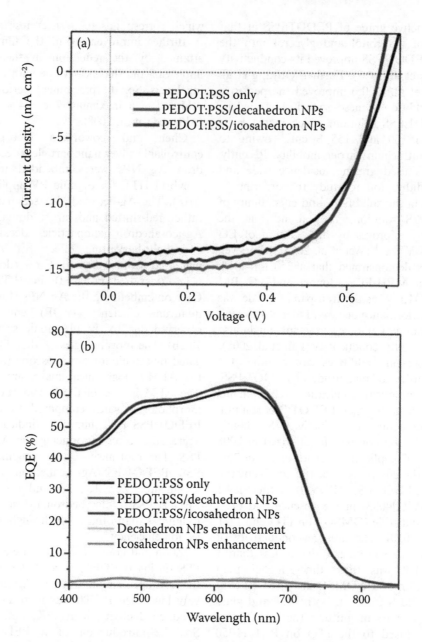

FIGURE 18.2 (a) *J–V* curves of the OPV device based on different HTL compositions under illumination with AM 1.5G solar-simulated sunlight illumination; (b) EQE spectra and EQE enhancement of the OPV devices. (Adapted from Chen et al. 2018.)

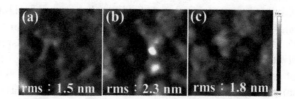

FIGURE 18.3 The AFM topographical images (1 μm × 1 μm) of HTLs (a) PEDOT:PSS; (b) PEDOT:PSS/Ag-decahedron; and (c) PEDOT:PSS/Ag-icosahedron films. (Adapted from Chen et al. 2018.)

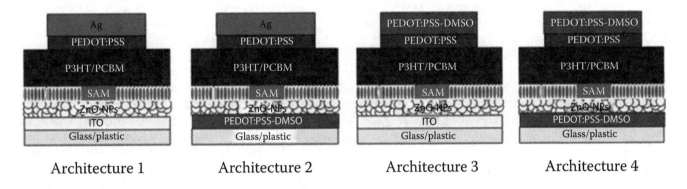

Architecture 1   Architecture 2   Architecture 3   Architecture 4

FIGURE 18.4 Device structures of the OSC using different electrode. (Adopted from Hau et al. 2009.)

glass substrate in OSC achieved a PCE of 2.99%, which is closed to the glass substrates. Architecture 3–4 (Figure 18.4) shows the replacement of the top metal electrode with highly conductive PEDOT:PSS, which was in fact fast, cost effective, energy efficient, and transparent when compared to the vacuum-deposited metal electrode.

After 2008, Heraeus introduced new formulations of PEDOT:PSS, namely, PH510, PH750, and PH1000 with superior conductivity when compared to that of PH500. The solvent additives, for example, DMSO, were combined with these formulations, which resulted in lower R below 100 X sq^{-1} and better PCE values about 3.5% (Do et al. 2009; Na et al. 2009). Zhu et al. (2015) demonstrated that blending of the film with Li salts can enhance its conductivity to 2610 times to 522 S cm^{-1}. Badre et al. (2012) established that the incorporation of 1-ethyl-3-methylimidazolium tetracyanoborate into PH1000 improved its transmittance up to 95%, sheet resistance was reduced to 31 X sq^{-1}, and its conductivity was enhanced up to 2084 S cm^{-1}. Furthermore, Mengistie et al. (2013) demonstrated that a PCE of 3.63% was achieved on P3HT-based OSC, showing the effect of polyethylene glycol molecular weight (MW) on the conductivity of PEDOT:PSS.

18.2.2 EFFECT OF POSTTREATMENT OF PEDOT:PSS FILMS IN AN ORGANIC SOLAR CELL

Some solvent additives are employed to improve the film conductivity of PEDOT:PSS, but still ITO has better conductivity. So nowadays, many researchers are working for the elimination of excessive amount of PSS for increasing the film conductivity. Some studies showed that the coulombic attractions can be weakened by using a dielectric between PSS counterion and PEDOT charge carrier (Kim et al. 2002; Martin et al. 2004; Ouyang, Xu, et al. 2004; Ashizawa et al. 2005; Gadisa et al. 2006). Hsiao et al. (2008) selected four different organic solvents including methoxy ethanol, EG, ethanol, and 1,2-dimethoxyethane for the treatment of as-prepared PEDOT:PSS films. These solvent were spin-coated top of

PEDOT:PSS membranes and allowed to dry at 150°C for 1 h inside the glove box. This treatment reduced the sheet resistance especially when EG and methoxy ethanol were used from 1.01×10^6 to 5.19×10^3 and 7.39×10^3 X sq^{-1}, respectively. Some combined studies of experimental techniques such as UV–Vis spectroscopy, tapping mode AFM, XPS, Raman, and scanning electron microscopy (SEM) showed that the enhancement of the conductivity depends on the following factors: PEDOT's phase segregation, PEDOT's chemical structure transformation, elimination of excess amount of PSS from the PEDOT's surface, and PSS domain. A very small OSC can show a PCE of 3.39% in spite of the large resistance of PEDOT:PSS TE (Hsiao et al. 2008).

18.3 APPROACHES TAILORING THE PEDOT:PSS HTL IN PEROVSKITE SOLAR CELLS

PEDOT:PSS conductivity depends mainly on two factors: one is PSS and PEDOT arrangement and the other is their ratio. In literature, additive methods are suggested for the rearrangement of PEDOT and PSS. Moreover, solvent posttreatments are also recommended to remove insulating PSS from films, which is a way to increase the PEDOT ratio to PSS (Na et al. 2008; Nardes et al. 2008). Both methods facilitate better orientation of PEDOT; thus in turn the high crystallinity of film efficiently improves the flow of charge (Takano et al. 2012). Huang and co-worker added DMSO solvent with different volume ratio into the aqueous solution of PEDOT:PSS. (Huang et al. 2017a) In this way, the charge extraction and photocurrent production were considerably improved in PSCs. The PCE of post treated PEDOT:PSS is reported to increase to 15.8% (with champion cell even to 16.7%) from 11.8% of untreated PEDOT:PSS. The reduction of leakage current in DMSO-treated PEDOS:PSS was attributed to the high value of shunt resistant and better PL quenching, particularly, indicating favorable charge transfer between the perovskite and DMSO-treated PEDOS:PSS (Arora et al. 2017; Sha et al. 2018). Moreover, DMSO also improved the crystallinity and morphology of PEDOT: PSS. This treatment enhanced the aggregation of PEDOT-rich particles by providing sites for the growth of crystal nucleus of perovskite films; thus, larger, compact, and highly crystalized grain are formed. Efficient transportation of charge carriers is attributed to improved morphology of perovskite (Fei et al. 2017; Chiang et al. 2017).

18.3.1   Composite Structure of PEDOT:PSS HTL

Although PEDOT:PSS has high conductivity and transparency, there are some drawbacks as well. PEDOT:PSS has a lower WK than that of perovskite. Moreover, the poor electron blocking ability, acidity, and hygroscopic nature are also observed in PEDOT:PSS-based devices (Kim et al. 2011; Lim et al. 2014; Ye et al. 2015). These drawback make PEDOT:PSS less interesting for the researcher, so the other alternative are more preferable in OSC and PSCs. (Jeng et al. 2014; Ye et al. 2015). Graphene-based derivatives are found to have potential applications for high-performance PSCs. The efficiency of more than 12% is found for graphene oxide (GO) applied as an HTL in a planar PSC (Wu et al. 2014). As GO is an insulator in nature, its performance highly depends on the thickness of HTL. The main discrepancy is the full coverage of ITO with a homogeneous and uniform dispersion of GO that causes, otherwise, direct contact of perovskite layer with ITO, thus resulting in poor hole transportation and collection. Because of the insulating property of GO, this nonuniform dispersion also causes poor repeatability and reliability of devices. This problem was solved by successively deposited layers of GO and PEDOT:PSS to get a composite structure (GO/PEDOT:PSS). (Lee, Na, and Kim 2016). This composite structure helps to fabricate efficient HTL, which help to design the high performance PSCs with improved repeatability of the device that was the outcome of smoother surface of GO, and archived by the sequential deposition of GO/PEDOT:PSS. GO is an insulator in nature as the functional groups at the carbon basal plan have oxygen that improves the conductivity of PEDOT:PSS. Graphene composites of PEDOT:PSS exhibit improved conductivity and charge extraction at perovskite/graphene–PEDOT:PSS interface (Chen et al. 2016). However, graphene composites have some drawbacks as well: an increase in PEDOT:PSS HTL roughness and shift of WK toward the vacuum level, which can be avoided for the Perovskite layer deposition. Therefore, GO is preferred instead of graphene in many reduction methods, that is, thermal (Wu et al. 2009), photoreduction (Cote et al. 2009), chemical, or electrochemical reduction methods (Li et al. 2008; Chen et al. 2008; Williams et al. 2008) are used to increase its conductivity. Giuri et al. (2016) used a simple and cost-effective method to obtain PEDOT:PSS/graphene HTL. They used UV radiations for the reduction of GO, and an enhancement in the device performance device

performance was observed owing to improved wettability of the HTL attributed to the better morphology of a perovskite film. The thermal reduction method was used by another group to reduce graphene oxide, also an environmental-friendly method (Huang et al. 2016). Both the above-mentioned reduction methods (thermal and UV reduction) helped to get HTLs with increased V_{OC}, which indicates the better alignment of rGO/PEDOT:PSS HTL energy levels with a valence band maximum (VBM) of perovskite. The synergistic effect of glucose and GO was also studied (Giuri et al. 2016). The composite structure showed an improved conductivity and better wettability of PEDOT:PSS. The conductivity is improved owing to the enhanced reduction process that results from the addition of glucose in the GO suspension (Zhu et al. 2010; Akhavan et al. 2012), while the presence of hydroxyl group in the functional group terminations causes better wettability of PEDOT:PSS (Perrozzi et al. 2014). These nanocomposites facilitate perovskite film deposition with improved morphology resulting PV devices showed better performance (12.8% PCE, 1.5 V V_{OC}) than the unmodified PEDOT:PSS-based cells (9.4% PCE). This improved performance infer in the reduction in charge-carriers recombination.

Jiang et al. (2017) demonstrated the transformation of monomer DBEDOT to polymer PEDOT, which changes the color from white to blue (Figure 18.5a). The optical microscopy images clearly showed that the colorless needle-shaped crystals of DBEDOT were transformed into black crystals of PEDOT (Figure 18.5b). The SEM image of the surface of PEDOT fabricated by solid-state synthesis is shown in Figure 18.5c. The transformation of DBEDOT to polymer PEDOT was further confirmed by X-ray diffraction (XRD) (Figure 18.5d). The crystalline DBEDOT thin film shows the main diffraction peaks at $2\theta = 8.5$, 16.8, 25.4, 27.0, and 34°, while a strong (The crystalline DBEDOT thin film shows the main diffraction peaks at $2\theta = 8.5$, 16.8, 25.4, 27.0, and 34°, while a strong and sharp (line width ca. 0.2°) diffraction peak at $2\theta = 30°$ was observed for the PEDOT thin film sample).

Electrochemical impedance spectroscopy (EIS) was carried out to gain more insight into electrical characteristics of PSCs with and without PEDOT as HTMs under one sun illumination (AM 1.5 G, 100 mW cm^{-2}) with varying bias voltages in the frequency range from 10^6 to 10^{-1} Hz (Figure 18.6a and b). The corresponding results are plotted in Figure 18.6c. The Nyquist plots exhibit two arcs: one arc in the high-frequency region is associated with the hole transport and extraction between the HTM and the carbon cathode (R_{HTM}); the main arc in the low-frequency region represents the characteristics of charge recombination (R_{rec}). From Figure 18.6c, it is clear that at a fixed bias potential, the PSC device employing PEDOT shows a lower R_{HTM} than the corresponding value for the device without any HTM, indicating a more sufficient hole transport/collection for the PEDOT-based device. This result accounts for the higher FF obtained for the device-containing PEDOT as an HTM. The larger R_{rec} observed

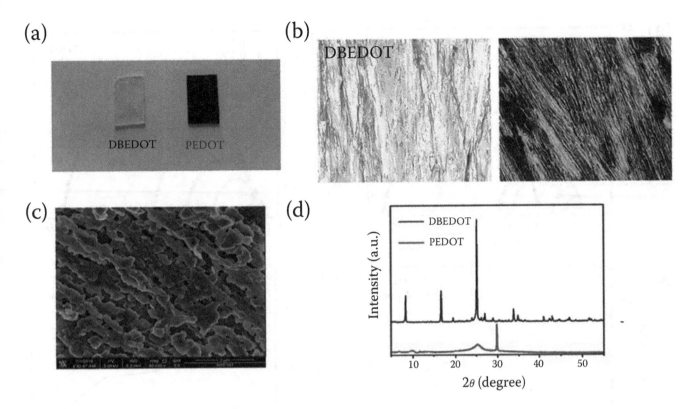

FIGURE 18.5 (a) Photographs of the monomer DBEDOT and PEDOT prepared on glass substrates. (b) Optical microscopic images of monomer DBEDOT (left) and PEDOT (right) formed by solid-state synthesis (magnification 200×). (c) The SEM image of the surface of PEDOT fabricated by solid-state synthesis. (d) XRD spectra of monomer DBEDOT (black) and PEDOT fabricated by the in situ polymerization method (red) on glass substrates. (Reproduced from Jiang et al. (2017); an open-access article licensed under a Creative Commons Attribution 4.0 International License (http://creative commons.org/licenses/by/4.0/).)

for the devices employing PEDOT as an HTM (Figure 18.6d) could explain the higher V_{oc} achieved from the J–V measurements. In another report, silver trifluoromethanesulfonate was used as an inorganic dopant (Liu et al. 2015). Electrical and optical properties of GO film directly depend on the doping concentration. With an increase in the doping concentration, film roughness and thickness also increase, resulting in a decrease in sheet resistant. Thus, the appropriate doping concentration is highly desirable for improving the performance of PSCs.

18.3.2 BILAYER STRUCTURE OF PEDOT:PSS HTL IN PEROVSKITE SOLAR CELL

Felicitous tailoring of energy levels can enhance the performance of the device. This tailoring enhances charge extraction and transfer and thus results in improved short current density (J_{sc}) and FF values. To reduce the loss of potential energy at the perovskite/PEDOT:PSS interface, the insertion of appropriate interfacial layer modifies the WF of ITO/PEDOT:PSS HTL and minimizes the loss. Malinkiewicz et al. (2013) deposited PEDOT:PSS/poly-TPD layer and Lim et al. (2014) employed double layers of PEDOT:PSS-conjugated polymers (DPP-DTT, PCDTBT, P3HT, and PCPDTBT). Although the energy gap was tailored between WF of HTLs and VBM of perovskite, poor wetting of DMF or DMSO (due to the hydrophobic surface of polymers) confined these structure only to vacuum-deposited PSCs. Xue et al. (2016) solved this issue by developing hydrophilic alcohol-soluble conjugated copolymers: PTPAF-SONa (HSL1) and PTPADCF3FSONa (HSL2). The use of HSL2 instead of pristine remarkably improves the PCEs of PSCs from 14.2% and 0.98 V to 16.6% and 1.07 V. This improvement is attributed to lower HOMO levels and higher LUMO levels that facilitate the hole transfer and block the electron flow, thus reducing charge recombination at interfaces. In the same way, QUPD and OTPD can be used to deposit cross-linked interlayers, suppresses electron–hole recombination and improves the efficiency of PSCs by blocking the electron (Jhuo et al. 2015). Li and colleagues (2016) fabricated the hybrid bilayered structure of GO and PEDOT:PSS. The GO layer was sandwiched between ITO and PEDOT:PSS, resulting in an extraordinary reduction in current

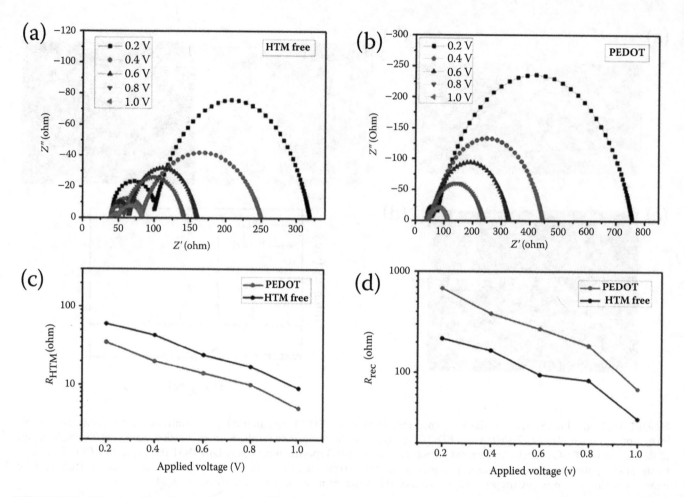

FIGURE 18.6 Nyquist plots of perovskite solar cell devices without hole-transporting materials (a) and with PEDOT (b) measured under one sun illumination with varied forward biases. (c) Hole-transport resistance and (d) charge recombination resistance obtained from EIS measurements at varied bias potentials. (Reproduced from Jiang et al. (2017); an open-access article licensed under a Creative Commons Attribution 4.0 International License (http://creativecommons.org/licenses/by/4.0/).)

leakage and charge recombination. On the other hand, the GO deposition on the PEDOT:PSS surface increased the work potential of film to 5.19 eV (from 5.10 eV for unprocessed PEDOT:PSS film). In another experiment, ammonia was introduced in the GO layer to form a modified PEDOT:PSS-GO:NH$_3$ film. About 1:03 (GO:NH$_3$) ratio was found to be more efficient (16.11%), which can be attributed to better optical absorption of perovskite films (Feng et al. 2016).

18.4 ROLES OF TAILORED PEDOT:PSS HTL IN PHOTOVOLTAICS

To improve the performance of PSCs using PEDOT:PSS as an HTL, the corresponding drawbacks should be appropriately addressed. In this section, we will discuss the ways to improve the electrical conductivity, energy level alignment, and trap passivation of PV devices by using PEDOT:PSS HTL.

18.4.1 ELECTRICAL CONDUCTIVITY OF TAILORED PEDOT:PSS HTLs

The difference of electric conductivity between perovskite and charge transport layers (CTL) results in a charge accumulation at the interface. This might be because of the creation of space charges that can restrain the effective carrier transport. Thus, the higher electric conductivity of CTLs is the key parameter for an effective charge transport to the relevant electrode. In PEDOT:PSSHTL, its conductivity can easily be enhanced by eliminating PSS chains (Vaagensmith et al. 2017). Moreover, the elimination of PSS chains changes an extended PEDOT structure from coil-like to an extended coil-like structure (Xia and Ouyang 2011). Therefore, the variation in the conductivity of PEDOT:PSS seriously influences the performance of PSCs. Cho and coworkers described that the supplementation of a little quantity of DMSO in the PEDOT:PSS before the spin coating can improve the conductivity without any considerable effect on the energy level alignment and can also help to improve grain size in PSCs (Sin et al. 2016). On the contrary, no major morphological effects were noticed between the perovskite films spin coated on varying conductive PEDOT:PSS films, thus verifying that the performance of PSCs is mainly afflicted by the properties of PEDOT:PSS films. Next, the incident light power was used to measure the charge accumulation and charge collection balance between perovskite and low- and high-conductive PEDOT:PSS film. The parameter S (~1 for the ideally fabricated device) for the high-conductive PEDOT is 0.969, while its value is 0.904 for low-conductive PEDOT, signifying that the PEDO:PSS film having higher conductivity reduces the charge accumulation between perovskite and HTL. However, S parameter at low effective applied voltage (0.4 V) also shows much higher value (0.953 compared to 0.893) for PEDOT-H based device, which means reduced nongeminate recombination with increased conductivity. Therefore, the PEDOT:PSS having higher conductivity helps to reduce the recombination losses and is also effective in the charge accumulation, resulting an enhancement in J_{sc}. This highly conductive PEDOT:PSS HTL improves the hole transport, which showed a 25% higher PCE in PSCs.

18.4.2 Energy Level Tailoring

Energy offset between perovskite and HTL reduces the performance of PSCs. Tailoring the energy level is beneficial to lessen the accumulation and recombination of charges at the interfaces, resulting in a better performance. Furthermore, a better energy level alignment between the HOMO (HTL) and LUMO (ETL) of CTLs and the valence and conduction band of perovskite, respectively, has a significant effect on the V_{oc} of the device. Cao and coworkers used double HTLs PEDOT:PSS/PTPAFSONa (HSL1) and PEDOT:PSS/PTPADCF3FSONa (HSL2) to measure the recombination lifetime of perovskite. The charge recombination lifetime increased for all the devices having HSL1 or HSL2 in comparison to that of the pristine PEDOT:PSS film, owing to the better energy level alignment with perovskite. This increased lifetime proposes that a proper energy level alignment efficiently suppressed charge recombination losses at the interfaces (Xue et al. 2016). Lin et al. observed a better performance of PSCs device by used of different p-type polymeric HTLs on the top of ITO/PEDOT:PSS film. These polymeric HTLs reduced the energy offset between VBM of perovskite and HOMO of the p-type layers, which always a key parameter for better performing device. (Lin et al. 2015). The higher value of V_{oc} (1.03) was achieved with the PCDTBT interlayer due to lower energy offset with perovskite, and on the

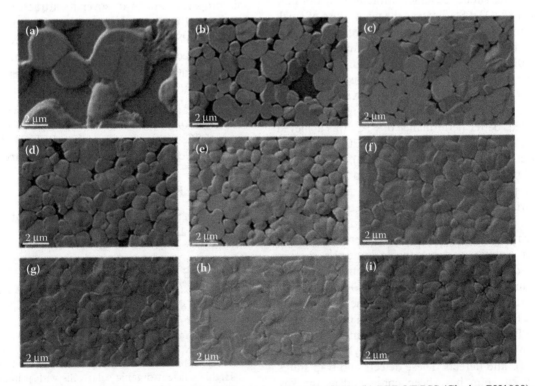

FIGURE 18.7 The SEM images of perovskite films coated on top of (a) glass/ITO; (b) PEDOT:PSS (Clevios PH1000); (c) PEDOT:PSS (Clevios PH1000) processed with 5% v/v DMSO; (d) PEDOT:PSS (Clevios PH1000) processed with 0.7% v/v Zonyl®FS-300; and (e) PEDOT:PSS (Clevios PH1000) processed with 0.7% v/v Zonyl® FS-300% and 2.5% v/v DMSO, (f) 5% v/v DMSO, (g) 10% v/v DMSO, (h) and 15% v/v DMSO, and (i) 20% v/v DMSO. (Reproduced with permission from Adam et al. 2016.)

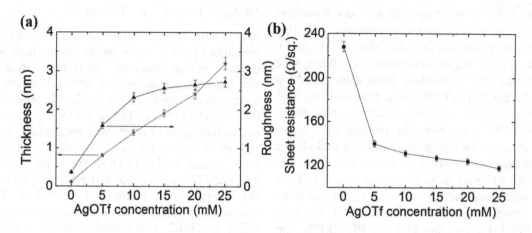

FIGURE 18.8

other hand, the lowest V_{oc} value (0.7) was achieved with P3HT due to its higher energy offset perovskite.

18.4.3 TRAPS PASSIVATION

It is noted that the higher conductivity of the PEDOT:PSS film all the time does not guarantee a better performance of PSCs. The surface properties of perovskite films critically depend on the surface morphology of HTL (PEDOT:PSS). Adverse morphology of the PEDOT:PSS surface persuaded the defects on the perovskite layer, causing a higher leakage current with lesser shunt resistance. Moreover, hydrophilic nature of the PEDOT:PSS film was the main cause of small grains and large grain boundaries of the perovskite film, which behaved as defect and trap states (Wang et al. 2014; Duan et al. 2015). However, in case of inferior performance of PSCs, these issues can be addressed by the surface treatment of PEDOT:PSS film. (Li et al. 2016; Xia et al. 2015; Liu, Li, Zhang, et al. 2017). However, as the majority of the groups reported, better performance of PSCs was achieved on the treated PEDOT:PSS film (Xue et al. 2016; Lee et al. 2016; Giuri et al. 2016; Wang et al. 2016b). Adam et al. (2016) used DMSO and zonyl additives into PEDOT:PSS, which improve its conductivity, resulting in a pinhole-free perovskite surface. They showed that the use of the additive in the PEDOT:PSS help to achieve a better quality perovskite film, which was the main cause for higher performance with a better stability and reasonable hysteresis (Figure 18.7). These additives help to decrease the number and size of grain boundaries of perovskite films (Figure 18.7c–f) when compared to the film deposited on the ITO (Figure 18.7a) and PEDOT:PSS (Figure 18.7b). The perovskite film on the PEDOT:PSS HTL having optimized additive concentration (5% DMSO and 0.7% Zonyl) showed no noticeable gaps and defects. These defects and trap states in the perovskite film have been considered as the key reason for photocurrent hysteresis of PSCs, which reduced the performance as well as the stability of PSCs devices (Shao et al. 2014; Noel et al. 2014; Heo et al. 2015; Liang

et al. 2015; Xu et al. 2015). Thus, tailoring PEDOT:PSS-based devices is helpful to reduce the charge accumulation and trap states, resulting a better performance of PSCs with less hysteresis (Shahbazi et al. 2016; Wang et al. 2016a; Liu, Li, Yuan, et al. 2017; Huang et al. 2017b).

18.5 CONCLUSION

In conclusion, PEDOT:PSS is an easily available and most growing conducting polymer of the present time in the PV community. Its outstanding processability and optoelectronic properties aid to discover its attractive energy conversion and PV applications. This chapter delineates that the PEDOT:PSS properties are very sensitive with regard to synthesis techniques, composite formation, dopant concentration, patterning, and so on. However, it is often required to modify its properties according to the specific application. With the persistent improvement in its properties or exploring newer applications, for example, buffer layer in PSCs, it is assumed that PEDOT:PSS will be a promising electronic material for emerging technologies.

ACKNOWLEDGMENTS

The authors are grateful to their representative institutes for providing literature facilities.

REFERENCES

Adam, Getachew, Martin Kaltenbrunner, Eric Daniel Głowacki, et al. 2016. Solution-processed perovskite solar cells using highly conductive PEDOT:PSS interfacial layer. *Solar Energy Materials and Solar Cells* 157:318–325.

Ahlswede, Erik, Wolfgang Mühleisen, Mohd Wahinuddin Bin Moh Wahi, Jonas Hanisch, and Michael Powalla. 2008. Highly efficient organic solar cells with printable low-cost transparent contacts. *Applied Physics Letters* 92 (14):127.

Akhavan, Omid, Elham Ghaderi, Samira Aghayee, Yasamin Fereydooni, and Ali Talebi. 2012. The use of a glucose-reduced graphene oxide suspension for photothermal cancer therapy. *Journal of Materials Chemistry* 22 (27):13773–13781.

Arias, A. C., M. Granström, K. Petritsch, and R. H. Friend. 1999. Organic photodiodes using polymeric anodes. *Synthetic Metals* 102 (1–3):953–954.

Arora, Neha M. Ibrahim Dar, Alexander Hinderhofer, et al. 2017. Perovskite solar cells with CuSCN hole extraction layers yield stabilized efficiencies greater than 20%. *Science* 358 (6364):768–771.

Ashizawa, S., R. Horikawa, and H. Okuzaki. 2005. Effects of solvent on carrier transport in poly (3, 4-ethylene dioxythiophene)/poly (4-styrenesulfonate). *Synthetic Metals* 153 (1–3):5–8.

Badre, Chantal, Ludovic Marquant, Ahmed M. Alsayed, and Lawrence A. Hough. 2012. Highly conductive poly (3, 4-ethylenedioxythiophene): poly (styrene sulfonate) films using the 1-ethyl-3-methylimidazolium tetracyanoborate ionic liquid. *Advanced Functional Materials* 22 (13):2723–2727.

Chen, Chih-Ping, I-Chan Lee, Yao-Yu Tsai, Cheng-Liang Huang, Yung-Chung Chen, and Guan-Wei Huang. 2018. Efficient organic solar cells based on PTB7/PC71BM blend film with embedded different shapes silver nanoparticles into PEDOT:PSS as hole transporting layers. *Organic Electronics* 62:95–101.

Chen, Haiqun, Marc B. Müller, Kerry J. Gilmore, Gordon G. Wallace, and Dan Li. 2008. Mechanically strong, electrically conductive, and biocompatible graphene paper. *Advanced Materials* 20 (18):3557–3561.

Chen, Qianli, Fatemeh Zabihi, and Morteza Eslamian. 2016. Improved functionality of PEDOT:PSS thin films via graphene doping, fabricated by ultrasonic substrate vibration-assisted spray coating. *Synthetic Metals* 222:309–317.

Chiang, Chien-Hung, Mohammad Khaja Nazeeruddin, Michael Grätzel, and Chun-Guey Wu. 2017. The synergistic effect of H2O and DMF towards stable and 20% efficiency inverted perovskite solar cells. *Energy & Environmental Science* 10 (3):808–817.

Chiang, Chwan K., C. R. Fincher, Yung W. Park, Jr, et al. 1977. Electrical conductivity in doped polyacetylene. *Physical Review Letters* 39 (17):1098.

Cote, Laura J., Rodolfo Cruz-Silva, and Jiaxing Huang. 2009. Flash reduction and patterning of graphite oxide and its polymer composite. *Journal of the American Chemical Society* 131 (31):11027–11032.

Do, Hung, Manuel Reinhard, Henry Vogeler, et al. 2009. Polymeric anodes from poly (3, 4-ethylenedioxythiophene): Poly (styrene sulfonate) for 3.5% efficient organic solar cells. *Thin Solid Films* 517 (20):5900–5902.

Duan, Hsin-Sheng, Huanping Zhou, Qi Chen, et al. 2015. The identification and characterization of defect states in hybrid organic-inorganic perovskite photovoltaics. *Physical Chemistry Chemical Physics* 17 (1):112–116.

Fan, Benhu, Xiaoguang Mei, and Jianyong Ouyang. 2008. Significant conductivity enhancement of conductive poly (3, 4-ethylenedioxythiophene): Poly (styrene sulfonate) films by adding anionic surfactants into polymer solution. *Macromolecules* 41 (16):5971–5973.

Fei, Chengbin, Bo Li, Rong Zhang, Haoyu Fu, Jianjun Tian, and Guozhong Cao. 2017. Highly efficient and stable perovskite solar cells based on monolithically grained CH3NH3PbI3 film. *Advanced Energy Materials* 7 (9):1602017.

Feng, Shanglei, Yingguo Yang, Meng Li, et al. 2016. High-performance perovskite solar cells engineered by an ammonia modified graphene oxide interfacial layer. *ACS Applied Materials & Interfaces* 8 (23):14503–14512.

Gadisa, Abay, Kristofer Tvingstedt, Shimelis Admassie, et al. 2006. Transparent polymer cathode for organic photovoltaic devices. *Synthetic Metals* 156 (16–17):1102–1107.

Giuri, Antonella, Sofia Masi, Silvia Colella, et al. 2016. Cooperative effect of GO and glucose on PEDOT:PSS for high VOC and hysteresis-free solution-processed perovskite solar cells. *Advanced Functional Materials* 26 (38):6985–6994.

Hau, Steven K., Hin-Lap Yip, Jingyu Zou, and Alex K. Y. Jen. 2009. Indium tin oxide-free semi-transparent inverted polymer solar cells using conducting polymer as both bottom and top electrodes. *Organic Electronics* 10 (7):1401–1407.

Heeger, Alan J. 2001. Semiconducting and metallic polymers: the fourth generation of polymeric materials (Nobel lecture). *Angewandte Chemie International Edition* 40 (14):2591–2611.

Heo, Jin Hyuck, Hye Ji Han, Dasom Kim, Tae Kyu Ahn, and Sang Hyuk Im. 2015. Hysteresis-less inverted CH 3 NH 3 PBI 3 planar perovskite hybrid solar cells with 18.1% power conversion efficiency. *Energy & Environmental Science* 8 (5):1602–1608.

Hsiao, Yu-Sheng, Wha-Tzong Whang, Chih-Ping Chen, and Yi-Chun Chen. 2008. High-conductivity poly (3, 4-ethylene-dioxythiophene): Poly (styrene sulfonate) film for use in ITO-free polymer solar cells. *Journal of Materials Chemistry* 18 (48):5948–5955.

Huang, Di, Tenghooi Goh, Jaemin Kong, et al. 2017a. Perovskite solar cells with a DMSO-treated PEDOT:PSS hole transport layer exhibit higher photovoltaic performance and enhanced durability. *Nanoscale* 9 (12):4236–4243.

Huang, Ju, Kai-Xuan Wang, Jing-Jing Chang, Yan-Yun Jiang, Qi-Shi Xiao, and Yuan Li. 2017b. Improving the efficiency and stability of inverted perovskite solar cells with dopamine-copolymerized PEDOT:PSS as a hole extraction layer. *Journal of Materials Chemistry A* 5 (26):13817–13822.

Huang, Xu, Heng Guo, Jian Yang, Kai Wang, Xiaobin Niu, and Xiaobo Liu. 2016. Moderately reduced graphene oxide/PEDOT:PSS as a hole transport layer to fabricate efficient perovskite hybrid solar cells. *Organic Electronics* 39:288–295.

Jeng, Jun-Yuan, Kuo-Cheng Chen, Tsung-Yu Chiang, et al. 2014. Nickel oxide electrode interlayer in CH3NH3PbI3 perovskite/PCBM planar-heterojunction hybrid solar cells. *Advanced Materials* 26 (24):4107–4113.

Jhuo, Hong-Jyun, Po-Nan Yeh, Sih-Hao Liao, Yi-Lun Li, Sunil Sharma, and Show-An Chen. 2015. Inverted perovskite solar cells with inserted cross-linked electron-blocking interlayers for performance enhancement. *Journal of Materials Chemistry A* 3 (17):9291–9297.

Jiang, Xiaoqing, Ze Yu, Yuchen Zhang, et al. 2017. High-performance regular perovskite solar cells employing low-cost poly (ethylene dioxythiophene) as a hole-transporting material. *Scientific Reports* 7:42564.

Kim, J. Y., J. H. Jung, D. E. Lee, and Jinsoo Joo. 2002. Enhancement of electrical conductivity of poly (3, 4-ethylene dioxythiophene)/poly (4-styrenesulfonate) by a change of solvents. *Synthetic Metals* 126 (2–3):311–316.

Kim, Yong Hyun, Christoph Sachse, Michael L. Machala, Christian May, Lars Müller-Meskamp, and Karl Leo. 2011. Highly conductive PEDOT:PSS electrode with optimized solvent and thermal post-treatment for ITO-free organic solar cells. *Advanced Functional Materials* 21 (6):1076–1081.

Kirchmeyer, Stephan, and Knud Reuter. 2005. Scientific importance, properties and growing applications of poly (3, 4-ethylenedioxythiophene). *Journal of Materials Chemistry* 15 (21):2077–2088.

Lee, Da-Young, Seok-In Na, and Seok-Soon Kim. 2016. Graphene oxide/PEDOT:PSS composite hole transport layer for efficient and stable planar heterojunction perovskite solar cells. *Nanoscale* 8 (3):1513–1522.

Li, Dan, Marc B. Müller, Scott Gilje, Richard B. Kaner, and Gordon G. Wallace. 2008. Processable aqueous dispersions of graphene nanosheets. *Nature Nanotechnology* 3:101.

Li, Xue-Yuan, Lian-Ping Zhang, Feng Tang, et al 2016. The solvent treatment effect of the PEDOT:PSS anode interlayer in inverted planar perovskite solar cells. *RSC Advances* 6 (29):24501–24507.

Liang, Po-Wei, Chu-Chen Chueh, Spencer T. Williams, and Alex K-Y Jen. 2015. Roles of fullerene-based interlayers in enhancing the performance of organometal perovskite thin-film solar cells. *Advanced Energy Materials* 5 (10):1402321.

Lim, Kyung-Geun, Hak-Beom Kim, Jaeki Jeong, Hobeom Kim, Jin Young Kim, and Tae-Woo Lee. 2014. Boosting the power conversion efficiency of perovskite solar cells using self-organized polymeric hole extraction layers with high work function. *Advanced Materials* 26 (37):6461–6466.

Lin, Qianqian, Ardalan Armin, Ravi Chandra Raju Nagiri, Paul L. Burn, and Paul Meredith. 2015. Electro-optics of perovskite solar cells. *Nature Photonics* 9 (2):106.

Liu, Dongyang, Yong Li, Jianyu Yuan, et al. 2017. Improved performance of inverted planar perovskite solar cells with F4-TCNQ doped PEDOT:PSS hole transport layers. *Journal of Materials Chemistry A* 5 (12):5701–5708.

Liu, Hongmei, Xueyuan Li, Lianping Zhang, et al. 2017. Influence of the surface treatment of PEDOT:PSS layer with high boiling point solvent on the performance of inverted planar perovskite solar cells. *Organic Electronics* 47:220–227.

Liu, Tongfa, Dongcheon Kim, Hongwei Han, Abd Rashid Bin Mohd Yusoff, and Jin Jang. 2015. Fine-tuning optical and electronic properties of graphene oxide for highly efficient perovskite solar cells. *Nanoscale* 7 (24):10708–10718.

Martin, Brett D., Nikolay Nikolov, Steven K. Pollack, et al. 2004. Hydroxylated secondary dopants for surface resistance enhancement in transparent poly (3, 4-ethylene dioxythiophene)–Poly (styrene sulfonate) thin films. *Synthetic Metals* 142 (1–3):187–193.

Mengistie, Desalegn Alemu, Pen-Cheng Wang, and Chih-Wei Chu. 2013. Effect of molecular weight of additives on the conductivity of PEDOT:PSS and efficiency for ITO-free organic solar cells. *Journal of Materials Chemistry A* 1 (34):9907–9915.

Na, Seok-In, Seok-Soon Kim, Jang Jo, and Dong-Yu Kim. 2008. Efficient and flexible ITO-free organic solar cells using highly conductive polymer anodes. *Advanced Materials* 20 (21):4061–4067.

Na, Seok-In, Gunuk Wang, Seok-Soon Kim, et al. 2009. Evolution of nanomorphology and anisotropic conductivity in solvent-modified PEDOT:PSS films for polymeric anodes of polymer solar cells. *Journal of Materials Chemistry* 19 (47):9045–9053.

Na, Seok-In, Byung-Kwan Yu, Seok-Soon Kim, et al. 2010. Fully spray-coated ITO-free organic solar cells for low-cost power generation. *Solar Energy Materials and Solar Cells* 94 (8):1333–1337.

Nardes, Alexandre Mantovani, Rene A. J. Janssen, and Martijn Kemerink. 2008. A morphological model for the solvent-enhanced conductivity of PEDOT:PSS thin films. *Advanced Functional Materials* 18 (6):865–871.

Noel, Nakita K., Antonio Abate, Samuel D. Stranks, et al. 2014. Enhanced photoluminescence and solar cell performance via Lewis base passivation of organic–Inorganic lead halide perovskites. *ACS Nano* 8 (10):9815–9821.

Ouyang, Jianyong, Chi-Wei Chu, Fang-Chung Chen, Qianfei Xu, and Yang Yang. 2004a. Polymer optoelectronic devices with high-conductivity poly (3, 4-ethylenedioxythiophene) anodes. *Journal of Macromolecular Science, Part A* 41 (12):1497–1511.

Ouyang, Jianyong, Chi-Wei Chu, Fang-Chung Chen, Qianfei Xu, and Yang Yang. 2005. High-conductivity poly (3, 4-ethylenedioxythiophene): poly (styrene sulfonate) film and its application in polymer optoelectronic devices. *Advanced Functional Materials* 15 (2):203–208.

Ouyang, Jianyong, Qianfei Xu, Chi-Wei Chu, Yang Yang, Gang Li, and Joseph Shinar. 2004. On the mechanism of conductivity enhancement in poly (3, 4-ethylenedioxythiophene): poly (styrene sulfonate) film through solvent treatment. *Polymer* 45 (25):8443–8450.

Peng, Bo, Xia Guo, Chaohua Cui, Yingping Zou, Chunyue Pan, and Yongfang Li. 2011. Performance improvement of polymer solar cells by using a solvent-treated poly (3, 4-ethylenedioxythiophene): poly (styrenesulfonate) buffer layer. *Applied Physics Letters* 98 (24):113.

Perrozzi, Francesco, Salvatore Croce, Emanuele Treossi, et al. 2014. Reduction dependent wetting properties of graphene oxide. *Carbon* 77:473–480.

Sha, Wei E. I., Hong Zhang, Zi Shuai Wang, et al. 2018. Quantifying efficiency loss of perovskite solar cells by a modified detailed balance model. *Advanced Energy Materials* 8 (8):1701586.

Shahbazi, S., F. Tajabadi, H. S. Shiu, et al. 2016. An easy method to modify PEDOT:PSS/perovskite interfaces for solar cells with efficiency exceeding 15%. *RSC Advances* 6 (70):65594–65599.

Shao, Yuchuan, Zhengguo Xiao, Cheng Bi, Yongbo Yuan, and Jinsong Huang. 2014. Origin and elimination of photocurrent hysteresis by fullerene passivation in CH3NH3PbI3 planar heterojunction solar cells. *Nature Communications* 5:5784.

Sin, Dong Hun, Hyomin Ko, Sae Byeok Jo, Min Kim, Geun Yeol Bae, and Kilwon Cho. 2016. Decoupling charge transfer and transport at polymeric hole transport layer in perovskite solar cells. *ACS Applied Materials & Interfaces* 8 (10):6546–6553.

Tait, Jeffrey G., Brian J. Worfolk, Samuel A. Maloney, et al. 2013. Spray coated high-conductivity PEDOT:PSS transparent electrodes for stretchable and mechanically-robust organic solar cells. *Solar Energy Materials and Solar Cells* 110:98–106.

Takano, Takumi, Hiroyasu Masunaga, Akihiko Fujiwara, Hidenori Okuzaki, and Takahiko Sasaki. 2012. PEDOT nanocrystal in highly conductive PEDOT:PSS polymer films. *Macromolecules* 45 (9):3859–3865.

Vaagensmith, Bjorn, Khan Mamun Reza, M. D. Nazmul Hasan, et al. 2017. Environmentally friendly plasma-treated PEDOT:PSS as electrodes for ITO-free perovskite solar cells. *ACS Applied Materials & Interfaces* 9 (41):35861–35870.

Wang, Qi, Yuchuan Shao, Qingfeng Dong, Zhengguo Xiao, Yongbo Yuan, and Jinsong Huang. 2014. Large fill-factor bilayer iodine perovskite solar cells fabricated by a low-temperature solution-process. *Energy & Environmental Science* 7 (7):2359–2365.

Wang, Qin, Chu-Chen Chueh, Morteza Eslamian, and Alex K-Y Jen. 2016a. Modulation of PEDOT:PSS pH for efficient inverted perovskite solar cells with reduced potential loss and enhanced stability. *ACS Applied Materials & Interfaces* 8 (46):32068–32076.

Wang, Zhao-Kui, Xiu Gong, Meng Li, et al. 2016b. Induced crystallization of perovskites by a perylene underlayer for high-performance solar cells. *ACS Nano* 10 (5):5479–5489.

Wei, Qingshuo, Masakazu Mukaida, Yasuhisa Naitoh, and Takao Ishida. 2013. Morphological change and mobility enhancement in PEDOT:PSS by adding co-solvents. *Advanced Materials* 25 (20):2831–2836.

Williams, Graeme, Brian Seger, and Prashant V. Kamat. 2008. TiO2-graphene nanocomposites. UV-assisted photocatalytic reduction of graphene oxide. *ACS Nano* 2 (7):1487–1491.

Wu, Zhong-Shuai, Wencai Ren, Libo Gao, et al. 2009. Synthesis of graphene sheets with high electrical conductivity and good thermal stability by hydrogen arc discharge exfoliation. *ACS Nano* 3 (2):411–417.

Wu, Zhongwei, Sai Bai, Jian Xiang, et al. 2014. Efficient planar heterojunction perovskite solar cells employing graphene oxide as hole conductor. *Nanoscale* 6 (18):10505–10510.

Xia, Yijie, and Jianyong Ouyang. 2011. PEDOT:PSS films with significantly enhanced conductivities induced by preferential solvation with cosolvents and their application in polymer photovoltaic cells. *Journal of Materials Chemistry* 21 (13):4927–4936.

Xia, Yijie, Kuan Sun, Jingjing Chang, and Jianyong Ouyang. 2015. Effects of organic-inorganic hybrid perovskite materials on the electronic properties and morphology of poly (3, 4-ethylenedioxythiophene): poly (styrene sulfonate) and the photovoltaic performance of planar perovskite solar cells. *Journal of Materials Chemistry A* 3 (31):15897–15904.

Xiao, Zeyun, Jegadesan Subbiah, Kuan Sun, David J. Jones, Andrew B. Holmes, and Wallace W. H. Wong. 2014. Synthesis and photovoltaic properties of thieno[3,2-b]thiophenyl substituted benzo[1,2-b:4,5-b']dithiophene copolymers. *Polymer Chemistry* 5 (23):6710–6717.

Xu, Jixian, Andrei Buin, Alexander H. Ip, et al. 2015. Perovskite–Fullerene hybrid materials suppress hysteresis in planar diodes. *Nature Communications* 6:7081.

Xue, Qifan, Guiting Chen, Meiyue Liu, et al. 2016. Improving film formation and photovoltage of highly efficient inverted-type perovskite solar cells through the incorporation of new polymeric hole selective layers. *Advanced Energy Materials* 6 (5):1502021.

Ye, Senyun, Weihai Sun, Yunlong Li, et al. 2015. CuSCN-based inverted planar perovskite solar cell with an average PCE of 15.6%. *Nano Letters* 15 (6):3723–3728.

Zhao, Baomin, Kuan Sun, Feng Xue, and Jianyong Ouyang. 2012. Isomers of dialkyl diketo-pyrrolo-pyrrole: electron-deficient units for organic semiconductors. *Organic Electronics* 13 (11):2516–2524.

Zhu, Chengzhou, Shaojun Guo, Youxing Fang, and Shaojun Dong. 2010. Reducing sugar: new functional molecules for the green synthesis of graphene nanosheets. *ACS Nano* 4 (4):2429–2437.

Zhu, Zhengyou, Haijun Song, Jingkun Xu, Congcong Liu, Qinglin Jiang, and Hui Shi. 2015. Significant conductivity enhancement of PEDOT:PSS films treated with lithium salt solutions. *Journal of Materials Science: Materials in Electronics* 26 (1):429–434.

19 Conducting Polymer-Based Ternary Composites for Supercapacitor Applications

Devalina Sarmah and Ashok Kumar
Department of Physics
Napaam, Tezpur
Assam, India

CONTENTS

19.1 INTRODUCTION: BACKGROUND AND DRIVING FORCES

Supercapacitors have extensively been studied for the last two decades owing to their instant power delivery when compared to those of rechargeable batteries and as they could sustain millions of fast charge–discharge cycles even at higher current densities so as to store energy and thus minimizing various environmental problems [1–3]. In 1957, H. Becker filed a patent titled "Low voltage electrolytic capacitor," in which for the first time a supercapacitor was discussed [4]. Becker described the behavior of charges present at the electric double layer (EDL) at electrode/electrolyte interfaces and the principle of electrical energy storage of porous carbon material in an aqueous electrolyte. With the evolution of electronic devices during 1990s, development of power sources with high energy as well as power density was essential to overcome the low energy density of conventional capacitors and knee-high power density of the lithium-ion batteries [5–7]. To solve the issue, in the first approach, the conventional capacitor with high power density was combined with lithium-ion battery having high energy density. However, the process required additional accessories, thereby increasing the weight of the device and also resulting in the decrease in the energy density. During the same time, Conway's group introduced a different charge storage mechanism using metal oxide RuO_2 films in H_2SO_4 electrolyte, which was nothing but pseudo-capacitance, associated with the electrochemical adsorption of charges with high specific capacitance and low internal resistance [8]. RuO_2 supercapacitors in H_2SO_4 aqueous electrolyte show almost the ideal capacitive response with appreciable cycling stability. The problem arose during the fabrication of capacitors at a large scale for use in hybrid systems such as electrical vehicles, as RuO_2 was very expensive. In the second approach, hybrid technology was developed based on traditional capacitors and electrochemical postulates called the hybrid electrochemical capacitors [9,10]. Electrochemical supercapacitors (ESs) can be commercialized with high specific capacitance, which can be achieved by using the available surface area of carbon materials and redox systems. Later, different groups reported about supercapacitor electrodes with composite materials incorporating both the charge storage mechanism, that is, electric double-layer capacitor (EDLC) and pseudocapacitance in a single electrode [11–15]. Today, conducting polymers, metal oxides, transition metal oxides, and graphene are the leading materials in the field of ESs because of their excellent electrochemical properties. Composites of reduced graphene oxide (rGO) with conducting polymers are potential candidates for supercapacitor electrodes, but the main challenges associated with these are as follows: (i) lower cycling stability of the conducting polymer, (ii) agglomeration of graphene during practical application, thus minimizing the utilization of total surface area, and (iii) lower energy density. These drawbacks of binary nanocomposites lead to the fabrication of ternary composites for a supercapacitor electrode so as to enhance the specific capacitance, cycling stability, and energy density as well as for maximum utilization of the total available surface area.

19.2 ELECTROCHEMICAL SUPERCAPACITORS

A capacitor is a charge storage device that stores energy by separating positive and negative charges with the help of an electric field and releases energy in the form of electrical energy (Figure 19.1a). The conventional capacitors are made up of two parallel metal plates separated by a dielectric like air, paper, ceramics, polyester, and so on. The capacitance of traditional capacitors is in the range of microfarad. The ESs are also a type of capacitor consisting of two current collectors with active material, that is, one is the cathode and another one is the anode separated by an electrolyte (Figure 19.1c). The electrolyte can be either solid or liquid that separates the positive electrode from the negative one. ESs are categorized in three different groups based on their charge storage mechanisms: (i) EDL capacitors where capacitances generated by the charges are accumulated at the electrode/electrolyte interfaces, (ii) pseudo-capacitors, corresponding to fast reversible redox reaction, and (iii) hybrid supercapacitors. In hybrid supercapacitors, both physical and chemical processes are responsible for charge storage, and they are the evolved form of EDLC and pseudo-capacitors to overcome various drawbacks of single mechanisms.

19.2.1 Electric Double-Layer Capacitors (EDLC)

The concept of EDL capacitance was first explained by Becker in 1957, where he used a porous carbon material and an aqueous electrolyte. The double-layer charge storage mechanism was first suggested by Helmholtz [16]. According to Helmholtz, oppositely charged ions accumulate at the electrode/electrolyte interfaces with a small separation depending on the size of the electrolyte's ions. These oppositely charged ions are developed near the electrode in the electrolyte to maintain the overall charge neutrality of the system by coulombic interaction. Thus, the EDL is formed in between the electrode and electrolyte, as shown in Figure 19.1b. The double-layer structure proposed by Helmholtz is quite similar to that of the conventional capacitors. Considering the thermal motion of the ions and assuming charges to be point charge, Gouy and Chapman described another model for double-layer formation. According to this model, the surface charge of the electrode is neutralized by the "diffuse layers" (scattered ions due to thermal fluctuation + assembled positive charges at electrode) of oppositely charged ions due to the thermal fluctuation of the solution. Stern [17] modified Gouy and Chapman's model by considering the finite size of the ions. Both the Helmholtz and the diffusion layers were taken into account by Stern to calculate the total capacitance of the double layer. Grahame [18] further refined Stern's model by considering the size difference of the cations and anions (cations are smaller than the anions); hence, the cation and anion possessed different distances from the electrode. The parameters affecting the EDLC component of the double layer are as follows: size and concentration of the ions, ion-to-solvent interaction, specific absorption of the ions, and the solvent used as the electrolyte. The EDLC electrodes

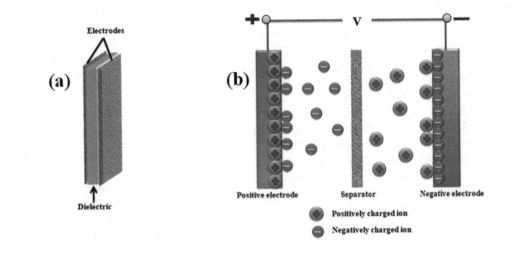

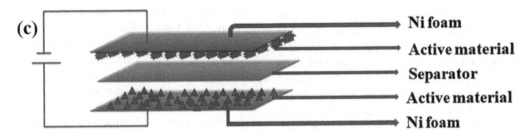

FIGURE 19.1 Schematic of (a) electrostatic capacitor, (b) charge storage in an electric double-layer capacitor (EDLC), and (c) different components of a supercapacitor by using Ni foam as the current collector.

made up of porous carbon material with a very high specific surface area ($\sim$2000 m^2 g^{-1}) and thin EDL give rise to an extremely high capacitance value. The pores in the porous carbon materials used in a supercapacitor electrode are categorized into three categories based on their size: (i) Micropores (pore size < 2 nm), (ii) mesopores (2 $\leq$ pore size $\geq$ 50 nm), and (iii) macropores (pore size > 50 nm) [19]. Dynamics of the electrolyte ions in these pores significantly affect the specific capacitance of ES.

If ES(1) and ES(2) are the surfaces of the two electrodes, C$^+$ and A$^-$ are the cations and anions, respectively, and // represents the EDL, then the reversible charging/discharging processes of EDLC can be explained as follows:

At the anode side:

$$\text{ES}(1) + \text{A}^- \rightleftharpoons \text{ES}(1)^+ // \text{A}^- + \text{e}^-$$

At the cathode side:

$$\text{ES}(2) + \text{e}^- + \text{C}^+ \rightleftharpoons \text{ES}(2)^- // \text{C}^+$$

The overall charging and discharging reactions at the cathode and anode are as follows:

$$\text{ES}(1) + \text{ES}(2) + \text{A}^- + \text{C}^+ \rightleftharpoons \text{ES}(1)^+ // \text{A}^- + \text{ES}(2)^- // \text{C}^+$$

During charging, electrons flow from the anode to cathode by an external power, whereas in the electrolyte the cations move toward the cathode and the anions move toward the anode. The membrane present between the cathode and anode (shown in Figure 19.1b) prevents a short circuit between them but allows the electrolyte ions to migrate from one side to another. As a result the cell voltage becomes very high and the energy is stored in the device. During discharge, electrons flow from the cathode to anode and the stored energy is utilized. In EDLC process, the charge transfer does not take place at the electrode/electrolyte interfaces. Therefore, during charging and discharging processes, EDLC electrodes do not undergo any structural and volumetric changes. Thus, these electrodes possess extremely high cycle life ($\sim$10^5 cycles) on repeated cycling. The supercapacitor following EDLC mechanism shows high power density (>500 W kg^{-1}) owing to quick and reversible charge storage-and-release processes at the electrode/electrolyte interfaces [20]. However, these physical charge transfer processes are confined only at the electrode surface, which results in low energy density of EDLC supercapacitors when compared to those of the batteries. Low energy density (<10 Wh kg^{-1}) is considered as one of the major drawbacks of the ESs and can be improved by using the pseudocapacitive charge storage mechanism in hybrid supercapacitors.

19.2.2 PSEUDOCAPACITORS

Pseudocapacitance is considered as the second mechanism to store charge in ES after EDLC. In pseudo-capacitive materials, fast charge transfer occurs by oxidation and reduction reactions at electrochemically active sites of the electrodes. During the redox reactions, the change in charge (dQ) depends almost linearly on the change in electrode potential (dV). This linear dependency gives a measurable capacitance of the electrodes, which is given as $C = dQ/dV$. As the capacitance measured does not result from charge separation or accumulation of charges near the electrode/electrolyte interfaces like EDLC and traditional capacitors, the capacitance generated by faradic redox reactions is termed *pseudocapacitance*. The interaction of the electrolyte ions with the electrode material strongly influences the pseudocapacitive response of ESs. Different types of pseudocapacitive processes of supercapacitors are as follows: (i) adsorption of the ions from the electrolyte at the electrode surface, (ii) ions involvement in the redox processes for some transition metal oxides, (iii) doping and de-doping in the case of conducting polymer systems, and (iv) lattice intercalations [21–24]. The adsorption pseudocapacitance is a result of the adsorption/desorption of electroactive species at the electrode surface from the electrolyte during the faradic charge transfer process. The second type of pseudocapacitance is a consequence of redox couple reactions at the electrode due to the variable oxidation states of electroactive materials. This kind of electrode possesses a rectangular-shaped geometry in cyclic voltammetry (CV) pattern during electrochemical measurements. Rapid doping and de-doping in the conducting polymers result from the intercalation of electrolyte ions in the three-dimensional (3D) array, thus providing a significant pseudocapacitance and energy density [25]. Fast intercalation/deintercalation of electrolyte ions at interfaces, lattice or van der Waals gaps of the electrode followed by faradic reactions, also exhibits pseudocapacitance in supercapacitors [26,27]. In the first two processes, the interaction of the electrolyte material is limited only to the surface. Therefore, it is also essential to use an electrode material with high available surface area to improve pseudocapacitance. The other two processes are associated with the penetration of electrolyte ions into the inner side of the bulk electrode, where the presence of interfaces or various gaps is taken into account. Pseudocapacitors possess much higher specific capacitance (10–100 times more) when compared to EDLC while using an electrode substrate of the same surface area. The pseudocapacitive charge storage process utilizes both the outer surface and the available inner surface area in the bulk of solid electrode through doping/de-doping and lattice intercalation process. This is the reason for higher specific capacitance in pseudocapacitors than that in EDL capacitors. At the same time, the penetration of the electrolyte ions to the bulk of the electrolyte and the redox couple reactions in faradic charge transfer processes result in low cycling stability in these materials. Therefore, pseudocapacitors provide elevated specific capacitance at the cost of cycle life. Improving the cycle life of pseudocapacitive materials is the biggest challenge in the field of supercapacitors and presently a number of researches are going on in this area.

19.2.3 HYBRID CAPACITORS

To improve the energy density of EDLC up to 20–30 Wh kg^{-1} in addition to the increasing demand for pseudo-capacitive material, the concept of hybrid supercapacitor was developed [28–30]. Both EDLC and pseudocapacitance storage mechanisms have some advantages as well as drawbacks. The drawback of EDLC can be compensated by using the advantages of pseudocapacitors and vice versa. The combination of both the mechanisms facilitates ESs with two to three times higher capacitive responses as compared to the conventional capacitors, and enhanced cycle life exhibiting wider potential window [31]. These are called hybrid supercapacitors where both EDLC and pseudocapacitive charge storage mechanisms are available. Hybrid supercapacitors can be designed as (i) symmetric supercapacitors by synthesizing electrodes with the composites of EDLC and pseudocapacitance. For example, the composites of graphene with conducting polymers, where graphene is the source of EDLC and conducting polymer provides pseudocapacitance [32]; (ii) asymmetric supercapacitors by preparing anode and cathode with different materials, that is, one with EDLC charge storage mechanism and the other electrode material follows pseudocapacitance.

19.3 ELECTRODE MATERIALS FOR SUPERCAPACITORS

The charge storage capacity and cycle life of ESs depend on the properties of electrode materials. It is essential to design and establish new materials for supercapacitors with elevated specific capacitance, energy density, and cycle life [33].

The capacitive response of the ES is dependent on the electroactive surface area present at the electrode. The size of the pores available in the electrode plays an important role while deciding the electroactive surface area. The pore size in the electrode materials significantly affects the double-layer capacitance depending upon the size of the electrolyte ions, as these sizes govern the kinetics of electrolyte ions at the pores while penetrating into the electrode [34]. Not only pore size, but also the distribution of pore size controls the movement of ions in the electrode during electrochemical processes in the ES. The brief overview of supercapacitor electrode materials is depicted in Figure 19.2. Mainly, the electrode materials for ESs are significantly divided into three categories, namely carbon materials, metal oxides, and conducting polymers [35,36].

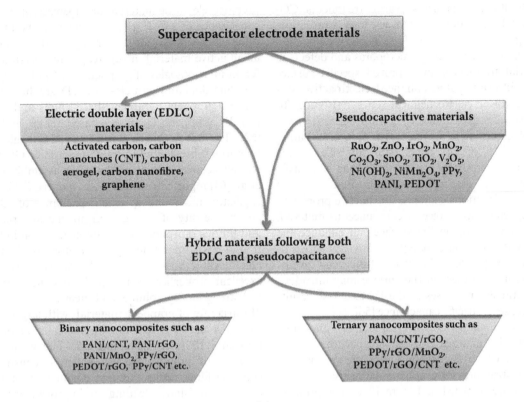

FIGURE 19.2 Brief overview of supercapacitor electrode materials.

19.3.1 CARBON MATERIALS

The double-layer capacitance was first discovered in porous carbon materials, and these electrode materials are continuously progressing to overcome the various limitations of ES. Carbon materials are extensively used in the electrodes of ES owing to high natural abundance, easy processing, low cost, good electronic conductivity, nontoxicity, large specific surface area, larger operating temperature, and chemical stability [37]. Crystal structures of different carbon materials such as three-dimensional (3D) diamond and graphite, two-dimensional (2D) graphene, one-dimensional (1D) carbon nanotube (CNT), and zero-dimensional (0D) carbon buckyballs are displayed in Figure 19.3.

Carbon materials are famous as an EDLC source for ES electrodes and charges are accumulated at the interface between the electrode and electrolyte. Carbon materials for EDL supercapacitors should possess high specific area (around 1000 m^2 g^{-1}), good intra- and inter-particle

conductivity, and easy electrolyte accessibility according to Conway [35,39]. Therefore, the parameters that influence the capacitive response and the electrochemical behavior of ES are specific surface area, electroactive surface area governed by pore shape, size, structure and, distribution, and electrical conductivity. Activated carbon (AC) [40,41], CNTs [42,43], carbon aerogel [44], carbon nanofiber [45], and graphene [38,46,47] are some of the famous carbon materials used in supercapacitor electrodes. EDLC is the main charge storage mechanism where charges are stored at the electrode and electrolyte interfaces. Owing to the absence of chemical involvement in charge storage process, carbon-based energy storage materials exhibit long cycle life (around 10^5 cycles) at the expense of specific capacity and energy density. The carbon materials possessing high surface area can accumulate more number of charges in the electrode/electrolyte interfaces and hence deliver higher specific capacitance [48]. To increase the capacitive response, carbon materials are

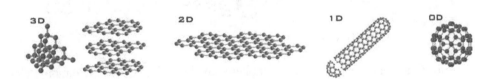

FIGURE 19.3 Different structures of carbon materials, (a) diamond, (b) graphite, (c) graphene, (d) carbon nanotube, and (e) carbon buckyballs are shown. (Reprinted from ref. [38], copyright (2007) with permission from Elsevier).

modified with alkaline treatments, plasma treatments, CO_2 activation, KOH activation, swift heavy ions' irradiation, and low energy ion implantation [49]. These modification techniques deliberately introduce micropores and defects in carbon material to furnish an improved specific surface area, shorter diffusion paths, enhanced electroactive surface area, and lower electrochemical series resistance. To increase the specific capacitance in porous carbon materials, it is not essential to increase the pore size, rather matching the sizes of the pores with the size of the electrolyte ions is important, so that the ions can easily penetrate into the electrode material. Besides these methods, surface functionalization is an effective process to increase the specific capacitance in carbonaceous material, as the functional groups on the surface can enhance the surface wettability of the electrode providing easy access to the electrolyte ions [50–55]. The surface functional groups (oxygenated groups) present in the surface may introduce redox charge transfer processes in the ES electrode resulting in the increase in specific capacitance [56].

Graphene is the single layer of graphite and the most versatile electrode material for supercapacitors under carbon material, where the carbon atoms are sp2 hybridized. The nanostructures of graphene nanosheets and its different forms are depicted in Figure 19.4. It is popular for its light weight, tunable surface area (up to 2675 m^2 g^{-1}), excellent electrical conductivity, strong mechanical strength (~1 TPa), and chemical stability [57,58]. The theoretical specific capacitance of graphene is ~21 μF cm^{-2} and experimentally ~550 F g^{-1} when the whole surface area is utilized [59]. Graphene can be used as a conductive network with transition metal oxides, conducting polymer to assist the redox reactions and these hybrid composites exhibit enhanced electrochemical performances due to the synergetic effects of metal oxides/conducting polymers. Ruoff et al. [60] pioneered chemical modification graphene as supercapacitor electrode with 135 F g^{-1} specific capacitance in aqueous electrolyte and specific capacitance of 99 F g^{-1} in organic electrolyte. Shi et al. [61] have reported one-step reduction process of GO and successfully prepared electrode material for supercapacitor with specific capacitance of 175 F g^{-1}. Graphene can be prepared with different morphologies by varying the dimension and used as an active material in supercapacitor electrodes. The different morphologies of graphene are (i) 0D graphene (freestanding dots and particles), (ii) 1D graphene (fiber-type), (iii) 2D graphene (graphene-based films and papers), and (iv) 3D graphene (graphene-based foams and hydrogels) [62]. The 0D graphene particles readily agglomerate and therefore it is essential to treat amphiphilic GO with selective surfactant to control the rGO assembly [63]. Zhang et al. [64] reported surfactant-assistant rGO film for supercapacitor and found specific capacitance of 194 F g^{-1} at current density of 1 A g^{-1}. Graphene material with yarn and fibrous morphology is gaining attention in the fabrication of next-generation supercapacitors for the use in portable devices, wearable and electric vehicles due to the excellent properties such as high flexibility, great mechanical strength, tiny volume, and great conductivity [65]. The 2D structure of graphene materials with larger surface area can offer easy intercalation/de-intercalation of electrolyte ions into the electrode materials. The great mechanical strength of graphene sheets helps to construct a freestanding film and its excellent conductivity minimizes the diffusion resistance leading to improved energy density, power density, high rate capability, and low electrode series resistance (ESR). The presence of all types of pores (micro, meso, and macropores) connected with each other in 3D porous graphene materials is desirable to fabricate ESs with high capacity, enhanced energy, and power density [66–69].

19.3.2 Metal Oxides

In order to improve the specific capacitance and energy density than that of carbonaceous material, transition metal oxides are employed as substitute material for supercapacitor electrodes. Among them, RuO_2 [70, 71, 76], ZnO [72], IrO_2 [73], MnO_2 [74], $NiCo_2O_4$ [75], and $MnFe_2O_4$ [77] are some popular supercapacitor electrode materials. Some metal oxides exhibit excellent pseudocapacitive

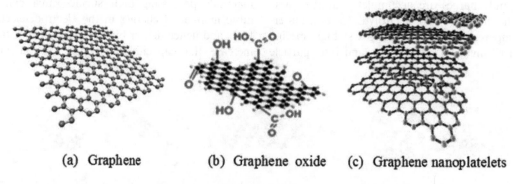

| (a) Graphene | (b) Graphene oxide | (c) Graphene nanoplatelets |

FIGURE 19.4 Schematic of graphene nanosheets, reduced graphene oxide and nanoplatelets, (reprinted from ref. [57], copyright (2017) with permission from Elsevier).

TABLE 19.1

Specific capacitance of a few metal oxides

Sl. no	Metal oxide	Specific capacitance	Ref.
1	Hydrous ruthenium oxide (RuO_2xH_2O)	786 F g^{-1}	[80]
2	Nickel Oxide (NiO)	1776 F g^{-1}	[78]
3	Cobalt oxide (Co_3O_4)	102 F g^{-1}	[85]
4	Vanadium oxide	350 F g^{-1}	[87]
5	Manganese oxide (MnO_2)	353 F g^{-1}	[81]

properties and hence deliver high specific capacitance. Unlike EDLC, electrolyte ions deeply penetrate into the lattice and interact at the redox centers due to the variable oxidation states of the metals in metal oxides. Grain boundaries, defects, and various surface-functionalized groups can act as redox center in metal oxides during charge storage processes. In spite of having large specific capacitance, metal oxides alone are not suitable in supercapacitor electrodes because conductivity of these metal oxides is very less except RuO_2. In the electrochemical measurements at higher current density, large IR drop appeared by increasing sheet resistance and large charge transfer resistance owing to the poor conductivity of metal oxides, hampering the practical measurements by lowering the energy density and rate capability. For example, supercapacitor electrode of NiO showed excellent specific capacitance of 1776 F g^{-1} at scan rate of 1 mV s^{-1} in 1 M KOH electrolyte but showed 23% of rate capability at 100 mV s^{-1} scan rate [78]. Pure metal oxide film develops strain in charge/discharge cycles causing mechanical degradation in the electrode leading to low cycling stability. Therefore, it is essential to prepare composite of metal oxides with carbon materials to mitigate the drawbacks and fruitful utilization of the merits to fabricate next-generation hybrid supercapacitor.

RuO_2 is considered to be the best pseudocapacitive material for supercapacitor electrodes due to very high specific capacitance, long cycle life, large potential window, with distinct oxidation states in the potential window of 1.2 V, higher rate capability, and remarkably high conductivity as compared to the other transition metal oxides [79,80]. With RuO_2, supercapacitor electrodes exhibit both EDLC and pseudocapacitive mechanisms in charge storage processes. The thin film of hydrous ruthenium oxide (RuO_2 xH_2O) exhibits specific capacitance of 786 F g^{-1}, high power density, excellent cycle life, and reversibility [76,80].

Owing to the high cost and environmental issues, RuO_2 was replaced with MnO_2 as electrode material for ESs. Mn oxide synthesized by hydrothermal route using 200°C exhibits high specific capacitance at higher scan rates and low specific capacitance at lower scan rates. This is because diffusion of the electrolyte ions is limited at higher scan rates due to the inaccessible voids and pores, which is attributed to the lower porosity and surface area of Mn oxides at 200°C [81]. Annealing only at appropriate

temperature can provide the improved capacitive behavior of Mn oxide supercapacitor electrodes. Also, modification of Mn oxides with carbon materials provides enhanced energy and power density owing to high available surface area with mesopores and improved electrical conductivity. The capacitive responses of some of the metal oxides are compared in Table 19.1.

Co_3O_4 is an alternative material for ES owing to its excellent reversible redox behavior, high available surface area, high conductivity, long cycle life, and high corrosion stability [82, 84]. The microspheres of Co_3O_4 having crater-like morphology can be synthesized with mesoporous silica template and exhibits specific capacitance of 102 F g^{-1} and intrinsic resistance of only 0.4 Ω [85]. Cobalt oxide thin film deposited on conducting current collector copper foil offered specific capacitance of 118 F g^{-1} [86]. Another electrode material, nickel oxide is famous as supercapacitor electrode material due to its easy synthesis, high capacitance (theoretical) of 3750 F g^{-1}, and low cost [83]. Vanadium oxide with variable oxidation states is famous as pseudocapacitive material due to the fast faradic charge storage processes and possesses specific capacitance of 350 F g^{-1} in 1 M KCl electrolyte [87].

19.3.3 CONDUCTING POLYMERS

Conducting polymers are a well-established source of pseudocapacitance and a deserving candidate for supercapacitor electrode material because of their high storage capacity, high conductivity when doped, high potential window, high energy density, environment-friendliness, low cost, and easy synthesis [88,89]. Conducting polymers show rapid doping and de-doping process during faradic charge transfer process. The electrolyte ions are moved to the polymer backbone in oxidation process and again released to the electrolyte during the reduction process in redox processes. These highly reversible redox reactions not only take place on the surface of the electrode but also transfers to the entire bulk of the electrode leading to structural alternation in the rapid charge–discharge processes [90]. Oxidation and reduction processes in the conducting polymer backbone can generate delocalized "π" electrons, and the electronic states of the π electrons determine the potentials of these states.

Polyaniline (PANI), an important pseudocapacitive material, is the most widely used conducting polymer for supercapacitor electrodes owing to high specific capacitance in acidic media, high electroactivity, good environmental stability, high doping level, and highly conducting matrix [91]. PANI is very selective for the electrolyte to be used in electrochemical measurements, as it needs protic solvent for proper charge and discharge processes [92,93]. Depending upon the synthesis process, morphology, binder type, and thickness of the film, PANI exhibits wide range of capacitance from 44 to 270 mA h g^{-1} [94]. Electrochemically deposited PANI shows higher specific capacitance than chemically synthesized due to the homogeneous distribution in electrochemical polymerization. Li-doped PANI possesses cycling stability of 70% after 5000 cycles, whereas $LiPF_6$-doped PANI exhibits cycling stability of 78% after 9000 cycles [97,98].

Polypyrrole (PPy) is famous among the conducting polymers for energy storage application owing to its high flexibility and easy synthesis [95]. The greater density of PPy minimizes the capacitive response of the polymer in thick coating electrodes due to the limited access of the electrolyte ions ascribed to high density. Suematsu et al. [96] stated that capacitance per unit volume of PPy is 100–200 F cm^{-3} but a typical PPy layer exhibited capacitance of about 400–500 F cm^{-3}.

Supercapacitor electrodes based on poly(3,4-ethylenedioxythiophene) (PEDOT) and its composite are growing due to the properties such as low oxidation potential, high energy density from wider potential window (1.2–1.5 V), low band gap (1–3 eV), thermal and chemical stability, conducting p-doped state, and fast electrode kinetics due to the high charge mobility [99]. The high surface area along with high conductivity of PEDOT provides fast kinetics of PEDOT electrodes and exhibits longer cycle life.

PEDOT do not dissolve easily in aqueous solvent and possesses low specific capacitance of about 90 F g^{-1} due to the higher molecular weight and specific capacitance of 180 F g^{-1} for very thin film of PEDOT electrode [100]. The chemical structures of PANI, PPy, and PEDOT are shown in Figure 19.5.

Unfortunately, conducting polymers show low cycling stability during rapid charge–discharge processes ascribed to swelling and shrinkage, leading to the mechanical degradation of the electrode material due to fast oxidation and reduction processes, which is a very serious issue of the pseudocapacitive material. For example, PPy electrode exhibits only 50% of capacitive retention after 1000 cycles at current density of 2 mA cm^{-2}. Polythiophene also shows poor cycling stability in its n-doped state, which minimizes the overall performance of ESs [101].

19.3.4 New Materials

Metal organic framework (MOF), covalent organic framework (COF), transition metal carbides and carbonitrides (MXenes), metal nitrides (MN), and black phosphorus (BP) are the new 2D energy storage electrode materials [102]. The porous coordinate polymer MOF is gaining attention owing to the diverse structure, enormous surface area (~6000 m^2 g^{-1}), and distinct pore size and volume. Like MOF, COFs are also considered as energy storage material with homogeneous pore size, flexibility, and large surface area. MXenes exhibit properties such as great mechanical strength, conductivity, and water affinity, which establishes them as potential candidate for supercapacitors' electrode materials. MNs where M stands for Mo, W, V, Ti, Nb, and Ga, exhibit very high electrical conductivity providing efficient transfer of the interfacial electrons, they can also be considered as high-performance

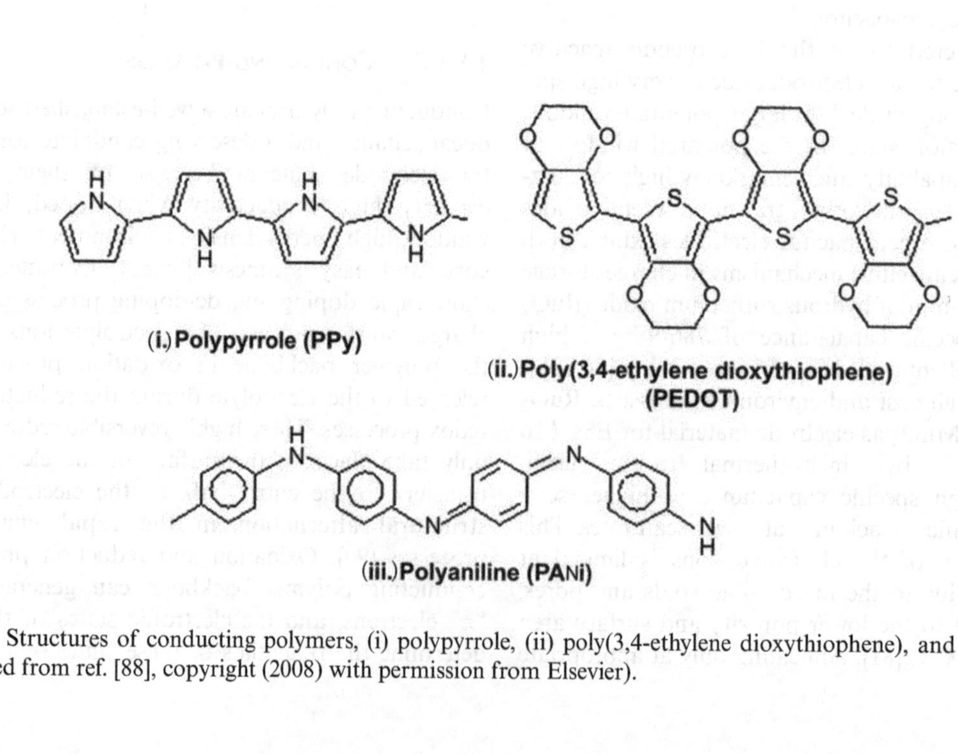

(i.) Polypyrrole (PPy)

(ii.) Poly(3,4-ethylene dioxythiophene) (PEDOT)

(iii.) Polyaniline (PANi)

FIGURE 19.5 Structures of conducting polymers, (i) polypyrrole, (ii) poly(3,4-ethylene dioxythiophene), and (iii) polyaniline are shown (reprinted from ref. [88], copyright (2008) with permission from Elsevier).

pseudocapacitive supercapacitor electrode materials. Owing to the shorter diffusion paths and closely packed structures, phosphorenes are also used as electrode material.

19.4 NANOCOMPOSITES FOR SUPERCAPACITORS

19.4.1 BINARY NANOCOMPOSITES

The performance of ESs is remarkably affected by the structure, electrode–electrolyte interaction, surface area, conductivity, interfaces, and the electrochemical behavior of the electrode materials. A flexible ideal supercapacitor electrode should possess: (i) excellent electrochemical properties, (ii) appreciable mechanical strength on bending, (iii) stable porous structure, and (iv) efficient accommodation of large amount of strain without losing the electrochemical performance. To fulfill the abovementioned requirements, it is necessary to choose suitable material for supercapacitor electrodes and design nanocomposites to satisfy the growing energy demand of modern society. As stated earlier, carbonaceous materials are the main source of EDLC, whereas conducting polymers and metal oxides provide pseudocapacitance in energy storage processes. Carbonaceous materials possess drawbacks such as low specific capacitance and low energy density but exhibit very high cycle life. Metal oxides possess low conductivity hence high value of ESR leading to high IR drop with lower rate capability. But metal oxides are established source of pseudocapacitance due to the fast reversible faradic reactions. Conducting polymers provide high specific capacitance due to the faradic reaction at the polymer backbone but undergo mechanical degradation and swelling during practical applications. Therefore, an idea was developed to mix carbonaceous materials (rGO, CNT, and AC) with pseudocapacitive materials (metal oxides and conducting polymers) and construct new composite materials to overcome the drawbacks of the individuals for the fabrication of high-performance hybrid supercapacitor electrodes. To utilize the synergetic effects of carbonaceous material and faradic materials, hybrid binary nanocomposites were designed for supercapacitors to achieve the desired properties in a single electrode. In this section, some of the binary nanocomposites with conducting polymers will be discussed. The comparisons of specific capacitance and cycling stability among the different binary composites of PANI, PPy, and PEDOT are displayed in Table 19.2.

19.4.1.1 PANI-based Binary Nanocomposites

Ryu et al. [98] have investigated PANI as symmetric redox supercapacitors and found only 70 F g^{-1} specific capacitance at the initial charge/discharge cycle, which decreased to 40 F g^{-1} at after 400 cycles only. To address such issues, composites of PANI have been prepared with

metal oxides and carbon materials. Ramaprabhu et al. [103] have prepared hybrid binary nanocomposite of PANI by *in-situ* polymerization method with α-MnO$_2$ nanotubes (MNTs). They have fabricated a symmetric supercapacitor and found enhanced specific capacitance of 626 F g^{-1} with energy density of 17.8 Wh kg^{-1} at current density of 2 A g^{-1} within the potential range of 0–0.7 V. Wang et al. [104] reported a flexible nanocomposite of CNTs and PANI nanowire array. The binary nanocomposite of PANI/CNT prepared from *in-situ* polymerization of aniline in CNT yarn delivered specific capacitance of 38 mF cm^{-2} at the current density of 0.01 mA cm^{-2} and cycling stability of 91% after 800 of continuous galvanostatic charge-discharge (GCD) cycles.

Binary nanocomposite of PANI nanowire and nitrophenyl-functionalized rGO (frGO) has been reported by Wang et al. [105] and they also investigated its electrochemical performances. The covalent bonding between the frGO and PANI offers specific capacitance of 590 F g^{-1} at current density of 0.1 A g^{-1} and cycling stability of 100% after 200 cycles at current density of 2 A g^{-1}. The synergetic effects between PANI and frGO provide improved capacitive response suggesting the binary composite as a well-known material of energy storage. The properties of graphene can be tuned by its structures by different processes. Meng et al. [106] reported 3D porous structure of graphene as flexible free-standing film and evaluated the electrochemical properties. Binary nanocomposite of 3D-rGO was prepared with PANI nanowire arrays by polymerizing aniline on the randomly stacked graphene sheets. The 3D-rGO film exhibits excellent rate capability of 89% at current density of 10 A g^{-1}. Binary 3D-rGO/PANI nanocomposite electrode exhibits specific capacitance of 385 F g^{-1} at 0.5-fold current density and rate capability of 94% at 10-fold current density. The interconnected porous structure of the binary electrode facilities connected charge transfer paths for the fast movement of the electrolyte ions in charging and discharging processes leading to superior rate capability of the electrodes.

19.4.1.2 Polypyrrole (PPy)-based Binary Nanocomposites

The conducting polymer PPy is attracting attention as a pseudocapacitive material for supercapacitor or battery electrodes due to the greater density, high volumetric capacitance (400–500 F cm^{-3}), rapid doping/de-doping, and appreciable degree of flexibility [111]. The limited excess of the doped ions to the interior sites of the polymer due to the dense growth in polymerization process reduces the specific capacitance (capacitance per gram) of the bulk electrodes. Composites of PPy can be synthesized to increase the number of interfaces in the polymer so that the ions can use those paths for easy movements in the electrode and hence provide a higher specific capacitance. The other parameters such as current density and active material loading in the bulk of the electrode can also

TABLE 19.2

Specific capacitance and cycling stability of some PANI-, PPy-, and PEDOT-based binary nanocomposites

Sl. no	Electrode material	Specific capacitance	Cycling stability	Ref.
1	PANI–MnO$_2$ nanotubes	626 F g^{-1}	-	[103]
2	CNT–PANI nanowire	38 mF cm^{-2}	91% after 800 cycles	[104]
3	PANI nanowire–rGO	590 F g^{-1}	100% after 200 cycles	[105]
4	3D rGO–PANI nanowire	385 F g^{-1}	-	[106]
5	PPy–CNT	390 F g^{-1}	-	[107]
6	PPy–NiCo(OH)$_2$	1469.25 F g^{-1}	95.2% after 10,000 cycles	[108]
7	PPy–rGO	466 F g^{-1}	85% after 600 cycles	[95]
8	PPy–melamine sponge	39.0 F g^{-1}	-	[96]
9	PEDOT/rGO	43.75 mF cm^{-2}	83.6% after 1000 cycles	[109]
10	MoS$_2$/PEDOT	149.8 mF cm^{-2}	85% after 1000 cycles	[110]

influence the capacitive response of a PPy-based supercapacitor electrode. The binary composite of PPy with CNTs was reported by Li et al. [107] with specific capacitance of 390 F g^{-1} at scan rate of 2 mV s^{-1}. Binary nanocomposite of nanoflower such as PPy with NiCo(OH)$_2$ reported by Wu et al. [108] exhibits specific capacitance of 1469.25 F g^{-1} (1 A g^{-1}) with cycling stability of 95.2% after 10,000 cycles (30 A g^{-1}). The extraordinary specific capacitance is due to the short electron transfer paths from the formation of hydrogen bonding and stable structure of PPy/NiCo(OH)$_2$ nanocomposite. The composite of PPy with graphene possessed specific capacitance of 466 F g^{-1} at scan rate of 10 mV s^{-1} in 1 M KCl electrolyte and cycling stability of 85% after 600 cycles, as reported by Das et al. [95]. The composites of PPy have also been synthesized by using artificial melamine sponge (aMS) by Chang et al. [96] with specific capacitance of 39.0 F g^{-1}.

19.4.1.3 Poly(3,4-ethylenedioxythiophene) (PEDOT)-based Binary Nanocomposites

PEDOT is gaining recent attention in supercapacitor applications due to the high conductivity, stability, and larger electrochemical window as compared to the other conducting polymers [112]. The composites of PEDOT have been reported with 2D materials such as graphene and MoS$_2$. These 2D materials exhibit remarkable EDLC mechanism with high surface area during charge storage process [113–115]. Xu et al. [109] investigated PEDOT/rGO nanocomposite by depositing PEDOT on Indium tin oxide (ITO) coated substrate through electrochemical polymerization method and then deposited GO nanosheets with spin coating. GO was reduced by laser writing method, which gave rGO nanosheets of very high quality and the maximum surface area of rGO can be utilized in practical application. The PEDOT/rGO nanocomposite exhibits nearly three times greater areal capacitance of 43.75 mF cm^{-2} than that of pure PEDOT electrode at a current density of 0.2 mA cm^{-2} and

cycling stability of 83.6% after 1000 cycles. The 2D materials such as MoS$_2$ and graphene possess layered structure bonded with van der Waals force. MoS$_2$ is a 2D transition metal dichalcogenides and offers excellent cycling stability in repetitive charge–discharge processes. The composites of MoS$_2$ with PEDOT can be synthesized by *in-situ* or electrode polymerization of EDOT monomer. A MoS$_2$/PEDOT nanocomposite offers specific capacitance of 405 F g^{-1} and cycling stability of about 90% after 1000 cycles, as reported by Wang et al. [110]. The capacitive response in MoS$_2$/PEDOT nanocomposite film is attributed to the synergetic effect between MoS$_2$ and PEDOT.

19.4.2 Ternary Nanocomposites for Supercapacitors

Energy density of EDL capacitors and pseudocapacitors is lower as compared to the batteries. To improve the energy density, ternary nanocomposites have been designed possessing both the charge storage mechanisms, which are EDLC and pseudocapacitance, without sacrificing the other important supercapacitor parameters, namely specific capacitance, cycle life, power density, rate capability, and coulombic efficiency. Supercapacitors with improved electrochemical responses prepared from these composites are named as hybrid supercapacitors. Binary composites have been prepared by mixing two materials using different routes such as chemical, electrochemical, or physical. Available surface area is an important parameter while synthesizing supercapacitor electrodes. The volume to surface ratio of the materials increases when moving from bulk to nanostructures. Therefore, it is important to use the nanostructures of the materials while preparing these composites for better utilization of surface area and more number of interfaces of the bulk electrode, so that the electrolyte ions can penetrate inside the active material. Conducting polymers are highly promising pseudocapacitive materials for

TABLE 19.3

Specific capacitance and cycling stability of some PANI-, PPy-, and PEDOT-based ternary nanocomposites

Sl. no	Electrode material	Specific capacitance	Cycling stability	Ref.
1	MoS_2/PANI/rGO	520 F g^{-1}	89.1% after 40,000 cycles	[119]
2	MoS_2/PANI/rGO aerogel	618 F g^{-1}	96% after 2000 cycles	[120]
3	MoS_2/rGO@PANI nanowire	1224 F g^{-1}	82.5% after 3000 loops	[121]
4	MWCNTs/PANI/MoS_2	542.56 F g^{-1}	73.71% after 3000 cycles	[122]
5	Fe_3O_4@ activated carbon @PANI symmetric cell	420 F g^{-1}	82% after 5000 cycles	[123]
6	lanthanum manganite ($LaMnO_3$)/rGO/PANI//rGO	111 F g^{-1}	117% after 100,000 cycles	[124]
7	Chitosan-ZnO/PANI	587.15 F g^{-1}	-	[125]
8	rGO/PANI/Co_3O_4	789.7 F g^{-1}	81.8% after 1000 GCD cycles	[126]
9	MnO_2/rGO/PANI	512 F g^{-1}	97% after 5000 cycles	[127]
10	Cobalt ferrite/graphene/Polyaniline	1133.3 F g^{-1}	96.6% after 5000 cycles	[128]
11	Sulfonated graphene/multi-walled carbon nanotubes/PANI//activated porous graphene	107 F g^{-1}	91% after 5000 cycles	[129]
12	Graphene/Au/PANI	572 F g^{-1}	88.54% after 10,000 cycles	[130]
13	Tin disulfide (SnS_2)/rGO/PANI	1021.67 F g^{-1}	78% after 5000 cycles	[131]
14	MnO_2/fFWNT/PEDOT-PSS	427 F g^{-1}	99% after 1000 cycles	[132]
15	CNT@PPy@MnO_2	325 F g^{-1}	90% after 1000 cycles	[133]
16	MoS_2–rGO/PPyNTs	1561 F g^{-1}	72% after 10,000 cycles	[32]

energy storage applications because of their low cost, easy synthesis, high specific capacitance, and energy density, and also provide higher specific capacitance due to faradic charge transfer process [116,117]. Conducting polymers also act as a matrix to the whole nanocomposite by dispersing the other elements and help to prepare a mechanically stable film. This homogeneous dispersion of the materials in the polymer matrix provides ordered charge transfer paths by minimizing the ESR. They contain alternate single and double bonds called conjugated system. Conjugated system reflects the delocalization of π-electrons in the back bonds of the polymer chain. These π-electrons in the polymer backbones are responsible for their remarkable electronic properties such as electrical conductivity, low energy optical transitions, high electron affinity, and low ionization potential. Conducting polymer shows relatively low cycling stability owing to the presence of unstable redox sites. The nanocomposites of conducting polymers with inorganic materials such as carbon material, metal oxides, transition metal oxides, and transition metal dichalcogenides show better cycle life, specific capacitance, energy density, power density, and mechanical stability, which are essential for supercapacitor electrodes [61,118]. Among the conducting polymers, PANI is a widely used supercapacitor material because of much higher specific capacitance (in the range of 600 F g^{-1}) as compared to the other conducting polymers PPy, polythiophene, and PEDOT. Herein, ternary nanocomposites prepared with PPy, PANI, and PEDOT for supercapacitor application

have been discussed and the capacitive responses are compared in Table 19.3.

19.4.2.1 Polyaniline (PANI)-based Ternary Nanocomposites

PANI is considered as one of the best pseudocapacitive material for supercapacitor electrode due to high flexibility, high conductivity, and higher specific capacitance of nearly 600 F g^{-1} because of multiple redox peaks that appeared in the electrochemical process in aqueous electrolyte [134,135]. During electrochemical measurements, PANI shows two pair of redox reactions because of faradic charge transfer process. These two pair of redox peaks appear due to the transformation from leucoemeraldine to emeraldine and from emeraldine to pernigraniline [136–139]. Leucoemeraldine, emeraldine, and pernigraniline are various oxidation states of PANI appeared due to oxidation and reduction processes in backbone of the polymer chain upon application of voltage in electrochemical measurements.

19.4.2.1.1 Polyaniline and MoS$_2$-based Ternary Nanocomposites

At present, extensive researches are going on in the ternary nanocomposite of MoS_2, rGO, and PANI for supercapacitor application and different groups are synthesizing the same ternary nanocomposite with different synthesis procedures in order to improve the capacitive response without losing the other supercapacitor parameters. Chao et al. [119] have synthesized ternary nanocomposite of MoS$_2$/polyaniline/rGO, where sandwiched structure of MoS$_2$/PANI is vertically arranged on rGO nanosheets as shown in Figure 19.6a, and investigated their electrochemical performances in 1 M aqueous H$_2$SO$_4$ electrolyte. They have synthesized MoS$_2$/rGO, MoS$_2$/PANI, and MoS$_2$/PANI/rGO nanocomposites and compared their electrochemical behavior by CV, GCD and electrochemical impedance spectroscopy (EIS).

The CV and GCD measurements of MoS$_2$/PANI, MoS$_2$/rGO, and MoS$_2$/PANI/rGO are depicted in Figure 19.6b, c. The ternary electrode exhibits highest capacitive response in CV curves and longest discharge duration for GCD patterns. The ternary electrode possessed specific capacitance of 520 F g^{-1}, 44.9% of rate capability in 30 A g^{-1} current density, and 89.1% of cyclic stability at after 40,000 cycles, as reported by the group. The ternary nanocomposite shows enhanced electrochemical behavior due to the sandwiched structure of MoS$_2$/PANI nanosheets vertically decorated on rGO nanosheets (Figure 19.6a). which forms a 3D conductive network of three nanomaterials with more number of interfaces and specific surface area (85.3 m^2 g^{-1}) accelerating the ion diffusion rates. PANI possesses lower cycling stability as compared to MoS$_2$ and rGO due to the involvement of chemical oxidation and reduction processes in faradic charge transfer in repeated cycling processes, which causes mechanical degradation and swelling in the PANI structure. MoS$_2$/rGO electrode possesses good cycling stability but at the same

time, the specific capacitance is low, whereas PANI contributes high specific capacitance at the cost of cycle life. To investigate the practical application of MoS$_2$/PANI/rGO nanocomposite, Chao's group have designed symmetric supercapacitor (MoS$_2$/PANI/rGO//MoS$_2$/PANI/rGO) using MoS$_2$/PANI/rGO electrodes both as cathode and anode, and asymmetric supercapacitor with MoS$_2$/PANI/rGO electrode as cathode and AC as anode (MoS$_2$/PANI/rGO-300//AC). The group found specific capacitance up to 97.8 and 73.3 F g^{-1} at two-fold of current density for symmetric and asymmetric configuration, respectively, and cycling stability of 84.2% and 87.9% after 20,000 cycles, for symmetric and asymmetric configuration, respectively. The synergetic effect of MoS$_2$, rGO, and PANI provides outstanding combination of specific capacitance, rate capability as well as long cycle life in MoS$_2$/PANI/rGO.

Another group Sha et al. [120] have reported ternary nanocomposite of MoS$_2$/PANI/rGO aerogel by *in-situ* polymerization method in presence of MoS$_2$ and then adding well-dispersed GO into MoS$_2$/PANI nanocomposite followed by reduction of rGO using urea as the reducing agent via hydrothermal treatment for supercapacitor application. The ternary MoS$_2$/PANI/rGO nanocomposite possesses high specific capacitance of 618 F g^{-1} at one-fold current density and capacitive retention of 96% at 20-fold current density. EDLC components MoS$_2$ and rGO with pseudocapacitive component PANI contribute to the overall specific capacitance of MoS$_2$/PANI/rGO aerogel electrode. In cycling stability test, MoS$_2$/PANI/rGO aerogel electrode loses only 2% of its initial specific capacitance

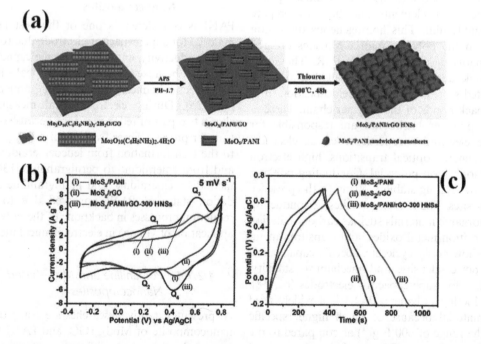

FIGURE 19.6 (a) Schematic of the synthesis of ternary MoS$_2$/PANI/rGO nanostructures, (b) CV patterns and (c) GCD measurements of MoS$_2$/PANI, MoS$_2$/rGO, and MoS$_2$/PANI/rGO electrodes (reprinted from ref. [119], copyright (2018) with permission from Elsevier).

after first 400 cycles and then hold fix at 96% from 600 to 2000 cycles. This remarkable cycle life arises due to the superior electrochemical stability of EDLC component graphene aerogel. The higher specific surface area of graphene aerogel strengthens the material compatibility and stability by dispersing the nanoparticle into it.

Li et al. [121] have fabricated a promising ternary nanocomposite of MoS_2/rGO@PANI by synthesizing MoS_2 on rGO nanosheets by hydrothermal route and *in-situ* polymerization of PANI nanowire on MoS_2/rGO nanocomposite. The schematic representation of the synthesis procedure of ternary MoS_2/rGO@PANI nanocomposite is depicted in Figure 19.7a. The CV measurements of PANI, MoS_2/rGO, and MoS_2/rGO@PANI electrodes were carried out in 1 M H_2SO_4 electrolyte at scan rates of 2, 5, 10, 20, 50, and 100 mV s^{-1} and depicted in Figure 19.7b, c. The GCD measurements of the electrodes are shown in Figure 19.7d, e. The synergetic effects of the three components in the electrode materials result in extraordinary electrochemical behavior in terms of excellent specific capacitance (1224 F g^{-1} at one-fold current density), rate capability (58% at 20-fold current density) (Figure 19.7f), and cycle life (Figure 19.7g) during repeated GCD cycles (82.5% after 3000 loops). The presence of good reversibility and efficient charge transfer process is suggested by 100% coulombic efficiency of the electrode material. The symmetric cell designed with MoS_2/RGO@PANI//MoS_2/RGO@PANI possesses good value of specific capacitance (160 F g^{-1} at one-fold current density), energy density (22.3 Wh kg^{-1}), and power density (5.08 kW kg^{-1}). The symmetric cell configuration shows excellent reversibility but potential drops in the GCD curve during discharge process arise due to the increased interfacial resistance in the cell as compared to the investigation done in three-electrode system, which is one of the biggest challenges in the area of supercapacitor. Pseudocapacitive element PANI nanowire helps in uniform dispersion of the EDLC components, that is, MoS_2 and rGO in the matrix of ternary nanocomposite providing improved interfacial contact and elevated electrochemical performances of the nanocomposite.

Ternary nanocomposites of MoS_2 nanoflower and PANI nanostem with multi-walled CNTs (MWCNTs) have been reported by Qian et al. [122] for supercapacitor application by synthesizing PANI on MWCNTs followed by incorporation of MoS_2 nanoflower hydrothermally. In the ternary nanocomposites of MWCNTs/PANI/MoS_2, MWCNTs act as framework by accommodating PANI in them to minimize the mechanical degradation of the conducting polymer and MoS_2 prevents the volumetric changes of PANI during repeated cycling measurements by participating as outer barrier of the ternary nanocomposites. MoS_2 also contributes to EDL capacitance during energy storage. The electrochemical performances of the ternary nanocomposite were analyzed with CV and GCD measurements in 1 M H_2SO_4 electrolyte and were compared with MoS_2, PANI, and MWCNTs/PANI

nanocomposites. MWCNTs/PANI/MoS_2 ternary electrodes exhibit highest specific capacitance of 542.56 F g^{-1} at 0.5 A g^{-1} current density and good cycling stability of 73.71% over 3000 cycles as compared to MoS_2 (128.75 F g^{-1} at 0.5 A g^{-1}, 38.6%), PANI (438.46 F g^{-1} at 0.5 A g^{-1}, 41.76%), and MWCNTs/PANI (480 F g^{-1} at 0.5 A g^{-1}, 61.82%) because of the synergistic effect from MoS_2 nanoflowers, PANI nanostem, and MWCNTs. The presence of MWCNTs offers high stability to the electrode and the ternary MWCNTs/PANI/MoS_2 electrodes show capacity retention of 62.5% at 10-fold current density.

19.4.2.1.2 Polyaniline and Metal Oxide-based Ternary Nanocomposites

Transition metal oxides have low electrical conductivity, which compromises the power density when used as a supercapacitor electrode. To overcome such issues, metal oxides are mixed with conductive network such as PANI. Moreover, ternary composites of metal oxides with PANI and AC have been established as a promising supercapacitor composite with elevated electrochemical behavior [140,141]. Ternary nanocomposites of Fe_3O_4 @C@PANI have been synthesized by Qiu et al. [123] and they also studied the electrochemical behavior with CV, GCD, and EIS as well as evaluated the behavior of Fe_3O_4@C@PANI//Fe_3O_4@C@PANI symmetric cell.

Fe_3O_4@C core@shell nanocomposites were synthesized by hydrothermal route followed by encapsulation with PANI by *in-situ* polymerization methods. Figure 19.8a–d depicts the transmission electron microscope (TEM) micrographs of Fe_3O_4 nanoparticles exhibited a spindle-like and the nanocomposites with SAED patterns. Configuration of the symmetric supercapacitor using the ternary Fe_3O_4@C@PANI electrodes is depicted in Figure 19.8e using 1 M KOH separator. CV curves of the symmetric cell exhibited ideal capacitive behavior with a rectangle-like geometry at 200 mV s^{-1} scan rate, as displayed in Figure 19.8f. During the electrochemical measurements, Fe_3O_4@C@PANI electrode shows 380 F g^{-1} of specific capacitance at 1 A g^{-1} current density in 1 M KOH electrolyte. The capacitive response increases for the ternary Fe_3O_4@C@PANI electrode as compared to binary Fe_3O_4@C core-shell electrode (120 F g^{-1}) after encapsulation with PANI. The charge transfer paths and charge storage properties enhance in the ternary Fe_3O_4@C@PANI electrode after the incorporation of PANI in Fe_3O_4@C. The charge transfer resistance of the ternary Fe_3O_4@C@PANI electrode (1.55 Ω) decreases as compared to the binary Fe_3O_4@C one (3.8 Ω) suggesting improved charge transfer kinetics and more number of interfaces from the EIS measurements. The symmetric cell Fe_3O_4@C@PANI//Fe_3O_4@C@PANI exhibits 420 F g^{-1} of specific capacitance at current density of 0.5 A g^{-1} and rate capability of 38% at current density of 50 A g^{-1} in (Figure 19.8h) 1 M KOH electrolyte due to smaller charge transfer resistance from

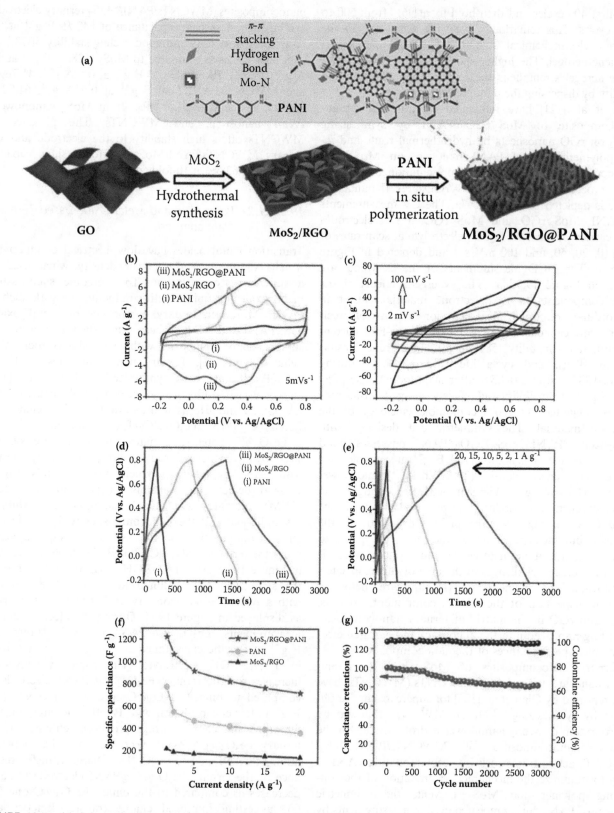

FIGURE 19.7 (a) Schematic representation of ternary $MoS_2/RGO@PANI$ nanocomposite synthesis process, (b) CV pattern of PANI, MoS_2/RGO, and $MoS_2/RGO@PANI$ electrodes in 1 M H_2SO_4 solution, and (c) CV pattern of ternary $MoS_2/RGO@PANI$ electrode at 2–100 mV s^{-1} scan rates in 1 M H_2SO_4 solution. GCD of (d) PANI, MoS_2/RGO, and $MoS_2/RGO@PANI$ electrodes at 1 A g^{-1} current density and (e) $MoS_2/RGO@PANI$ electrode at 1–20 A g^{-1} current density. (f) Specific capacitances vs. current density of PANI, MoS_2/RGO, and $MoS_2/RGO@PANI$ electrodes and (g) capacitive retention and coulombic efficiency vs. cycle numbers of $MoS_2/RGO@PANI$ electrode after 3000 cycles at current density of 10 A g^{-1} (reprinted with permission from ref. [121], copyright (2016) American Chemical Society).

the 3D conducting network of PANI. The symmetric cell shows 82% of cycling stability at 10 A g^{-1} current density after 5000 GCD cycles and possesses energy density and power density of 32.7 Wh kg^{-1} and 500 W kg^{-1} at 0.5 A g^{-1} current density, respectively. Fabrication of ternary Fe$_3$O$_4$@C@PANI nanocomposite is a new strategy to design efficient supercapacitors with ternary nanocomposites. Another group from India, Shafi et al. [124], has reported ternary nanocomposites synthesized from lanthanum manganite (LnMnO$_3$), rGO, and PANI for supercapacitor application. This group has reported the fabrication of ternary LaMnO$_3$/rGO/PANI composite by *in-situ* polymerization of PANI in rGO/LaMnO$_3$ suspension for efficient supercapacitor electrode with improved energy density. Symmetric supercapacitors were fabricated with ternary nanocomposites by assembling electrodes as LaMnO$_3$/rGO/PANI//LaMnO$_3$/rGO/PANI and asymmetric supercapacitors as LaMnO$_3$/rGO/PANI//rGO. The asymmetric supercapacitor showed better electrochemical performances with specific capacitance of 111 F g^{-1} at 2.5 A g^{-1} and 50% of capacitive retention at 20 A g^{-1} (55.5 F g^{-1}) when compared with symmetric device (89 F g^{-1} at 2.5 A g^{-1} current density). The electrochemical potential window for the asymmetric device is 1.8 V, whereas the symmetric device exhibits a potential window of 1.4 V. The asymmetric device delivers 50 Wh kg^{-1} energy density while power density is 2.25 kW kg^{-1}. LaMnO$_3$/rGO/PANI//rGO asymmetric device was quite stable during repeated GCD cycles with 117% of capacitive retention after 100,000 cycles. All the above results strongly suggest the ternary LaMnO$_3$/rGO/PANI composite as an efficient cathode material for the fabrication of supercapacitor device with very high cycle life. In 2014, Pandiselvi et al. [125] have reported a ternary nanocomposite of chitosan-ZnO/ PANI for supercapacitor electrodes. The ternary composite was synthesized by *in-situ* polymerization in the presence of chitosan-ZnO composite and electrochemical performances were evaluated in 1 M aqueous H$_2$SO$_4$ electrolyte within the potential range of 0.2–1.0 V with CV, GCD, and EIS measurements. The ternary chitosan-ZnO/ PANI electrode possesses specific capacitance 587.15 F g^{-1} at 175 mA cm^{-2} current density within the potential window of 0–0.8 V. After the incorporation of chitosan in ZnO/PANI nanocomposite, specific capacitance increases up to 25% in the ternary nanocomposite, because of the homogeneous distribution of mesopores facilitates faster diffusion of the electrolyte ions and shorter solid transportation path length [142]. Among the various transition metal oxides for energy storage application, Co$_3$O$_4$ has attracted recent attention due to the extraordinary specific capacitance (3560 F g^{-1}), good stability, and relatively low cost [143,144]. Wang et al. [126] have synthesized ternary nanocomposite of graphene/polyaniline/Co$_3$O$_4$ and investigated the supercapacitor properties with CV, GCD, and EIS in 6 M KOH in the potential window of 0– 0.4 V. The ternary nanocomposite was synthesized by *in-situ* polymerization of PANI on rGO nanosheets followed by synthesis of Co$_3$O$_4$ on rGO/PANI nanocomposite via hydrothermal route. The ternary nanocomposite graphene/polyaniline/Co$_3$O$_4$ possesses 789.7 F g^{-1} specific capacitance at a current density of 1 A g^{-1} and 81.8% of cycling stability at 10 A g^{-1} after 1000 GCD cycles. The good electrochemical performance of rGO/PANI/Co$_3$O$_4$ ternary electrodes comes from the synergistic effects of Co$_3$O$_4$ and PANI along with the beneficial cross-linker graphene nanosheets. MnO$_2$ is also a promising electrode material for the next-generation supercapacitors due to the cycling stability, environmental friendliness, and low cost [145,146]. Ternary nanocomposites of MnO$_2$ nanorods with PANI and GO were reported by Han et al. [127], where PANI was synthesized from *in-situ* polymerization process on GO nanosheets and presynthesized MnO$_2$ nanorods were added into PANI GO suspension followed by the addition of aniline for the second time. The ternary nanocomposites were finally ready after polymerization of PANI for two times. The ternary MnO$_2$/rGO/PANI electrode exhibits 70% increase in specific capacitance (512 F g^{-1}) after incorporation of MnO$_2$ as compared to the binary rGO/PANI electrode. The MnO$_2$/rGO/PANI electrode possesses outstanding cycling stability of 97% specific capacitance after 5000 cycles. On the other hand, rGO, MnO$_2$, and PANI ternary nanocomposites were synthesized by *in-situ* polymerization of PANI on the MnO$_2$-garnished graphene sheets by Mu group [147]. MnO$_2$/rGO nanocomposites were prepared simultaneously via hydrothermal route from MnO$_2$ precursors and GO. In the electrochemical tests for energy storage, rGO/MnO$_2$/PANI ternary nanocomposite exhibit specific capacitance of 395 F g^{-1} at current density of 10 mA cm^{-2} in 1 M H$_2$SO$_4$ electrolyte and 92% of cycling stability over 1200 CV cycles at a scan rate of 100 mV S^{-1}. The graphene–MnO$_2$–polyaniline-based ternary composites are considered as promising electrode materials to fabricate the next generation of hybrid supercapacitors [132,145,148–151].

Cobalt ferrite (CoFe$_2$O$_4$) is a source of pseudocapacitors according to Deng et al. [152], but the common problem associated with the metal oxides is low conductivity. To overcome the low conductivity of CoFe$_2$O$_4$ for utilizing the pseudocapacitive properties, Wang et al. [128] have synthesized composites with PANI and rGO. The ternary nanocomposites of cobalt ferrite/graphene/PANI are prepared by Wang's group by loading cobalt ferrite on rGO nanosheets hydrothermally with simultaneous reduction of GO followed by *in-situ* polymerization of PANI. The ternary nanocomposite possesses specific capacitance of 1133.3 F g^{-1} (1 mV s^{-1}) and 767.7 F g^{-1} (0.1 A g^{-1}), respectively. In two-electrode system, the ternary cobalt ferrite/graphene/PANI electrode exhibits specific capacitance of 716.4 F g^{-1} and 392.3 F g^{-1} at a scan rate of 1 mV s^{-1} and a current density of 0.1 A g^{-1}, respectively, as compared to CoFe$_2$O$_4$ (39.4 F g^{-1}), PANI (24.6 F g^{-1}), graphene (171.5 F g^{-1}), binary nanocomposite of CoFe$_2$O$_4$ and graphene (183.2 F g^{-1}), CoFe$_2$O$_4$ and PANI (30.0 F g^{-1}), and

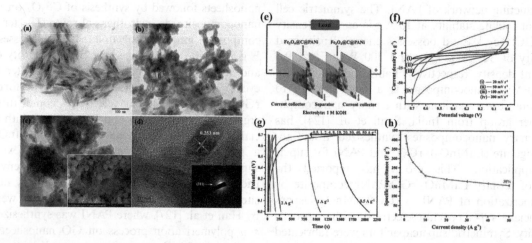

FIGURE 19.8 Transmission electron microscope (TEM) images of (a) Fe_3O_4, (b) Fe_3O_4@C, and (c) Fe_3O_4@C@PANI. Scale bars are all 100 nm. Inset of (c) is a corresponding scanning electron microscope (SEM) image, with a scale bar of 500 nm, (d) Representative high-resolution TEM image of Fe_3O_4, with a scale bar of 10 nm. Inset is the corresponding Selected-area electron diffraction (SAED) patterns. (e) Schematic diagram of a symmetric Fe_3O_4@C@PANI supercapacitor device. (f) CV curves of the device at various scan rates, (g) GCD pattern of the device at different current densities and (h) specific capacitance vs. current density plot of Fe_3O_4@C@PANI (reprinted with permission from ref. [123], copyright (2018) Springer Nature).

graphene and PANI ($578.8 \ F \ g^{-1}$). The elevated capacitive response of the ternary nanocomposites is due to the interaction between the molecular chain of PANI and graphene from electrostatic interaction, $\pi - \pi$ stacking, and hydrogen bonding. $CoFe_2O_4$ nanoparticle provides pseudocapacitance with fast reversible redox reaction in charge storage processes and also prevents the restacking of graphene by decorating itself on the rGO nanosheets during the hydrothermal process. The ternary cobalt ferrite/rGO/PANI electrode is highly reversible in repetitive charge–discharge measurements with 96.6% (5000 cycles) of cycling stability attributed to the well-designed structures of the nanocomposite and synergetic effects between the three components.

19.4.2.1.3 PANI and Graphene-based Ternary Nanocomposites

PANI is a promising pseudocapacitive positive electrode material because of its relatively high electrical conductivity and graphene is considered as an excellent EDLC component due to the high available surface area, mechanical strength, and excellent conductivity due to zero band gap. To prevent the mechanical degradation of PANI and agglomeration of rGO during practical application, Shen et al. [129] have synthesized ternary nanocomposite of PANI with sulfonated graphene and multi-walled CNTs by interfacial polymerization method.

Figure 19.9a shows the schematic representation of ASC device based on a sGNS/cMWCNTs/PANI composite as the positive electrode and activated porous graphene (aGNS) as the negative electrode in 1 M H_2SO_4 electrolyte. Moreover, asymmetric supercapacitor devices have been designed with sGNS/cMWCNT/PANI as cathode

and AC as anode using 1 M H_2SO_4 electrolyte. The energy density vs. power density plot of the symmetric and asymmetric devices is shown in Figure 19.9b. The asymmetric device exhibits higher energy density as well as power density than that of the symmetric device, as observed from the Ragone plot, which is attributed to the larger potential window of the sGNS/cMWCNT/PANI//aGNS asymmetric device. The sGNS/cMWCNT/PANI//aGNS asymmetric device delivers energy density of $20.5 \ Wh \ kg^{-1}$ and power density of $25 \ kW \ kg^{-1}$. The superior electrochemical performance of the asymmetric device is attributed to the unique architecture of the sGNS/cMWCNT/PANI-positive electrode with excellent interfacial contact providing better migration of electrolyte ions, shorter diffusion path, and the porous structure of the sGNS-negative electrode provides higher storage ability with larger potential window of 1.6 V. The asymmetric device also possesses excellent cycle life of 91% after 5000 cycles (Figure 19.9c). The synergetic effects and $\pi - \pi$ interaction between PANI and sGNS/cMWCNTs can accommodate the volumetric changes of PANI during rapid cycling due to the good elasticity of the ternary electrode along with the porous structure of the negative electrode providing enhanced cycling stability of the asymmetric device. On the other hand, ternary graphene/Au/PANI (GAP) nanocomposites have been designed by Wei group as supercapacitor electrode [130]. They synthesized Au nanoparticles on rGO sheets by hydrothermal method and then *in-situ* polymerization of PANI to coat the rGO/Au nanocomposite. The ternary graphene/Au/PANI electrode possesses specific capacitance of $572 \ F \ g^{-1}$ at current density of $0.1 \ A \ g^{-1}$ in saturated Ag/AgCl reference electrode in 2 M H_2SO_4 electrolyte. 88.54% of cycling stability of the ternary electrode after

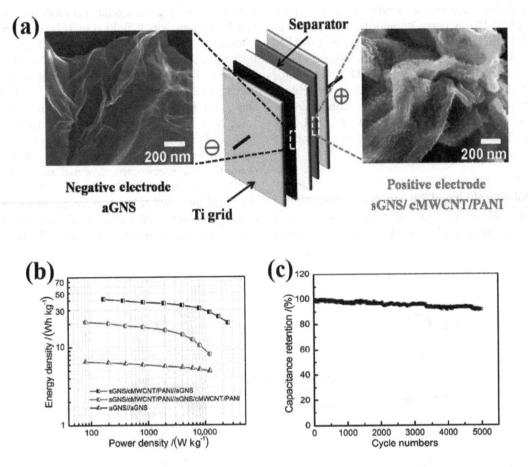

FIGURE 19.9 (a) Schematic representation of Asymmetric Supercapacitor (ASC) device in 1 M H_2SO_4 electrolyte with sGNS/ cMWMCNT/PANI nanocomposite as positive electrode and aGNS as negative electrode, (b) energy density vs. current density plots of symmetric and asymmetric devices, and (c) cycling stability measurements of the asymmetric device in 1 A g^{-1} current density for 5000 cycles (reprinted with permission from ref. [129], copyright (2013) American Chemical Society).

10,000 GCD cycles suggests extraordinary electrochemical behavior of the graphene/Au/PANI electrode due to the synergetic effects of the three components.

Another ternary nanocomposite of PANI and graphene has been reported by Lin et al. [131] with tin disulfide (SnS_2) exhibiting specific energy of 69.53 Wh kg^{-1} and specific power of 575.46 W kg^{-1}. Binary nanocomposite of SnS_2 and rGO was synthesized by hydrothermal route. The ternary nanocomposites can be synthesized by *in-situ* polymerization of aniline monomers on a massage cushion-like SnS_2/rGO binary nanocomposite by varying the mass percentage of PANI in the ternary nanocomposite and obtained maximum capacitive response for 60% mass load of PANI in the ternary SnS_2/rGO/PANI nanocomposite electrode. The ternary nanocomposite exhibits specific capacitance of 1021.67 F g^{-1} and cycling stability of 78% after 5000 consecutive GCD cycles at current density of 1 A g^{-1} in 6 M KOH electrolyte. The elevated capacitive behavior of the ternary SnS_2/rGO/PANI nanocomposite electrodes are ascribed to (i) the surface interaction between PANI nanofiber and SnS_2/rGO nanosheets and (ii) the

homogeneously distributed electron transfer paths provided by the conductive network of PANI.

19.4.2.2 PEDOT–PSS-based Ternary Nanocomposite

To overcome the poor conductivity of metal oxides, Hou et al. [132] synthesized ternary nanocomposites with MnO_2, CNT, and PEDOT–poly(4-styrenesulfonate) (PSS). For efficient utilization of the properties of MnO_2, binary nanocomposites were synthesized with conducting polymers and CNTs [145,148–151]. To achieve high cycling stability and overcome the mechanical instability of MnO_2 and conducting polymers, ternary nanocomposites were designed with CNTs. MnO_2 functionalized few-walled CNTs (fFWCNTs) were synthesized by functionalizing oxygenated groups in the large surface area of CNT followed by addition of the MnO_2 precursors and commercially available PEDOT–PSS conducting polymer was incorporated in the composites for effective utilization of the properties and potentials of the three components. The electrochemical performances were analyzed in 1 M Na_2SO_4 electrolyte and compared with MnO_2 electrode,

MnO_2/PEDOT–PSS composite electrode, and MnO_2/fFWNT/PEDOT–PSS ternary composite electrode. The MnO_2/fFWNT/PEDOT–PSS ternary electrode exhibited specific capacitance of 427 F g^{-1}, excellent rate capability, and 99% of cycling stability after 1000 GCD cycles. All the electrochemical behavior demonstrated that the charge storage properties of MnO_2 are effectively utilized in the ternary nanocomposites with fFWNTs and PEDOT-PSS. Such ternary electrodes are also considered as a promising candidate for the next-generation ESs.

19.4.2.3 Polypyrrole-based Ternary Nanocomposites

PPy, the potential pseudocapacitive conducting polymer possesses wide potential window (1.8 V), specific capacitance

(220–450 F g^{-1}), good mechanical properties, solubility, and tunable conductivity [95]. Li et al. [133] have reported a very interesting way to synthesize PPy-based ternary nanocomposite of core-double-shell structure with CNTs and MnO_2. Figure 19.10a depicts the schematic that illustrates the synthesis process of ternary CNT@PPy@MnO_2 nanocomposites, and morphology of the core-double-shell structure from high-resolution TEM images is displayed in Figure 19.10b. The ternary CNT@PPy@MnO_2 nanocomposites were synthesized by considering CNT as a framework and depositing the pseudocapacitive materials, that is, PPy and MnO_2 on it. On the conductive network of CNT, PPy was deposited by *in-situ* polymerization method followed by decorating a MnO_2 shell on the PPy shell and therefore it is a two core-shell structure. The outer MnO_2 layer provides

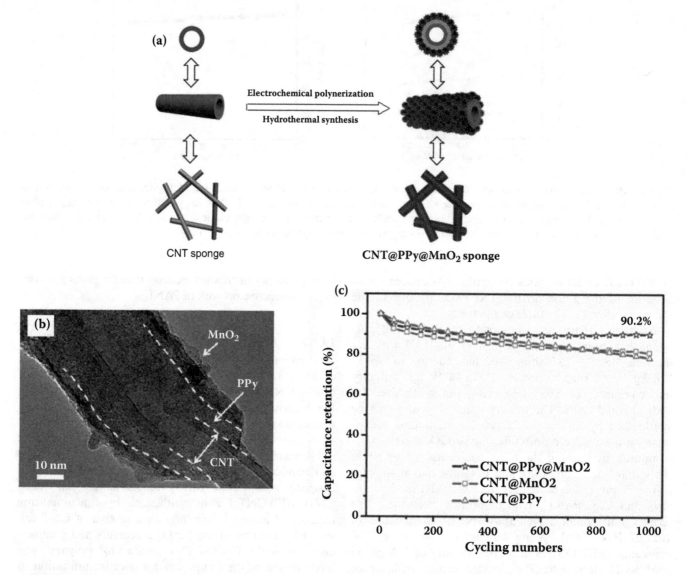

FIGURE 19.10   (a) Schematic representation for synthesis of CNT@PPy@MnO_2 core-double-shell sponge, (b) TEM image of CNT@PPy@MnO_2 core-double-shell structure, and (c) cycling stability of the synthesized electrodes for 1000 cycles at 100 mV s^{-1} scan rate (reprinted with permission from ref. [133], copyright (2014) American Chemical Society).

enhanced surface area and the intermediate PPy layer maintains good interfaces in between for the movement of electrolyte ions.

The CNT@PPy@MnO$_2$ core-double-shell sponge provides highest specific capacitance of 325 F g^{-1} at a scan rate of 2 mV s^{-1}. The cycling stability of the synthesized electrodes were evaluated for 1000 CV cycles at a scan rate of 100 mV s^{-1} (Figure 19.10c). The CNT@PPy@MnO$_2$ (90%) ternary electrode offers higher cycling stability after 1000 cycles at a scan rate of 100 mV s^{-1} than that for the CNT@MnO$_2$ (81%) electrode and CNT@PPy (78.5%) electrode. The enhanced electrochemical performance of the ternary sponge is attributed to the porous conductive network of CNTs that maintains the structural stability by reducing the internal resistance and to the synergetic response of PPy and MnO$_2$.

19.5 MOS$_2$-RGO@PPYNTS NANOCOMPOSITE FOR HIGH-PERFORMANCE SUPERCAPACITOR ELECTRODE

For effective utilization of surface area and to prevent the agglomeration of rGO nanosheets, binary nanocomposite of MoS$_2$/rGO was prepared from layer-by-layer self-assembly of negatively charged GO nanosheet with positively charged MoS$_2$ nanosheets and solvothermal

reduction of GO [32]. Ternary nanocomposites of MoS$_2$–rGO/PPyNTs were synthesized by adding the previously prepared PPy nanotubes (PPyNTs) in binary MoS$_2$-GO heterostructures and solvothermal reduction of GO at 200°C for 24 h. The MOS$_2$–rGO heterostructure provides high surface area, good electronic conductivity, and more electro-active sites from increased interfaces. rGO–PPyNTs, MoS$_2$–PPyNTs, and PPyNTs electrodes were also prepared on Indium tin oxide (ITO) coated substrate to compare their electrochemical performances.

19.5.1 SCHEMATIC REPRESENTATION

Schematic diagram of ternary MoS$_2$–rGO@PPyNTs fabrication process is depicted in Figure 19.11. The MoS$_2$–rGO heterostructures were obtained from the layer-by-layer assembly process in between negatively charged GO and positively charged MoS$_2$ nanosheets after exfoliation followed by annealing. The ternary nanocomposites of MoS$_2$–rGO@PPyNTs were synthesized by adding PPyNTs into MoS$_2$–GO heterostructures and solvothermal reduction of GO.

19.5.2 MORPHOLOGICAL CHARACTERIZATION

Morphology of the MoS$_2$, rGO, MoS$_2$–rGO, PPyNTs, MoS$_2$–rGO@PPyNTs was evaluated with TEM and

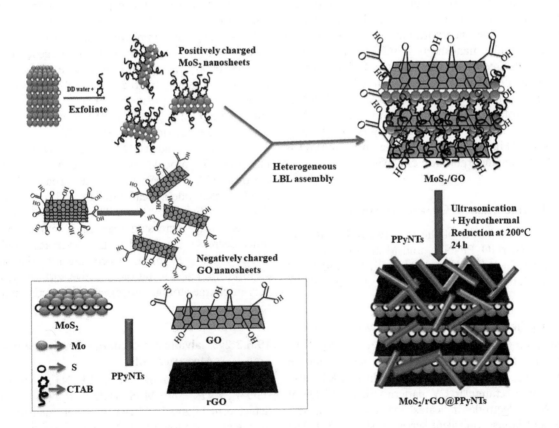

FIGURE 19.11 Schematic representation for synthesis of MoS$_2$–GO and ternary MoS$_2$–rGO@PPyNTs nanocomposite (reprinted from ref. [32] with permission from Elsevier).

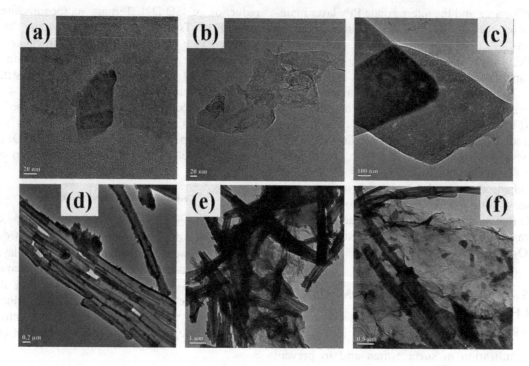

FIGURE 19.12 TEM micrographs of (a) MoS$_2$, (b) rGO, (c) MoS$_2$/rGO, (d) PPyNTs, (e) rGO/PPyNTs, and (f) MoS$_2$–rGO/PPyNTs (reprinted from ref. [32] with permission from Elsevier).

depicted in Figure 19.12. The single-layer MoS$_2$ and rGO nanosheets are observed from Figures 19.12a and b, respectively. After the incorporation of MoS$_2$ in rGO nanosheets, the folded nanosheets of rGO are turned into plane nanosheets free of wrinkles (Figure 19.12c). Micrograph in Figure 19.12d is corresponding to PPyNTs where the nanotubes are observed. rGO nanosheets and PPyNTs are observed from Figure 19.12e for the rGO/PPyNTs nanocomposite. In the ternary MoS$_2$–rGO@PPyNTs nanocomposite (Figure 19.12f), agglomeration of rGO is minimized after the incorporation of MoS$_2$.

19.5.3 Electrochemical Characterizations

Electrochemical properties of the ternary MoS$_2$–rGO@PPyNTs, PPyNTs, rGO–PPyNTs, and MoS$_2$–PPyNTs electrodes were conducted and compared with CV and GCD in 1 M KCl solution in the potential range of −0.2 to 0.6 V.

19.5.3.1 Cyclic Voltammetry (CV) Analysis

Cyclic voltammograms of MoS$_2$–rGO@PPyNTs, PPyNTs, rGO–PPyNTs, MoS$_2$–PPyNTs electrodes are shown in Figure 19.13(i) at a scan rate of 20 mV s^{-1} in 1 M KCl electrolyte. The nonrectangular CV pattern (Figure 19.13) of pure PPyNTs electrode indicates the pseudocapacitive properties and larger equivalent series resistance [32]. The rectangular geometry in the CV curve is observed upon incorporation of rGO in rGO–PPyNTs electrode indicating

the contribution of EDL capacitance. Larger area of CV curve for MoS$_2$–PPyNTs electrode as compared to rGO–PPyNTs suggests improved capacitive response after incorporation of MoS$_2$ due to the homogeneous dispersion of the binary nanocomposites. The highest capacitive response is shown by the ternary MoS$_2$–rGO@PPyNTs electrode with highest surface area (625.456 m^2 g^{-1} specific surface area) leading to high current vs. potential response with enhanced capacitive response [32].

Here MoS$_2$ and rGO are the EDLC contributors and PPyNTs provide pseudocapacitance in the ternary electrode [32]. CV patterns of the ternary MoS$_2$–rGO@-PPyNTs electrode at scan rates from 5 to 100 mV s^{-1} are depicted in Figure 19.13(ii). The current–voltage response increases with increasing scan rates, but specific capacitance decreases attributed to the lower accessibility of the electrolyte ions at higher sweep rate into the bulk of the electrode and as a result outer active surface area can take part in charge storage processes of the active material due to the time constraint.

19.5.3.2 Galvanostatic Charge–Discharge (GCD) Measurements

The GCD measurements of PPyNTs, MoS$_2$–PPyNTs, rGO–PPyNTs, and MoS$_2$–rGO@PPyNTs electrodes are depicted in Figure 19.14(i) at a current density of 1 A g^{-1} in potential window of −0.2 to 0.6 V. Specific capacitance (C_{sp}) and coulombic efficiency (η) are calculated using the following equation:

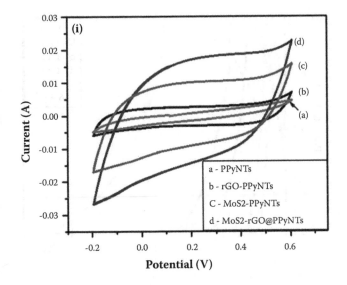 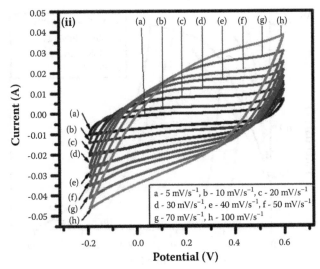

FIGURE 19.13 CV curves of (i) MoS_2–rGO@PPyNTs, rGO–PPyNTs, MoS_2–PPyNTs, and PPyNTs electrodes at a scan rate of 20 mV s^{-1} and (ii) ternary MoS_2–rGO@PPyNTs at different scan rates from 5 to 100 mV s^{-1} (reprinted from ref. [32] with permission from Elsevier).

$$C_{sp} = \frac{I \times \Delta T_d}{m \times \Delta V} \tag{19.1}$$

$$\eta = \frac{\Delta T_d}{\Delta T_c} \times 100 \tag{19.2}$$

where I, ΔT_d, ΔT_c, m, and ΔV indicate discharge current, galvanostatic discharge period, charging period, mass of the active material, and potential range of discharge process, respectively. The GCD pattern of PPyNTs electrode deviates from perfect linearity owing to the faradic charge transfer processes at the electrode/electrolyte interfaces [32]. The rGO–PPyNTs electrode exhibits a nonlinear pattern may be due to the agglomeration of rGO in the PPyNTs matrix and pseudocapacitive dominance from PPyNTs. MoS_2–PPyNTs electrode shows perfectly linear triangle-shaped GCD pattern suggesting the EDLC dominance over pseudocapacitance as potential changes linearly with time. The ternary MoS_2–rGO@PPyNTs electrode shows longest discharge duration and maximum capacitive response. The first two parts are perfectly linear in the discharge curve with two slopes corresponding to synergetic effects of MoS_2 and rGO. The distinct plateaus suggested the electrolyte ion rich/poor phase or due to the intercalation and de-intercalation of the electrolyte ions into the electrode material [32]. PPyNTs, rGO–PPyNTs, MoS_2–PPyNTs, and MoS_2–rGO@PPyNTs electrodes exhibit specific capacitance of 313.75, 625, 1249.36, and 1561.25 F g^{-1}, respectively, at current density of 1 A g^{-1}.

The ternary MoS_2–rGO@PPyNTs/ITO electrodes possess a large specific capacitance owing to: (i) pseudocapacitance from PPyNTs; (ii) number of interfaces from MoS_2/rGO self-assembly, and (iii) higher surface area. The charge–discharge curves of MoS_2–rGO@PPyNTs, MoS_2–PPyNTs, rGO–PPyNTs, and PPyNTs at current densities of 1–6 A g^{-1} are shown in Figure 19.14 (ii–v). The discharge duration of MoS_2–rGO@PPyNTs, rGO–PPyNTs, MoS_2–PPyNTs, and PPyNTs electrodes decreases with increasing current densities and the specific capacitance, which is due to the participation of outer layers in charge storage. The rate capability of the MoS_2–rGO@PPyNTs, MoS_2–PPyNTs, rGO–PPyNTs, and PPyNTs electrodes are shown in Figure 19.14(vi) with different current densities. The PPyNTs, rGO–PPyNTs, MoS_2–PPyNTs, and MoS_2–rGO@PPyNTs electrodes exhibit specific capacitance of 112.5, 251.25, 823.7, and 1145.2 F g^{-1} at the current density of 6 A g^{-1} and retain 35.86%, 40.2%, 65.92%, and 73.35% of the initial specific capacitance of 1-fold current density, respectively, suggesting good rate capability of the electrode. Almost 100% coulombic efficiency for the ternary electrode suggests the high reversibility of the electrode kinetics and efficient charge–discharge process.

Ragone plot (energy density vs. power density) is displayed in Figure 19.15(a). Energy and power density are calculated using the following equations:

Energy density,

$$E = \frac{1}{2} C_{sp} (\Delta V)^2 \tag{19.3}$$

Power density,

$$P = \frac{E \times 3600}{T_{dis}} \tag{19.4}$$

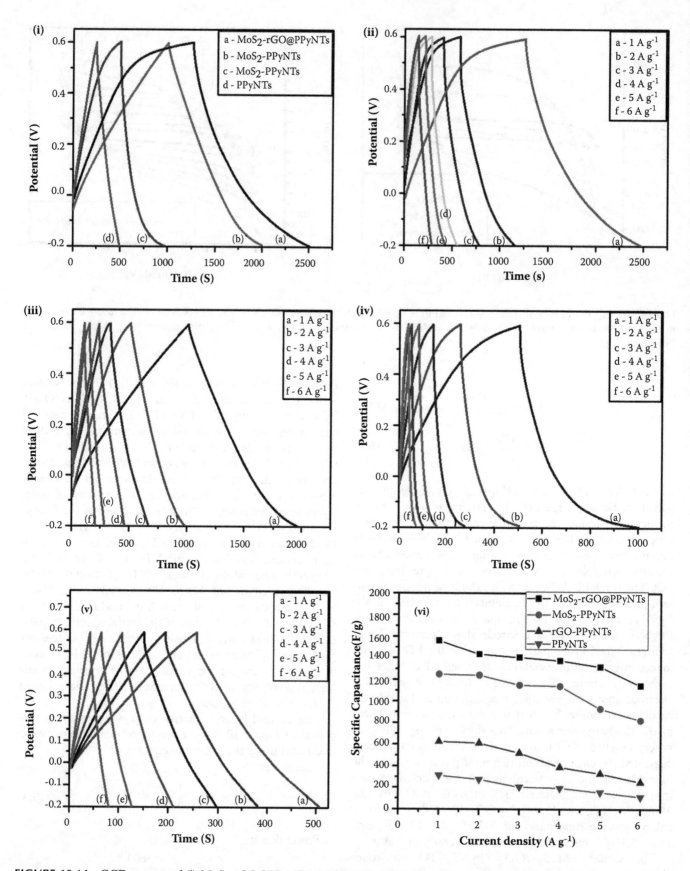

FIGURE 19.14 GCD pattern of (i) MoS$_2$–rGO@PPyNT, rGO/PPyNT, MoS$_2$/PPyNT, and PPyNT at current density of 1 A g^{-1}, (ii) MoS$_2$–rGO@PPyNTs, (iii) MoS$_2$–PPyNTs, (iv) rGO–PPyNTs, and (v) PPyNTs electrode at different current densities and (vi) specific capacitance with increasing current density (reprinted from ref. [32] with permission from Elsevier).

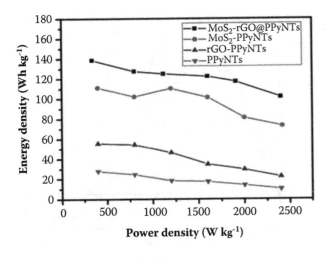
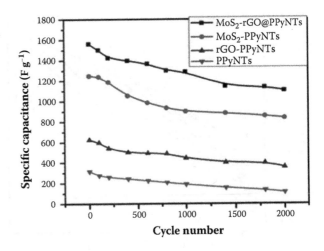

FIGURE 19.15 (a) Energy density vs. power density (Ragone plot) and (b) specific capacitance vs. cycle numbers (reprinted from ref. [32] with permission from Elsevier).

where C_{sp}, ΔV, and T_{dis} are the specific capacitance, potential drop, and discharge period, respectively. PPyNTs/ITO electrode exhibits energy density of 27.89 Wh kg^{-1} and power density of 400.032 W kg^{-1} at current density of 1 A g^{-1} and energy density of 10.08 Wh kg^{-1} with power density of 2400.192 W kg^{-1} at current density of 6 A g^{-1}. MoS$_2$/PPyNTs electrode possesses energy density of 111.062 Wh kg^{-1} with power density of 400.029 W kg^{-1}, and energy density of 73.16 Wh kg^{-1} with power density of 2416.34 W kg^{-1} at current densities of 1 and 6 A g^{-1}, respectively. Ternary MoS$_2$–rGO@PPyNTs electrode exhibits enormous power range of 399.94 to 2400.15 W kg^{-1} and energy density of 101.78–138.76 Wh kg^{-1} at one- to six-fold of current density.

19.5.3.3 Cycling Stability Study

Cycling stability is a crucial parameter for ES electrodes for their possible application in practical use. Pseudocapacitive materials are rich in specific capacitance and energy density by scarifying cycle life and EDLC material exhibit extremely high cycling stability at the cost of specific capacitance. Cyclic stability of PPyNTs, rGO–PPyNTs, MoS$_2$–PPyNTs, and MoS$_2$–rGO@PPyNTs electrodes was studied with GCD measurements at current density of 1 A g^{-1} up to 2000 cycles. The calculated value of specific capacitance vs. cycle numbers is displayed in Figure 19.15(b). PPyNTs, rGO–PPyNTs, and MoS$_2$–PPyNTs electrode possess capacitive retention of 30.5%, 54.24%, and 75.2% after 2000 cycles. The ternary MoS$_2$–rGO@PPyNTs electrode exhibits cycling stability of 76.68% after 2000 cycles attributed to the excellent cycle life of MoS$_2$ and minimum degradation due to the MoS$_2$–rGO heterostructure. PPyNT electrode possesses poor cycling stability in faradic processes leading to volume instability and incorporation of rGO enhances the cycling performances of the rGO–PPyNTs

electrode. Highest cycling stability is observed for the ternary MoS$_2$–rGO@PPyNTs electrodes corresponding to the synergetic effects of MoS$_2$–rGO heterostructure and PPyNTs.

To understand the operating potential range of MoS$_2$–rGO@PPyNT electrode in 1 M KCl solution CV was carried out in potential window of −1 to 1.5 V [32]. A pair of redox peaks has appeared at 1.32 V (oxidation peak) and −0.34 V (reduction peak). The capacitive potential range (CPR) is the upper (oxidation) and lower (reduction) potential limit. In CPR, only the electrodes exhibit good capacitive properties [32].

For ternary MoS$_2$–rGO@PPyNTs electrodes −0.3 to 1.3 V is considered as CPR. CV and charge–discharge cycles were carried out in CPR to check the stability of ternary MoS$_2$–rGO@PPyNTs electrodes. CV patterns displayed in Figure 19.16(i) suggest the stability of ternary MoS$_2$–rGO@PPyNTs electrode in the potential window of −0.3 to 1.3 V. GCD (Figure 19.16(ii)) curves of the ternary electrode at current density of 1 A g^{-1} in the potential window of −0.3 to 1.3 V are almost symmetric and maintaining specific capacitance of 1561 F g^{-1} for all potential windows of (−0.3–0.6 V), (−0.3–0.7 V), (−0.3–1.1 V), (−0.3–1.2 V), (−0.3–1.3 V), (−0.3–1.4 V), (−0.3–1.5 V), and (−0.3–1.6 V). GCD measurements of the ternary MoS$_2$–rGO@PPyNTs/ITO electrode within −0.3 to 1.3 V at current density range of 1–6 A g^{-1} as depicted in Figure 19.16(iii). MoS$_2$–rGO@PPyNTs/ITO electrode exhibits capacitive retention of 71.4% at current density of 6 A g^{-1}. Larger potential window gives higher energy at the cost of capacitive retention, attributes to easy degradation of the electrode at higher potential range.

The cycling stability performances of MoS$_2$–rGO/PPyNTs electrode were evaluated with GCD measurements within the potential range of −0.3 to 1.3 V at current density of 10 A g^{-1} for 10,000 consecutive cycles. The calculated

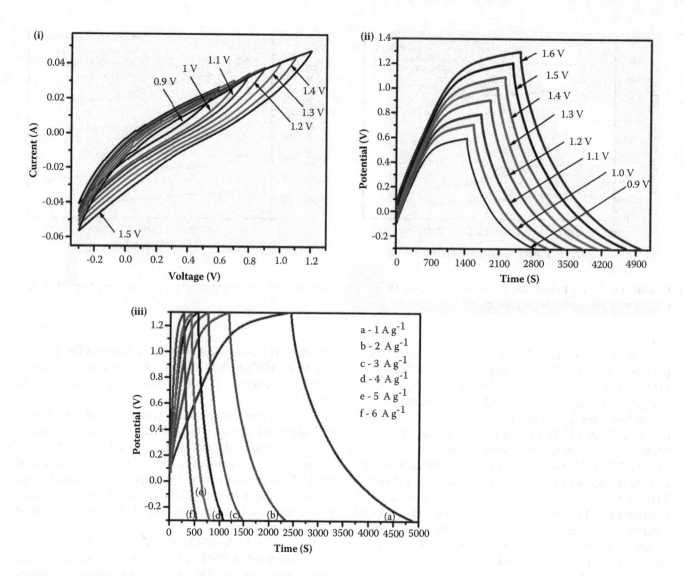

FIGURE 19.16 (i) CV patterns of MoS$_2$–rGO@PPyNTs electrode at different potential window at a scan rate of 50 mV s^{-1} (inset the CV pattern within CPR), (ii) GCD pattern of MoS$_2$–rGO@PPyNTs electrode at different potential window at current density of 1 A g^{-1}, and (iii) GCD pattern of MoS$_2$–rGO@PPyNTs electrode at current densities 1–6 A g^{-1} (reprinted from Ref. [32] with permission from Elsevier).

capacitive retention vs. GCD cycle numbers are depicted in Figure 19.17. The ternary MoS$_2$–rGO/PPyNTs electrode offers 100% of cycling stability up to first 82 cycles and 83.21%, 75.48%, and 72 % of cycling stability after 1000, 2000, and 10,000 GCD cycles, respectively.

The scanning electron microscope (SEM) images of the ternary MoS$_2$–rGO/PPyNTs electrode were captured before and after 10,000 GCD cycles to compare the surface morphology and are shown in Figure 19.18. Before the cycling process, PPyNTs coated the MoS$_2$–rGO-layered structures and after the cycling process, the nanosheets are also visible in the SEM images (Figure 19.18a and b). Pseudo-capacitive materials, PPyNTs, undergo swelling/shrinkage (Figures 19.18c and d) in repetitive GCD cycles causing degradation in capacitive performance

19.5.3.4 Electrochemical Impedance Spectroscopy (EIS) Analysis

EIS is performed to understand the electrode kinetics in terms of faradic and diffusion reactions in the frequency range of 10 μHz–10^5 Hz. The Nyquist plots of the electrodes, that is, MoS$_2$–rGO@PPyNTs, MoS$_2$–PPyNTs, rGO–PPyNTs, and PPyNTs are depicted in Figure 19.19. EIS spectra are showing a straight line in lower frequency side and a semi-circle in high-frequency side.

The intercept of semi-circle at x axis is the equivalent series resistance (R_s), which implies the intrinsic electrode resistance and electrolyte's ionic resistance and diameter of the semi-circle gives charge transfer resistance (R_{ct}) of the electrochemical cell. Nyquist plot clearly indicates that the R_s of rGO–PPyNTs (2.59 Ω) is smaller than that of

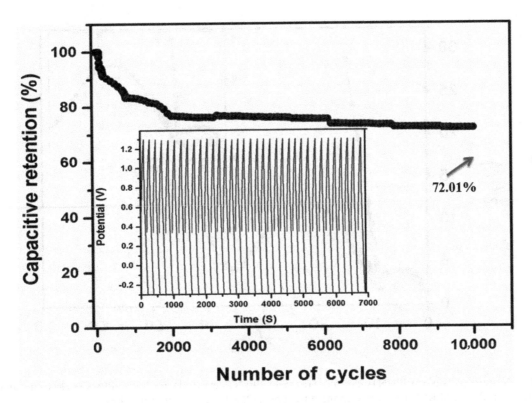

FIGURE 19.17 Capacitive retention vs. GCD cycle number up to 10,000 cycles (reprinted from ref. [32] with permission from Elsevier).

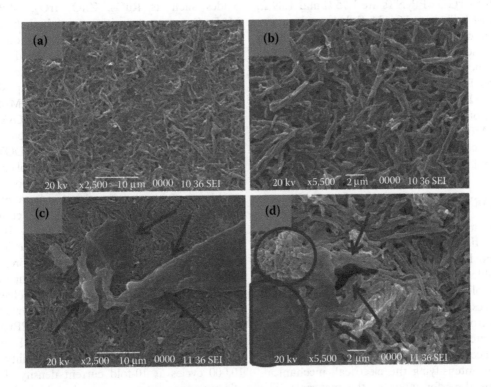

FIGURE 19.18 SEM images of MoS_2–rGO@PPyNTs electrode (a), (b) before GCD cycles and (c), (d) after GCD cycles at two different magnifications (reprinted from ref. [32] with permission from Elsevier).

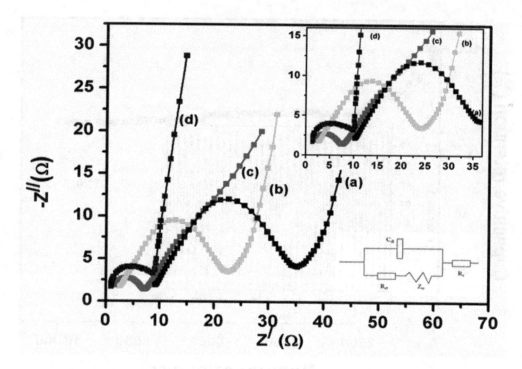

FIGURE 19.19 Nyquist plots of (a) PPyNTs electrode, (b) rGO–PPyNTs electrode, (c) MoS$_2$–PPyNTs electrode, and (d) MoS$_2$–rGO@PPyNTs electrode in 1 M KCl electrolyte (reprinted from ref. [32] with permission from Elsevier).

PPyNTs (7.01 Ω). The calculated values of R_s for MoS$_2$–PPyNTs and MoS$_2$–rGO@PPyNTs are 1.75 Ω and 1.09 Ω, respectively in 1 M KCl electrolyte. MoS$_2$ helps in homogeneous dispersion of all the components in the ternary nanocomposite and rGO reduces the intrinsic electrode resistance. As ionic resistance is constant, PPyNTs, rGO–PPyNTs, MoS$_2$–PPyNTs, and MoS$_2$–rGO@PPyNTs exhibit R_{ct} of 27 Ω, 11 Ω, 4.13 Ω, and 5.4 Ω, respectively. Interestingly, R_{ct} of the ternary electrode is slightly higher as compared to MoS$_2$–PPyNTs electrode, attributed to the presence of three components that hinder the charge flow to some extent [32].

19.6 SUMMARY AND PERSPECTIVES

Supercapacitors as alternate energy storage devices are gaining attention due to their versatile applications in both low- and high-power devices, including e-paper, wearable, bendable/stretchable displays, different environmental and biological sensors, and other health-monitoring devices. It is imperative to fabricate high-performance electrodes and hybrid supercapacitor strategies to overcome the barriers of energy density, power density, mechanical flexibility, capacitive response, and cycle life by intensifying the electrical, mechanical, and electrochemical performance of the materials. The growth of supercapacitor electrode materials has been discussed in this chapter, primarily focusing on the conducting polymer-based ternary nanocomposites with

carbon nanomaterials, such as graphene, CNTs, metal oxides such as RuO$_2$, ZnO, IrO$_2$, MnO$_2$, NiCo$_2$O$_4$, MnFe$_2$O$_4$, and transition metal dichalcogenides like MoS$_2$. The most widely studied conducting polymers PPy, PANI, and PEDOT are outlined in this chapter to discuss the ternary nanocomposites. In particular, recent studies on ternary nanocomposites including PANI/MoS$_2$/rGO, MWCNTs/PANI/MoS$_2$, Fe$_3$O$_4$@activatedcarbon@PANI, LaMnO$_3$/rGO/PANI, and graphene/PANI/Co$_3$O$_4$ for supercapacitor applications are summarized. PPy-, PANI-, and PEDOT-based binary nanocomposites for supercapacitor application have also been discussed. On the basis of the above discussions, it is clear that to overcome the drawback of single materials designing of composite systems by combining two or three suitable components strategically is important to fabricate supercapacitors with high specific capacitance, energy and power density. Furthermore, MoS$_2$–rGO@PPyNTs nanocomposites for hybrid supercapacitors are included and their electrochemical behaviors are discussed with CV, GCD and electrochemical impedance spectroscopy. Ternary MoS$_2$–rGO@PPyNTs electrode exhibits specific capacity of 1561.25 F g^{-1} at current density of 1 A g^{-1}, and cycling stability of 72% after 10,000 cycles at 10-fold current density. The advantages of ternary nanocomposites such as high electrical conductivity, increased electroactive sites, high energy density, and stable structures facilitate in tailoring high-performance supercapacitor electrodes.

Hybrid material design and synthesis strategy have provided a lot of hope in the field of ESs, yet the growth of next-generation supercapacitors is in the initial phase and much improvement in specific capacitance, rate capability, energy density, coulombic efficiency, and cycling stability is needed before commercialization. Some of the issues in the field of supercapacitor technology that need to be further explored are mentioned below:

1. The ternary material should be so designed that the charge transfer resistance (R_s) is minimized. The hierarchical structure of the ternary nanocomposites arrests the movement of ions and electrons at the electrode/electrolyte interface leading to the increase in ESR of the electrode degrading the rate performance.
2. Architecture, morphology, and dimensions of the materials of ternary supercapacitor electrodes show distinct effects on the specific capacitance and charge transfer kinetics. Designing special structures with high specific surface area such as core-shell structures, 3D nanostructures, nanostructured arrays, and heterostructures should be developed to maximize interfacial area to enhance the storage capacity of the electrode.
3. The conducting polymer-based electrodes undergo swelling and shrinkage during the repeated charge–discharge cycles and exhibit poor cycling performance. Coating of conducting polymers with metal oxides and transition metal dichalcogenide nanostructures may prevent the degradation of the conducting polymers in cycling processes.
4. The ternary nanocomposites of conducting polymers with MOFs, COFs, MXenes, MNs, and BP can be fabricated, as these materials exhibit properties such as very high surface area, homogeneous pore distribution, flexibility, wettability, and shorter ion diffusion lengths that are beneficial for supercapacitor electrode materials.
5. New supercapacitor devices such as flow supercapacitors, AC line-filtering supercapacitors, redox electrolyte-enhanced supercapacitor, micro-supercapacitors, metal-ion hybrid supercapacitors, and self-healing supercapacitors can be fabricated with the ternary electrodes to integrate many functions in a single system.

REFERENCES

1. Purkait, T., G. Singh, D. Kumar, M. Singh, and R. S. Dey. High-performance flexible supercapacitors based on electrochemically tailored three-dimensional reduced graphene oxide networks. *Scientific Reports* 8, no. 1 (2018): 640.
2. Chiam, S. L., H. N. Lim, S. M. Hafiz, A. Pandikumar, and N. M. Huang. Electrochemical performance of supercapacitor with stacked copper foils coated with graphene nanoplatelets. *Scientific Reports* 8, no. 1 (2018): 3093.
3. Zhou, Y., P. Jin, Y. Zhou, and Y. Zhu. High-performance symmetric supercapacitors based on carbon nanotube/ graphite nanofiber nanocomposites. *Scientific Reports* 8, no. 1 (2018): 9005.
4. Becker, H. I. Low voltage electrolytic capacitor. U.S. Patent 2,800,616, issued July 23, 1957.
5. Wang, D. W., H. T. Fang, F. Li, Z. G. Chen, Q. S. Zhong, G. Q. Lu, and H. M. Cheng. Aligned titania nanotubes as an intercalation anode material for hybrid electrochemical energy storage. *Advanced Functional Materials* 18, no. 23 (2008): 3787–3793.
6. Zuo, X., Z. Jin, P. Müller-Buschbaum, and Y. J. Cheng. Silicon based lithium-ion battery anodes: A chronicle perspective review. *Nano Energy* 31 (2017): 113–143.
7. Chen, D., F. Zheng, L. Li, M. Chen, X. Zhong, W. Li, and L. Lu. Effect of Li3PO4 coating of layered lithium-rich oxide on electrochemical performance. *Journal of Power Sources* 341 (2017): 147–155.
8. Juodkazis, K., J. Juodkazytė, V. Šukienė, A. Grigucevičienė, and A. Selskis. On the charge storage mechanism at $RuO_2/0.5$ M H_2SO_4 interface. *Journal of Solid State Electrochemistry* 12, no. 11 (2008): 1399–1404.
9. Aida, T., K. Yamada, and M. Morita. An advanced hybrid electrochemical capacitor that uses a wide potential range at the positive electrode. *Electrochemical and Solid-state Letters* 9, no. 12 (2006): A534–A536.
10. Ding, J., H. Wang, Z. Li, K. Cui, D. Karpuzov, X. Tan, A. Kohandehghan, and D. Mitlin. Peanut shell hybrid sodium ion capacitor with extreme energy–power rivals lithium ion capacitors. *Energy & Environmental Science* 8, no. 3 (2015): 941–955.
11. Cherusseri, J., K. S. Kumar, N. Choudhary, N. Nagaiah, Y. Jung, T. Roy, and J. Thomas. Novel mesoporous electrode materials for symmetric, asymmetric and hybrid supercapacitors. *Nanotechnology* 30, no. 20 (2019): 202001.
12. Dong, G., H. Fan, K. Fu, L. Ma, S. Zhang, M. Zhang, J. Ma, and W. Wang. The evaluation of super-capacitive performance of novel g-C_3N_4/PPy nanocomposite electrode material with sandwich-like structure. *Composites Part B: Engineering* 162 (2019): 369–377.
13. Cheng, Y., M. Zhai, M. Guo, Y. Yu, and J. Hu. A novel electrode for supercapacitors: Spicules-like Ni3S2 shell grown on molybdenum nanoparticles doped nickel foam. *Applied Surface Science* 467 (2019): 1113–1121.
14. Luo, G., K. S. Teh, Y. Xia, Y. Luo, S. Li, S. Wang, L. Zhao, and Z. Jiang. A novel three-dimensional spiral CoNi LDHs on Au@ ErGO wire for high performance fiber supercapacitor electrodes. *Materials Letters* 236 (2019): 728–731.
15. Huang, K. J., L. Wang, Y. J. Liu, H. B. Wang, Y. M. Liu, and L. L. Wang. Synthesis of polyaniline/2-dimensional graphene analog MoS2 composites for high-performance supercapacitor. *Electrochimica Acta* 109 (2013): 587–594.
16. Helmholtz, V. H. On the conservation of force. *Annals of Physics (Leipzig)* 89 (1853): 21.
17. Stern, O. The theory of the electrolytic double-layer. *Z. Elektrochem.* 30, no. 508 (1924): 1014–1020.
18. Grahame, D. C. The electrical double layer and the theory of electrocapillarity. *Chemical Reviews* 41, no. 3 (1947): 441–501.
19. Zhang, L., T. D. Zhang, R. Gao, D. Y. Tang, J. Y. Tang, and Z. L. Zhan. Preparation and characterization of mesoporous carbon materials of chinese medicine residue with high specific surface areas. *Chemical Engineering Transactions* 55 (2016): 79–84.

20. Kötz, R., M. Hahn, and R. Gallay. Temperature behavior and impedance fundamentals of supercapacitors. *Journal of Power Sources* 154, no. 2 (2006): 550–555.

21. Augustyn, V., P. Simon, and B. Dunn. Pseudocapacitive oxide materials for high-rate electrochemical energy storage. *Energy & Environmental Science* 7, no. 5 (2014): 1597–1614.

22. Conway, B. E. *Electrochemical supercapacitors: Scientific fundamentals and technological applications.* Springer Science & Business Media, New York, 2013.

23. Conway, B. E., V. Birss, and J. Wojtowicz. The role and utilization of pseudocapacitance for energy storage by supercapacitors. *Journal of Power Sources* 66, no. 1–2 (1997): 1–14.

24. Lu, P., D. Xue, H. Yang, and Y. Liu. Supercapacitor and nanoscale research towards electrochemical energy storage. *International Journal of Smart and Nano Materials* 4, no. 1 (2013): 2–26.

25. Bryan, A. M., L. M. Santino, Y. Lu, S. Acharya, and J. M. D'Arcy. Conducting polymers for pseudocapacitive energy storage. *Chemistry of Materials* 28, no. 17 (2016): 5989–5998.

26. Rauda, I. E., V. Augustyn, B. Dunn, and S. H. Tolbert. Enhancing pseudocapacitive charge storage in polymer templated mesoporous materials. *Accounts of Chemical Research* 46, no. 5 (2013): 1113–1124.

27. Liu, H., H. Zhou, L. Chen, Z. Tang, and W. Yang. Electrochemical insertion/deinsertion of sodium on NaV_6O_{15} nanorods as cathode material of rechargeable sodium-based batteries. *Journal of Power Sources* 196, no. 2 (2011): 814–819.

28. Burke, A. R&D considerations for the performance and application of electrochemical capacitors. *Electrochimica Acta* 53, no. 3 (2007): 1083–1091.

29. Rangom, Y., X. Tang, and L. F. Nazar. Carbon nanotube-based supercapacitors with excellent ac line filtering and rate capability via improved interfacial impedance. *ACS Nano.* 9, no. 7 (2015): 7248–7255.

30. Zhu, Y., X. Ji, C. Pan, Q. Sun, W. Song, L. Fang, Q. Chen, and C. E. Banks. A carbon quantum dot decorated RuO_2 network: Outstanding supercapacitances under ultrafast charge and discharge. *Energy & Environmental Science* 6, no. 12 (2013): 3665–3675.

31. Muzaffar, A., M. B. Ahamed, K. Deshmukh, and J. Thirumalai. A review on recent advances in hybrid supercapacitors: Design, fabrication and applications. *Energy & Environmental Science* 101 (2019): 123–145.

32. Sarmah, D., and A. Kumar. Layer-by-layer self-assembly of ternary MoS 2-rGO@ PPyNTs nanocomposites for high performance supercapacitor electrode. *Synthetic Metals* 243 (2018): 75–89.

33. Theerthagiri, J., K. Karuppasamy, G. Durai, A. Rana, P. Arunachalam, K. Sangeetha, P. Kuppusami, and H. S. Kim. Recent advances in metal chalcogenides (MX; X= S, Se) nanostructures for electrochemical supercapacitor applications: A brief review. *Nanomaterials* 8, no. 4 (2018): 256.

34. Largeot, C., C. Portet, J. Chmiola, P. L. Taberna, Y. Gogotsi, and P. Simon. Relation between the ion size and pore size for an electric double-layer capacitor. 130, no. 9 (2008): 2730–2731.

35. Conway, B. E. *Electrochemical Supercapacitors.* Kluwer Academic/Plenum Press, New York, 1999.

36. Lee, H., M. S. Cho, I. H. Kim, J. D. Nam, and Y. Lee. RuOx/polypyrrole nanocomposite electrode for electrochemical capacitors. *Synthetic Metals* 160, no. 9–10 (2010): 1055–1059.

37. Zhang, Y., H. Feng, X. Wu, L. Wang, A. Zhang, T. Xia, H. Dong, X. Li, and L. Zhang. Progress of electrochemical capacitor electrode materials: A review. *International Journal of Hydrogen Energy* 34, no. 11 (2009): 4889–4899.

38. Katsnelson, M. I. Graphene: Carbon in two dimensions. *Materials Today* 10, no. 1–2 (2007): 20–27.

39. Wang, G., L. Zhang, and J. Zhang. A review of electrode materials for electrochemical supercapacitors. *Chemical Society Reviews* 41, no. 2 (2012): 797–828.

40. Du, X., C. Wang, M. Chen, Y. Jiao, and J. Wang. Electrochemical performances of nanoparticle Fe3O4/activated carbon supercapacitor using KOH electrolyte solution. *The Journal of Physical Chemistry C* 113, no. 6 (2009): 2643–2646.

41. Farzana, R., R. Rajarao, B. R. Bhat, and V. Sahajwalla. Performance of an activated carbon supercapacitor electrode synthesised from waste compact discs (CDs). *Journal of Industrial and Engineering Chemistry* 65 (2018): 387–396.

42. Wu, G., P. Tan, D. Wang, Z. Li, L. Peng, Y. Hu, C. Wang, W. Zhu, S. Chen, and W. Chen. High-performance supercapacitors based on electrochemical-induced vertical-aligned carbon nanotubes and polyaniline nanocomposite electrodes. *Scientific Reports* 7 (2017): 43676.

43. Aval, L. F., M. Ghoranneviss, and G. Behzadi Pour. High-performance supercapacitors based on the carbon nanotubes, graphene and graphite nanoparticles electrodes. *Heliyon.* 4, no. 11 (2018): e00862.

44. Fischer, U., R. Saliger, V. Bock, R. Petricevic, and J. Fricke. Carbon aerogels as electrode material in supercapacitors. *Journal of Porous Materials* 4, no. 4 (1997): 281–285.

45. Daraghmeh, A., S. Hussain, I. Saadeddin, L. Servera, E. Xuriguera, A. Cornet, and A. Cirera. A study of carbon nanofibers and active carbon as symmetric supercapacitor in aqueous electrolyte: A comparative study. *Nanoscale Research Letters* 12, no. 1 (2017): 639.

46. Yang, H., S. Kannappan, A. S. Pandian, J. H. Jang, Y. S. Lee, and W. Lu. Graphene supercapacitor with both high power and energy density. *Nanotechnology* 28, no. 44 (2017): 445401.

47. Liu, C., Z. Yu, D. Neff, A. Zhamu, and B. Z. Jang. Graphene-based supercapacitor with an ultrahigh energy density. *Nano Letters* 10, no. 12 (2010): 4863–4868.

48. Frackowiak, E. Carbon materials for supercapacitor application. *Physical Chemistry Chemical Physics* 9, no. 15 (2007): 1774–1785.

49. Natter, N., N. Kostoglou, C. Koczwara, C. Tampaxis, T. Steriotis, R. Gupta, O. Paris, C. Rebholz, and C. Mitterer. Plasma-derived graphene-based materials for water purification and energy storage. *C.* 5, no. 2 (2019): 16.

50. Regisser, F., M. A. Lavoie, G. Y. Champagne, and D. Bélanger. Randomly oriented graphite electrode. Part 1. Effect of electrochemical pretreatment on the electrochemical behavior and chemical composition of the electrode. *Journal of Electroanalytical Chemistry* 415, no. 1–2 (1996): 47–54.

51. Béguin, F., K. Szostak, G. Lota, and E. Frackowiak. A self-supporting electrode for supercapacitors prepared by one-step pyrolysis of carbon nanotube/polyacrylonitrile blends. *Advanced Materials* 17, no. 19 (2005): 2380–2384.

52. Momma, T., X. Liu, T. Osaka, Y. Ushio, and Y. Sawada. Electrochemical modification of active carbon fiber

electrode and its application to double-layer capacitor. *Journal of Power Sources* 60, no. 2 (1996): 249–253.

53. Lota, G., B. Grzyb, H. Machnikowska, J. Machnikowski, and E. Frackowiak. Effect of nitrogen in carbon electrode on the supercapacitor performance. *Chemical Physics Letters* 404, no. 1–3 (2005): 53–58.

54. Frackowiak, E., G. Lota, J. Machnikowski, C. Vix-Guterl, and F. Béguin. Optimisation of supercapacitors using carbons with controlled nanotexture and nitrogen content. *Electrochimica Acta* 51, no. 11 (2006): 2209–2214.

55. Leitner, K., A. Lerf, M. Winter, J. O. Besenhard, S. Villar-Rodil, F. Suarez-Garcia, A. Martinez-Alonso, and J. M. D. Tascon. Nomex-derived activated carbon fibers as electrode materials in carbon based supercapacitors. *Journal of Power Sources* 153, no. 2 (2006): 419–423.

56. Kötz, R., and M. J. E. A. Carlen. Principles and applications of electrochemical capacitors. *Electrochimica Acta* 45, no. 15–16 (2000): 2483–2498.

57. Amaral, C., R. Vicente, P. A. A. P. Marques, and A. Barros-Timmons. Phase change materials and carbon nanostructures for thermal energy storage: A literature review. *Renewable and Sustainable Energy Reviews* 79 (2017): 1212–1228.

58. Lee, C., X. Wei, J. W. Kysar, and J. Hone. Measurement of the elastic properties and intrinsic strength of monolayer graphene. *Science* 321, no. 5887 (2008): 385–388.

59. Xia, J., F. Chen, J. Li, and N. Tao. Measurement of the quantum capacitance of graphene. *Nature Nanotechnology* 4, no. 8 (2009): 505.

60. Stoller, M. D., S. Park, Y. Zhu, J. An, and R. S. Ruoff. Graphene-based ultracapacitors. *Nano Letters* 8, no. 10 (2008): 3498–3502.

61. Shi, W., J. Zhu, D. H. Sim, Y. Y. Tay, Z. Lu, X. Zhang, Y. Sharma et al. Achieving high specific charge capacitances in Fe 3 O 4/reduced graphene oxide nanocomposites. *Journal of Materials Chemistry* 21, no. 10 (2011): 3422–3427.

62. Ke, Q., and J. Wang. Graphene-based materials for supercapacitor electrodes–A review. *Journal of Materiomics* 2, no. 1 (2016): 37–54.

63. Byon, H. R., S. W. Lee, S. Chen, P. T. Hammond, and Y. Shao-Horn. Thin films of carbon nanotubes and chemically reduced graphenes for electrochemical micro-capacitors. *Carbon.* 49, no. 2 (2011): 457–467.

64. Zhang, K., L. Mao, L. L. Zhang, H. S. O. Chan, X. S. Zhao, and J. Wu. Surfactant-intercalated, chemically reduced graphene oxide for high performance supercapacitor electrodes. *Journal of Materials Chemistry* 21, no. 20 (2011): 7302–7307.

65. Meng, Y., Y. Zhao, C. Hu, H. Cheng, Y. Hu, Z. Zhang, G. Shi, and L. Qu. All-graphene core-sheath microfibers for all solid state, stretchable fibriform supercapacitors and wearable electronic textiles. *Advanced Materials* 25, no. 16 (2013): 2326–2331.

66. Chen, H., M. B. Müller, K. J. Gilmore, G. G. Wallace, and D. Li. Mechanically strong, electrically conductive, and biocompatible graphene paper. *Advanced Materials* 20, no. 18 (2008): 3557–3561.

67. Tung, V. C., J. Kim, L. J. Cote, and J. Huang. Sticky interconnect for solution-processed tandem solar cells. *Journal of the American Chemical Society* 133, no. 24 (2011): 9262–9265.

68. Liu, F., and T. S. Seo. A controllable self-assembly method for large-scale synthesis of graphene sponges and free-standing graphene films. *Advanced Functional Materials* 20, no. 12 (2010): 1930–1936.

69. Yin, S., Y. Zhang, J. Kong, C. Zou, C. M. Li, X. Lu, J. Ma, F. Y. C. Boey, and X. Chen. Assembly of graphene sheets into hierarchical structures for high-performance energy storage. *Acs Nano* 5, no. 5 (2011): 3831–3838.

70. Jiang, Q., N. Kurra, M. Alhabeb, Y. Gogotsi, and H. N. Alshareef. All pseudocapacitive MXene-RuO2 asymmetric supercapacitors. *Advanced Energy Materials* 8, no. 13 (2018): 1703043.

71. Ahn, Y. R., C. R. Park, S. M. Jo, and D. Y. Kim. Enhanced charge-discharge characteristics of RuO2 supercapacitors on heat-treated TiO2 nanorods. *Applied Physics Letters* 90, no. 12 (2007): 122106.

72. Wang, Y., X. Xiao, H. Xue, and H. Pang. Zinc oxide based composite materials for advanced supercapacitors. *Chemistry Select* 3, no. 2 (2018): 550–565.

73. Korkmaz, S., F. M. Tezel, and İ. A. Kariper. Synthesis and characterization of GO/IrO2 thin film supercapacitor. *Journal of Alloys and Compounds* 754 (2018): 14–25.

74. Chen, S., J. Zhu, X. Wu, Q. Han, and X. Wang. Graphene oxide− MnO2 nanocomposites for supercapacitors. *ACS Nano* 4, no. 5 (2010): 2822–2830.

75. Zhang, G., and X. W. Lou. General solution growth of mesoporous NiCo2O4 nanosheets on various conductive substrates as high-performance electrodes for supercapacitors. *Advanced Materials* 25, no. 7 (2013): 976–979.

76. Sakiyama, K., S. Onishi, K. Ishihara, K. Orita, T. Kajiyama, N. Hosoda, and T. Hara. Deposition and properties of reactively sputtered ruthenium dioxide films. *Journal of the Electrochemical Society* 140, no. 3 (1993): 834–839.

77. Sankar, K. V., and R. K. Selvan. The ternary MnFe2O4/graphene/polyaniline hybrid composite as negative electrode for supercapacitors. *Journal of Power Sources* 275 (2015): 399–407.

78. Liang, K., X. Tang, and W. Hu. High-performance three-dimensional nanoporous NiO film as a supercapacitor electrode. *Journal of Materials Chemistry* 22, no. 22 (2012): 11062–11067.

79. Kim, I. H., and K.-B. Kim. Electrochemical characterization of hydrous ruthenium oxide thin-film electrodes for electrochemical capacitor applications. *Journal of the Electrochemical Society* 153, no. 2 (2006): A383–A389.

80. Zheng, Y. Z., H. Y. Ding, and M. L. Zhang. Hydrous–ruthenium–oxide thin film electrodes prepared by cathodic electrodeposition for supercapacitors. *Thin Solid Films* 516, no. 21 (2008): 7381–7385.

81. Wei, J., N. Nagarajan, and I. Zhitomirsky. Manganese oxide films for electrochemical supercapacitors. *Journal of Materials Processing Technology* 186, no. 1–3 (2007): 356–361.

82. Kim, H. K., T. Y. Seong, J. H. Lim, W. I. Cho, and Y. S. Yoon. Electrochemical and structural properties of radio frequency sputtered cobalt oxide electrodes for thin-film supercapacitors. *Journal of Power Sources* 102, no. 1–2 (2001): 167–171.

83. Wu, M. S., Y. A. Huang, J. J. Jow, W. D. Yang, C. Y. Hsieh, and H. M. Tsai. Anodically potentiostatic deposition of flaky nickel oxide nanostructures and their electrochemical performances. *International Journal of Hydrogen Energy* 33, no. 12 (2008): 2921–2926.

84. Kulesza, P. J., S. Zamponi, M. A. Malik, M. Berrettoni, A. Wolkiewicz, and R. Marassi. Spectroelectrochemical

characterization of cobalt hexacyanoferrate films in potassium salt electrolyte. *Electrochimica Acta* 43, no. 8 (1998): 919–923.

85. Wang, L., X. Liu, X. Wang, X. Yang, and L. Lu. Preparation and electrochemical properties of mesoporous Co_3O_4 crater-like microspheres as supercapacitor electrode materials. *Current Applied Physics* 10, no. 6 (2010): 1422–1426.

86. Kandalkar, S. G., D. S. Dhawale, C.-K. Kim, and C. D. Lokhande. Chemical synthesis of cobalt oxide thin film electrode for supercapacitor application. *Synthetic Metals* 160, no. 11–12 (2010): 1299–1302.

87. Kudo, T., Y. Ikeda, T. Watanabe, M. Hibino, M. Miyayama, H. Abe, and K. Kajita. Amorphous V_2O_5/carbon composites as electrochemical supercapacitor electrodes. *Solid State Ionics* 152 (2002): 833–841.

88. Green, R. A., N. H. Lovell, G. G. Wallace, and L. A. Poole-Warren. Conducting polymers for neural interfaces: Challenges in developing an effective long-term implant. *Biomaterials* 29, no. 24–25 (2008): 3393–3399.

89. Zhou, Y., B. L. He, W. Zhou, J. Huang, X. H. Li, B. Wu, and H. L. Li. Electrochemical capacitance of well-coated single-walled carbon nanotube with polyaniline composites. *Electrochimica Acta* 49, no. 2 (2004): 257–262.

90. Fan, L. Z., and J. Maier. High-performance polypyrrole electrode materials for redox supercapacitors. *Electrochemistry Communications* 8, no. 6 (2006): 937–940.

91. Wang, H., J. Lin, and Z. X. Shen, Polyaniline (PANi) based electrode materials for energy storage and conversion. *Journal of science: Advanced materials and devices* 1, no. 3 (2016): 225–255.

92. Villers, D., D. Jobin, C. Soucy, D. Cossement, R. Chahine, L. Breau, and D. Bélanger. The influence of the range of electroactivity and capacitance of conducting polymers on the performance of carbon conducting polymer hybrid supercapacitor. *Journal of the Electrochemical Society* 150, no. 6 (2003): A747–A752.

93. Vol'fkovich, Y. M., and T. M. Serdyuk. Electrochemical capacitors. *Russian Journal of Electrochemistry* 38, no. 9 (2002): 935–959.

94. Snook, G. A., P. Kao, and A. S. Best. (2011). Conducting-polymer-based supercapacitor devices and electrodes. *Journal of power sources* 196, no. 1 (2011): 1–12.

95. Sahoo, S., S. Dhibar, G. Hatui, P. Bhattacharya, and C. K. Das. Graphene/polypyrrole nanofiber nanocomposite as electrode material for electrochemical supercapacitor. *Polymer* 54, no. 3 (2013): 1033–1042.

96. Chang, Y., N. Wang, G. Han, M. Li, Y. Xiao, and H. Li. The properties of highly compressible electrochemical capacitors based on polypyrrole/melamine sponge-carbon fibers. *Journal of Alloys and Compounds* 786 (2019): 668–676.

97. Ryu, K. S., K. M. Kim, Y. J. Park, N. G. Park, M. G. Kang, and S. H. Chang. Redox supercapacitor using polyaniline doped with Li salt as electrode. *Solid State Ionics* 152 (2002): 861–866.

98. Ryu, K. S., K. M. Kim, N. G. Park, Y. J. Park, and S. H. Chang. Symmetric redox supercapacitor with conducting polyaniline electrodes. *Journal of Power Sources* 103, no. 2 (2002): 305–309.

99. Shown, I., A. Ganguly, L. C. Chen, and K. H. Chen. Conducting polymer-based flexible supercapacitor. *Energy Science & Engineering* 3, no. 1 (2015): 2–26.

100. Lei, C., P. Wilson, and C. Lekakou. Effect of poly (3, 4-ethylenedioxythiophene)(PEDOT) in carbon-based composite electrodes for electrochemical supercapacitors. *Journal of Power Sources* 196, no. 18 (2011): 7823–7827.

101. Fu, C., H. Zhou, R. Liu, Z. Huang, J. Chen, and Y. Kuang. Supercapacitor based on electropolymerized polythiophene and multi-walled carbon nanotubes composites. *Materials Chemistry and Physics* 132, no. 2–3 (2012): 596–600.

102. Wang, F., X. Wu, X. Yuan, Z. Liu, Y. Zhang, L. Fu, Y. Zhu, Q. Zhou, Y. Wu, and W. Huang. Latest advances in supercapacitors: From new electrode materials to novel device designs. *Chemical Society Reviews* 46, no. 22 (2017): 6816–6854.

103. Jafri, R. I., A. K. Mishra, and S. Ramaprabhu. Polyaniline–MnO_2 nanotube hybrid nanocomposite as supercapacitor electrode material in acidic electrolyte. *Journal of Materials Chemistry* 21, no. 44 (2011): 17601–17605.

104. Wang, K., Q. Meng, Y. Zhang, Z. Wei, and M. Miao. High-performance two-ply yarn supercapacitors based on carbon nanotubes and polyaniline nanowire arrays. *Advanced Materials* 25, no. 10 (2013): 1494–1498.

105. Wang, L., Y. Ye, X. Lu, Z. Wen, Z. Li, H. Hou, and Y. Song. Hierarchical nanocomposites of polyaniline nanowire arrays on reduced graphene oxide sheets for supercapacitors. *Scientific Reports* 3 (2013): 3568.

106. Meng, Y., K. Wang, Y. Zhang, and Z. Wei. Hierarchical porous graphene/polyaniline composite film with superior rate performance for flexible supercapacitors. *Advanced Materials* 25, no. 48 (2013): 6985–6990.

107. Li, X., and I. Zhitomirsky. Electrodeposition of polypyrrole–carbon nanotube composites for electrochemical supercapacitors. *Journal of Power Sources* 221 (2013): 49–56.

108. Wu, X., M. Lian, and Q. Wang. A high-performance asymmetric supercapacitors based on hydrogen bonding nanoflower-like polypyrrole and NiCo (OH)2 electrode materials. *Electrochimica Acta* 295 (2019): 655–661.

109. Mao, X., W. Yang, X. He, Y. Chen, Y. Zhao, Y. Zhou, Y. Yang, and J. Xu. The preparation and characteristic of poly (3, 4-ethylenedioxythiophene)/reduced graphene oxide nanocomposite and its application for supercapacitor electrode. *Materials Science and Engineering: B* 216 (2017): 16–22.

110. Wang, J., Z. Wu, H. Yin, W. Li, and Y. Jiang. Poly (3, 4-ethylenedioxythiophene)/MoS_2 nanocomposites with enhanced electrochemical capacitance performance. *RSC Advances* 4, no. 100 (2014): 56926–56932.

111. Huang, Y., H. Li, Z. Wang, M. Zhu, Z. Pei, Q. Xue, Y. Huang, and C. Zhi. Nanostructured polypyrrole as a flexible electrode material of supercapacitor. *Nano Energy* 22 (2016): 422–438.

112. Ryu, K. S., Y. G. Lee, Y. S. Hong, Y. J. Park, X. Wu, K. M. Kim, M. G. Kang, N. G. Park, and S. H. Chang. Poly (ethylenedioxythiophene)(PEDOT) as polymer electrode in redox supercapacitor. *Electrochimica Acta* 50, no. 2–3 (2004): 843–847.

113. Ge, Y., R. Jalili, C. Wang, T. Zheng, Y. Chao, and G. G. Wallace. A robust free-standing MoS2/poly (3, 4-ethylenedioxythiophene): Poly (styrenesulfonate) film for supercapacitor applications. *Electrochimica Acta* 235 (2017): 348–355.

114. Ranjusha, R., K. M. Sajesh, S. Roshny, V. Lakshmi, P. Anjali, T. S. Sonia, A. Sreekumaran Nair et al. Supercapacitors based on freeze dried MnO_2 embedded PEDOT: PSS hybrid sponges. *Microporous and Mesoporous Materials* 186 (2014): 30–36.

115. Alamro, T., and M. K. Ram. Polyethylenedioxythiophene and molybdenum disulfide nanocomposite electrodes for supercapacitor applications. *Electrochimica Acta* 235 (2017): 623–631.

116. Alvi, F., M. K. Ram, P. A. Basnayaka, E. Stefanakos, Y. Goswami, and A. Kumar. Graphene–polyethylenedioxythiophene conducting polymer nanocomposite based supercapacitor. *Electrochimica Acta* 56, no. 25 (2011): 9406–9412.

117. Yoo, D., J. Kim, and J. H. Kim. Direct synthesis of highly conductive poly (3, 4-ethylenedioxythiophene): Poly (4-styrenesulfonate)(PEDOT: PSS)/graphene composites and their applications in energy harvesting systems. *Nano Research* 7, no. 5 (2014): 717–730.

118. Xiao, J., D. Choi, L. Cosimbescu, P. Koech, J. Liu, and J. P. Lemmon. Exfoliated MoS2 nanocomposite as an anode material for lithium ion batteries. *Chemistry of Materials* 22, no. 16 (2010): 4522–4524.

119. Chao, J., L. Yang, J. Liu, R. Hu, and M. Zhu. Sandwiched MoS2/polyaniline nanosheets array vertically aligned on reduced graphene oxide for high performance supercapacitors. *Electrochimica Acta* 270 (2018): 387–394.

120. Sha, C., B. Lu, H. Mao, J. Cheng, X. Pan, J. Lu, and Z. Ye. 3D ternary nanocomposites of molybdenum disulfide/polyaniline/reduced graphene oxide aerogel for high performance supercapacitors. *Carbon* 99 (2016): 26–34.

121. Li, X., C. Zhang, S. Xin, Z. Yang, Y. Li, D. Zhang, and P. Yao. Facile synthesis of MoS2/reduced graphene oxide@ polyaniline for high-performance supercapacitors. *ACS Applied Materials & Interfaces* 8, no. 33 (2016): 21373–21380.

122. Zhang, R., Y. Liao, S. Ye, Z. Zhu, and J. Qian. Novel ternary nanocomposites of MWCNTs/PANI/MoS2: Preparation, characterization and enhanced electrochemical capacitance. *Royal Society Open Science* 5, no. 1 (2018): 171365.

123. Qiu, Z., Y. Peng, D. He, Y. Wang, and S. Chen. Ternary Fe3O4@ C@ PANi nanocomposites as high-performance supercapacitor electrode materials. *Journal of Materials Science* 53, no. 17 (2018): 12322–12333.

124. Shafi, P. M., V. Ganesh, and A. C. Bose. LaMnO3/RGO/ PANI ternary nanocomposites for supercapacitor electrode application and their outstanding performance in all-solid-state asymmetrical device design. *ACS Applied Energy Materials* 1, no. 6 (2018): 2802–2812.

125. Pandiselvi, K., and S. Thambidurai. Chitosan-ZnO/polyaniline ternary nanocomposite for high-performance supercapacitor. *Ionics* 20, no. 4 (2014): 551–561.

126. Wang, H., Z. Guo, S. Yao, Z. Li, and W. Zhang. Design and synthesis of ternary graphene/polyaniline/Co3O4 hierarchical nanocomposites for supercapacitors. *International Journal of Electrochemical Science* 12 (2017): 3721–3731.

127. Han, G., Y. Liu, L. Zhang, E. Kan, S. Zhang, J. Tang, and W. Tang. MnO 2 nanorods intercalating graphene oxide/ polyaniline ternary composites for robust high-performance supercapacitors. *Scientific Reports* 4 (2014): 4824.

128. Xiong, P., H. Huang, and X. Wang. Design and synthesis of ternary cobalt ferrite/graphene/polyaniline hierarchical nanocomposites for high-performance supercapacitors. *Journal of Power Sources* 245 (2014): 937–946.

129. Shen, J., C. Yang, X. Li, and G. Wang. High-performance asymmetric supercapacitor based on nanoarchitectured polyaniline/graphene/carbon nanotube and activated graphene electrodes. *ACS Applied Materials & Interfaces* 5, no. 17 (2013): 8467–8476.

130. Wang, L., T. Wu, S. Du, M. Pei, W. Guo, and S. Wei. High performance supercapacitors based on ternary graphene/ Au/polyaniline (PANI) hierarchical nanocomposites. *RSC Advances* 6, no. 2 (2016): 1004–1011.

131. Xu, Z., Z. Zhang, L. Gao, H. Lin, L. Xue, Z. Zhou, J. Zhou, and S. Zhuo. Tin disulphide/nitrogen-doped reduced graphene oxide/polyaniline ternary nanocomposites with ultra-high capacitance properties for high rate performance supercapacitor. *RSC Advances* 8, no. 70 (2018): 40252–40260.

132. Hou, Y., Y. Cheng, T. Hobson, and J. Liu. Design and synthesis of hierarchical MnO2 nanospheres/carbon nanotubes/conducting polymer ternary composite for high performance electrochemical electrodes. *Nano Letters* 10, no. 7 (2010): 2727–2733.

133. Li, P., Y. Yang, E. Shi, Q. Shen, Y. Shang, S. Wu, J. Wei, K. Wang, H. Zhu, Q. Yuan, and A. Cao. Core-double-shell, carbon nanotube@ polypyrrole@ MnO2 sponge as freestanding, compressible supercapacitor electrode. *ACS Applied Materials & Interfaces* 6, no. 7 (2014): 5228–5234.

134. Zhou, H., H. Chen, S. Luo, G. Lu, W. Wei, and Y. Kuang. The effect of the polyaniline morphology on the performance of polyaniline supercapacitors. *Journal of Solid State Electrochemistry* 9, no. 8 (2005): 574–580.

135. Li, H., J. Wang, Q. Chu, Z. Wang, F. Zhang, and S. Wang. Theoretical and experimental specific capacitance of polyaniline in sulfuric acid. *Journal of Power Sources* 190, no. 2 (2009): 578–586.

136. Mi, H. X., Z. S. Yang, X. Ye, and J. Luo. Polyaniline nanofibers as the electrode material for supercapacitors. *Materials Chemistry and Physics* 112, no. 1 (2008): 127–131.

137. Wang, Y. G., H. Q. Li, and Y. Y. Xia. Ordered whiskerlike polyaniline grown on the surface of mesoporous carbon and its electrochemical capacitance performance. *Advanced Materials* 18, no. 19 (2006): 2619–2623.

138. Kim, M., Y. K. Kim, J. Kim, S. Cho, G. Lee, and J. Jang. Fabrication of a polyaniline/MoS2 nanocomposite using self-stabilized dispersion polymerization for supercapacitors with high energy density. *RSC Advances* 6, no. 33 (2016): 27460–27465.

139. Luo, J., W. Zhong, Y. Zou, C. Xiong, and W. Yang. Preparation of morphology-controllable polyaniline and polyaniline/graphene hydrogels for high performance binder-free supercapacitor electrodes. *Journal of Power Sources* 319 (2016): 73–81.

140. Wang, K., H. Wu, Y. Meng, and Z. Wei. Conducting polymer nanowire arrays for high performance supercapacitors. *Small* 10, no. 1 (2014): 14–31.

141. Rana, U., K. Chakrabarti, and S. Malik. Benzene tetracarboxylic acid doped polyaniline nanostructures: Morphological, spectroscopic and electrical characterization. *Journal of Materials Chemistry* 22, no. 31 (2012): 15665–15671.

142. Yan, L. J., and Y. Y. Xia. Effect of pore structure on the electrochemical capacitive performance of MnO2. *Journal of the Electrochemical Society* 154, no. 11 (2007): A987–A992.

143. Guan, Q., J. Cheng, B. Wang, W. Ni, G. Gu, X. Li, L. Huang, G. Yang, and F. Nie. Needle-like Co3O4 anchored on the graphene with enhanced electrochemical performance for aqueous supercapacitors. *ACS Applied Materials & Interfaces* 6, no. 10 (2014): 7626–7632.

144. Dong, X. C., H. Xu, X. W. Wang, Y. X. Huang, M. B. Chan-Park, H. Zhang, L. H. Wang, W. Huang, and

P. Chen. 3D graphene–Cobalt oxide electrode for high-performance supercapacitor and enzymeless glucose detection. *ACS Nano* 6, no. 4 (2012): 3206–3213.

145. Ma, S. B., K. W. Nam, W. S. Yoon, X. Q. Yang, K. Y. Ahn, K. H. Oh, and K. B. Kim. Electrochemical properties of manganese oxide coated onto carbon nanotubes for energy-storage applications. *Journal of Power Sources* 178, no. 1 (2008): 483–489.

146. Dong, X., W. Shen, J. Gu, L. Xiong, Y. Zhu, H. Li, and J. Shi. MnO_2-embedded-in-mesoporous-carbon-wall structure for use as electrochemical capacitors. *The Journal of Physical Chemistry B* 110, no. 12 (2006): 6015–6019.

147. Mu, B., W. Zhang, S. Shao, and A. Wang. Glycol assisted synthesis of graphene–MnO_2–polyaniline ternary composites for high performance supercapacitor electrodes. *Physical Chemistry Chemical Physics* 16, no. 17 (2014): 7872–7880.

148. Bordjiba, T., and D. Bélanger. Direct redox deposition of manganese oxide on multiscaled carbon nanotube/microfiber carbon electrode for electrochemical capacitor.

Journal of the Electrochemical Society 156, no. 5 (2009): A378–A384.

149. Yan, D., P. X. Yan, G. H. Yue, J. Z. Liu, J. B. Chang, Q. Yang, D. M. Qu et al. Self-assembled flower-like hierarchical spheres and nanobelts of manganese oxide by hydrothermal method and morphology control of them. *Chemical Physics Letters* 440, no. 1–3 (2007): 134–138.

150. Liu, R., and S. B. Lee. MnO_2/poly (3, 4-ethylenedioxythiophene) coaxial nanowires by one-step coelectrodeposition for electrochemical energy storage. *Journal of the American Chemical Society* 130, no. 10 (2008): 2942–2943.

151 Sun, L. J., and X. X. Liu. Electrodepositions and capacitive properties of hybrid films of polyaniline and manganese dioxide with fibrous morphologies. *European Polymer Journal* 44, no. 1 (2008): 219–224.

152. Deng, D. H., H. Pang, J. M. Du, J. W. Deng, S. J. Li, J. Chen, and J. S. Zhang. Fabrication of cobalt ferrite nanostructures and comparison of their electrochemical properties. *Crystal Research and Technology* 47, no. 10 (2012): 1032–1038.

Index